# Handbook of Potato Production, Improvement, and Postharvest Management

J. Gopal
S. M. Paul Khurana
Editors

CRC Press
Taylor & Francis Group
Boca Raton London New York

CRC Press is an imprint of the
Taylor & Francis Group, an **informa** business

CRC Press
Taylor & Francis Group
6000 Broken Sound Parkway NW, Suite 300
Boca Raton, FL 33487-2742



**Visit the Taylor & Francis Web site at**
**http://www.taylorandfrancis.com**

**and the CRC Press Web site at**
**http://www.crcpress.com**

# CONTENTS

## ABOUT THE EDITORS

**Jai Gopal, PhD,** is Principal Scientist in the Division of Crop Improvement at the Central Potato Research Institute in Shimla, India. His research experience includes nearly three decades in potato genetics, breeding, and biotechnology. He has studied genetics and floral biology of the potato, has bred a number of potato varieties and parental lines, and has developed selection procedures combining in vitro and in vivo approaches. His work has appeared in *Theoretical & Applied Genetics, Plant Breeding, Plant Cell Reports, Euphytica,* and *Potato Research.*

**S. M. Paul Khurana, PhD,** is Vice-Chancellor at Rani Durgawati University in Jabalpur, India. He was National Coordinator of Potato for ten years and Director of the Central Potato Research Institute for three years. His research experience includes over three decades of work on plant viruses and viroids, and he is an expert in virus detection techniques. He standardized use of ELISA and ISEM for detection of potato viruses and developed simple techniques for screening resistance to potato viruses. His published works have appeared in various international journals. Dr. Khurana has also edited several books, conference proceedings, and journals.

## CONTRIBUTORS

**Hanafi Abdelhaq,** Department of Plant Protection, Complexe Horticole, Cite Balneaire, Agadir, Morocco.

**H. Barker,** Scottish Crop Research Institute, Invergowrie, Dundee, United Kingdom.

**John E. Bradshaw,** Scottish Crop Research Institute, Invergowrie, Dundee, United Kingdom.

**Warren K. Coleman,** Agriculture and Agri-Food Canada, Potato Research Centre, Fredericton, New Brunswick, Canada.

**R. Ezekiel,** Division of Post-Harvest Technology, Central Potato Research Institute, Himachal Pradesh, India.

**G. A. Forbes,** International Potato Center (CIP), Lima, Peru.

**Klaus Gottschalk,** Institut für Agrartechnik Bornim e.V. ATB, Potsdam, Germany.

**P. M. Govindakrishnan,** Division of Crop Production, Central Potato Research Institute, Himachal Pradesh, India.

**A. J. Haverkort,** Wageningen University, Plant Research International, Wageningen, The Netherlands.

**Kazuto Iwama,** Department of Botany and Agronomy, Graduate School of Agriculture, Hokkaido University, Sapporo, Japan.

**Colin Jeffries,** Scottish Agricultural Science Agency, East Craigs, Edinburgh, United Kingdom.

**J. A. Landeo,** International Potato Center (CIP), Lima, Peru.

**R. Lebecka,** Plant Breeding and Acclimatization Institute (IHAR), Mlochow Research Center, Mlochow, Poland.

**Xiu-Qing Li,** Agriculture and Agri-Food Canada, Potato Research Centre, Fredericton, New Brunswick, Canada.

**Qiang Liu,** Agriculture and Agri-Food Canada, Food Research Program, Guelph, Ontario, Canada.

**E. Łojkowska,** Department of Plant Protection and Biotechnology, Intercollegiate Faculty of Biotechnology, University of Gdansk and Medical University of Gdansk, Gdansk, Poland.

**R. Ortiz,** International Maize and Wheat Improvement Center (CIMMYT), Apdo, Mexico, D.F.

**R. D. Peters,** Agriculture and Agri-Food Canada, Crops and Livestock Research Centre, Charlottetown, Prince Edward Island, Canada.

**H. W. (Bud) Platt,** Agriculture and Agri-Food Canada, Crops and Livestock Research Centre, Charlottetown, Prince Edward Island, Canada.

**Alberto Salas,** International Potato Center (CIP), Lima, Peru.

**Martin G. Scanlon,** Department of Food Science, Faculty of Agricultural and Food Sciences, University of Manitoba, Winnipeg, Manitoba, Canada.

**David M. Spooner,** U.S. Department of Agriculture, Agricultural Research Service, Department of Horticulture, University of Wisconsin, Madison, Wisconsin, USA.

**G. C. C. Tai,** Agriculture and Agri-Food Canada, Potato Research Centre, Fredericton, New Brunswick, Canada.

**T. R. Tarn,** Agriculture and Agri-Food Canada, Potato Research Centre, Fredericton, New Brunswick, Canada.

**Gerhard Wenzel,** Technical University of Munich, Freising-Weihenstephan, Germany.

**Junichi Yamaguchi,** Department of Botany and Agronomy, Graduate School of Agriculture, Hokkaido University, Sapporo, Japan.

**E. Zimnoch-Guzowska,** Plant Breeding and Acclimatization Institute (IHAR), Mlochow Research Center, Mlochow, Poland.

# Foreword

Potato is a significant crop and an important staple food in temperate countries. It is rapidly gaining importance as a major food crop in tropical and subtropical countries, where it is overtaking staple grains in volume of production and consumption. Potato cropping in Asia, Africa, and Latin America has grown rapidly and continues to expand as a result of better varieties, improved access to inputs, integrated crop management, enhanced postharvest handling, and optimum irrigation. Potato is valuable both as a fresh product and as a processed product, and the current trend is for rapid growth in processed forms, mainly chips, French fries, and powder. The area under production is expected to more than double in the future with the increasing demand for fresh and processed potato products arising from the fast-changing food habits in many developing countries.

Accompanying this rapid increase in potato production and consumption, especially in developing countries, is the need to constantly improve varieties and methodologies for higher productivity. Improved postharvest management of potato is vital in terms of total production; conservation systems that make food available after harvest will form a significant requirement. As a semiperishable product, potato needs special care in storage, transport, and processing. It is imperative to acquire specific scientific knowledge to develop and select varieties, to improve the management of inputs and crop protection, and to select efficient methods of storage, handling, and processing.

The *Handbook of Potato Production, Improvement, and Postharvest Management* is a text-cum-reference book covering all aspects of potatoes from biosystematics through breeding, agronomy, diseases, and insect pests to storage, processing, and value addition. Its publication is particularly timely as potato becomes a significant crop in many tropical and subtropical countries. It collects valuable contributions from renowned scientists associated with potato production and postharvest handling. The well-illustrated chapters provide up-to-date reviews on both basic and applied aspects and include references at the end for readers who want to obtain further details on each topic.

The book is a valuable resource for potato researchers, teachers, and students.

*Pamela Anderson, Director General*
*International Potato Center*
*Lima, Peru*

# Preface

Potato is an important component of the human diet in both developed and developing countries. Originally domesticated by Andean farmers, probably in the Lake Titicaca basin, potato is now grown in nearly 150 countries and is the world's fourth most important crop after rice, wheat, and maize. Hence, the production, improvement, and postharvest management of potato are of great interest. Many research institutes and universities, including the International Potato Center (CIP), Lima, Peru, have generated a vast amount of knowledge about potatoes, and in 1978 Professor Paul Harris edited an excellent book entitled *The Potato Crop,* which was revised in 1992. Since then much information has been generated and a few subject- and region-specific compilations have been produced by well-known potato scientists, yet no other comprehensive book on potato has been published.

The *Handbook of Potato Production, Improvement, and Postharvest Management* is a handbook in 15 chapters authored by authorities in their respective fields. It is a text-cum-reference book aiming to provide state-of-the-art information to scientists, students, and industrialists so that it can be effectively used for future research into, and development and utilization of, potato.

The book begins with a contribution from the leading taxonomists and germplasm experts David M. Spooner and Alberto Salas, who provide an introduction to the structure, distribution, and habitats of potato. Their chapter deals critically with the taxonomy of wild and cultivated potatoes and the present taxonomic status, highlighting the problems associated with previous classifications and discussing how these are being resolved.

The second chapter, "Genetics of Agrihorticultural Traits" by John E. Bradshaw, provides comprehensive information on the genetics of flowering, yield and its components, dormancy, tuber shape, pigmentation, and other topics.

"Considerations for Successful Breeding" by Jai Gopal, the third chapter, highlights theoretical and practical strategies for consideration in the selection of superior parents, crosses, and clones from segregating progenies. In addition, it proposes novel and realistic selection procedures based on the combined use of in vitro and in vivo methods.

Chapter 4, "Biotechnology in Potato Improvement" by Gerhard Wenzel, reviews the role of biotechnology in the conservation and characterization of potato germplasm and breeding. It also covers haploids, somatic cell fusion, molecular markers, genetic maps, transgenics, and high-throughput DNA techniques.

In Chapter 5, "Quality Improvement," T. R. Tarn, G. C. C. Tai, and Qiang Liu review the genetics of texture, dry matter, sugar, cold-induced sweetening, starch, proteins, vitamins, antioxidants, flavor, glycoalkaloids, and enzymatic browning from the point of view of the progress being made in improving these characteristics through biotechnology and other methods.

In Chapter 6, "Ecophysiology and Agronomic Management," P. M. Govindakrishnan and A. J. Haverkort analyze the phases of crop growth and the factors important in the development of cultural practices in diverse environments. In addition, in dealing with agrotechniques for optimum resource use, they illustrate how these principles can be embodied into decision support systems.

Under the title "Abiotic Stresses" in Chapter 7 Kazuto Iwama and Junichi Yamaguchi cover the physiobiochemical basis of tuberization and topics relevant to the understanding and management of stresses due to such factors as drought, heat, salinity, frost, and ozone.

The next two chapters deal with fungal and oomycete diseases of potato. Chapter 8 is devoted to late blight disease: G. A. Forbes and J. A. Landeo describe the morphology, taxonomy, origin, and spread of the pathogen and the epidemiology, symptoms, disease cycle, and management through cultural, chemical, and host-mediated resistance. Progress in breeding *R*-gene-free populations and in understanding the molecular aspects of host-pathogen interaction are also reviewed. H. W. (Bud) Platt and R. D. Peters systematically describe the causal agents, distribution and impact, symptoms, epidemiology, and management aspects of another 14 important fungal and oomycete diseases in Chapter 9.

In Chapter 10, R. Lebecka, E. Zimnoch-Guzowska, and E. Łojkowska describe six bacterial diseases of potato, including the four major ones—bacterial wilt, ring rot, black leg, and common scab. They also provide an extensive review of host resistance and its utilization in developing resistant varieties.

Chapter 11 on "Viruses and Viroids," by Colin Jefferies, H. Barker, and S. M. Paul Khurana, covers the distribution, morphology, nucleic acid types, and mode of transmission of viruses and viroids naturally infecting potato. In addition to detection and management methods, the authors analyze seed certification, phytosanitary regulations, cultural/chemical control measures, and the use of host- and pathogen-derived resistance.

Chapter 12, "Insects, Nematodes, and Integrated Pest Management" by Hanafi Abdelhaq, covers all important pest species, their distribution and the damage they cause, their biology and management, and integrated pest management and biopesticides.

Chapters 13 and 14 address the problems associated with postharvest management of potatoes, which are of great importance to growers and also to industry. Chapter 13, "Storage" by Klaus Gottschalk and R. Ezekiel, addresses both temperate and tropical problems in potato storage, providing information on storage methodology and factors affecting storage as well as mathematical models governing storage conditions. It also reviews the basics of such topics as physiological age and dormancy.

Chapter 14, "Processing and Value Addition" by Xiu-Qing Li, Martin G. Scanlon, Qiang Liu, and Warren K. Coleman, starts with the food value of potatoes, before moving on to review the industrial technology for processed products, including molecular farming and the production of vaccines, pharmaceutical proteins, and so on. This chapter complements the discussion of quality improvement in Chapter 5.

The final chapter is on "True Potato Seed," once a hot research topic. In this chapter, Jai Gopal and R. Ortiz review the past and present status of TPS and analyze its future prospects. Although these prospects are not very encouraging, the desire to use TPS as an alternative to the traditional method of cropping through seed tubers is not going to die down quickly, as research is continuing to overcome its shortcomings.

The help and support of the authors by way of their contributions is gratefully acknowledged. We hope that this team effort will benefit the cause of the potato. As editors, we have tried to make the presentation comprehensive and error free. Should readers find technical lacunae, we would be happy to be informed of them and also to receive comments for improving future editions.

*Jai Gopal*
*S. M. Paul Khurana*

# Chapter 1

# Structure, Biosystematics, and Genetic Resources

David M. Spooner
Alberto Salas

The potato is one of the world's most important food crops and the world's most important vegetable crop. Potato produces more carbohydrate per acre per year than any other crop except sugarcane. It has a higher quality protein than any other vegetable, and only soybean yields more protein per acre (Rhoades, 1982; Anonymous, 1984; Ortiz, 1998). Cultivated and wild potatoes are both tuber-bearing members of the genus *Solanum*. The only difference in the designation of a potato as "cultivated" or "wild" is cultural, that is, whether it is intentionally grown for food or is growing naturally. The "modern" cultivated potatoes of world commerce are collectively designated under the name *S. tuberosum*. L. Potato has a rich gene pool of nearly 200 tuber-bearing species that represent a huge and only partially explored reservoir of germplasm useful for potato breeding. Wild species of potato have known desirable traits such as resistance to heat and frost, fungi, bacteria, viruses, nematodes, and insects. Most of these species are cross-compatible with the cultivated potato either directly or through the use of 2*n* gametes.

---

The authors thank Mercedes Ames and Diego Fajardo for taxonomic insights provided by their unpublished morphological data.

This chapter provides an introduction to the structure, distribution, habitats, and taxonomy of wild and cultivated potatoes. It also discusses the collection, conservation, and use of potato genetic resources. Potato continues to be the focus of intense collection and taxonomic research; hence our understanding of the number of species and their interrelationships continues to change. Therefore an effort has been made to give the latest taxonomic summary of wild and cultivated potatoes.

## *STRUCTURE*

One of the earliest descriptions of the potato was by Gerard (1597), who briefly outlined the outward morphology of the plant. In stark contrast, Cutter (1992) provided a 97-page treatise on both the external morphology and internal anatomy of potato and stated, "it is clear that the structure and development of the potato, admittedly a very complex plant, are by no means fully understood" (p. 65). This chapter covers the basic details of the external morphology of potato.

In potato, the stems arise at the beginning of each growing season from a tuber of the previous year, referred to as the mother tuber (shown at the base of the stem in Figure 1.1). A potato tuber is a modified underground stem, not a swollen root. The "eyes" of potato tubers are stem buds, similar to buds on the nodes of a stem. Each potato tuber has several eyes, and each eye is capable of producing a separate stem.

The young plant draws its initial nutrition from the mother tuber, which soon withers. This plant produces (1) roots that develop from the base of the plant, (2) aerial stems, and (3) underground horizontal stolons originating above the roots near the base of the aerial stems. As the plant grows, the stolons thicken at their ends, or in some wild species thicken like beads on a string along their length (see section "*Solanum* series *Piurana*"), and produce new tubers.

A tuber is a starch storage organ that perpetuates the plant into the next year by nonsexual means, also called vegetative propagation. In contrast, plants can also arise from "sexual" or "true" seeds produced from the flowers (see Chapter 15, "True Potato Seed"). Tubers of various cultivars/genotypes vary considerably in skin and flesh color,

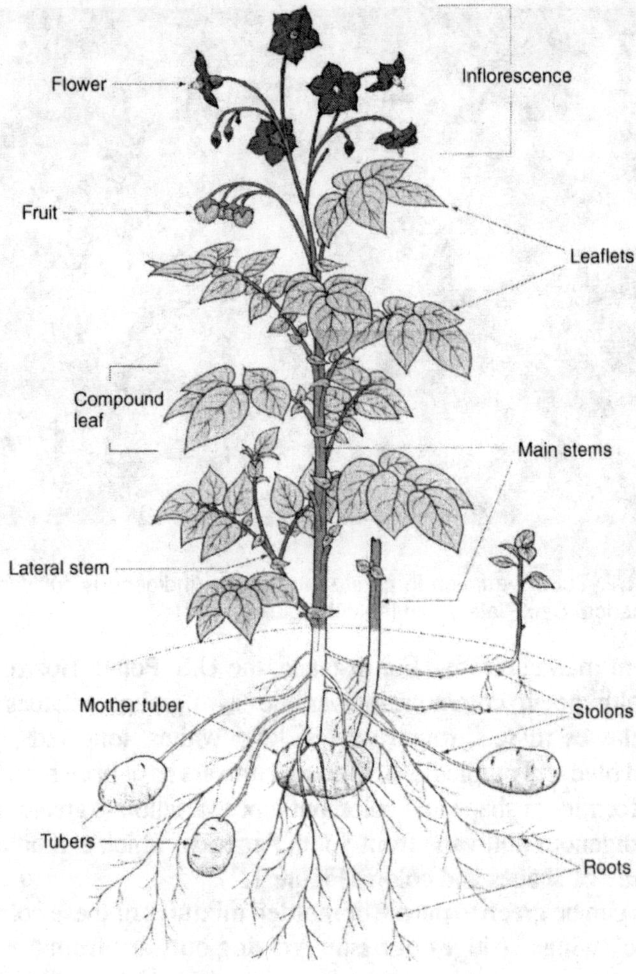


FIGURE 1.1. Stylized illustration of a cultivated potato plant. Drawing courtesy of the International Potato Center. See Plate 1.1 in the color gallery.

shape, and size. In modern cultivars skin colors vary from white to yellow to red to tan, but most have white tubers (some cream to yellow). Tuber shapes vary from round to elongate, and tuber skin surface varies from smooth to netted (russets) (Hils and Pieterse, 2005). These colors and shapes are commonly grouped into

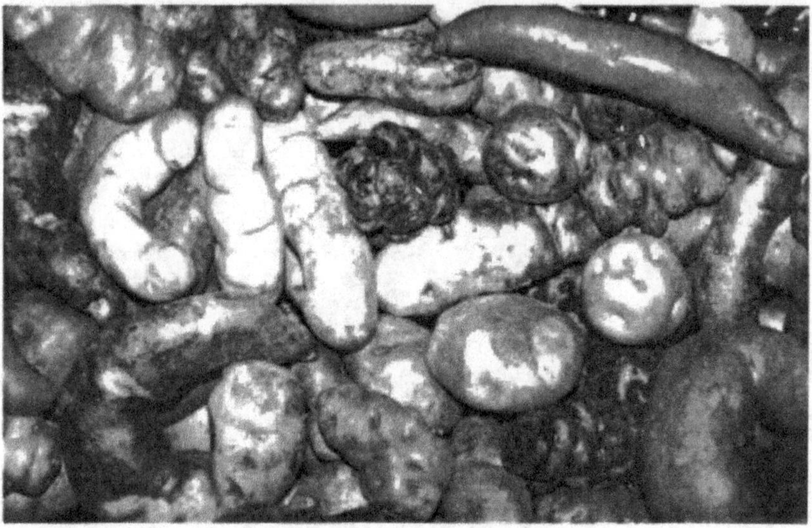

FIGURE 1.2. Tuber variation in potato land races (indigenous cultivars) from South America. See Plate 1.2 in the color gallery.

convenient market classes. For instance, the U.S. Potato Board (http://www.healthypotato.com/nutrition/varieties.asp) groups potatoes by the market classes russets, round whites, long whites, long reds, yellow flesh, and blue and purple flesh. Modern cultivars have been bred for relative uniformity in shape and color, but tuber variation is greater in land races (indigenous cultivars) from South America, which exhibit a much wider array of shapes and colors (Figure 1.2).

Stems can be green to purple or mottled mixtures of these colors and may have "wings" (ridges of tissue growing outward from the stem). Compound leaves arise in a spiral pattern on the aerial stems. In the angle between the stems and leaves (the leaf axil), branches or flower clusters (inflorescences) are produced. Typically, at the base of the leaves, flaps of tissue called "pseudostipular leaflets" are produced (Figure 1.3A). Sometimes these fall off or are not produced at all.

Leaves of cultivated and wild potatoes vary tremendously in relative degree of dissection and shape. Figure 1.3A (one leaf form of *S. tuberosum*) illustrates parts of a potato leaf. Except that secondary lateral leaflets are not very common, the leaf in Figure 1.3A looks

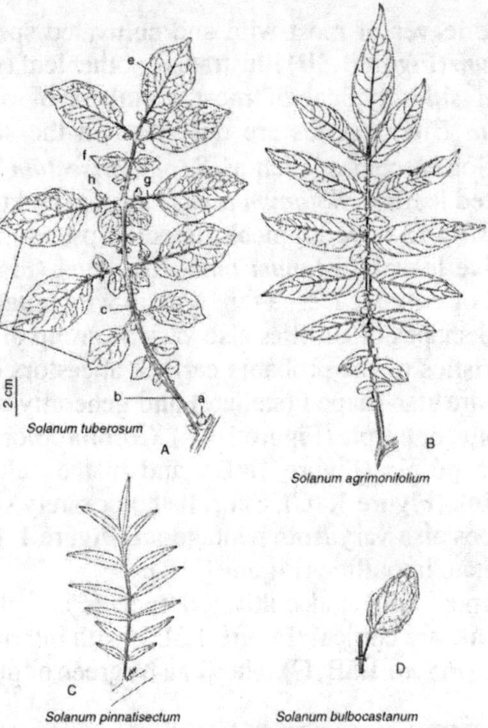


FIGURE 1.3. Variation in leaf morphology in wild and cultivated potatoes (*Solanum* section *Petota*). *(A)* The most dissected leaf type is illustrated in a collection of *S. tuberosum,* showing *a.* pseudostipular leaflets, here appearing as ovate flaps but sometimes shaped as half-moons (lunate); *b.* petiole (base of the leaf lacking lateral leaflets); *c.* rachis; *d.* lateral leaflets; *e.* apical leaflet; *f.* interjected leaflets; *g.* petiolule (base of the lateral and apical leaflets without leaf tissue); *h.* secondary lateral leaflets (separate leaf tissue arising on the petiolule). The designation of the small structures near the base of the leaf axis as interjected leaflets or small lateral leaflets is open to interpretation. *(B) Solanum agrimonifolium* shows a leaf type more characteristic of a group of approximately 40 wild potatoes in *Solanum* series *Conicibaccata* with leaves having a somewhat parallel-sided morphology, with leaflet pairs more subequal or diminishing gradually toward the base. *(C) Solanum pinnatisectum* and only one other species *(S. jamesii)* have pseudostipular leaflets that are not ovate or lunate but pinnate; this species, and a few others, such as *S. infundibuliforme,* have narrow lateral and terminal leaflets without petiolules and no interjected leaflets. *(D) Solanum bulbocastanum* and a few other species have entire leaves; this species has lunate pseudostipules. There is a wide range of variation among these types.

similar to the leaves of most wild and cultivated species. *Solanum agrimonifolium* (Figure 1.3B) illustrates another leaf type with somewhat parallel sides typical of most members of *Solanum* series *Conicibaccata* Bitter (series are described in the section "Genus *Solanum*"). Some species, such as *S. pinnatisectum* (Figure 1.3C), lack interjected leaflets. *Solanum pinnatisectum* and the related species *S. jamesii* also have atypical dissected pseudostipules shaped like diminutive leaves. *Solanum bulbocastanum* (Figure 1.3D) has nondissected or entire leaves, a rare type in wild potatoes.

Colors and shapes of corollas also vary. A group of species showing characteristics of the probable earliest ancestors of potato have corollas that are star-shaped (stellate) and generally white, or white tinged with blue or purple (Figure 1.4A). Corolla colors are more typically blue to purple (Figure 1.4B), and in the cultivated species sometimes pink (Figure 1.4C), a corolla color rarely seen in the wild species. Shapes also vary from pentagonal (Figure 1.4C, D) to more circular or rotate in outline (Figure 1.4B).

Fruits of most wild and cultivated species are globose (Figure 1.5A), but some are conical (Figure 1.5D), with intermediate shapes that are ovoid (Figure 1.5B, C). They can be green or purple or almost

FIGURE 1.4. Wild and cultivated potato flowers. See Plate 1.4 in the color gallery.

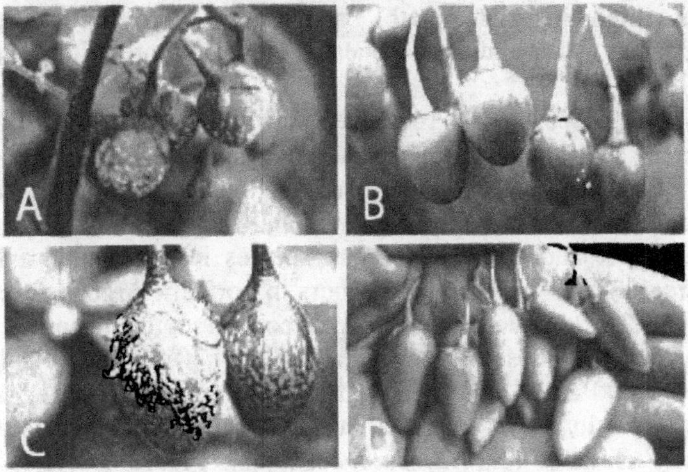


FIGURE 1.5. Wild potato fruits. See Plate 1.5 in the color gallery.

white or have mottled or dotted combinations of these colors (Figure 1.5A, C).

## *ORIGIN, SPREAD, AND HABITAT*

The cultivated potato was believed to have originated somewhere in the Andes; different ideas were advanced for its wild species progenitors and specific place of origin. Spooner et al. (2005) supported a single origin of potato from a wild species progenitor in the *S. brevicaule* complex in southern Peru. The Spaniards found cultivated potatoes in Peru at the time of their conquest beginning in 1536 (Hawkes, 1990), and records exist of potatoes crated for shipment from Chile to Europe in 1587 (Glendinning, 1983). Cultivated potato first appeared outside South America in the Canary Islands (Spain) in 1567 (Hawkes and Francisco-Ortega, 1992). Between 1650 and 1840 potatoes had become a vital part of the basic food supply in Ireland. When late blight disease wiped out the crop in the 1840s, famine forced many Irish people to immigrate to America. Because of its early food use and importance in Ireland, the potato plant is sometimes called the Irish potato. From Europe, potato soon became established as a popular and cheap food crop worldwide (Hawkes, 1990).

Wild potatoes grow in 16 countries, from the southwestern United States to central coastal Chile and adjacent Argentina. Most of the 188 species of wild potatoes (see section "Taxonomy of section *Petota*") are rare and narrowly endemic. In general, they occur between 38°N and 41°S, with more species in the Southern Hemisphere. High species richness occurs in northern Argentina, central Bolivia, central Ecuador, central Mexico, and south and north-central Peru. Wild potatoes grow from sea level to 4,300 m but typically between 2,000 and 4,000 m altitudes (Hijmans and Spooner, 2001). They grow in an amazing variety of habitats from very seasonal wet/dry climates (e.g., *S. bulbocastanum,* Figure 1.6A) to high-altitude grasslands (e.g., *S. colombianum,* Figure 1.6B), to beach margins (e.g., *S. tuberosum,* Figure 1.6C), to upland rain forests (e.g., *S. candolleanum,* Figure 1.6D).

Native cultivated potatoes, or "land race" cultivars, grow throughout the Andes from Venezuela to northern Argentina, and then again in

FIGURE 1.6. Wild potato habitats. See Plate 1.6 in the color gallery.

central Chile (in the Chonos Archipelago). No formal distributional analysis like that for the wild species exists for the cultivated species, but most populations in the Andes are planted in mid (2,000 m) to high elevations (4,000 m), and in Chile in low elevations near sea level.

The origin of "modern" cultivated potatoes has been very controversial. Juzepczuk and Bukasov (1929) proposed that modern potatoes originated from the tetraploid land races from Chile (Chilotanum Group), whereas Salaman and Hawkes (1949), Salaman (1946, 1954), and Hawkes and Francisco-Ortega (1992) suggested an initial origin from tetraploid land races from the Andes (Andigenum Group). The different arguments relate to day-length adaptation of these types in Europe, morphology of early herbarium specimens, shipping records, and times when the early cultivars were grown in Europe. For example, the Chilean introduction hypothesis rests on the fact that Chilean land races form tubers under long days, but potatoes from the Andes form tubers only in the short days in the tropics and thus would form tubers only very late in the year or not at all in Europe. The Andean introduction hypothesis argues that early cultivated herbarium specimens from Europe show a form similar to members of the Andigenum Group, the first cultivated potatoes in Europe were harvested late in the year, as expected for this short-day crop, and potatoes from Chile would not survive the long transport time to Europe.

Whatever the source of the early introductions to Europe, modern cultivars were later bred with Andean and Chilean land races and wild species. Hosaka et al. (1988) showed that tetraploid Chilean land races are primarily distinguished from most populations of tetraploid Andean land races and other cultivated and wild species by an approximately 400 base pair deletion in chloroplast DNA (cpDNA). Chloroplast DNA restriction site data documented five chloroplast genotypes (A, C, S, T, W types) in *S. tuberosum* (including the Andigenum Group and the Chilotanum Group). Subspecies *andigenum* has all five types, and the Chilotanum Group had three types, A, T, and W (Hosaka and Hanneman, 1988). The most frequently observed type in the Chilotanum Group is T, which was more precisely found to have a 241 base pair deletion, not a 400 base pair deletion (Kawagoe and Kikuta, 1991).

Hosaka (1995) further studied the origin of the cultivated tetraploid potato species using cpDNA restriction site data. He studied 132 accessions of the diploid Stenotomum Group and of six related diploid wild species. He showed extensive cpDNA polymorphism in all taxa except *S. brevicaule* and concluded that Andean diploid land races were domesticated many times from the wild species, followed by sexual polyploidization to form Andean land races. However, a recent study by Spooner et al. (2005) supported a single origin of cultivated potato. Hosaka (1995) also concluded that Chilean land races were selected from a limited subset of the Andean tetraploid land races somewhere near the Bolivian and Argentinean border. Hosaka (2003) showed an identical cpDNA deletion in some populations of the wild species *S. tarijense* and suggested that the Chilean populations arose from a history of hybridization with this species.

On the basis of cytoplasmic sterility factors, geographical isolation, and ecological differences, Grun (1990) suggested that the Chilotanum Group was distinct from the Andigenum Group. Hawkes (1990) distinguished these two Groups by the Chilotanum Group having fewer stems, with foliage aligned at a broad angle to the stem, and having less-dissected leaves with wider leaflets and thicker pedicels. Huamán and Spooner (2002) corroborated Hawkes's conclusion of morphological differences between the potatoes from these two regions but showed that they could be distinguished only with great difficulty by using many characters that individually were not always diagnostic (polythetic support).

## *GENUS* SOLANUM

*Solanum* L. is one of the largest genera of flowering plants, including perhaps 1,500 species of herbs, shrubs, vines, and trees. It is one of the world's most economically important genera, including potato (*S. tuberosum* L.), tomato (*S. lycopersicum* L.), eggplant (*S. melongena* L.), and minor tropical fruits such as pepino (*S. muricatum* Aiton). The inclusion of tomato in *Solanum* (rather than in the genus *Lycopersicon* Miller) is novel to some, but it is well supported by many modern morphological and molecular studies.

Dunal (1852) wrote the first species-level worldwide monograph on *Solanum*. Later, Seithe (1962) and Danert (1970) attempted to use selected characters such as hair morphology and branching patterns to infer relationships within *Solanum*. D'Arcy (1972, 1991) divided *Solanum* into seven subgenera. Hunziker (2001) modified D'Arcy's (1972) system slightly and provided descriptions, nomenclature, and commentary for each of D'Arcy's sections. D'Arcy (1972, 1991) and Hunziker (2001) recognized potatoes in the subgenus *Potatoe* (G. Don), containing various sections, including the section *Petota* Dumort (covering all the wild and cultivated potatoes). A species-level revision of the genus *Solanum* is currently being undertaken by Solanaceae colleagues Lynn Bohs, Sandra Knapp, Michael Nee, and David Spooner.

Until Hawkes's (1990) classification, most classifications of wild and cultivated potatoes used overall similarity (phenetics) to delimit species and to infer their interrelationships. Sections and series are taxonomic ranks below genus that group related species (section is the higher rank and contains various series). Hawkes (1990) divided section *Petota* into subsection *Potatoe* Hawkes, with 19 tuber-bearing series, and subsection *Estolonifera* Hawkes, with 2 non-tuber-bearing series: *Etuberosa* Buk. and Kameraz and *Juglandifolia* (Rydb.) Hawkes.

Spooner et al. (1993) used cpDNA restriction site data and morphological data to investigate the relationships between members of potatoes (section *Petota*) and *Solanum* and other genera (including the genus *Lycopersicon*) of the Solanaceae. Their results confirmed placement of all members of Hawkes's (1990) tuber-bearing species into section *Petota*. However, the non-tuber-bearing species that Hawkes (1990) had placed into section *Petota* (series *Juglandifolia* and series *Etuberosa*) were excluded from the group. Members of series *Juglandifolia* were related to tomatoes, and, quite surprisingly, the branch of the phylogenetic tree that included tomatoes and series *Juglandifolia* was most closely related to the branch that included potatoes (i.e., they were sister clades). Members of series *Etuberosa* formed a sister clade to the combined tomato and potato clade, showing a need to modify Hawkes's (1990) classification of section *Petota* to exclude series *Juglandifolia* and series *Etuberosa* (Figure 1.7).

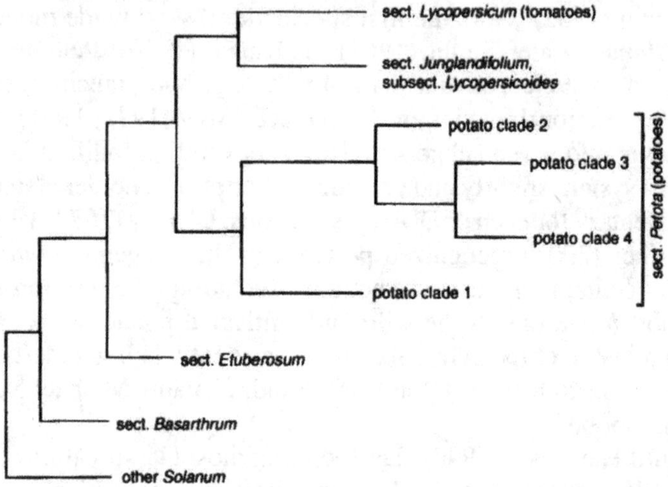


FIGURE 1.7. Cladogram of potato *(Solanum* section *Petota)* and immediate outgroups (see text).

On the basis of cpDNA and morphological data and a classification method that relied on phylogenetic relationships (cladistics, in contrast to phenetics), Spooner et al. (1993) placed tomatoes in the genus *Solanum* (not *Lycopersicon*), and Contreras and Spooner (1999) recognized Hawkes's series *Etuberosa* at the sectional level as section *Etuberosum* (Buk. and Kameraz) Child, at the same taxonomic rank as potatoes.

Recent molecular studies (Spooner et al., 1993; Olmstead and Palmer, 1997; Bohs and Olmstead, 2001; Bohs, 2005) have clarified the definition of *Solanum* and led to new insights into infrageneric classification. These studies established more natural groups, based on branches of a phylogenetic tree (monophyletic "clades") of *Solanum* that includes the previously segregate genera *Lycopersicon, Cyphomandra, Normania,* and *Triguera* (recognized as separate genera by D'Arcy and Hunziker). Bohs (2005) analyzed cpDNA sequence data from more than 100 species of *Solanum* representing the majority of infrageneric groups recognized by previous workers as well as a number of taxa thought to represent distinct clades based on morphology. She recognized 12 major clades within *Solanum,* gave them unranked informal clade designations, and speculated on potential nonmolecular characters that might be used to identify the major clades. She placed potatoes in the "potato" clade that included tomatoes and their relatives, a treatment distinctly different from those of D'Arcy (1972, 1991) and Hunziker (2001) and providing a more natural classification.

## *RELATIONSHIPS WITHIN* SOLANUM *SECTION* PETOTA

Wild potatoes constitute approximately 10 percent of the total species diversity of *Solanum*. Hosaka et al. (1984) analyzed 37 species of section *Petota*, including members of section *Etuberosum* and tomatoes. They used cpDNA, digested with eight endonucleases, and analyzed the data by comparing the banding patterns of the entire cpDNA molecule. They interpreted four clades: (1) the South American species, Mexican polyploids, and *S. verrucosum;* (2) the Mexican diploids (including *S. bulbocastanum* and *S. cardiophyllum*); (3) *S. etuberosum* (section *Etuberosum*); and (4) tomatoes. (The authors of potato species are mentioned in Table 1.A1 at the end this chapter.)

This cpDNA study was extended by Spooner and Sytsma (1992) and Spooner and Castillo (1997). They used 19 cloned cpDNA probes and 22 restriction endonucleases with 86 species representing 17 of the 19 tuber-bearing series of Hawkes (1990; germplasm of series *Cuneoalata* Hawkes and *Olmosiana* Ochoa was not available). These studies defined four exclusively tuber-bearing clades in section *Petota:* (1) the U.S., Mexican, and Central American diploid species, exclusive of *S. bulbocastanum, S. cardiophyllum,* and *S. verrucosum;* (2) *S. bulbocastanum,* and *S. cardiophyllum;* (3) all examined members of the South American series *Piurana* and some South American species classified to other series; and (4) all remaining South American species and the U.S., Mexican, and Central American polyploid species (Figure 1.7). These studies redefined our understanding of outgroup relationships through the definition of *S. bulbocastanum* and *S. cardiophyllum* as a distinct clade and by the definition of a distinct *Piurana* clade.

The separation of *S. bulbocastanum* and *S. cardiophyllum* (clade 2) from the other species of clade 1 was unexpected based on all prior taxonomic interpretations. Rodríguez and Spooner (1997) studied many accessions of all subspecies of these two species and showed that *S. cardiophyllum* subsp. *ehrenbergii* fell in clade 1, while all other subspecies fell in clade 2. These results suggested misclassification of subspecies *ehrenbergii* or introgressive hybridization with members of this clade. Spooner et al. (2004) used these results to justify separate species status for *S. cardiophyllum* subsp. *ehrenbergii* (Tables 1.1 and 1.A1). Because cpDNA is predominately or entirely maternally inherited in *Solanum* (Corriveau and Coleman, 1988), the interpretation of cpDNA "gene trees" into "species trees" awaits confirmation from biparentally inherited nuclear markers (Wendel and Doyle, 1998).

Nuclear markers were used for phylogenetic reconstruction in section *Petota* by Debener et al. (1990; nuclear restriction fragment length polymorphisms [RFLPs]), Bonierbale et al. (1990; nuclear RFLPs), Kardolus et al. (1998; amplified fragment length polymorphisms [AFLPs] that included nuclear and organellar markers), and Volkov et al. (2003; external transcribed spacer regions of ribosomal DNA). The first three studies corroborated the cpDNA studies above in placing the outgroups (*Etuberosa* and/or tomatoes) sister to the Mexican diploids but did not include representatives of cpDNA clade 2 *(S. bulbocastanum* and *S. cardiophyllum)* or clade 3 (series *Piurana* clade) to corroborate the cpDNA results. Volkov et al. (2003) included representatives *S. bulbocastanum* of clade 2 but not *S. cardiophyllum* subsp. *cardiophyllum* to address this relationship, and they did not include members of clade 3. Their results failed to support section *Etuberosum* as an outgroup. The results of Bonierbale et al. (1990) and Volkov et al. (2003) are significant in that unlike the cpDNA results, they grouped a representative of series *Circaeifolia* sister to the Mexican diploids found in clade 1, not with members of clade 4 as expected. Species in series *Circaeifolia* are atypical in that they are South American species with white stellate corollas like the Mexican diploid species. This striking discordance of organellar and nuclear datasets is well documented in other groups (Rieseberg and Soltis, 1991) and suggests a history of hybridization in the evolution of this species. Clearly, a well-supported phylogeny

TABLE 1.1. New determinations of synonymy, new combinations, and newly resurrected names in wild and cultivated potato species *(Solanum* section *Petota)* subsequent to Hijmans and Spooner (2001).

**Huamán and Spooner (2002)**

*Solanum tuberosum* L.

- *S. ajanhuiri* Juz. and Bukasov (Ajanhuiri Group, diploid)
- *S. chaucha* Juz. and Bukasov (Chaucha Group, triploid)
- *S. curtilobum* Juz. and Bukasov (Curtilobum Group, pentaploid)
- *S. juzepczukii* Bukasov (Juzepczukii Group, triploid)
- *S. phureja* Juz. and Bukasov
  - subsp. *hygrothermicum* (not understood well enough yet to assign to a Group)
  - subsp. *phureja* (Phureja Group, diploid)
- *S. stenotomum* Juz. and Bukasov
  - subsp. *goniocalyx* (Juz. and Bukasov) Hawkes (Stenotomum Group, diploid)
  - subsp. *stenotomum* (Stenotomum Group, diploid)
- *S. tuberosum* L.
  - subsp. *andigenum* (Juz. and Bukasov) Hawkes (Andigenum Group, tetraploid)
  - subsp. *tuberosum* (Chilotanum Group—the Chilean land races only; the modern cultivars have yet to be assigned Groups, tetraploid)

**Spooner et al. (2004)**

*S. bulbocastanum*

- *S. bulbocastanum* subsp. *bulbocastanum*
- *S. bulbocastanum* subsp. *dolichophyllum* (Bitter) Hawkes
- *S. bulbocastanum* subsp. *partitum* (Correll) Hawkes

*S. cardiophyllum* Bitter

- *S. cardiophyllum* subsp. *cardiophyllum*
- *S. cardiophyllum* subsp. *lanceolatum* (P. Berthault) Bitter

*S. demissum* Lindley

- *S.* × *semidemissum* Juz.

S. × *edinense* P. Berthault

- *S. edinense* subsp. *edinense* Hawkes
- *S. edinense* subsp. *salamanii* (Hawkes) Hawkes

*S. ehrenbergii* (Bitter) Rydberg

- *S. cardiophyllum* subsp. *ehrenbergii* Bitter

*S. hjertingii* Hawkes

- *S. hjertingii* var. *hjertingii*
- *S. fendleri* var. *physaloides* Correll
- *S. matehualae* Hjerting & T. R. Tarn

TABLE 1.1 *(continued)*

- *S. iopetalum* (Bitter) Hawkes
  - *S. brachycarpum* (Correll) Correll
- *S. stenophyllidium* Bitter
  - *S. brachistotrichium* (Bitter) Rydberg
  - *S. nayaritense* (Bitter) Rydberg
- *S. stoloniferum*
  - *S. fendleri* A. Gray
    - *S. fendleri* subsp. *arizonicum* Hawkes
    - *S. fendleri* subsp. *fendleri*
  - *S. leptosepalum* Correll
  - *S. papita* Rydberg
  - *S. polytrichon* Rydberg
  - *S. stoloniferum* subsp. *moreliae* Hawkes
  - *S. stoloniferum* subsp. *stoloniferum*
- *S. verrucosum* Schlechtendal
  - *S. macropilosum* Correll

*Note:* The listed name is the new one proposed by the authors in the heading, and the indented names are the old, superseded (synonymized) taxa.

of section *Petota* awaits the use of more classes of molecular markers and further morphological studies with a wider range of species.

## *TAXONOMY OF SECTION* PETOTA

Taxonomic research is partly justified on its ability to be used as a predictive tool (Rollins, 1965; Daly et al., 2001). For plant breeders, prediction means that germplasm can be chosen or avoided based on the past positive or negative evaluations of related species. Germplasm evaluations for resistance or agronomic traits, organized taxonomically, are common in the literature. For example, species-specific statements about the breeding value of wild potato germplasm are found in Ross (1986), Hawkes (1990), and Ruiz de Galerreta et al. (1998). Clearly, not all accessions of a species will share all traits, but when one lacks prior evaluation data, taxonomy should provide a useful guide for making inferences about unevaluated germplasm based on present knowledge. For this reason, potato has been the subject of intensive taxonomic research.

The taxonomy of wild and cultivated potatoes continues to be notoriously difficult. Many potato species, sometimes even those that look very different, maintain the ability to hybridize when they come into contact, which further blurs species boundaries. In addition, different taxonomists have applied different taxonomic concepts to recognize and to group species (Spooner and Van den Berg, 1992a). Most accepted species are distinguished from similar ones by a "polythetic morphological species concept"; that is, species are defined by the greatest number of shared features, no single one of which is essential for group membership or sufficient to make an organism a member of a group (Stuessy, 1990). Stated otherwise, species are distinguished only by a complex of morphological features, all or most of which overlap in extent with other species. A wide range of molecular markers continue to be applied to questions of the validity and interrelationships of potato species, and their taxonomic interpretation surely will be modified.

Taxonomic changes of wild and cultivated potatoes are the subject of a continuing series of reviews (Grun, 1990; Spooner and Van den Berg, 1992a; Hawkes, 1997; Hijmans and Spooner, 2001). The last review (Hijmans and Spooner, 2001) updated the taxonomy of Hawkes (1990) through publications from the period 1990-2000. The present review updates Hijmans and Spooner (2001) through changes in the taxonomy of the cultivated species by Huamán and Spooner (2002) and changes in the taxonomy of the North and Central American species by Spooner et al. (2004). Relative to the account in Spooner and Hijmans (2001), the papers by Huamán and Spooner (2002) and Spooner et al. (2004) account for a net loss of 6 species and 6 subspecies for the cultivated potatoes and 9 species, 12 subspecies, and 2 varieties for the wild species. Updated species diversity estimates are 188 wild and 1 cultivated species for section *Petota,* plus 3 species in section *Etuberosa*. The present review also provides speculation on future taxonomic changes through literature research and field collections. Species in Table 1.A1 are indicated as "stable" (likely to be maintained by future work) or placed into groups where future work is needed.

Descriptions of the groups mentioned in Table 1.A1 follow. Because of unresolved questions, and because of the many species involved, designation of species in Table 1.A1 as *S. brevicaule* complex north or south, series *Conicibaccata,* or series *Piurana* is based on the publications mentioned here. For the species not yet studied in these publications, designation of a species to a group is based on speculation from original species descriptions, the excellent descriptions and illustrations in Ochoa (1999), or inferences based on geography for the *S. brevicaule* north and south subsets. Table 1.A1 is bibliographic in nature and should *not* be construed as final acceptance of names or groups, but rather as a working hypothesis. The series are used purely to connect these names to the literature; they do not imply acceptance of these ranks.

## Solanum brevicaule *Complex*

The *S. brevicaule* complex represents a group of approximately 20 morphologically similar species, distributed from the south of Peru to Argentina, that are difficult to distinguish from some land races of *S. tuberosum*. All members of the *S. brevicaule* complex

1. have pinnately dissected leaves
2. have round fruits
3. have rotate to rotate-pentagonal corollas
4. are largely sexually compatible
5. are hypothesized to form occasional natural hybrid swarms
6. have endosperm balance numbers (a phenomenon of sexual compatibility based on ratios of maternal/paternal genomes in the endosperm) matching their ploidy levels
7. are frequently confused in the literature, herbarium, and germplasm collections regarding identifications

They vary in ploidy from diploid ($2n = 2x = 24$), through tetraploid ($2n = 4x = 48$), to hexaploid ($2n = 6x = 72$) (Van den Berg et al., 1998).

The species boundaries were studied using morphological phenetics by Van den Berg et al. (1998), and species boundaries and relationships with molecular marker data by Miller and Spooner (1999). All data were concordant in showing the following:

1. No species-specific morphological characters existed, and any species that may be valid would be distinguished only by use of a complex of widely overlapping traits (polythetic support).
2. At best, two "species" were defined with a north-south geographical partitioning, one from Peru and northwestern Bolivia and another from northwestern Bolivia to northern Argentina. AFLP data (Kardolus et al., 1998; Spooner et al., 2005) further confirmed the north-south split of the two geographical groups of the *S. brevicaule* complex.
3. Within these two geographical groups were contained morphologically distinct species that were not previously considered to be part of the complex, suggesting that the complex was "unnatural" in a taxonomic philosophy that relied on relationships to define and group species.
4. Morphological similarity was not always a good indicator of phylogenetic relationships.
5. The cultivated accessions fell into the north *S. brevicaule* clade.

These results suggest that some of the species in the complex will eventually be placed into synonymy, further reducing the number of names in sect. *Petota*.

## Solanum *Series* Circaeifolia

*Solanum* series *Circaeifolia* has been treated as containing three species and two subspecies (Hawkes, 1990) or two species and two subspecies (Ochoa, 1990). Despite this confusion at the species and subspecies level, all taxa in series *Circaeifolia* are clearly related based on morphological and molecular data. All species of the series have white stellate corollas, entire to little-dissected leaves, and narrow elliptic-lanceolate fruits unlike any other in section *Petota*. Van den Berg and Groendijk-Wilders (1999) analyzed morphological phenetics for the four taxa of series *Circaeifolia* and showed difficulty in distinguishing *S. capsicibaccatum* and *S. circaeifolium* subsp. *quimense*. Van den Berg et al. (2001), however, showed that all four taxa of this series were well distinguished by AFLP data, suggesting that the taxa were valid at some rank (species or subspecies). Further studies are needed to resolve these discrepancies and make final decisions on the species status.

### Solanum *Series* Conicibaccata

Many species of *Solanum* series *Conicibaccata* usually are easily grouped morphologically by leaves with a somewhat parallel-sided morphology and narrowly ovate to elliptical leaflets (Figure 1.3B) and by conical fruits (Figure 1.5D). Most grow in rich organic soil of rain forests. However, the leaf and fruit characters are not consistently distinctive as the leaf morphology can intergrade with those of members from other groups. Some members of the series have fruits only slightly elongate (elliptical), and some apparently unrelated species also have conical fruits (e.g., *S. hintonii, S. iopetalum*).

Castillo and Spooner (1997) examined species status and phylogenetic relationships of 23 out of 40 species of series *Conicibaccata* (as determined by Hawkes, 1990) using cpDNA restriction site variation and morphology. The results defined three main groups within the series: (1) polyploids from central Mexico to southern Ecuador, (2) diploids from northern Peru to Bolivia, and (3) diploids and tetraploids cladistically related to members of series *Piurana* and suggesting misclassification in series *Conicibaccata*. Some species within these ploidy groups were not clearly distinct from others within the group based on morphology. These results suggest that either prior designations of the series were not correct or some species may have to be regarded as synonymous.

### Solanum *Series* Piurana

*Solanum* series *Piurana* includes 15 species, distributed from southern Ecuador to central Peru (Hawkes, 1990). It contains some of the morphologically most diverse species in section *Petota,* but what constitutes a species remains controversial. The most distinctive features of series *Piurana,* used by all major taxonomists of section *Petota* (Correll, 1962; Hawkes, 1990; Ochoa, 1999), are the combination of globose to ovoid fruits and coriaceous glossy leaves.

The problem is that this fruit shape and coriaceous and shiny leaves are present in species placed in other series, and it is difficult to clearly decide what to include in series *Piurana*. All of the above authors have expressed doubt as to the limits of the series, but perhaps the clearest statement of this difficulty was made by Correll (1962, p. 139): "This series, more than any of the others, may be considered a catchall. Paradoxically, its component species are held together not so much by their similarity as by their differences."

The cpDNA restriction site phylogenetic studies of section *Petota* by Spooner and Castillo (1997) and Castillo and Spooner (1997) defined only four clades, with all members of series *Piurana* in a well-defined clade (Figure 1.7, but this clade also included species placed in series *Conicibaccata, Megistacroloba, Yungasensia,* and *Tuberosa*). Field studies in Peru (Salas et al., 2001), combined with insights into tuber morphology from Ochoa (1999), suggested a third character defining the series: many members possess "moniliform" tubers, or tubers arranged like beads along the stolons, unlike the more typical arrangement of single tubers placed at stolon ends. However, this trait is present in some species in series *Conicibaccata* (e.g., *S. irosinum*), so even this trait is difficult to use to absolutely define series.

## TAXONOMY OF CULTIVATED POTATO

Hawkes (1990) recognized seven cultivated species and seven subspecies: *S. ajanhuiri, S. chaucha, S. curtilobum, S. juzepczukii, S. phureja* subsp. *phureja, S. phureja* subsp. *estradae, S. phureja* subsp. *hygrothermicum, S. stenotomum* subsp. *stenotomum, S. stenotomum* subsp. *goniocalyx, S. tuberosum* subsp. *andigenum,* and *S. tuberosum* subsp. *tuberosum*. However, this seven cultivated species taxonomy is not universally accepted (Spooner and Van den Berg, 1992a). For example, the Russian potato taxonomists Bukasov (1971) and Lechnovich (1971) recognized 21 species, including separate species status for *S. tuberosum* subsp. *andigenum* (as *S. andigenum* Juz. and Bukasov) and subsp. *tuberosum* (as *S. tuberosum*). Ochoa (1990, 1999) recognized nine species and 141 infraspecific taxa (subspecies, varieties, and forms, including his unlisted autonyms) for the Bolivian cultivated species alone.

Dodds (1962) had a radically different view of cultivated potato taxonomy. He contended that the morphological characters used by Hawkes (1956) to separate the cultivated species exaggerated the consistency of qualitative and quantitative characters. He also showed that Andean farmers grow land races of all ploidy levels together in the same field and that these can all potentially hybridize. He showed no genetic differentiation of the cultivated diploids (Dodds and Paxman, 1962). In contrast to the above treatments of potatoes as distinct "species," named by nomenclature rules governed by the International Code of Botanical Nomenclature (ICBN; Greuter et al., 2000), Dodds (1962) treated the cultivated species as three species, *S.* ×*curtilobum, S.* ×*juzepczukii,* and *S. tuberosum,* with five "Groups" recognized in the latter. Groups are taxonomic categories used by the International Code of Nomenclature for Cultivated Plants (ICNCP; Brickell et al., 2004) solely to associate cultivated plants with traits that are of use to agriculturists, not to group phylogenetically related organisms. The ICNCP recognizes the complex hybrid origins of most crops and focuses on a classification of convenience to users and the nomenclature stability needed for trade (Hetterscheid and Brandenburg, 1995).

Huamán and Spooner (2002) studied morphological differences of all land race potato species (i.e., all cultivated potatoes except the modern cultivars) in a field plot in Peru. Their results showed some degree of morphological support for *S. ajanhuiri, S. chaucha, S. curtilobum, S. juzepczukii,* and *S. tuberosum* subsp. *tuberosum,* but little for *S. phureja* subsp. *phureja, S. stenotomum* subsp. *goniocalyx, S. stenotomum* subsp. *stenotomum,* and *S. tuberosum* subsp. *andigenum.* Most characters overlapped extensively with those of other species (polythetic support). These results led Huamán and Spooner (2002) to recognize all land race populations of cultivated potatoes as a single species, *S. tuberosum,* with the eight Groups: Ajanhuiri Group, Andigenum Group, Chaucha Group, Chilotanum Group, Curtilobum Group, Juzepczukii Group, Phureja Group, and Stenotomum Group.

Some phenetic support means that a reasonable argument could be made to recognize *S. ajanhuiri, S. chaucha, S. curtilobum, S. juzepczukii,* and *S. tuberosum* subsp. *tuberosum* as separate species or subspecies and all the other taxa as Groups under a separate cultivated species *S. andigenum.* Support for a separate taxon treatment was provided by Raker and Spooner (2002), who demonstrated that most of the land race populations of *S. tuberosum* subsp. *tuberosum* can be distinguished with microsatellite data from most populations of *S. tuberosum* subsp. *andigenum,* and we expect that molecular support will be provided for *S. ajanhuiri, S. curtilobum,* and *S. juzepczukii.* No final "correct" taxonomic treatment exists in such situations, and Huamán and Spooner (2002) used the Group classification for cultivated potatoes because of predominant polythetic morphological support, reticulate origins (Huamán et al., 1982, 1983; Schmiediche et al., 1982; Cribb and Hawkes, 1986; Hawkes, 1990), evolutionary dynamics or continuing hybridization, and their classification philosophy of the appropriateness of the ICNCP for cultivated species.

## *COLLECTION, CONSERVATION, AND USE OF GERMPLASM*

Wild and cultivated potato genetic resources have proven value in breeding programs for disease resistance, environmental tolerances, and other agronomic traits of interest (Ross, 1986; Plaisted and Hoopes, 1989; Hawkes, 1990; Spooner and Bamberg, 1994; Ruiz de Galerreta et al., 1998; Jansky, 2000). Plaisted and Hoopes (1989) documented the parentage of North American cultivars, citing the importance of both wild species and the *S. tuberosum* Phureja Group, Andigenum Group, and Tuberosum Group. The following 14 wild species have entered into the parentage of European and North American cultivars: *S. acaule, S. chacoense, S. commersonii, S. demissum, S. kurtzianum, S. maglia, S. microdontum, S. raphanifolium, S. sparsipilum, S. spegazzinni, S. stoloniferum, S. megistacrolobum* subsp. *toralapanum, S. vernei,* and *S. verrucosum. Solanum demissum,* a hexaploid Mexican species with late blight resistance, has been most widely used, but the other wild species have conferred resistance to a variety of viral, fungal, and bacterial diseases as well as nematode and insect pests of potatoes. Wild species and primitive cultivars have improved agronomic traits such as yield, specific gravity, chipping quality, and suppression of enzymatic browning. Current research is discovering the disease resistance and improved horticultural traits present in many other wild species, and this list is sure to increase.

The economic value of genetic resources has stimulated more than 40 potato-collecting expeditions over the past 40 years by various international organizations. Correll (1962), Hawkes and Hjerting (1969, 1989), Ochoa (1990), and Spooner et al. (2004) have summarized accounts of these potato collections. Further germplasm needs to be collected, particularly from Peru.

Seeds and tubers of these collections have been sent to potato genebanks worldwide. The major genebanks for potatoes are Instituto Nacional de Tecnología Agropecuaria, Balcarce, Argentina; Universidad Austral de Chile, Valdivia, Chile; Colección Central Colombiana, Bogotá, Colombia; Institüt fur Pflanzenbau und Pflanzenzüchtung, Germany; Landwirtschaftswissenschaften, Gross-Lüsewitz, Germany; International Potato Center, Lima, Peru; Commonwealth Potato Collection, Dundee, United Kingdom; The United States Potato Genebank, Sturgeon Bay, United States; and N.I. Vavilov Research Institute of Plant Industry, Leningrad, Russia (Hawkes, 1990). An intergenebank catalogue documents common and unique holdings of these collections (Huamán et al., 2000).

Most potato germplasm accessions are preserved as botanical seeds. Seeds are easier to maintain in a disease-free state, require less labor and materials for storage and shipping, contain greater genetic diversity, and have much greater longevity than vegetative propagules (tubers and in vitro shoot tips). Clonal stocks are maintained for collections that cannot be increased sexually, for cultivated collections where the genotype must be maintained, and for certain mapping populations.

In conclusion, wild and cultivated potatoes represent an amazing diversity of forms of great use for potato improvement. The taxonomy of wild and cultivated potatoes is complex, owing in part to the ability of many species to easily hybridize. Taxonomic research is continuing to refine our understanding of what constitutes a species and how these species are interrelated, to aid in the organization and planned use of these genetic resources by potato breeders.

TABLE 1.A1. Wild and cultivated potato species *(Solanum* section *Petota)* and wild species of close outgroup relatives in section *Etuberosa (S. etuberosum, S. fernandezianum, S. palustre).*

| Taxon (putative hybrid origins)[a] | Code[b] | Country[c] | Ploidy and (EBN)[d] | Status of name[e] |
|---|---|---|---|---|
| **Wild species** | | | | |
| *Solanum acaule* Bitter | acl | ARG, BOL, PER | 4*x* (2EBN) | A |
| subsp. *acaule* | | | | A |
| f. *acaule* | | | | |
| f. *incuyo* Ochoa | inc | PER | 4*x* | |
| subsp. *aemulans* (Bitter and Wittm.) Hawkes and Hjert. | aem | ARG | 4*x* (2EBN) | A |
| subsp. *palmirense* J. Kardolus | pal | ECU | 6*x* | A |
| subsp. *punae* (Juz.) Hawkes and Hjert. | pne | PER | 4*x* (2EBN) | A |
| *S. achacachense* Cárdenas | ach | BOL | 2*x* | B |
| *S. acroglossum* Juz. | acg | PER | 2*x* (2EBN) | D |
| *S. acroscopicum* Ochoa | acs | PER | 2*x* | B |
| *S. agrimonifolium* Rydb. | agf | GUA, HON, MEX | 4*x* (2EBN) | E, H |
| *S. alandiae* Cárdenas | aln | BOL | 2*x* | |
| *S. albicans* (Ochoa) Ochoa | alb | ECU, PER | 6*x* (4EBN) | A |
| *S. albornozii* Correll | abz | ECU | 2*x* (2EBN) | D |
| *S. amayanum* Ochoa | amy | PER | 2*x* (2EBN) | B |
| *S. ambosinum* Ochoa | amb | PER | 2*x* (2EBN) | B |
| *S. anamatophilum* Ochoa | amp | PER | 2*x* (2EBN) | D |
| *S. ancophilum* (Correll) Ochoa | acp | PER | 2*x* (2EBN) | I |
| *S. ancoripae* Ochoa | anp | PER | 2n | B |
| *S. andreanum* Baker | adr | COL, ECU | 2*x* (2EBN) | D |
| *S.* × *arahuayum* Ochoa (med × wtm) | ara | PER | 2*x* | |
| *S. ariduphilum* Ochoa | adp | PER | 2*x* (2EBN) | D |
| *S. arnezii* Cárdenas | arz | BOL | | |
| *S. augustii* Ochoa | agu | PER | 2*x* (1EBN) | D |
| *S. avilesii* Hawkes and Hjert. | avl | BOL | 2*x* | C |
| *S. ayacuchense* Ochoa | ayc | ***PER*** | *2x (2EBN)* | E |
| *S. aymaraesense* Ochoa | aym | PER | 2*x* | B |
| *S. berthaultii* Hawkes | ber | BOL | 2*x* (2EBN) | F |
| *S. bill-hookeri* Ochoa | bhk | PER | 2*x* | B |
| *S.* × *blanco-galdosii* Ochoa (amp × plq) | blg | PER | 2*x* (2EBN) | D |

TABLE 1.A1 *(continued)*

| Taxon (putative hybrid origins)[a] | Code[b] | Country[c] | Ploidy and (EBN)[d] | Status of name[e] |
|---|---|---|---|---|
| *S. boliviense* Dunal | blv | BOL | 2*x* (2EBN) | G |
| subsp. *astleyi* (Hawkes and Hjert.) D. M. Spooner, M. Ugarte, and P. M. Skroch | ast | BOL | 2*x* (2EBN) | G |
| subsp. *boliviense* | | | | G |
| *S. bombycinum* Ochoa | bmb | BOL | 4*x* | E |
| *S. brevicaule* Bitter | brc | BOL | 2*x* (2EBN) | C |
| *S.* × *bruecheri* Correll (*S. gourlayi* [= lph by Ochoa, 1990] × inf) | bru | ARG | | |
| *S. buesii* Vargas | bue | PER | 2*x* (2EBN) | E |
| *S. bukasovii* Juz. | buk | PER | 2*x* (2EBN) | B |
| f. *bukasovii* | | | | B |
| f. *multidissectum* (Hawkes) Ochoa | mlt | PER | 2*x* (2EBN) | B |
| *S. bulbocastanum* Dunal | blb | MEX, GUA, HON | 2*x* (1EBN) | H |
| *S. burkartii* Ochoa | brk | PER | 2*x* | E |
| *S. burtonii* Ochoa | brt | ECU | 3*x* | |
| *S. cajamarquense* Ochoa | cjm | PER | 2*x* (1EBN) | D |
| *S. calacalinum* Ochoa | cln | ECU | 2*x* | E |
| *S. calvescens* Bitter | clv | BRA | 3*x* | |
| *S. candolleanum* P. Berthault | cnd | BOL, PER | 2*x* (2EBN) | C |
| *S. cantense* Ochoa | cnt | PER | 2*x* (2EBN) | D |
| *S. cardiophyllum* Lindl. | cph | MEX | 2*x* (1EBN), 3*x* | H |
| *S. chacoense* Bitter | chc | ARG, BOL, PAR, URU | 2*x* (2EBN) | I (species only) |
| subsp. *chacoense* | chc | | | |
| subsp. *muelleri* (Bitter) Hawkes and Hjert. | mue | ARG, BRA | 2*x* (2EBN) | |
| *S. chancayense* Ochoa | chn | PER | 2*x* (1EBN) | D |
| *S. chilliasense* Ochoa | chl | ECU | 2*x* (2EBN) | D |
| *S. chillonanum* Ochoa | chi | PER | 2*x* | B |
| *S. chiquidenum* Ochoa | chq | PER | 2*x* (2EBN) | D |
| var. *chiquidenum* | chq | PER | | D |
| f. *amazonense* Ochoa | ama | PER | 2*x* | D |
| f. *chiquidenum* | chq | | | D |
| var. *gracile* Ochoa | gra | PER | 2*x* | D |

TABLE 1.A1 *(continued)*

| Taxon (putative hybrid origins)[a] | Code[b] | Country[c] | Ploidy and (EBN)[d] | Status of name[e] |
|---|---|---|---|---|
| var. *robustum* Ochoa | rob | PER | 2*x* | D |
| *S. chomatophilum* Bitter | chm | | | D (species only) |
| var. *chomatophilum* | chm | ECU, PER | 2*x* (2EBN) | |
| f. *angustifolium* Correll | ang | ECU | | |
| f. *chomatophilum* | chm | PER | | |
| f. *sausianense* Ochoa | sau | PER | | |
| var. *subnivale* Ochoa | sbn | PER | 2*x* | |
| *S. circaeifolium* Bitter | crc | BOL | 2*x* (1EBN) | J |
| var. *capsicibaccatum* (Cárdenas) Ochoa | cap | BOL | 2*x* (1EBN) | J |
| var. *circaeifolium* | crc | | | J |
| *S. clarum* Correll | clr | MEX, GUA | 2*x* | H |
| *S. coelestipetalum* Vargas | cop | PER | 2*x* (2EBN) | B |
| *S. colombianum* Bitter | col | COL, ECU, VEN | 4*x* (2EBN) | E |
| *S. commersonii* Dunal | cmm | ARG, BRA, URU | 2*x* (1EBN) | I (species only) |
| subsp. *commersonii* | cmm | | | |
| subsp. *malmeanum* (Bitter) Hawkes and Hjert. | mlm | ARG, BRA, PAR, URU | 2*x* (1EBN) | |
| *S. contumazaense* Ochoa | ctz | | 2*x* (2EBN) | |
| *S. demissum* Lindl. | dms | GUA, MEX | 6*x* (4EBN) | H |
| *S.* × *doddsii* Correll (aln × chc) | dds | BOL | 2*x* (2EBN) | |
| *S. dolichocremastrum* Bitter | dcm | PER | 2*x* (1EBN) | D? |
| *S. donachui* (Ochoa) Ochoa | dnc | COL | | E |
| *S.* × *edinense* P. Berthault | edn | MEX | 5*x* | H |
| *S. ehrenbergii* (Bitter) Rydb. | ehr | MEX | 2*x* (1EBN) | H |
| *S. etuberosum* Lindl. | etb | CHL | 2*x* (1EBN) | K |
| *S. fernandezianum* Phil. | frn | CHL | 2*x* (1EBN) | K |
| *S. flahaultii* Bitter | flh | COL | 4*x* | E |
| *S. flavoviridens* Ochoa | flv | BOL | | |
| *S. gandarillasii* Cárdenas | gnd | BOL | 2*x* (2EBN) | I |
| *S. garcia-barrigae* Ochoa | gab | COL | | E |
| *S. gracilifrons* Bitter | grc | PER | 2*x* | |
| *S. guerreroense* Correll | grr | MEX | 6*x* (4EBN) | H |
| *S. guzmanguense* Whalen and Sagást. | gzm | PER | 2*x* (1EBN) | I |
| *S. hastiforme* Correll | hsf | PER | 2*x* (2EBN) | I |
| *S. hintonii* Correll | hnt | MEX | | H |
| *S. hjertingii* Hawkes | hjt | MEX | 4*x* (2EBN) | H |

TABLE 1.A1 *(continued)*

| Taxon (putative hybrid origins)[a] | Code[b] | Country[c] | Ploidy and (EBN)[d] | Status of name[e] |
|---|---|---|---|---|
| *S. hoopesii* Hawkes and K. A. Okada | hps | BOL | 4*x* | C |
| *S. hougasii* Correll | hou | MEX | 6*x* (4EBN) | H |
| *S. huancabambense* Ochoa | hcb | PER | 2*x* (2EBN) | |
| *S. huancavelicae* Ochoa | hcv | PER | 2*x* (2EBN) | B |
| *S. huarochiriense* Ochoa | hro | PER | 2*x* (2EBN) | D |
| *S. humectophilum* Ochoa | hmp | PER | 2*x* (1EBN) | D |
| *S. hypacrarthrum* Bitter | hcr | PER | 2*x* (1EBN) | D |
| *S. immite* Dunal | imt | PER | 2*x* (1EBN), 3*x* | D |
| var. *immite* | | | | |
| var. *vernale* Correll | vrl | PER | | |
| *S. incahuasinum* Ochoa | inh | PER | 2*x* (1EBN) | |
| *S. incamayoense* K. A. Okada and A. M. Clausen | inm | ARG | 2*x* | C |
| *S. incasicum* Ochoa | ins | PER | 2*x* (2EBN) | I |
| *S.* × *indunii* K. A. Okada and A. M. Clausen (acl × mga) | ind | ARG | 36 | |
| *S. infundibuliforme* Phil. | ifd | ARG, BOL | 2*x* (2EBN) | I |
| *S. ingifolium* Ochoa | igf | PER | 2*x* (1EBN) | D? |
| *S. iopetalum* (Bitter) Hawkes | iop | MEX | 6*x* (4EBN) | H |
| *S. irosinum* Ochoa | irs | PER | 2*x* (2EBN) | D or E |
| var. *irosinum* | irs | | | |
| var. *tarrosum* Ochoa | trr | PER | 2*x* | |
| *S. jaenense* Ochoa | jnn | PER | 6*x* (4EBN) | D or E |
| *S. jalcae* Ochoa | jlc | PER | 2*x* (2EBN) | D |
| var. *jalcae* | | | | |
| var. *pubescens* Correll | pub | PER | | |
| *S. jamesii* Torr. | jam | MEX, USA | 2*x* (1EBN) | H |
| *S. kurtzianum* Bitter and Wittm. | ktz | ARG | 2*x* (2EBN) | I |
| *S. laxissimum* Bitter | lxs | PER | 2*x* (2EBN) | E |
| *S. leptophyes* Bitter | lph | BOL, PER | 2*x* (2EBN) 4*x* (4EBN) | C |
| *S. lesteri* Hawkes and Hjert. | les | MEX | 2*x* | H |
| *S. lignicaule* Vargas | lgl | PER | 2*x* (1EBN) | I |
| *S. limbaniense* Ochoa | lmb | PER | 2*x* (2EBN) | E |
| *S.* × *litusinum* Ochoa (ber × tar) | lit | BOL | 2*x* (2EBN) | F |
| *S. lobbianum* Bitter | lbb | COL | 4*x* (2EBN) | E |
| *S. longiconicum* Bitter | lgc | CRI, PAN | 4*x* | E, H |
| *S. longiusculus* Ochoa | lgs | PER | 2*x* | B |
| *S. lopez-camarenae* Ochoa | lpc | PER | 2*x* (1EBN) | D |
| *S. maglia* Schltdl. | mag | CHL | 2*x*, 3*x* | I |

TABLE 1.A1 *(continued)*

| Taxon (putative hybrid origins)[a] | Code[b] | Country[c] | Ploidy and (EBN)[d] | Status of name[e] |
|---|---|---|---|---|
| *S. marinasense* Vargas | mrn | PER | 2*x* (2EBN) | B |
| *S. medians* Bitter | med | PER | 2*x* (2EBN), 3*x* | I (species only) |
| var. *autumnale* Correll | aut | PER | 2*x* (2EBN) | |
| var. *medians* | med | PER | | |
| *S. megistacrolobum* Bitter | mga | ARG, PER, BOL | 2*x* (2EBN) | L |
| subsp. *megistacrolobum* | mga | | | L |
| f. *megistacrolobum* | mga | | | |
| f. *purpureum* Ochoa | prp | PER | 2*x* | |
| subsp. *toralapanum* (Cárdenas and Hawkes) R. B. Giannattasio and D. M. Spooner | tor | ARG, PER, BOL | 2*x* (2EBN) | L |
| *S.* × *michoacanum* (Bitter) Rydb. (blb × pnt) | mch | MEX | 2*x* | H |
| *S. microdontum* Bitter | mcd | ARG, BOL | 2*x* (2EBN), 3*x* | M (species only) |
| var. *microdontum* | mcd | | | |
| var. *montepuncoense* Ochoa (mcd × vio) | mon | BOL | 2*x* | |
| *S. minutifoliolum* Correll | min | ECU | 2*x* (1EBN) | I |
| *S. mochiquense* Ochoa | mcq | PER | 2*x* (1EBN) | I |
| *S. morelliforme* Bitter and G. Muench | mrl | MEX, GUA, HON | 2*x* | H |
| *S. moscopanum* Hawkes | msp | COL | 6*x* (4EBN) | E |
| *S. multiinterruptum* Bitter | mtp | PER | 2*x* (2EBN) | B |
| var. *machaytambinum* Ochoa | mac | PER | | |
| var. *multiinterruptum* | mtp | | | |
| f. *albiflorum* Ochoa | alf | PER | 2*x* | |
| f. *multiinterruptum* | mtp | | | |
| *S. nemorosum* Ochoa | nmr | PER | 6*x* (4EBN) | E |
| *S. neocardenasii* Hawkes and Hjert. | ncd | BOL | 2*x* | I |
| *S. neorosii* Hawkes and Hjert. | nrs | ARG | 2*x* | |
| *S. neovalenzuelae* L. López | nvz | COL | 4*x* | E |
| *S. neovargasii* Ochoa | nvg | PER | 2*x* | |
| *S. neovavilovii* Ochoa | nvv | PER | 2*x* (2EBN) | |
| *S.* × *neoweberbaueri* Wittm. (med × chc) | nwb | PER | 3*x* | |
| *S. nubicola* Ochoa | nub | PER | 4*x* (2EBN) | E |
| *S. okadae* Hawkes and Hjert. | oka | ARG, BOL | 2*x* | C |
| *S. olmosense* Ochoa | olm | ECU, PER | 2*x* (2EBN) | D? |

TABLE 1.A1 *(continued)*

| Taxon (putative hybrid origins)[a] | Code[b] | Country[c] | Ploidy and (EBN)[d] | Status of name[e] |
|---|---|---|---|---|
| *S. oplocense* Hawkes | opl | ARG, BOL | 2*x* (2EBN)<br>4*x* (4EBN)<br>6*x* (4EBN) | N |
| *S. orocense* Ochoa | oro | COL | | E |
| *S. orophilum* Correll | orp | PER | 2*x* (2EBN) | B |
| *S. ortegae* Ochoa | ort | PER | 2*x* | B |
| *S. otites* Dunal | oti | COL, VEN | | E |
| *S. oxycarpum* Schiede | oxc | MEX | 4*x* (2EBN) | E, H |
| *S. palustre* Poepp. | pls | ARG, CHL | 2*x* (1EBN) | K |
| *S. pampasense* Hawkes | pam | PER | 2*x* (2EBN) | B |
| *S. pamplonense* L. López | ppl | COL | 4*x* | E |
| *S. pascoense* Ochoa | psc | PER | 2x | D |
| *S. paucijugum* Bitter | pcj | ECU | 4*x* (2EBN) | D |
| *S. paucissectum* Ochoa | pcs | PER | 2*x* (2EBN) | D |
| *S. peloquinianum* Ochoa | plq | PER | 2*x* (2EBN) | D |
| *S. pillahuatense* Vargas | pll | PER | 2*x* (2EBN) | E |
| *S. pinnatisectum* Dunal | pnt | MEX | 2*x* (1EBN) | H |
| *S. piurae* Bitter | pur | PER | 2*x* (2EBN) | D |
| *S. polyadenium* Greenm. | pld | MEX | 2*x* | H |
| *S. puchupuchense* Ochoa | pch | BOL, PER | 2*x* | B |
| *S. raphanifolium* Cárdenas and Hawkes | rap | PER | 2*x* (2EBN) | O |
| *S. raquialatum* Ochoa | raq | PER | 2*x* (1EBN) | D |
| *S.* × *rechei* Hawkes and Hjert. (ktz × mcd) | rch | ARG | 2*x*, 3*x* | P |
| *S. regularifolium* Correll | rgf | ECU | 2*x* | |
| *S. rhomboideilanceolatum* Ochoa | rhl | PER | 2*x* (2EBN) | |
| *S.* × *ruiz-lealii* Brücher (chc × ktz) | rzl | ARG | | |
| *S. salasianum* Ochoa | sls | PER | 2*x* | E |
| *S.* ×sambucinum Rydb. (ehr × pnt) | smb | MEX | 2*x* | H |
| *S. sanctae-rosae* Hawkes | sct | ARG | 2*x* (2EBN) | I |
| *S. sandemanii* Hawkes | snd | PER | 2*x* (2EBN) | I |
| *S. santolallae* Vargas | san | PER | 2*x* (2EBN) | E |
| *S. sarasarae* Ochoa | srs | PER | 2*x* (2EBN) | B |
| *S. sawyeri* Ochoa | swy | PER | 2*x* (2EBN) | B |
| *S. saxatilis* Ochoa | sax | PER | 2*x* (2EBN) | B |
| *S. scabrifolium* Ochoa | scb | PER | 2*x* | |
| *S. schenckii* Bitter | snk | MEX | 6*x* (4EBN) | H |
| *S.* ×*setulosistylum* Bitter (chc ×spg) | stl | ARG | 2*x* | |
| *S. simplicissimum* Ochoa | smp | PER | 2*x* (1EBN) | D |

TABLE 1.A1 *(continued)*

| Taxon (putative hybrid origins)[a] | Code[b] | Country[c] | Ploidy and (EBN)[d] | Status of name[e] |
|---|---|---|---|---|
| *S. soestii* Hawkes and Hjert. | sst | BOL | 2*x* | J |
| *S. sogarandinum* Ochoa | sgr | PER | 2*x* (2EBN) | I |
| *S. solisii* Hawkes | sol | ECU | | D |
| *S. sparsipilum* (Bitter) Juz. and Bukasov | spl | BOL, PER | 2*x* (2EBN) | C |
| *S. spegazzinii* Bitter | spg | ARG | 2*x* (2EBN) | I |
| *S. stenophyllidium* Bitter | sph | MEX | 2*x* (1EBN) | H |
| *S. stoloniferum* Schltdl. and Bouchet | sto | MEX, USA | 4*x* (2EBN) | H |
| *S. subpanduratum* Ochoa | sup | VEN | 4*x* | E |
| *S.* ×*sucrense* Hawkes (adg × opl) | scr | BOL | 4*x* (4EBN) | |
| *S. sucubunense* Ochoa | suc | COL | | E |
| *S. tacnaense* Ochoa | tcn | PER | 2*x* (2EBN) | |
| f. *decurrentialanum* (Ochoa) Correll | dec | PER | 2*x* | |
| f. *tacnaense* | tcn | | | |
| *S. tapojense* Ochoa | tpj | PER | 2*x* (2EBN) | B |
| *S. tarapatanum* Ochoa | trp | PER | 2*x* | I |
| *S. tarijense* Hawkes | tar | ARG, BOL | 2*x* (2EBN) | F |
| *S. tarnii* Hawkes and Hjert. | trn | MEX | 2*x* | H |
| *S. taulisense* Ochoa | tau | PER | 2*x* (2EBN) | |
| *S. trifidum* Correll | trf | MEX | 2*x* (1EBN) | H |
| *S. trinitense* Ochoa | trt | PER | 2*x* (1EBN) | I |
| *S. tundalomense* Ochoa | tnd | ECU | 6*x* (4EBN) | E |
| *S. tuquerrense* Hawkes | tuq | COL, ECU | 4*x* (2EBN) | D |
| *S. ugentii* Hawkes and K. A. Okada | ugt | BOL | 4*x* | C |
| *S. urubambae* Juz. | uru | PER | 2*x* (2EBN) | |
| f. *chakchabambense* Ochoa | chk | PER | 2*x* | |
| f. *urubambae* | | | | |
| f. *velutinum* (Correll) Ochoa | vel | PER | | |
| *S.* ×*vallis-mexici* Juz. (sto × ver) | vll | MEX | 3*x* | H |
| *S. velardei* Ochoa | vlr | PER | 2*x* | B |
| *S. venturii* Hawkes and Hjert. | vnt | ARG | 2*x* (2EBN) | |
| *S. vernei* Bitter and Wittm. | vrn | | | I |
| subsp. *ballsii* (Hawkes) Hawkes and Hjert. | bal | ARG | 2*x* (2EBN) | |
| subsp. *vernei* | vrn | ARG | 2*x* (2EBN) | |
| *S. verrucosum* Schltdl. | ver | | 2*x* (2EBN) | H |
| *S. vidaurrei* Cárdenas | vid | ARG, BOL | 2*x* (2EBN) | I |
| *S.* ×*viirsoii* K. A. Okada and A. M. Clausen (acl × ifd) | vrs | ARG | 3*x* | |

TABLE 1.A1 *(continued)*

| Taxon (putative hybrid origins)[a] | Code[b] | Country[c] | Ploidy and (EBN)[d] | Status of name[e] |
|---|---|---|---|---|
| *S. violaceimarmoratum* Bitter | vio | BOL | 2*x* (2EBN) | |
| *S. virgultorum* (Bitter) Cárdenas and Hawkes | vrg | BOL | 2*x* | |
| *S. wittmackii* Bitter | wtm | PER | 2*x* (1EBN) | I |
| *S. woodsonii* Correll | wds | PAN | | E, H |
| *S. yamobambense* Ochoa | ymb | PER | 2*x* | D |
| *S. yungasense* Hawkes | yun | BOL, PER | 2*x* (2EBN) | I |
| **Cultivated species** | | | | |
| *S. tuberosum* L., a single cultivated species with eight Groups | | | | |
| Ajanhuiri Group | ajh | BOL, PER | 2*x* | |
| Andigenum Group | adg | widespread in the Andes | 4*x* (4EBN) | |
| Chaucha Group | cha | BOL, PER | 3*x* | |
| Chilotanum Group (the Chilean land races only; the modern cultivars traditionally classified as *S. tuberosum* subsp. *tuberosum* have yet to be assigned Groups) | chl | CHL | 4*x* (4EBN) | |
| Curtilobum Group | cur | BOL, PER | 4*x* (4EBN), most 5*x* | |
| Juzepczukii Group | juz | BOL, PER | 3*x* | |
| Phureja Group | phu | widespread in the Andes | 2*x* (2EBN) | |
| Stenotomum Group | stn | BOL, PER | 2*x* (2EBN) | |

*Source:* Based on Spooner and Hijmans (2001), with subsequent changes as outlined in Table 1.1.

*Note:* The country code, EBN, and ploidy are listed only at the species level in the case of autonyms.

[a]Author abbreviations are as standardized by Brummitt and Powell (1992). Putative hybrid origins are from Hawkes (1990) and from Ochoa (1999) for *S. ×arahuayum*, *S. ×blanco-galdosii*, and *S. ×neoweberbaueri*.

[b]Species and subspecies codes follow Spooner and Hijmans (2001).

[c]Country abbreviations are ARG, Argentina; BOL, Bolivia; BRA, Brazil; CHL, Chile; COL, Colombia; CRI, Costa Rica; ECU, Ecuador; GUA, Guatemala; HON, Honduras; MEX, Mexico; PAN, Panama; PER, Peru; URU, Uruguay; USA, United States; VEN, Venezuela.

[d]See Spooner and Hijmans (2001) for references to ploidy and EBN determinations.

[e]This column represents, through the publications cited below or through literature research, field collections, and our speculation, the future taxonomic status of most wild and cultivated species in this list. The status of some of these names will change in rank (from species to sub-

species or subspecies to species), or more likely some names will be placed in synonymy with other species. The status of some names is speculated to be stable; that is, they will remain as a good species (e.g., G, H, I, K, L, M, N, O, P). In other cases putative group relationships are provided (e.g., B, C, D, E, F, J), and some reductions in names within these groups is possible. For other names we do not know enough to speculate about species status or relationships and leave these cells empty.

A. The phenetic analysis of Kardolus (1999) suggested that these names are stable, although we find it very difficult to distinguish *S. acaule* subsp. *acaule* from subsp. *punae*, and subsp. *acaule* f. *incuyo* needs evaluation.
B. Possible member of *S. brevicaule* complex north (see text).
C. Possible member of *S. brevicaule* complex south (see text).
D. Possible member of *Solanum* series *Piurana* (see text).
E. Possible member of *Solanum* series *Conicibaccata* (see text).
F. The morphological phenetic analysis of Spooner and Van den Berg (1992b) suggests that the names *S. berthaultii*, *S. litusinum*, and *S. tarijense* may need to be combined into one.
G. The RAPD analysis of Spooner et al. (1997) suggests these names are stable.
H. Species from North and Central America have been well researched in different phenetic and molecular studies (Spooner and Hijmans, 2001), were recently documented in a monograph (Spooner et al., 2004) and are likely to remain stable.
I. A phenetically very distinctive species from South America that likely will remain as a good species.
J. Member of *Solanum* series *Circaeifolia* (see text).
K. The three species in *Solanum* section *Etuberosum* were studied using field and herbarium collections by Contreras and Spooner (1999) and likely will remain stable.
L. The two subspecies in *S. megistacrolobum* were studied using morphological and molecular marker data by Giannattasio and Spooner (1994a,b) and likely will remain stable.
M. The species *S. microdontum* was studied using morphological phenetics by Van den Berg and Spooner (1992) and likely will remain stable.
N. *Solanum oplocense* was shown to be a well-defined species using morphological phenetics by Van den Berg et al. (1998) and using molecular marker data by Miller and Spooner (1999), but it was not defined in the AFLP study of Spooner et al (2005).
O. *Solanum raphanifolium* is a phenetically well-defined species and well characterized by cpDNA data (Spooner et al., 1991; Spooner and Castillo, 1997) and likely will remain stable. Clausen and Spooner (1998) supported the orgin of *S.* × *rechi* from *S. kurtzianum* and *S. microdontum*.
P. *Solanum* ×*rechei* is of clear hybrid origin between *S. kurtzianum* and *S. microdontum* (Clausen and Spooner, 1998). Although it may be hard to distinguish from them it is clearly worthy of its hybrid designation and likely will remain stable.

## REFERENCES


Anonymous. (1984). *Potatoes for the Developing World*. Lima: International Potato Centre (CIP).

Bohs, L. (2005). Major clades in *Solanum* based on *ndh*F sequence data. *Monographs in Systematic Botany from the Missouri Botanical Garden, St. Louis, Missouri 104*:27-49.

Bohs, L. and Olmstead, R.G. (2001). A reassessment of *Normania* and *Triguera* (Solanaceae). *Plant Systematics and Evolution 228*:33-48.

Bonierbale, M., Ganal, W., and Tanksley, S.D. (1990). Applications of restriction fragment length polymorphisms and genetic mapping in potato breeding and molecular genetics. In Vayda, M.E. and Park, W.D. (Eds.), *The Molecular and Cellular Biology of the Potato* (pp. 13-24). Wallingford, U.K.: CAB International.

Brickell, C.D., Baum, B.R., Hetterscheid, W.L.A., Leslie, A.C., McNeill, J., Trehane, P., Vrugtman, F., and Wiersema, J.H. (2004). International code of nomenclature of cultivated plants. *Acta Horticulturae 647*:1-123.

Brummitt, R.K. and Powell, C.E. (1992). *Authors of Plant Names: A List of Authors of Scientific Names of Plants, with Recommended Standard Forms of Their Names, Including Abbreviations.* Kew, U.K.: Royal Botanic Gardens.

Bukasov, S.M. (1971). Cultivated potato species. In Bukasov, S.M. (Ed.), *Flora of Cultivated Plants*, vol. 9 (pp. 5-40). Leningrad, Russia: Kolos.

Castillo, R. and Spooner, D.M. (1997). Phylogenetic relationships of wild potatoes, *Solanum* series *Conicibaccata* (section *Petota*). *Systematic Botany 22*:45-83.

Clausen, A.M. and Spooner, D.M. (1998). Molecular support for the hybrid origin of the wild potato species *Solanum* ×*rechei (Solanum* sect. *Petota). Crop Science 38*:858-865.

Contreras, A.M. and Spooner, D.M. (1999). Revision of *Solanum* section *Etuberosum*. In Nee, M., Symon, D.E., and Jessop, J.P. (Eds.), *Solanaceae IV: Advances in Biology and Utilization* (pp. 227-245). Kew, U.K.: Royal Botanic Gardens.

Correll, D.S. (1962). The potato and its wild relatives. *Contributions from Texas Research Foundation: Botanical Studies 4*:1-606.

Corriveau, J.L. and Coleman, A.W. (1988). Rapid screening method to detect potential biparental inheritance of plastid DNA and results for over 200 angiosperm species. *American Journal of Botany 75*:1443-1458.

Cribb, P.J. and Hawkes, J.G. (1986). Experimental evidence for the origin of *Solanum tuberosum* subspecies *andigenum*. In D'Arcy, W.G. (Ed.), *Solanaceae: Biology and Systematics* (pp. 384-404). New York: Columbia University Press.

Cutter, E.G. (1992). Structure and development of the potato plant. In Harris, P.M. (Ed.), *The Potato Crop* (pp. 65-161). London: Chapman and Hall.

Daly, D.C., Cameron, K.M., and Stevenson, D.M. (2001). Plant systematics in the age of genomics. *Plant Physiology 127*:1328-1333.

Danert, S. (1970). Infragenerische Taxa der Gattung *Solanum* L. *Kulturpflanze 18*:253-297.

D'Arcy, W.G. (1972). Solanaceae studies II: Typification of subdivisions of *Solanum*. *Annals of the Missouri Botanical Garden 59*:262-278.

———. (1991). The Solanaceae since 1976, with a review of its biogeography. In Hawkes, J.G., Lester, R.N., Nee, M., and Estrada, N. (Eds.), *Solanaceae III: Taxonomy, Chemistry, Evolution* (pp. 75-137). Kew, U.K.: Royal Botanic Gardens.

Debener, T., Salamini, F., and Gebhardt, C. (1990). Phylogeny of wild and cultivated *Solanum* species based on nuclear restriction fragment length polymorphisms (RFLPs). *Theoretical and Applied Genetics 79:*360-368.

Dodds, K.S. (1962). Classification of cultivated potatoes. In Correll, D.S. (Ed.), *The Potato and Its Wild relatives. Contributions from Texas Research Foundation: Botanical Studies 4:*517-539.

Dodds, K.S. and Paxman, G.J. (1962). The genetic system of cultivated potatoes. *Evolution 16:*154-167.

Dunal, M.F. (1852). *Solanum*. In DeCandolle, A.P. (Ed.), *Prodromus*, vol. 13 (pp. 27-387). Paris: Treuttel and Würtz.

Gerard, J. (1597). *The Herbal; or General Historie of Plantes*, ed. John Norton. London.

Giannattasio, R. and Spooner, D.M. (1994a). A reexamination of species boundaries and hypotheses of hybridization concerning *Solanum megistacrolobum* and *S. toralapanum (Solanum* section *Petota*, series *Megistacroloba):* Molecular data. *Systematic Botany 19:*106-115.

———. (1994b). A reexamination of species boundaries between *Solanum megistacrolobum* and *S. toralapanum (Solanum* section *Petota*, series *Megistacroloba):* Morphological data. *Systematic Botany 19:*89-105.

Glendinning, D.R. (1983). Potato introductions and breeding up to the early 20th century. *The New Phytologist 94:*479-505.

Greuter, W., McNeill, J., Barrie, F.R., Burdett, H.M., Demoulin, V., Filgueiras, T.S., Nicolson, D.H., Silva, P.C., Skog, J.E., Trehane, P., et al. (Eds. and Compilers.) (2000). International Code of Botanical Nomenclature (St. Louis Code). *Regnum Vegetabile 138:*1-474.

Grun, P. (1990). The evolution of cultivated potatoes. In Bretting, P.K. (Ed.), *New Perspectives on the Origin and Evolution of New World Domesticated Plants. Economic Botany 44*(3 Suppl.):39-55.

Hawkes, J.G. (1956). A revision of the tuber-bearing solanums. *Annual Report of the Scottish Plant Breeding Station*, pp. 37-109.

———. (1990). *The Potato: Evolution, Biodiversity and Genetic Resources.* Washington, DC: Belhaven Press.

———. (1997). Cultivated plant diversity and taxonomy. In Claridge, M.F., Dawah, H.A., and Wilson, M.R. (Eds.), *Species: The Units of Biodiversity* (pp. 191-198). New York: Chapman and Hall.

Hawkes, J.G. and Francisco-Ortega, J. (1992). The potato in Spain during the late 16th century. *Economic Botany 46:*86-97.

Hawkes, J.G. and Hjerting, J.P. (1969). *The Potatoes of Argentina, Brazil, Paraguay and Uruguay: A Biosystematic Study.* Oxford: Oxford University Press.

———. (1989). *The Potatoes of Bolivia: Their Breeding. Value and Evolutionary Relationships.* Oxford: Oxford University Press.

Hetterscheid, W.L.A. and Brandenberg, W.A. (1995). Culton vs. taxon: Conceptual issues in cultivated plant systematics. *Taxon 44:*161-175.

Hijmans, R. and Spooner, D.M. (2001). Geographic distribution of wild potato species. *American Journal of Botany 88:*2101-2112.

Hils, U and L. Pieterse (Eds.). (2005). *World Catalogue of Potato Varieties.* Bergen/Dumme, Germany: Agrimedia GmbH.

Hosaka, K. (1995). Successive domestication and evolution of the Andean potatoes as revealed by chloroplast DNA restriction endonuclease analysis. *Theoretical and Applied Genetics 90:*356-363.

———. (2003). T-type chloroplast DNA in *Solanum tuberosum* L. subsp. *tuberosum* was conferred by some populations of *S. tarijense* Hawkes. *American Journal of Potato Research 80:*21-32.

Hosaka, K., de Zoeten, G.A., and Hanneman, R.E., Jr. (1988). Cultivated potato chloroplast DNA differs from the wild type by one deletion: Evidence and implications. *Theoretical and Applied Genetics 75:*741-745.

Hosaka, K. and Hanneman, R.E., Jr. (1988). The origin of the cultivated tetraploid potato based on chloroplast DNA. *Theoretical and Applied Genetics 76:*172-176.

Hosaka, K., Ogihara, Y., Matsubayashi, M., and Tsunewaki, K. (1984). Phylogenetic relationship between the tuberous *Solanum* species as revealed by restriction endonuclease analysis of chloroplast DNA. *Japanese Journal of Genetics 59:*349-369.

Huamán, Z., Hawkes, J.G., and Rowe P.R. (1982). A biosystematic study of the origin of the diploid potato, *Solanum ajanhuiri. Euphytica 31:*665-675.

———. (1983). Chromatographic studies on the origin of the cultivated potato *Solanum ajanhuiri. American Potato Journal 60:*361-367.

Huamán, Z., Hoekstra, R., and Bamberg. J.B. (2000). The Intergenebank Potato Database and the dimensions of available wild potato germplasm. *American Journal of Potato Research 77:*353-362.

Huamán, Z. and Spooner, D.M. (2002). Reclassification of landrace populations of cultivated potatoes *(Solanum* section *Petota). American Journal of Botany 89:*947-965.

Hunziker, A.T. (2001). *Genera Solanacearum, the Genera of Solanaceae Illustrated, Arranged according to a New System.* Ruggell, Germany: A.R.G. Gantner Verlag Kommanditgesellschaft.

Jansky, S.H. (2000). Breeding for disease resistance in potato. *Plant Breeding Reviews 19:*69-155.

Juzepczuk, S.W. and Bukasov, S.M. (1929). A contribution to the question of the origin of the potato. *Trudy Vsesoyuznogo Szeda po Genetike i Selektsii 3:*593-611 (in Russian, English summary).

Kardolus, J.P. (1999). Morphological variation within series *Acaulia* Juz. *(Solanum* section *Petota).* In Nee, M., Symon, D.E., Lester, R.N., and Jessop, J.P. (Eds.), *Solanaceae IV, Advances in Biology and Utilization* (pp. 257-274). Kew, U.K.: Royal Botanic Gardens.

Kardolus, J.P., van Eck, H.J., and Van den Berg, R.G. (1998). The potential of AFLPs in biosystematics: A first application in *Solanum* taxonomy *(Solanaceae). Plant Systematics and Evolution 210:*87-103.

Kawagoe, Y. and Kikuta, Y. (1991). Chloroplast DNA evolution in potato (*Solanum tuberosum* L.). *Theoretical and Applied Genetics 81:*13-20.

Lechnovich, V.S. (1971). Cultivated potato species. In Bukasov, S.M. (Ed.), *Flora of Cultivated Plants*, vol. 9 (pp. 41-304). Leningrad, Russia: Kolos.

Miller, J. and Spooner, D.M. (1999). Collapse of species boundaries in the wild potato *Solanum brevicaule* complex (Solanaceae, *S.* section *Petota*): Molecular data. *Plant Systematics and Evolution 214:*103-130.

Ochoa, C.M. (1990). *The Potatoes of South America: Bolivia.* Cambridge, MA: Cambridge University Press.

———. (1999). *Las papas de Sudamerica: Pert.* Lima, Peru: International Potato Centre.

Olmstead, R.G. and Palmer, J.D. (1997). Implications for the phylogeny, classification, and biogeography of *Solanum* from cpDNA restriction site variation. *Systematic Botany 22:*19-29.

Ortiz, R. (1998). Potato breeding via ploidy manipulations. *Plant Breeding Reviews 16:*15-86.

Plaisted, R.L. and Hoopes, R.W. (1989). The past record and future prospects for the use of exotic potato germplasm. *American Potato Journal 66:*603-627.

Raker, C. and Spooner, D.M. (2002). The Chilean tetraploid cultivated potato, *Solanum tuberosum*, is distinct from the Andean populations; microsatellite data. *Crop Science 42:*1451-1458.

Rhoades, R.E. (1982). The incredible potato. *National Geographic*, May, pp. 668-694.

Rieseberg, L.H. and Soltis, D.E. (1991). Phylogenetic consequences of cytoplasmic gene flow in plants. *Evolutionary Trends in Plants 5:*65-84.

Rodríguez, A. and Spooner, D.M. (1997). Chloroplast DNA analysis of *Solanum bulbocastanum* and *S. cardiophyllum*, and evidence for the distinctiveness of *S. cardiophyllum* subsp. *ehrenbergii* (sect. *Petota*). *Systematic Botany 22:*31-43.

Rollins, R.C. (1965). On the basis of biological classification. *Taxon 14:*1-6.

Ross, H. (1986). Potato breeding-problems and perspectives. *Advances in Plant Breeding* (Suppl. 13). Berlin: Verlag Paul Parey.

Ruiz de Galerreta, J.I., Carrasco, A., Salazar, A., Barrena, I., Iturritxa, E., Marquinez, R., Legorburu, F.J., and Ritter, E. (1998). Wild *Solanum* species as resistance sources against different pathogens of potato. *Potato Research 41:*57-68.

Salaman, R. (1946). The early European potato; its character and place of origin. *Journal of the Linnean Society of Botany 53:*1-27.

———. (1954). The origin of the early European potato. *Journal of the Linnean Society of Botany 55:*185-190.

Salaman, R. and Hawkes, J.G. (1949). The character of the early European potato. *Proceedings of the Linnean Society 161:*71-84.

Salas, A.R., Spooner, D.M., Huamán, Z., Torres-Maita, R.V., Hoekstra, R., Schüler, K., and Hijmans, R.J. (2001). Taxonomy and new collections of wild potato species in central and southern Peru in 1999. *American Journal of Potato Research 78:*197-207.

Schmiediche, P.E., Hawkes, J.G., and Ochoa, C.M. (1982). The breeding of the cultivated potato species *Solanum ×juzepczukii* and *S. ×curtilobum*. II. The resynthesis of *S. ×juzepczukii* and *S. ×curtilobum*. *Euphytica 31:*395-707.

Seithe, A. (1962). Die Haararten der Gattung *Solanum* L. und ihre taxonomische Verwertung. *Botanische Jahrbücher für Systematik, Pflanzengeschichte und Pflanzengeographie 81:*261-335.

Spooner, D.M., Anderson, G.J., and Jansen, R.K. (1993). Chloroplast DNA evidence for the interrelationships of tomatoes, potatoes, and pepinos *(Solanaceae)*. *American Journal of Botany 80:*676-688.

Spooner, D.M. and Bamberg, J.B. (1994). Potato genetic resources: Sources of resistance and systematics. *American Potato Journal 71:*325-337.

Spooner, D.M. and Castillo, R.T. (1997). Reexamination of series relationships of South American wild potatoes (Solanaceae: *Solanum* sect. *Petota*): Evidence from chloroplast DNA restriction site variation. *American Journal of Botany 84:*671-685.

Spooner, D.M. and Hijmans, R. (2001). Potato systematics and germplasm collecting, 1989-2000. *American Journal of Potato Research 78:*237-268.

Spooner, D.M., McLean, K., Ramsay, G., Waugh, R., and Bryan, G. (2005). A single domestication for potato based on multilocus amplified fragment length polymorphism genotyping. *Proceedings of the National Academy of Sciences (USA) 102:*14694-14699.

Spooner, D.M. and Sytsma, K.J. (1992). Reexamination of series relationships of Mexican and Central American wild potatoes *(Solanum* sect. *Petota):* Evidence from chloroplast DNA restriction site variation. *Systematic Botany 17:*432-448.

Spooner, D.M., Sytsma, K.J., and Smith, J.F. (1991). A molecular reexamination of diploid hybrid speciation of *Solanum raphanifolium*. *Evolution 45:*757-764.

Spooner, D.M., Ugarte, M.L., and Skroch, P.W. (1997). Species boundaries and interrelationships of two closely related sympatric diploid wild potato species, *Solanum astleyi* and *S. boliviense* based on RAPDs. *Theoretical and Applied Genetics 95:*764-771.

Spooner, D.M. and Van den Berg, R.G. (1992a). An analysis of recent taxonomic concepts in wild potatoes *(Solanum* section *Petota)*. *Genetic Resources and Crop Evolution 39:*23-37.

———. (1992b). Species limits and hypotheses of hybridization of *Solanum berthaultii* Hawkes and *S. tarijense* Hawkes: Morphological data. *Taxon 41:*685-700.

Spooner, D.M., Van den Berg, R.G., Rodríguez, A., Bamberg, J., Hijmans, R.J., and Lara-Cabrera, S.I. (2004). Wild potatoes *(Solanum* section *Petota)* of North and Central America. *Systematic Botany Monographs 68:*1-209.

Stuessy, T.F. (1990). *Plant Taxonomy: The Systematic Evaluation of Comparative data*. New York: Columbia University Press.

Van den Berg, R.G. and Groendijk-Wilders, N. (1999). Numerical analysis of the taxa of series *Circaeifolia (Solanum* section *Petota)*. In Nee, M., Symon, D.E., Lester, R.N., and Jessop, J.P. (Eds.), *Solanaceae IV, Advances in Biology and utilization* (pp. 213-226). Kew, U.K.: Royal Botanic Gardens.

Van den Berg, R.G., Groendijk-Wilders, N., Zevenbergen, M.J., and Spooner, D.M. (2001). Molecular systematics of *Solanum* series *Circaeifolia (Solanum* section *Petota)* based on AFLP and RAPD markers. In Van den Berg, R.G., Barendse, G., Van der Weerden, G.W., and Mariani, C. (Eds.), *Solanaceae V: Progress in Taxonomy and Utilization* (pp. 72-84). Nijmegen, The Netherlands: Nijmegen University Press.

Van den Berg, R.G., Miller, J.T., Ugarte, M.L., Kardolus, J.P., Villand, J., Nienhuis, J., and Spooner, D.M. (1998). Collapse of morphological species in the wild potato *Solanum brevicaule* complex (section *Petota*). *American Journal of Botany 85*:92-109.

Van den Berg, R.G. and Spooner, D.M. (1992). A reexamination of infraspecific taxa of a wild potato, *Solanum microdontum* Bitter (*Solanum* section *Petota*: Solanaceae). *Plant Systematics and Evolution 182*:239-252.

Volkov, R.A., Komarova, N.Y., Panchuk, I.I., and Hemblen, V. (2003). Molecular evolution or rDNA external transcribed spacer and phylogeny of section *Petota* (genus *Solanum*). *Molecular Phylogenetics and Evolution 29*:187-202.

Wendel, J.F. and Doyle, J.J. (1998). Phylogenetic incongruence: Window into genome history and molecular evolution. In Soltis, D.E., Soltis, P.S., and Doyle, J.J. (Eds.), *Molecular Systematics of Plants II: DNA Sequencing* (pp. 265-296). Dordrecht, The Netherlands: Kluwer Academic Publishers.

# Chapter 2

# Genetics of Agrihorticultural Traits

John E. Bradshaw

Today, potatoes are grown in 149 countries from latitudes 65°N to 47°S and from sea level to altitudes of approximately 4,000 m (Hijmans, 2001). Plant breeders are continuing their efforts for the better adaptation of the potato to a wide range of environments while seeking cultivars that meet current consumer preferences and the demanding quality requirements of processors and supermarkets. To achieve these goals, they need an increasing amount of genetic knowledge, as well as a wealth of germplasm and modern molecular breeding techniques.

Cultivated potato *Solanum tuberosum* L. is a tetraploid, displays tetrasomic inheritance, and comprises highly heterozygous individuals, which suffer inbreeding depression on selfing. Hence, in potato genetics, it has not been possible to achieve the same degree of sophistication as in the genetic analysis of crosses between true-breeding inbred lines that display disomic inheritance. In this chapter, genetic methods applicable to potato are briefly considered before reviewing the genetics of flowering, yield and its components, maturity, dormancy, tuber shape, eye depth, and pigmentation.

## *GENETIC METHODS*

Traits are either qualitative or quantitative in nature. For qualitative traits (e.g., flower and tuber color and presence or absence of molecular markers), individuals can be classified into distinct categories and the number of individuals in each category can be counted. One can look for simple Mendelian ratios in order to understand the inheritance of the trait, whether in diploid or tetraploid potatoes. Furthermore, one can look for associations between traits and build up linkage maps of the underlying genes. For example, Jacobs et al. (1995) used a diploid backcross population to construct a genetic map of the 12 chromosomes of potato integrating molecular and classical markers. The total length of the combined female and male maps was 1120 cM.

For quantitative traits, measurements are made and individual genes cannot be recognized by Mendelian analysis because discrete classes cannot be discerned. Nevertheless, genetic information can be inferred from measurements on related individuals on the supposition of Mendelian inheritance. Biometrical methods have been developed and refined to undertake genetic analyses of quantitative traits (for potatoes see reviews by Bradshaw, 1994; Ortiz and Peloquin, 1994).

A statistically significant offspring-midparent regression is evidence of heritable variation, and the slope of the regression line is a measure of heritability. A detailed genetic analysis can be performed and an explanation of heritability provided when both full-sib and half-sib families (a group of full-sib families with a common parent) are produced in a hierarchical, factorial, or diallel set of crosses. Combining-ability analysis has become popular in both diploid and tetraploid potatoes because no random mating population in equilibrium has to be assumed. The general combining ability (GCA) of a parental clone provides an assessment of its genetic output, as judged by the mean performance of its progenies from crosses with other clones. In order to make valid inferences, these latter clones should be either all those of interest or a random sample of them. Two types of factorial crossing design are commonly used to determine combining abilities. In the first, one set of clones is crossed with another set that complements it for desirable traits. In the second, a diallel set of crosses is made among clones showing a range of values for the trait(s) of interest. Where possible, reciprocal crosses should be made and, ideally, the subsequent progenies assessed in more than one environment in order to study genotype-environment interactions. The departure of a progeny mean from that expected on the basis of the GCAs of its parents is called specific combining ability (SCA). Hence, the progeny mean of the cross between parents *i* and *j*, as a departure from the population mean $\mu$, is

$$\text{progeny}_{ij} - \mu = \text{GCA}_i + \text{GCA}_j + \text{SCA}_{ij}$$

The total genetic variation between families for a factorial design is

$$\sigma^2_G = \sigma^2_{GCA}(\text{set 1}) + \sigma^2_{GCA}(\text{set 2}) + \sigma^2_{SCA}$$

and for a diallel design it is

$$\sigma^2_G = 2\sigma^2_{GCA} + \sigma^2_{SCA}$$

Should the parents come from a random mating population in equilibrium, the total genetic variation can be partitioned into uncorrelated components that are attributable to different causes and can be used to determine the covariance of relatives (FS: full-sib, HS: half-sib) and to predict the response to selection.

$$\sigma^2_{GCA} = \text{cov(HS)}$$

$$\sigma^2_{SCA} = \text{cov(FS)} - 2\,\text{cov(HS)}$$

In the absence of epistasis (non-allelic interactions)

In diploids: $\text{cov(HS)} = \frac{1}{4}\sigma^2_A$

$\text{cov(FS)} - 2\,\text{cov(HS)} = \frac{1}{4}\sigma^2_D$

$\text{cov(parent-offspring)} = \frac{1}{2}\sigma^2_A$

In tetraploids: $\text{cov(HS)} = \frac{1}{4}\sigma^2_A + \frac{1}{36}\sigma^2_D$

$$\text{cov(FS)} - 2\,\text{cov(HS)} = \tfrac{1}{6}\sigma^2_D + \tfrac{1}{12}\sigma^2_T + \tfrac{1}{36}\sigma^2_Q$$
$$\text{cov(parent-offspring)} = \tfrac{1}{2}\sigma^2_A + \tfrac{1}{6}\sigma^2_D$$

where $\sigma^2_A$ and $\sigma^2_D$ are the additive and dominance (diallelic interactions) components of variance and $\sigma^2_T$ and $\sigma^2_Q$ are those for triallelic and tetra-allelic interactions in tetraploids. However, it must be remembered that we are dealing with gene and genotype frequencies in populations and that diallelic interactions (and triallelic and tetra-allelic ones in tetraploids) can contribute to the additive genetic variance. A few examples exist of variance component analysis in potato. Thompson and Mendoza (1984), for example, used 4cov(HS) and 6[cov(FS) − 2cov(HS)] to estimate $\sigma^2_A + \tfrac{1}{9}\sigma^2_D$ and $\sigma^2_D + \tfrac{1}{2}\sigma^2_T + \tfrac{1}{6}\sigma^2_Q$, respectively, for 11 traits including yield in a heterogeneous potato population propagated from true seed. More recently, Ortiz and Golmirzaie (2002) compared hierarchical and factorial mating designs (also known as North Carolina mating designs I and II, respectively) for the quantitative genetic analysis of tuber yield in tetraploid potato. They recommended the hierarchical design because of the ease of generating the families and sampling a large number of parents, but they conceded that the design does not allow a direct and precise estimate of $\sigma^2_D$. A decrease in population mean on selfing is further evidence of nonadditive gene action, but the situation in tetraploids is more complex than that in diploids (Bradshaw, 1994).

Although biometrical genetic analyses provide valuable information about the variation present in a breeding program for traits displaying continuous variation, they do not provide any clues as to the number of genes segregating, their chromosomal locations, or the magnitude of their effects. More detailed genetic analyses of quantitative trait loci (QTLs) can now be performed through their linkage to molecular markers. A recent summary of the potato genetic maps developed using diploid populations and of the gene mapping and genetic analyses performed using molecular markers can be found in the review by Ortiz (1998). Schäfer-Pregl et al. (1996) developed a model for interval mapping of QTLs in offspring of noninbred diploid parents. Hackett et al. (2001) have published a method for QTL interval mapping in autotetraploid species for a full-sib family derived by crossing two parents. QTL mapping provides information on the minimum number and approximate genomic positions of the factors controlling a quantitative trait but does not identify their molecular basis.

For quantitative and qualitative traits, the ultimate goal is to understand the gene function at the biochemical level and to know how this relates to the observed phenotype. Genetic and biochemical information is available on metabolic pathways for carbohydrates, anthocyanins, and carotenoids in potatoes. Candidate genes for QTLs can be postulated and hypotheses tested, as described in later sections of this chapter. Furthermore, large "gene catalogs" now exist for many species in the form of expressed sequence tag (EST) databases. An EST is a partial cDNA (complementary or copy DNA) molecule that has been sequenced. As the cDNA is synthesized from messenger RNA, an EST represents a gene being expressed in the target cells of interest. By 2005, there were 236,717 potato ESTs in GenBank (http://www.ncbi.nlm.nih.gov), a public database that includes potato and tomato. Their value will increase as they are located on physical maps, which in turn are integrated with genetic maps. Further, ESTs will be more useful as the extent of the colinearity of different genomes is established and exploited for comparative genomics, where known gene position and function in one genome are used to draw inferences for other genomes (e.g., from tomato to potato, pepper, and eggplant in the Solanaceae).

Finally, as the potato can be transformed with relative ease using *Agrobacterium*-mediated systems, the advent of "targeted" gene transfer technology has resulted in the use of transgenic potato to manipulate biochemical pathways by antisense suppression or by overexpression of key genes and to provide proof of function of cloned genes at the molecular level.

## *FLOWERING*

Though breeding new cultivars and genetic experiments are both dependent upon making successful crosses, little is known about the genetics of flowering in potato. However, summaries of reproduction biology can be found in Tarn et al. (1992) and in Chapter 3 "Considerations for Successful Breeding."

Gopal (1994) conducted a survey of the flowering behavior, male sterility, and berry setting in 676 accessions of tetraploid *S. tuberosum* germplasm (cultivars and parental lines) from 25 countries outdoors at Kufri in India (32°N, 70°E, 2,500 m above sea level) during summer (12.1-14.1 hours day length). Flowering intensity ranged from the dropping of floral buds just after initiation to profuse blooming. The majority of accessions (58.3 percent) bloomed profusely, but 20.4 percent did not bloom at all. Time to flowering ranged from 6 to 15 weeks and duration of flowering from 1 to 10 weeks. Pollen stainability (as an indication of fertility) ranged from 0 to 90 percent, with 23 percent of accessions completely male sterile. Berry setting ranged from zero to more than five berries per plant, with none in 31.8 percent of the flowering accessions. Thus, only 54.3 percent of the accessions were found to be fertile in all respects and could be used as male and female parents. Premature bud abscission was the major cause of sterility. Genotypes from Peru and Poland were the best sources of profuse and early flowering, respectively, whereas those from Mexico were the best for long-duration flowering and good berry setting. If, as seems likely, this survey is typical of Tuberosum germplasm, much variation is available in flowering traits for future genetic studies.

## *TUBER YIELD AND ITS COMPONENTS*

Although tuber yields of 120 $t \cdot ha^{-1}$ FW (tonnes per hectare fresh weight) have been achieved experimentally in the absence of pests and pathogens and with adequate inputs of water and fertilizers (Mackay, 1996), average world yields are only 16.7 $t \cdot ha^{-1}$ (Lang, 2001), with some European countries achieving 45 $t \cdot ha^{-1}$. Hence, great potential exists for improvement, as has happened, for example, in Great Britain. British Potato Council statistics show that over the 44-year period from 1960 to 2003, the average yield in Great Britain doubled from 22 to 45 $t \cdot ha^{-1}$, and the area planted halved. Hence potato production and consumption have remained almost constant, albeit with the proportion of processed potatoes increasing from 30 to 50 percent—a worldwide trend. Better disease control, more irrigation, and other improvements in agronomy, together with new cultivars, have contributed to the increase in yield, but the contribution of each has not been quantified. Similar advances can be anticipated in developing countries, particularly as larger areas are being planted with healthy (i.e., disease-free) seed tubers at the right physiological stage, and new cultivars with better inbuilt resistance to abiotic and biotic stresses are being grown.

Breeding programs, and hence genetic studies on yield and its components, tuber number and weight, are always conducted in the context of potato production in specific environments and for specific end uses rather than in the context of maximum potential yield. For example, Frusciante et al. (1999) reviewed the breeding and physiological aspects of potato cultivation in the Mediterranean region, where tuber yields in 1997 were 15.6 and 15.0 $t \cdot ha^{-1}$ in the north and south and 24.5 $t \cdot ha^{-1}$ in the east. Potato breeders are less interested in dry matter yield than in high fresh-weight yields with a dry matter content (estimated by specific gravity) appropriate for the desired end use. For example, high dry matter content is desirable for chips (crisps) because it is associated with an increased yield of chips and lower oil absorption.

Exploiting heterosis in crosses of Tuberosum with Neotuberosum, Andigena, and Phureja/Stenotomum offers potential to increase tuber yield. Under long days, Tuberosum × long-day Andigena/Neotuberosum hybrids have shown yield advantages of 13-17 percent over Tuberosum × Tuberosum hybrids (Glendinning, 1969; Cubillos and Plaisted, 1976; Tarn and Tai, 1983; Maris, 1989). Under a short-day subtropical environment, Tuberosum × Andigena hybrids had a yield advantage of 19-56 percent over Tuberosum × Tuberosum hybrids (Gopal et al., 2000). It appears that the extent of heterotic advantage is dependent on day length (Cubillos and Plaisted, 1976). Likewise, the tetraploid offspring from Tuberosum (4*x*) × Tuberosum/Phureja hybrids (2*x*) have shown yield advantages of 27 percent over their 4*x* parents and controls (Tai and De Jong, 1980).

Bonierbale et al. (1993) used codominant molecular markers (restriction fragment length polymorphisms [RFLPs]) at 62 loci distributed across all 12 linkage groups in tetraploid potatoes to test the hypothesis that hybrid vigor is maximized by maximum heterozygosity. Three contrasting sexual progenies from the Cornell University potato breeding program were used in their study: 120

clones from Tuberosum cultivar × Tuberosum cultivar (TT), 80 clones from Neotuberosum-Tuberosum hybrid × bulk pollen from Neotuberosum population (NT), and 80 clones from (4*x* Tuberosum × 2*x* Phureja–dihaploid Tuberosum hybrid) × Tuberosum selection (PT). From within each family, 20 "less heterozygous" and 20 "more heterozygous" clones were selected for yield trials. Tuber yields were similar for the three families, but TT had fewer and heavier tubers than NT and PT. The TT family had the highest level of inbreeding (homozygosity at marker loci) and the PT family the highest degree of heterozygosity, as expected. The correlations found within families between heterozygosity and yield were small and positive in TT, not significant in NT, and negative in PT. In other words, maximum heterozygosity influences tuber yield and its components in a favorable direction when only adapted germplasm sources are combined. Furthermore, stronger correlations were found with specific combinations of individual RFLP fragments (additive effects) than with indices of heterozygosity. This suggests that the heterosis for tuber yield when wild species are crossed with elite material may be largely due to the additive combination of genes for reduced tuber set (number) and larger tuber size in the resulting progeny. It is pertinent to mention here that tuber set, rather than tuber initiation, determines the total tuber number (Struik et al., 1990). Fragments whose segregation patterns showed significant associations with the measured yield characteristics corresponded to 26 mapped loci from seven of the potato linkage groups (1, 2, 4, 5, 7, 9, 12).

Gopal (2002) compared 36 Tuberosum × Tuberosum (TT) families with 36 Tuberosum × Andigena (TA) families involving 18 common female parents of the Tuberosum Group to elucidate the nature of gene action leading to heterosis in TA families. Additive gene action was higher in TA families than in TT families, although the relative significance of additive and nonadditive gene actions in both families varied with the character and the generation. Gopal (2002) concluded that TA families differed from TT families in terms of the qualitative rather than quantitative nature of heterozygosity as a result of the introduction of different types of alleles from Andigena.

Most published work on the genetics of yield and its components has come from breeders conducting combining-ability analyses to help them decide their hybridization strategy (i.e., their choice of parents and crosses). However, the information on GCA and SCA is usually specific to the particular population of genotypes under investigation and, hence, must be sought afresh when new material is introduced into a breeding program. GCA effects are also known to vary from generation to generation even for the same set of progenies (Gopal, 1998). Tetraploid examples for the relative importance of GCA and SCA, rather than an exhaustive list, are given in Table 2.1. It can be seen that GCA and SCA contribute in varying proportions. Neele et al. (1991) suggested that SCA tends to be more important than GCA in crossing schemes involving closely related parents (e.g., Killick, 1977), and Tai (1976) suggested the same for traits, such as tuber yield, that have been subjected to continuous directional selection. Gopal (1998) felt that both these explanations were tenable for his results with parents used in the National Potato Breeding Program in India. Bradshaw et al. (2000) provided an example of the value of including the parents and the parents' self-progenies in any assessment of a diallel set of crosses. They confirmed that inbreeding depression was marked for total tuber yield, with values of 14.53, 20.16, and 22.61 kg per plot for selfs, $F_1$s, and the parents, providing additional evidence for nonadditive gene action.

Although these combining-ability analyses provide valuable information about the genetic variation present in a breeding program for traits displaying continuous variation, they do not provide any clues as to the number of genes segregating, their chromosomal locations, or the magnitudes of their effects. For this, we need to consider the results of QTL analysis.

Schäfer-Pregl et al. (1998) performed a QTL mapping experiment over more than 1 year for tuber starch content (percentage) and tuber yield using RFLP markers on two different crosses (K31 and LH) among dihaploid breeding lines. The environments in the study were five and two combinations of sites in Germany and Italy (Mediterranean), respectively. Population K31 was fully genotyped, whereas population LH was selectively genotyped according to high and low tuber starch content, high content being associated with lower yield. Eighteen putative QTLs for tuber starch content were identified, covering all 12 potato linkage groups, and eight putative QTLs for tuber

TABLE 2.1. Relative importance of general and specific combining ability for tuber yield, tuber number, and tuber weight, as reported in some studies.

| Trait and reference | Combining ability | GCA-parental value correlation | Midparent-progeny correlation |
|---|---|---|---|
| Tuber number | | | |
| Tai (1976) | SCA | | |
| Killick (1977) | GCA and SCA | | |
| Veilleux and Lauer (1981) | GCA | | |
| Brown and Caligari (1989) | GCA | | 0.71 |
| Maris (1989) | GCA (>recip.) | 0.69 | 0.59 |
| Neele et al. (1991) | GCA > SCA | | 0.57 |
| Gopal (1998) | GCA < SCA | | |
| Tuber weight (mean) | | | |
| Tai (1976) | GCA (>GCA × year) | | |
| Killick (1977) | GCA and SCA | | |
| Brown and Caligari (1989) | GCA and SCA (and GCA × sites) | | 0.40 |
| Maris (1989) | GCA (>recip.) | 0.96 | 0.78 |
| Neele et al. (1991) | GCA > SCA (>recip.) | | 0.81 |
| Gopal (1998) | GCA < SCA | | |
| Tubers < 30 mm by weight (%) | | | |
| Neele et al. (1991) | GCA = SCA (>recip.) | | 0.73 |
| Tuber yield | | | |
| Plaisted et al. (1962) | GCA < SCA (both interact with locations) | | |
| Tai (1976) | SCA | | |
| Killick (1977) | SCA | | |
| Veilleux and Lauer (1981) | GCA and SCA | | |
| Brown and Caligari (1989) | GCA (and GCA × sites) | | 0.55 |
| Maris (1989) | GCA (= recip.) | 0.78 | 0.51 |
| Neele et al. (1991) | GCA = SCA (>recip.) | | 0.14 |
| Gopal (1998) | GCA < SCA | | |
| Bradshaw et al. (2000) | GCA > SCA > GCA × years | 0.85 | 0.61 |

*Notes:* The symbols >, <, and = are used where components of variance are estimated; recip. = reciprocal difference.

yield were identified on 8 linkage groups. Twenty of the twenty-six putative QTLs were reproducibly detected in at least two environments or mapping populations, but none of the QTLs was detected in all environments and in both populations. Most QTLs for tuber yield were linked with QTLs for tuber starch content, suggesting that the effects on both traits are controlled by the same genetic factors.

Schäfer-Pregl et al. (1998) also pointed out that time to plant maturity and tuberization are related physiological traits controlled by genetic factors and day length. They provided some evidence from their research and that of others that gene(s) with pleiotropic effects on tuberization, plant maturity, tuber starch content, and tuber yield are located on potato chromosome 5. Thus, although tuber yield and starch content may be truly polygenic traits, QTLs of particular interest may be found through this kind of analysis. However, the molecular and genetic factors underlying the initiation of tuberization and the partitioning of photosynthates to the tuber are still poorly understood. Schäfer-Pregl et al. (1998) further suggested that quantitative trait alleles in tetraploid potatoes might be tagged when large numbers of amplified fragment length polymorphism (AFLP) fragments are scored either for presence or absence and tested for effect using a single-marker statistical test such as the t-test. AFLP fragments with a significant effect on the trait of interest can be physically isolated and mapped as conventional RFLPs on existing diploid mapping populations.

Chen et al. (2001) developed a molecular-function map for carbohydrate metabolism and transport in potato as a prerequisite for a candidate gene approach to identifying the genes responsible for tuber starch content at the molecular level. The map has 85 loci based on 69 genes involved in the amount of starch accumulated in mature potato tubers. Starch accumulated is the net result of photosynthetic carbon fixation, the synthesis of transient starch and its conversion into sucrose in photosynthetically active source leaves, the vascular transport of sucrose from the leaves to the developing sink tuber, and starch synthesis and degradation in the tuber during the growth period. The map was developed by searching public databases for genes of interest from potato, tomato, or other plant species. DNA sequence information was used to develop polymerase chain reaction–based marker assays (mainly cleaved amplified polymorphic sequences [CAPSs] and a few sequence characterized amplified regions [SCARs]) that allowed the localization of the corresponding potato genes on existing RFLP linkage maps. Potato cDNA and EST clones coding for relevant enzymes were also used as marker probes for RFLP analysis. The molecular-function map for genes operating in carbohydrate metabolism and transport in the K31 population of Schäfer-Pregl et al. (1998) allowed 14 putative candidate genes for tuber starch content to be identified. Particularly intriguing was the *Sut2* locus on linkage group 5, which was identified using a sucrose transporter–like gene as an RFLP probe. *Sut2* is tightly linked to the anchor marker GP179, which itself is most closely linked to QTLs for tuber starch content, tuber yield, plant maturity, and vigor (Schäfer-Pregl et al., 1998; Oberhagemann et al., 1999). Hence, Chen et al. (2001) posited that alleles of either a sucrose transporter or a sucrose sensor might affect the efficiency with which sucrose is translocated from source to sink tissues and hence might have an effect on the above traits.

"Targeted" gene transfer technology has been tried with the objective of improving carbon partitioning (yield) and dry matter content and its allocation within tubers (even distribution of starch) (Davies, 1998). Sonnewald et al. (1997) expressed a gene encoding a yeast invertase in tubers, and the invertase protein was targeted to the cell wall. This resulted in a substantial increase in tuber glucose content, a two- to threefold decrease in tuber number per plant, and a 30 percent increase in tuber fresh weight. Davies (1998) concluded that the role of specific sucrose transporters on the plasma membrane in regulating overall unloading rates now becomes an important issue for potato as well as for other sink organs well known to have apoplastic unloading mechanisms. Once inside the cells of developing tubers, sucrose is cleaved by the enzyme sucrose synthase, which is thought to be an important regulator of sink strength. Zrenner et al. (1995) used an antisense approach to down-regulate the expression of the sucrose synthase gene and enzyme. This resulted in a significant reduction in total dry matter production, thus providing evidence that the enzyme plays a key role in establishing sink strength potential. Within the amyloplast, the conversion of glucose-1-P to ADP glucose by the enzyme ADGP pyrophosphorylase (ADGP PPase) is the first essentially irreversible reaction in starch synthesis in the plastid. When Stark et al. (1992) expressed an *Escherichia coli glg C16* gene encoding a bacterial ADGP PPase in tubers of cultivar Russet Burbank, starch content was elevated by up to 60 percent (35 percent average), but total tuber yield appeared unaffected. Hence the transgenic manipulation of carbohydrate metabolism holds promise for the understanding and improvement of yield and quality traits, but much research remains to be done.

## *MATURITY*

Where growing seasons are short, farmers and plant breeders must have practiced selection for faster tuber development and earlier maturity. Even where growing seasons are relatively long, such as in northern temperate latitudes, first and second early cultivars, as well as maincrop ones, are grown to meet the requirements for fresh produce over many months. Compared with the late-maturing cultivars, early (short-duration) potato cultivars divert a larger part of the assimilates to the tubers early in the growing season, resulting in shorter crop periods and also lower yields (Kooman and Rabbinge, 1996).

Combining-ability and regression analyses have both pointed to high heritability and the importance of GCA for maturity, as illustrated by the following three examples. Killick (1977) analyzed the genetic variation for nine traits of importance in Tuberosum potatoes, including foliage maturity, in a half-diallel set of crosses with six tetraploid parents. All of the genetic variation for maturity was due to GCA, with two parents conferring early maturity and four late maturity. Maris (1989) studied the behavior of seven long-day-adapted Andigena clones and three Tuberosum varieties in an incomplete diallel set of crosses in the Netherlands in 1980. There was variation among the parents in maturity, but the means for the Andigena and Tuberosum parents showed little difference. This was because Andigena had been selected for earliness in long days. The observed variation was due to GCA and reciprocal differences, the former accounting for more variation than the latter. There was a high correlation between GCA and parental values ($r = 0.84$) and between offspring means and midparent values ($r = 0.72$). Hutten et al. (1996) estimated the relative effects of diploid and tetraploid parents on the vine maturity of tetraploid progeny from $4x \times 2x$ crosses in a trial in the Netherlands. Three early- and three late-maturing tetraploid varieties were crossed with three early- and three late-maturing diploids. The multiple regression relationship between progeny means ($y$) and tetraploid ($x_1$) and diploid ($x_2$) parents was

$$y = 1.38 + 0.38x_1 + 0.30x_2$$

The coefficient of determination ($R^2$) was 0.87 and the slope for the tetraploid parents (0.38) was significantly larger than the slope for the diploid parents (0.30), indicating the greater impact of the tetraploids. Progeny means were toward those of the tetraploid parents. The vine maturity of the $4x \times 2x$ progeny was well predicted from the performance of their parents.

Resistance to diseases such as late blight (Umaerus and Umaerus, 1994; Bradshaw et al., 1995) and early blight (Caligari and Nachmias, 1988; Boiteux et al., 1995) is often associated with late maturity. Work on late blight and maturity showed that a QTL of large effect for early maturity was present on linkage group 5 in the region of marker GP179 (Collins et al., 1999; Oberhagemann et al., 1999; Visker et al., 2003; Bradshaw et al., 2004). As already mentioned, Chen et al. (2001) thought that this gene might code for a sucrose transporter or sensor.

The wild Bolivian species *S. berthaultii* carries trichome-mediated insect resistance but is very late maturing in northern climates because it requires short days for tuberization. Van den Berg et al. (1996a) detected QTLs affecting tuberization in reciprocal backcrosses between *S. tuberosum* and *S. berthaultii*. Eleven distinct loci on seven linkage groups were associated with variation in tuberization. Most of the loci had small effects but, again, a QTL explaining 27 percent of the variance was found on linkage group 5. Subsequently a QTL affecting leaf Gibberellin A1 levels was also detected in this region of linkage group 5, with high levels associated with late tuberization (Ewing et al., 2004).

## *DORMANCY*

Dormancy is the obligate period of nonsprouting after harvest, even under conditions favorable to sprouting. Where one crop is grown annually, tubers are stored both for planting in the next season and to provide continuity of supply to end users. In such cases long dormancy may be desirable, provided it does not delay the sprouting of seed tubers beyond the planting time for the next season. In contrast, with multiple crops per year, short dormancy is required, as found in *S. phureja*.

Simmonds (1964) showed that true seed and tuber dormancies were highly correlated in cultivated diploids and that the former was under polygenic control. Later, Thompson et al. (1980) crossed Phureja clones (20 days' dormancy at 21-27°C) with Stenotomum clones (75 days' dormancy) in a North Carolina II design and found a high additive genetic variance estimate with high ($h^2 = 0.73$) narrow-sense heritability, indicating the importance of additive gene action for tuber dormancy. As *S. berthaultii* is noted for its long tuber dormancy, Van den Berg et al. (1996b) used reciprocal backcrosses between diploid *S.tuberosum* and *S. berthaultii* to study the genetics of tuber dormancy. QTLs were detected on nine linkage groups that affected tuber dormancy at 13°C in the dark, either alone or through epistatic interactions. Recessive alleles from the wild parent promoted dormancy, with the largest effect at a QTL on linkage group 2. At least three QTLs for dormancy were mapped to markers previously found to be associated with tuberization in these crosses. Subsequently, hormonal analyses were carried out to identify candidate genes, and a combination of alleles was found at QTLs on linkage groups 2 and 4 in which long dormancy was associated with high levels of tuber abscisic acid (ABA) (Ewing et al., 2004).

Previously, Freyre and Douches (1994b) had performed a similar study on two populations derived from hybrids of diploid *(S. tuberosum × S. chacoense)* × *S. phureja* (two different clones). Here, *S. phureja* parents had a much shorter dormancy (10 days to sprouting at 10°C) than *S. tuberosum* (80 days for hybrids with *S. chacoense*) and the variation in the populations was highly skewed toward lack of dormancy, suggesting that *S. phureja* contributed dominant alleles for short dormancy. Multiple linear regression estimated that 36.9 percent of the total phenotypic variation (broad-sense heritability 93.8 percent) for tuber dormancy was due to the effect of six loci in one population, and for the other population 8.5 percent of the phenotypic variation (broad-sense heritability 92.6 percent) could be explained by the effect of two loci. However, for this QTL analysis the authors had used only 11 isozyme markers distributed among 7 of the 12 linkage groups. Subsequently, the first population was analyzed with the addition of RFLP and random amplified polymorphic DNA (RAPD) markers, and the phenotypic variation for tuber dormancy explained by QTLs and epistatic interactions rose to 72.1 percent (Freyre et al., 1994). Six QTLs were identified, one on each of linkage groups 2, 3, 4, 5, 7, and 8. The isozyme marker *Got-2* on linkage group 7 had the largest effect of all loci. Interestingly, *Got-2* also had the largest effect on specific gravity in the same population when averaged across three environments (Freyre and Douches, 1994a). A total of ten putative QTLs for specific gravity were identified over environments, located on linkage groups 1, 2, 3, 5, 7, and 11.

## *TUBER SHAPE AND EYE DEPTH*

Although tuber shape is mainly a varietal characteristic, it can be influenced by environmental conditions and cultural practices. Tuber shape is described in qualitative terms as compressed, round, ovate, obovate, elliptic, oblong, long-oblong, and long, but the length to breadth ratio shows continuous variation from less than 1.4 to more than 2.0 (Ortiz and Huamán, 1994). Unusual tuber shapes are found in Andean potato cultivars; these have been described by Huamán et al. (1977) and summarized as flattened, clavate, reniform, fusiform, falcate, coiled, digitate, concertina-shaped, and tuberosed. In today's markets, regular shapes with shallow eyes are preferred for processing and table use in order to reduce wastage. Round potatoes are preferred for producing chips (crisps) and long ovals for French fries.

The two most recent reviews of the inheritance of tuber shape in cultivated diploid potatoes are by De Jong and Burns (1993) and Ortiz and Huamán (1994). Despite the continuous nature of the length to breadth ratio, the former concluded that one major gene controls the round (compressed plus round) and long (oblong and long) shapes, with round dominant over the long. This gene, *Ro*, was shown to be linked to locus *I*, which controls the expression of anthocyanin in the tuber skin, with a recombination frequency of 11.8 percent. Subsequently, van Eck, Jacobs, Stam, et al. (1994) were able to map *Ro* on chromosome 10 using RFLPs in a diploid potato cross between two round parents. When tuber shape was studied as a quantitative trait, using the length/width ratio, the estimated broad-sense heritability was 0.80 and the morphologically mapped *Ro* locus explained 75 percent of the genotypic variation, with evidence for different *Ro* alleles in the two parents. Genes for unusual tuber shapes have not yet been mapped.

Li et al. (2005) found that in diploid cultivated potatoes, the *Ro* locus is closely linked (4 cM) to a locus for tuber eye depth *(Eyd)*, with deep eye dominant over shallow. In tetraploid potatoes, Neele et al. (1991) found that for eye depth GCA was more important than SCA (1.178 versus 0.223) and the midparent-progeny correlation was 0.78. They also found GCA to be more important than SCA for tuber shape and regularity of tuber shape, with midparent-progeny correlations of 0.50 and 0.85, respectively. Gopal (1998) too reported GCA to be more important than SCA for tuber shape. The diploid results should aid understanding of the inheritance of eye depth in tetraploid potatoes.

## *PIGMENTATION*

The pink, red, blue, and purple colors in potato tubers and flowers are due to anthocyanins whose antioxidant properties are now appreciated and have been demonstrated in red- and purple-fleshed potatoes (Brown et al., 2003). Yellow and orange flesh are due to carotenoids, which are also antioxidants with health-promoting attributes such as protection against certain cancers (Colditz et al., 1985), cardiovascular diseases (Gaziano et al., 1995), and particularly macular eye degeneration (Snodderly, 1995). Short-term feeding of foods rich in lutein and zeaxanthin can substantially increase macula pigment density in human subjects (Hammond et al., 1997). Hence, the enhancement of anthocyanins and carotenoids by breeding provides an opportunity to improve the nutritive value of potatoes and processed foods made from potatoes (Brown et al., 2003).

### *Anthocyanins*

The genetics and biochemistry of anthocyanin biosynthesis were reviewed by Holton and Cornish (1995) for maize *(Zea mays)*, snapdragon *(Antirrhinum majus)*, and the Solanaceous ornamental species petunia *(Petunia hybrida)*, which has become a model system for isolating flavonoid biosynthetic genes and studying their interactions and regulation. The colorless dihydroflavonols (dihydrokaempferol [DHK], dihydroquercetin [DHQ], and dihydromyricetin [DHM]) represent a branch point in flavonoid biosynthesis as they are the intermediates in the production of both colored anthocyanins (Figure 2.1), through the action of the enzyme dihydroflavonol 4-reductase (DFR), and the colorless flavonols, produced by flavonol synthase (FLS). DHK can be hydroxylated by flavonoid 3′-hydroxylase (F3′H) to produce DHQ, and both DHK and DHQ can be hydroxylated by flavonoid 3′,5′-hydroxylase (F3′5′H) to produce DHM. The action of DFR and further oxidation, dehydration, and glycosylation of the different leucoanthocyanidins produce the corresponding brick-red pelargonidin, red cyanidin, and blue delphinidin pigments. In petunia and potato, anthocyanidin-3-glycosides (stable anthocyanins) are further modified by rhamnosylation to anthocyanidin-3-rutinosides, by glycosylation to anthocyanidin-3-rutinoside-5-glycosides, and by methylation of the anthocyanidin (0-methyl-transferases) at the 3′ and 3′,5′ positions. The rutinosides (rha) can be acylated with a p-coumaroyl group.

Lewis et al. (1998b) confirmed that petunidin-3-(p-coumaroyl-rutinoside)-5-glucoside was the major anthocyanin in the tuber skin and flowers of the wild *Solanum* species they examined, tubers of which were mostly white or light purple. In contrast, a much wider range of anthocyanins is found in cultivated species. Harborne (1960) described the isolation and characterization of ten anthocyanins from cultivated diploid and tetraploid potatoes: the six anthocyanidin-3-(p-coumaroyl)-rutinoside-5-glucosides shown in Figure 2.1, plus the 3-rutinoside derivatives of pelargonidin, cyanidin, delphinidin, and petunidin. In cultivated diploids, blue or purple flowers and tubers

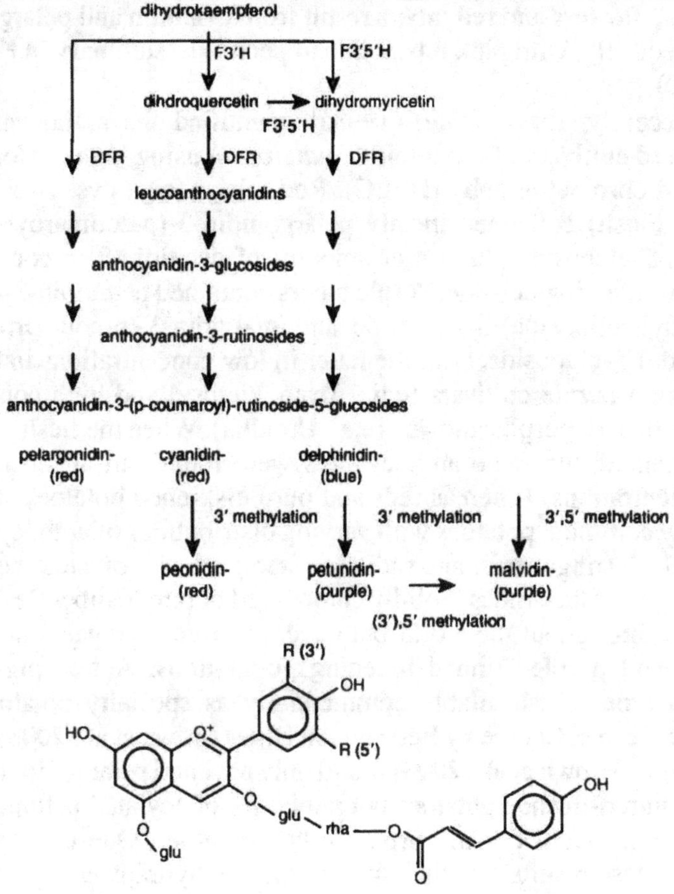


| anthocyanidin | R (3') | R (5') | Color |
|---|---|---|---|
| pelargonidin | H | H | red |
| cyanidin | OH | H | red |
| delphinidin | OH | OH | blue |
| peonidin | $OCH_3$ | H | red |
| petunidin | $OCH_3$ | OH | purple |
| malvidin | $OCH_3$ | $OCH_3$ | purple |

FIGURE 2.1. Anthocyanin biosynthetic pathway.

result from the production of delphinidin and petunidin, and reddish purple flowers and red tubers result from cyanidin and pelargonidin, respectively, with pink tubers due to peonidin (summary in Howard, 1970).

Recently, Lewis et al. (1998a) quantified the anthocyanins in colored cultivars of tetraploid *S. tuberosum* using high-performance liquid chromatography (HPLC). Red tubers (e.g., cvs. Desiree and Red Flush) contained mostly pelargonidin-3-(p-coumaroyl-rutinoside)-5-glucoside plus lesser amounts of peonidin-3-(p-coumaroyl-rutinoside)-5-glucoside. Purple tubers contained petunidin-3-(p-coumaroyl-rutinoside)-5-glucoside and malvidin-3-(p-coumaroyl-rutinoside)-5-glucoside, with the latter in low concentrations in light to medium purple cultivars (e.g., Arran Victory) and high concentrations in dark purple cultivars (e.g., Urenika). When the flesh was also pigmented, the same anthocyanins were found but at much lower concentrations. Whereas red- and purple-skinned potatoes are relatively common, potatoes with varying distributions of anthocyanin in the flesh (rings, arcs, and radiating stars) are rare outside the native cultivars of the Andes. Solidly colored red or purple tuber flesh is unusual throughout the world but occurs as rare segregants in certain red- and purple-skinned breeding populations. A few pigmented flesh types are available commercially as specialty potatoes, for example cvs. Cranberry Red and All Blue (Brown et al., 2003). Interestingly, Brown et al. (2003) found only acylated pelargonidin glycoside in red-fleshed potatoes but a mixture of acylated petunidin and peonidin glycosides in purple-fleshed potatoes. One can conclude from these results that the anthocyanin methyltransferase system is very active in purple tubers.

The major anthocyanins in flowers are rutinosides or other glycosides of pelargonidin, petunidin, and malvidin, but glycosides of cyanidin and delphinidin were found in some flowers (Lewis et al., 1998b).

The inheritance of anthocyanin pigmentation in cultivated potatoes was reviewed by De Jong (1991), who attempted to resolve some of the confusion that had arisen from previous reviews. Separate work had been done on diploid and tetraploid potatoes and different nomenclatures used for the postulated genes. However, it is reasonable to assume that the same loci affect pigmentation in both diploids and tetraploids, and van Eck et al. (1993) and van Eck, Jacobs, van den Berg, et al. (1994) proceeded to map the flower and tuber skin color loci *R, P, F,* and *I* in diploid populations, some of which involved dihaploids derived from tetraploid *S. tuberosum.* More recently, De Jong et al. (2004) mapped, in tomato, 13 genes known to be involved in anthocyanin biosynthesis or its regulation in *Petunia* and found associations with classical and QTL loci that had been mapped in the closely related tomato, potato, pepper, and eggplant genomes. In several cases, including the *R, P,* and *I* loci in potato, the associations were strong enough for candidate genes to be identified. In summary, the genotypes and phenotypes shown in Table 2.2 were recognized. In their first paper, van Eck et al. (1993) classified pale blue as BLUE and pink, red, and purple as RED, whereas in common English usage, purple is a mixture of blue and red.

The locus *R* (referred to as *D* by van Eck et al., 1993; van Eck, Jacobs, van den Berg, et al. 1994) mapped onto chromosome 2 and was involved in the biosynthesis of red anthocyanins throughout the plant, including the tips of sprouts. Likewise, locus *P* on chromosome 11 was involved in the production of blue anthocyanins throughout the plant, including the tips of sprouts. The dominant allele *P* appeared epistatic to *R* in the tubers but not in the flowers, although it is not always clear in the literature which anthocyanins are produced in genotype *R- P- F- I-*. The loci *F* and *I* were closely linked on chromosome 10, near locus *Ro* (*Ro* at 42 cM, *F* at 57 cM, and *I* at 62 cM), and determine whether or not *R* and *P* are expressed in the flowers (petals) and tubers (both skin and flesh), respectively. Sprout tips, stems, and inflorescences are not affected. Two alleles were recognized at the *I* locus: *Iep* giving pigmentation in the epidermis of the tuber and *Ico* in the outer cortex. Van Eck, Jacobs, van den Berg, et al. (1994) considered the *PSC* locus of Gebhardt et al. (1989) to be *I*. There appeared to be three alleles at the *R* locus: *R*, $R^{pw}$, and *r*, with $R^{pw}$ giving pink tubers and white flowers. As van Eck, Jacobs, van den Berg, et al. (1994) give the genotype of clone 07506-1 as $R^{pw}R^{pw}$ and De Jong (1987) gives it as *rr*, the literature shows some confusion over which alleles are present in certain clones.

TABLE 2.2. Genotypes and phenotypes for anthocyanin pigmentation.

| Flowers (petals) | | Tubers (skin) | |
|---|---|---|---|
| *R- P- ff* | White | *R- P- ii* | White |
| *rr pp F-* | White | *rr pp I-* | White |
| $R^{pw}r$ *pp F-* | White | $R^{pw}r$ *pp I-* | Pink |
| *R- pp F-* | Red | *R- pp I-* | Red |
| *rr P- F-* | Blue | *rr P- I-* | Purple |
| *R- P- F-* | Purple | *R- P- I-* | Purple |

Jung et al. (2005) have recently confirmed that the dominant allele at the *P* locus codes for F3′5′H activity, whereas the recessive *p* allele lacks this activity. They introduced a potato *f3′5′h* cDNA clone under the control of a doubled CaMV 35S promoter into the red-skinned cultivar Desiree and found that tuber and stem tissues were purple in 9 out of 17 independently transformed lines. Furthermore, sequencing the *f3′5′h* allele expressed in the red-skinned cultivar Rideau revealed 10 amino acid differences compared with the functional allele of the purple-skinned clone W5281.2, including one within a highly conserved motif. Hence the difference between dominant (functional) and recessive (nonfunctional) alleles at *P* results from mutations that affect protein function rather than mutations that affect gene transcription. Given the precedent of *Petunia an2* gene encoding an MYB domain transcriptional regulator of the anthocyanin pathway and the mapping of *an2, F,* and *I* close to marker *TG63* in the tomato and potato genomes, they considered potato genes related to *an2* good candidates for *F* and *I*. Finally, W. S. De Jong, De Jong, De Jong, et al. (2003) postulated that locus *R* encodes DFR and demonstrated that an identically sized *dfr* gene restriction fragment lacking the *Bam*HI restriction site was present in all potato clones that produced red anthocyanin pigments but that the same fragment was absent in many potato clones with white tuber skin and flowers. Their data were consistent with a 48-year-old hypothesis by Dodds and Long (1955) that *R* was selected just once during the domestication of potato.

W. S. De Jong, De Jong, and Bodis (2003) were able to distinguish four alleles at the *R* locus, of which just *R* gave a red color. They were also able to show that although the dosage of *R* in tetraploids contributes to the intensity of skin color, it is not the major factor. Furthermore, the DNA sequences of *dfr* associated with *r* in clone 07506-01 (genotype *rr ii*, white tubers) and with $R^{pw}$ in clone 79-48 (genotype $RR^{pw}$ II, red tubers) were indistinguishable, which means that an explanation is required for pink tubers in genotype $R^{pw}r$ *I-* (W. S. De Jong, De Jong, De Jong, et al., 2003), the pink color being derived from peonidin.

A plausible explanation offered by W. S. De Jong, De Jong, De Jong, et al. (2003) was that the observed variation in tuber color results from the substrate specificity of DFR and that superimposed on this is natural variation for tuber expression of F3′H (Figure 2.1). In this model, the *dfr* allele present in all red-skinned potatoes encodes an isozyme that is able to utilize DHK as a substrate, and this permits the synthesis of pelargonidins (the *R* phenotype). In genetic backgrounds that lack the red allele of *dfr* but do express F3′H, DHQ will be synthesized and, because it can be utilized by all alleles of *dfr*, this permits the synthesis of pink peonidin-based anthocyanins (the $R^{pw}$ phenotype). When neither DHK-competent DFR nor F3′H activities are present, no pink or red anthocyanins are produced (the *r* phenotype). Although Johnson et al. (2001) showed that alteration of a single amino acid changes the substrate specificity of DFR in petunia, more work is required to test the model in potato. Indeed, further research is required to demonstrate that the expression of combinations of alleles at the *R, P, F,* and *I* loci can explain all of the observed phenotypes in terms of both color and the anthocyanins present.

In tetraploid Tuberosum potatoes, all anthocyanins found so far have been acylated, whereas in diploids segregation has been found at an unmapped locus *Ac*; that is, not all diploids have acylated anthocyanins. Locus *Ac* is associated with the attachment of a glucose residue in the 5 position and methylation, as well as acylation of the anthocyanin molecule with p-coumaric acid.

Other genes listed by De Jong (1991) as affecting the distribution of pigments and reviewed by Ortiz and Huamán (1994) were *B* (series of alleles controlling distribution to floral abscission layer, tuber eyebrow, cotyledonary node of the embryo [seed spot], and nodal band at the base of the petiole), *Ow* (distributes pigment to ovary wall), *Pf* (distributes pigment to tuber flesh), *Pw* (controls accumulation of anthocyanins in the youngest leaves of the plant: pigmented whorl), *Ul* (distributes pigment to underside of leaf), *Pd* (pigmentation of dorsal side of leaf), *Pv* (pigmentation of ventral side of leaf), and *M* (in tetraploids, restricts pigmentation of tuber periderm to areas around the eye). *B, Pf, Pw,* and *Ow* are linked to *F* and *I* on chromosome 10. Combinations of the genes that affect the distribution of pigments in the tuber skin result in pigmented eyes, pigmented eyebrows, and splashed, scattered, spectacled, or stippled tubers (Ortiz and Huamán, 1994).

### *Carotenoids*

Structurally, carotenoids are terpenoids, derived by condensation of prenyl pyrophosphates (Sandmann, 2001), whose conjugated double-bond system determines their color and is responsible for their biological actions as antioxidants. The conversion of geranylgeranyl pyrophosphate to phytoene by phytoene synthase is the first dedicated step in the formation of carotenoids (Figure 2.2).

Tuber flesh color in potatoes can be classified as white, yellow, or orange, but different degrees of yellowness exist. Brown et al. (1993) provided evidence in diploid cultivated potatoes for orange, yellow, and white flesh all being controlled by alleles at a single locus, with orange *(Or)* dominant over yellow *(Y)* and white *(y)*, and yellow dominant over white. The yellow flesh locus has been mapped to chromosome 3 (Bonierbale et al., 1988).

Brown et al. (1993) showed that the orange flesh color was associated with large amounts of zeaxanthin (2055 μg per 100 g FW). The orange-fleshed parent also had 120 μg per 100 g FW of lutein, which was also present in the yellow diploid standard (140 μg per 100 g FW) and white tetraploid standard (65 μg per 100 g FW), whereas zeanxanthin was not found in these standards. When averaged over a number of backcrosses, lutein levels were identical (146 μg per 100 g FW) in orange and white segregants, whereas zeaxanthin levels were much higher in the former (1,426 versus 304 μg per 100 g FW). Zeaxanthin also accounted for slightly more than half (51 percent) of

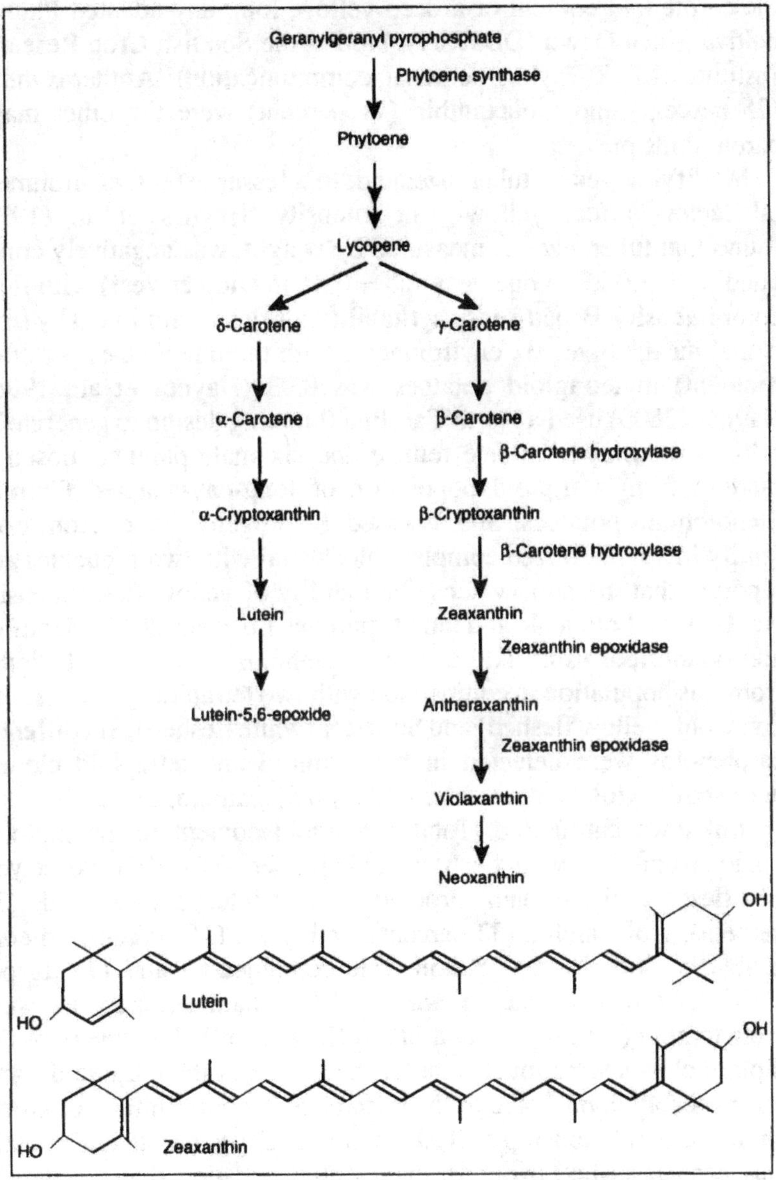


FIGURE 2.2. Carotenoid biosynthetic pathway.

the carotenoid content of a deep yellow, long-day-adapted Phureja cultivar, Inca Dawn (DB375/1), bred at the Scottish Crop Research Institute (Mark Taylor, personal communication). Antheraxanthin (25 percent) and violaxanthin (11 percent) were the other major carotenoids present.

Modifying genes, tuber size, and, to a lesser extent, environmental factors affect yellow-flesh intensity. Haynes et al. (1994) found that tuber size, as measured by weight, was negatively correlated ($r_s = -0.27$ in one year and $-0.31$ in another year) with flesh color intensity. Broad-sense heritability for flesh color intensity from clonal means (over six environments with three replicates per environment) in tetraploid potatoes was 0.93 (Haynes et al., 1996). Haynes (2000) used a North Carolina II mating design to generate 30 full-sib families from five female and six male parents chosen at random from a diploid population of long-day-adapted Phureja/Stenotomum potatoes. She assessed 12 progeny clones from each family in a randomized complete block trial with two replicates and reported that the narrow-sense heritability of yellow-flesh intensity was 0.99 for both male and female parents. Lu et al. (2001) identified and quantified, using HPLC, the carotenoids present in 11 clones from this population in comparison with two tetraploid cultivars, Yukon Gold (yellow fleshed) and Superior (white fleshed). Six different carotenoids were detected in both diploid and tetraploid clones: neoxanthin, violaxanthin, lutein-5,6-epoxide, lutein, zeaxanthin, and an unknown carotenoid. Total carotenoid content in the diploids ranged from 136 (white flesh) to 1,435 µg per 100 g FW (intense yellow flesh), and the main carotenoids were lutein-5,6-expoxide (34 percent), violaxanthin (33 percent), and lutein (26 percent). In contrast, cvs. Superior and Yukon Gold contained 64 and 111 µg per 100 g FW of total carotenoids, and the main components were violaxanthin (40 percent) and lutein (22 percent). The tubers of the diploid clones were much smaller than those of the tetraploids, and this probably contributed to the difference in concentration of carotenoids. The relationship of flesh color intensity to carotenoid content was best described using an exponential equation that resulted in correlation coefficients (*r*) of 0.95 for total carotenoids and 0.92, 0.90, and 0.87 for lutein-5,6-epoxide, violaxanthin, and lutein, respectively. Hence, in practical potato breeding, selection for yellow flesh should result in high contents of both total and individual carotenoids such as lutein. Lutein, like neoxanthin, is stable under normal cooking conditions, whereas violaxanthin is easily destroyed by heat (Khachik et al., 1992).

Breithaupt and Bamedi (2002) determined the carotenoid pattern of four yellow- and four white-fleshed potato cultivars common on the German market, but they could not differentiate between the two groups. Violaxanthin, antheraxanthin, lutein, and zeaxanthin were present in varying amounts in all eight cultivars. However, total carotenoid content was higher in the yellow cultivars (58-175 μg per 100 g FW) than in the white ones (38-62 μg per 100 g FW). Carotenoid esters were also found (41-131 μg per 100 g FW) in all eight cultivars.

Nesterenko and Sink (2003) determined the carotenoid content of 15 tetraploid breeding lines from the Michigan State University breeding program along with orange-fleshed OR-4 and cvs. Atlantic, Spunta, and Yukon Gold. Violaxanthin (3.1-124.3 μg per 100 g FW) and lutein (19.8-119.0 μg per 100 g FW) were the prominent carotenoids detected in all genotypes studied. Neoxanthin and antheraxanthin were present in some genotypes, but zeaxanthin was present in high concentrations only in OR-4 (476 μg per 100 g FW).

Thorup et al. (2000) mapped ten structural genes from the *Capsicum* (pepper) carotenoid biosynthetic pathway on a *Capsicum annum* × *Capsicum chinense* $F_2$ genetic map anchored in *Lycopersicon* (tomato). The positions of these genes were compared with the positions of the same genes in tomato when known and with loci from pepper, potato, and tomato that affect carotenoid levels in different tissues. They suggested *CrtZ-2* (β-carotene hydroxylase) or a linked regulatory element as a candidate gene for the *Y* locus on chromosome 3 in potato (Bonierbale et al., 1988) but could not rule out Psy (phytoene synthase) in view of its map position in tomato and the existence in pepper of an unmapped Psy homolog. Furthermore, maize shows evidence of yellow endosperm being the result of two independent gain-of-function events associated with insertions into the promoter of the *Y1* gene that encodes phytoene synthase and up-regulation of expression in the endosperm (Palaisa et al., 2003). Romer et al. (2002) discovered that tuber-specific down-regulation of the zeaxanthin epoxidase gene in *S. tuberosum* not only increased the amount of zeaxanthin that accumulated but also increased total carotenoid level by up to 5.7 times that of the controls. At the Scottish Crop Research Institute, an inverse trend has been found between zeaxanthin epoxidase transcript level and tuber carotenoid content in a wide range of germplasm (Mark Taylor, personal communication). Clearly, the control of carotenoid biosynthesis still requires much elucidation, but the results should be of value to potato improvement programs.

In conclusion, considerable genetic variation exists in cultivated potatoes for agrihorticultural traits. Analysis of this variation and that in wild relatives should aid future crop improvement. During the past ten years, potato genetics has advanced beyond Mendelian analysis of qualitative traits and combining-ability analysis of quantitative traits. Dense linkage maps of molecular markers for all 12 linkage groups (chromosomes) of potato have allowed the development of QTL analysis and its application to more and more traits, and hence the meiotic (genetic) mapping of more and more genes of economic importance and quantification of their effects, although not always that precisely. Coupled with biochemical information on metabolic pathways, such as those for carbohydrates, anthocyanins, and carotenoids, progress has been made in identifying candidate genes (genetic loci) for QTLs and, hence, in functional analysis. Genetic transformation of potato has proved useful in testing hypotheses about function. The partial sequences of large numbers of expressed genes (ESTs) are now held in public databases and are expected to increase in use as they are located on physical maps that, in turn, are integrated with genetic maps. The challenge for breeders is going to be in identifying the most desirable alleles at the genetic loci of economic importance and getting them into farmers' fields as quickly as possible through new cultivars.

## REFERENCES

Boiteux, L.S., Reifschneider, F.J.B., Fonseca, M.E.N., and Buso, J.A. (1995). Search for sources of early blight *(Alternaria solani)* field resistance not associated with vegetative late maturity in tetraploid potato germplasm. *Euphytica 83*:63-70.

Bonierbale, M.W., Plaisted, R.L., and Tanksley, S.D. (1988). RFLP maps based on a common set of clones reveal modes of chromosomal evolution in potato and tomato. *Genetics 120:*1095-1103.

———. (1993). A test of the maximum heterozygosity hypothesis using molecular markers in tetraploid potatoes. *Theoretical and Applied Genetics 86:*481-491.

Bradshaw, J.E. (1994). Quantitative genetics theory for tetrasomic inheritance. In Bradshaw, J.E. and Mackay, G.R. (Eds.), *Potato Genetics* (pp. 71-99). Wallingford, U.K.: CAB International.

Bradshaw, J.E., Pande, B., Bryan, G.J., Hackett, C.A., McLean, K., Stewart, H.E., and Waugh, R. (2004). Interval mapping of quantitative trait loci for resistance to late blight [*Phytophthora infestans* (Mont.) de Bary], height and maturity in a tetraploid population of potato *(Solanum tuberosum* subsp. *tuberosum). Genetics 168:*983-995.

Bradshaw, J.E., Stewart, H.E., Wastie, R.L., Dale, M.F.B., and Phillips, M.S. (1995). Use of seedling progeny tests for genetical studies as part of a potato *(Solanum tuberosum* subsp. *tuberosum)* breeding programme. *Theoretical and Applied Genetics 90:*899-905.

Bradshaw, J.E., Todd, D., and Wilson, R.N. (2000). Use of tuber progeny tests for genetical studies as part of a potato *(Solanum tuberosum* subsp. *tuberosum)* breeding programme. *Theoretical and Applied Genetics 100:*772-781.

Breithaupt, D.E. and Bamedi, A. (2002). Carotenoids and carotenoid esters in potatoes (*Solanum tuberosum* L.): New insights into an ancient vegetable. *Journal of Agricultural and Food Chemistry 50:*7175-7181.

Brown, C.R., Edwards, C.G., Yang, C.-P., and Dean, B.B. (1993). Orange flesh trait in potato: Inheritance and carotenoid content. *Journal of the American Society for Horticultural Science 118:*145-150.

Brown, C.R., Wrolstad, R., Durst, R., Yang, C.-P., and Clevidence, B. (2003). Breeding studies in potatoes containing high concentrations of anthocyanins. *American Journal of Potato Research 80:*241-250.

Brown, J. and Caligari, P.D.S. (1989). Cross prediction in a potato breeding programme by evaluation of parental material. *Theoretical and Applied Genetics 77:*246-252.

Caligari, P.D.S. and Nachmias, A. (1988). Screening for field resistance to early blight *(Alternaria solani)* in potatoes. *Potato Research 31:*451-460.

Chen, X., Salamini, F., and Gebhardt, C. (2001). A potato molecular-function map for carbohydrate metabolism and transport. *Theoretical and Applied Genetics 102:*284-295.

Colditz, G.A., Branch, L.G., Lipnick, R.J., Willett, W.C., Rosner, B., Posner, B.M., and Hennekens, C.H. (1985). Increased green and yellow vegetable intake and lowered cancer deaths in an elderly population. *American Journal of Clinical Nutrition 41:*32-36.

Collins, A., Milbourne, D., Ramsay, L., Meyer, R., Chatot-Balandras, C., Oberhagemann, P., De Jong, W., Gebhardt, C., Bonnel, E., and Waugh, R. (1999). QTL for field resistance to late blight in potato are strongly correlated with maturity and vigour. *Molecular Breeding 5:*387-398.

Cubillos, A.G. and Plaisted, R.L. (1976). Heterosis for yield in hybrids between *S. tuberosum* spp. *tuberosum* and *S. tuberosum* spp. *andigena*. *American Potato Journal 53:*143-150.

Davies, H.V. (1998). Prospects for manipulating carbohydrate metabolism in potato tuber. *Aspects of Applied Biology 52:*245-254.

De Jong, H. (1987). Inheritance of pigmented tuber flesh in cultivated diploid potatoes. *American Potato Journal 64:*337-343.

———. (1991). Inheritance of anthocyanin pigmentation in the cultivated potato: A critical review. *American Potato Journal 68:*585-593.

De Jong, H. and Burns, V.J. (1993). Inheritance of tuber shape in cultivated diploid potatoes. *American Potato Journal 70:*267-283.

De Jong, W.S., De Jong, D.M., and Bodis, M. (2003). A fluorogenic 5′ nuclease (TaqMan) assay to assess dosage of a marker tightly linked to red skin color in autotetraploid potato. *Theoretical and Applied Genetics 107:*1384-1390.

De Jong, W.S., De Jong, D.M., De Jong, H., Kalazich, J., and Bodis, M. (2003). An allele of dihydroflavonol 4-reductase associated with the ability to produce red anthocyanin pigments in potato (*Solanum tuberosum* L.). *Theoretical and Applied Genetics 107:*1375-1383.

De Jong, W.S., Eannetta, N.T., De Jong, D.M., and Bodis, M. (2004). Candidate gene analysis of anthocyanin pigmentation loci in the Solanaceae. *Theoretical and Applied Genetics 108:*423-432.

Dodds, K.S. and Long, D.H. (1955). The inheritance of colour in diploid potatoes. 1. Types of anthocyanidins and their genetic loci. *Journal of Genetics 53:*136-149.

Ewing, E.E., Simko, I., Omer, E.A., and Davies, P.J. (2004). Polygene mapping as a tool to study the physiology of potato tuberization and dormancy. *American Journal of Potato Research 81:*281-289.

Freyre, R. and Douches, D.S. (1994a). Development of a model for marker-assisted selection of specific gravity in diploid potato across environments. *Crop Science 34:*1361-1368.

———. (1994b). Isoenzymatic identification of quantitative traits in crosses between heterozygous parents: Mapping tuber traits in diploid potato (*Solanum* spp.). *Theoretical and Applied Genetics 87:*764-772.

Freyre, R., Warnke, S., Sosinski, B., and Douches, D.S. (1994). Quantitative trait locus analysis of tuber dormancy in diploid potato (*Solanum* spp.). *Theoretical and Applied Genetics 89:*474-480.

Frusciante, L., Barone, A., Carputo, D., and Ranalli, P. (1999). Breeding and physiological aspects of potato cultivation in the Mediterranean region. *Potato Research 42:*265-277.

Gaziano, J.M., Manson, J.E., Branch, L.G., Colditz, G.A., Willett, W.C., and Buring, J.E. (1995). A prospective study of consumption of carotenoids in fruits and vegetables and decreased cardiovascular mortality in the elderly. *Annals of Epidemiology 5:*255-260.

Gebhardt, C., Ritter, E., Schachtschabel, U., Walkemeier, B., Uhrig, H., and Salamini, F. (1989). RFLP analysis and linkage mapping in *Solanum tuberosum. Theoretical and Applied Genetics 78:*65-75.

Glendinning, D.R. (1969). The performance of progenies obtained by crossing Groups Andigena and Tuberosum of *Solanum tuberosum. European Potato Journal 12:*13-19.

Gopal, J. (1994). Flowering behaviour, male sterility and berry setting in tetraploid *Solanum tuberosum* germplasm. *Euphytica 72:*133-142.

———. (1998). General combining ability and its repeatability in early generations of potato breeding programmes. *Potato Research 41:*21-28.

———. (2002). Nature of gene action in Tuberosum × Tuberosum and Tuberosum × Andigena families in potato. *Bangladesh Journal of Agricultural Research 27:*543-551.

Gopal, J., Chahal, G.S., and Minocha, J.L. (2000). Progeny mean, heterosis and heterobeltiosis in *Solanum tuberosum* × *tuberosum* and *S. tuberosum* × *andigena* families under a short day sub-tropic environment. *Potato Research 43:*61-70.

Hackett, C.A., Bradshaw, J.E., and McNicol, J.W. (2001). Interval mapping of quantitative trait loci in autotetraploid species. *Genetics 159:*1819-1832.

Hammond, B.R., Johnson, E.J., Russell, R.M., Krinsky, N.I., Yeum, K.J., Edwards, R.B., and Snodderly, D.M. (1997). Dietary modification of human macular pigment density. *Investigative Ophthalmology and Visual Science 38:*1795-1801.

Harborne, J.B. (1960). Plant polyphenols. I: Anthocyanin production in the cultivated potato. *Biochemical Journal 74:*262-269.

Haynes, K.G. (2000). Inheritance of yellow-flesh intensity in diploid potatoes. *Journal of the American Society for Horticultural Science 125:*63-65.

Haynes, K.G., Potts, W.E., Chittams, J.L., and Fleck, D.L. (1994). Determining yellow-flesh intensity in potatoes. *Journal of the American Society for Horticultural Science 119:*1057-1059.

Haynes, K.G., Sieczka, J.B., Henninger, M.R., and Fleck, D.L. (1996). Clone × environment interactions for yellow-flesh intensity in tetraploid potatoes. *Journal of the American Society for Horticultural Science 121:*175-177.

Hijmans, R.J. (2001). Global distribution of the potato crop. *American Journal of Potato Research 78:*403-412.

Holton, T.A. and Cornish, E.C. (1995). Genetics and biochemistry of anthocyanin biosynthesis. *Plant Cell 7:*1071-1083.

Howard, H.W. (1970). *Genetics of the Potato.* London: Logos Press Limited.

Huamán, Z., Williams, J.T., Salhuana, W., and Vicent, L. (1977). *Descriptors for the Cultivated Potato and for the Maintenance and Distribution of Germplasm Collections.* Rome: International Board for Plant Genetic Resources.

Hutten, R.C.B., Schippers, M.G.M., Eising, J., van Til, P.M., Hermsen, J.G.T., and Jacobsen, E. (1996). Analysis of parental effects on mean vine maturity and chip colour of *4x.2x* potato progenies. *Euphytica 88:*175-179.

Jacobs, J.M.E., van Eck, H.J., Arens, P., Verkerk-Bakker, B., te Lintel Hekkert, B., Bastiaanssen, H.J.M., El-Kharbotly, A., Pereira, A., Jacobsen, E., and Stiekema, W.J. (1995). A genetic map of potato *(Solanum tuberosum)* integrating molecular markers, including transposons, and classical markers. *Theoretical and Applied Genetics 91:*289-300.

Johnson, E.T., Ryu, S., Yi, H., Shin, B., Cheong, H., and Choi, G. (2001). Alteration of a single amino acid changes the substrate specificity of dihydroflavonol 4-reductase. *Plant Journal 25:*325-333.

Jung, C.S., Griffiths, H.M., De Jong, D.M., Cheng, S., Bodis, M., and De Jong, W.S. (2005). The potato *P* locus codes for flavonoid 3′,5′-hydroxylase. *Theoretical and Applied Genetics 110:*269-275.

Khachik, F., Goli, M.B., Beecher, G.R., Holden, J., Lusby, W.R., Tenorio, M.D., and Barrera, M.R. (1992). Effect of food preparation on qualitative and quantitative distribution of major carotenoid constituents of tomatoes and several green vegetables. *Journal of Agricultural and Food Chemistry 40:*390-398.

Killick, R.J. (1977). Genetic analysis of several traits in potatoes by means of a diallel cross. *Annals of Applied Biology 86:*279-289.

Kooman, P.L. and Rabbinge, R. (1996). An analysis of the relation between dry matter allocation to the tuber and earliness of a potato crop. *Annals of Botany 77:*235-242.

Lang, J. (2001). *Notes of a Potato Watcher*. College Station: Texas A&M University Press.

Lewis, C.E., Walker, J.R.L., Lancaster, J.E., and Sutton, K.H. (1998a). Determination of anthocyanins, flavonoids and phenolic acids in potatoes. I: Coloured cultivars of *Solanum tuberosum* L. *Journal of the Science of Food and Agriculture 77:*45-57.

———. (1998b). Determination of anthocyanins, flavonoids and phenolic acids in potatoes. II: Wild, tuberous *Solanum* species. *Journal of the Science of Food and Agriculture 77:*58-63.

Li, X.Q., De Jong, H., De Jong, D.M., and De Jong, W.S. (2005). Inheritance and genetic mapping of tuber eye depth in cultivated diploid potatoes. *Theoretical and Applied Genetics 110:*1068-1073.

Lu, W., Haynes, K., Wiley, E., and Clevidence, B. (2001). Carotenoid content and color in diploid potatoes. *Journal of the American Society for Horticultural Science 126:*722-726.

Mackay, G.R. (1996). An agenda for future potato research. *Potato Research 39:*387-394.

Maris, B. (1989). Analysis of an incomplete diallel cross among three ssp. *tuberosum* varieties and seven long-day adapted ssp. *andigena* clones of the potato (*Solanum tuberosum* L.). *Euphytica 41:*163-182.

Neele, A.E.F., Nab, H.J., and Louwes, K.M. (1991). Identification of superior parents in a potato breeding programme. *Theoretical and Applied Genetics 82:*264-272.

Nesterenko, S. and Sink, K.C. (2003). Carotenoid profiles of potato breeding lines and selected cultivars. *Horticultural Science 38:*1173-1177.

Oberhagemann, R., Chatot-Balandras, C., Schäfer-Pregl, R., Wegener, D., Palomino, C., Salamini, F., Bonnel, E., and Gebhardt, C. (1999). A genetic analysis of quantitative resistance to late blight in potato: Towards marker-assisted selection. *Molecular Breeding 5:*399-415.

Ortiz, R., (1998). Potato breeding via ploidy manipulations. In Janick, J. (Ed.), *Plant Breeding Reviews*, vol. 16 (pp. 15-86). New York: John Wiley & Sons.

Ortiz, R. and Golmirzaie, A. (2002). Hierarchical and factorial mating designs for quantitative genetic analysis in tetrasomic potato. *Theoretical and Applied Genetics 104:*675-679.

Ortiz, R. and Huamán, Z. (1994). Inheritance of morphological and tuber characteristics. In Bradshaw, J.E. and Mackay, G.R. (Eds.), *Potato Genetics* (pp. 263-283). Wallingford, U.K.: CAB International.

Ortiz, R. and Peloquin, S.J. (1994). Use of 24-chromosome potatoes (diploids and dihaploids) for genetical analysis. In Bradshaw, J.E. and Mackay, G.R. (Eds.), *Potato Genetics* (pp. 133-154). Wallingford, U.K.: CAB International.

Palaisa, K.A., Morgante, M., Williams, M., and Rafalski, A. (2003). Contrasting effects of selection on sequence diversity and linkage disequilibrium at two phytoene synthase loci. *Plant Cell 15:*1795-1806.

Plaisted, R.L., Sanford, L., Federer, W.T., Kehr, A.E., and Peterson, L.C. (1962). Specific and general combining ability for yield in potatoes. *American Potato Journal 39:*185-197.

Romer, S., Lubeck, J., Kauder, F., Steiger, S., Adomat, C., and Sandmann, G. (2002). Genetic engineering of a zeaxanthin-rich potato by antisense inactivation and co-suppression of carotenoid epoxidation. *Metabolic Engineering 4:*263-272.

Sandmann, G. (2001). Genetic manipulation of carotenoid biosynthesis: Strategies, problems and achievements. *Trends in Plant Science 6:*14-17.

Schäfer-Pregl, R., Ritter, E., Concilio, L., Hesselbach, J., Lovatti, L., Walkemeier, B., Thelen, H., Salamini, F., and Gebhardt, C. (1998). Analysis of quantitative trait loci (QTLs) and quantitative trait alleles (QTAs) for potato tuber yield and starch content. *Theoretical and Applied Genetics 97:*834-846.

Schäfer-Pregl, R., Salamini, F., and Gebhardt, C. (1996). Models for mapping quantitative trait loci (QTLs) in progeny of non-inbred parents and their behaviour in the presence of distorted segregation ratios. *Genetical Research 67:* 43-54.

Simmonds, N.W. (1964). The genetics of seed and tuber dormancy in the cultivated potatoes. *Heredity 19:*489-504.

Snodderly, D.M. (1995). Evidence for protection against age-related macular degeneration by carotenoids and antioxidant vitamins. *American Journal of Clinical Nutrition* 62(Suppl.):1448-1461.

Sonnewald, U., Hajirezaei, M.H., Kossmann, J., Heyer, A., Trethewey, R.N., and Willmitzer, L. (1997). Increased potato tuber size resulting from apoplastic expression of a yeast invertase. *Nature Biotechnology 15:*794-797.

Stark, D.M., Timmerman, K.P., Barry, G.F., Preiss, J., and Kishore, G.M. (1992). Regulation of the amount of starch in plant tissues by ADPglucose pyrophosphorylase. *Science 258:*287-292.

Struik, P.C., Haverkort, A.J., Vreugdenhil, D., Bus, C.B., and Dankert, R. (1990). Manipulation of tuber-size distribution of a potato crop. *Potato Research 33:*417-432.

Tai, G.C.C. (1976). Estimation of general and specific combining abilities in potato. *Canadian Journal of Genetics and Cytology 18:*463-470.

Tai, G.C.C. and De Jong, H. (1980). Multivariate analyses of potato hybrids. 1. Discrimination between tetraploid-diploid hybrid families and their relationship to cultivars. *Canadian Journal of Genetics and Cytology 22:*277-235.

Tarn, T.R. and Tai, G.C.C. (1983). Tuberosum × Tuberosum and Tuberosum × Andigena potato hybrids: Comparisons of families and parents, and breeding strategies for Andigena potatoes in long-day temperate environments. *Theoretical and Applied Genetics 66:*87-91.

Tarn, T.R., Tai, G.C.C., De Jong, H., Murphy, A.M., and Seabrook, J.E.A. (1992). Breeding potatoes for long-day, temperate climates. In Janick, J. (Ed.), *Plant Breeding Reviews,* vol. 9 (pp. 217-332). New York: John Wiley & Sons.

Thompson, P.G., Haynes, F.L., and Moll, R.H. (1980). Estimation of genetic variance components and heritability for tuber dormancy in diploid potatoes. *American Potato Journal 57:*39-46.

Thompson, P.G. and Mendoza, H.A. (1984). Genetic variance estimates in a heterogenous potato population propagated from true seed (TPS). *American Potato Journal 61:*697-702.

Thorup, T.A., Tanyolac, B., Livingstone, K.D., Popovsky, S., Paran, I., and Jahn, M. (2000). Candidate gene analysis of organ pigmentation loci in the Solanaceae. *Proceedings of the National Academy of Sciences (USA) 97:* 11192-11197.

Umaerus, V. and Umaerus, M. (1994). Inheritance of resistance to late blight. In Bradshaw, J.E. and Mackay, G.R. (Eds.), *Potato Genetics* (pp. 365-401). Wallingford, U.K.: CAB International.

Van den Berg, J.H., Ewing, E.E., Plaisted, R.L., McMurry, S., and Bonierbale, M.W. (1996a). QTL analysis of potato tuberization. *Theoretical and Applied Genetics 93:*307-316.

———. (1996b). QTL analysis of potato tuber dormancy. *Theoretical and Applied Genetics 93:*317-324.

van Eck, H.J., Jacobs, J.M.E., Stam, P., Ton, J., Stiekema, W.J., and Jacobsen, E. (1994). Multiple alleles for tuber shape in diploid potato detected by qualitative and quantitative genetic analysis using RFLPs. *Genetics 137:*303-309.

van Eck, H.J., Jacobs, J.M.E., van den Berg, P.M.M.M., Stiekema, W.J., and Jacobsen, E. (1994). The inheritance of anthocyanin pigmentation in potato (*Solanum tuberosum* L.) and mapping of tuber skin colour loci using RFLPs. *Heredity 73:*410-421.

van Eck, H.J., Jacobs, J.M.E., van Dijk, J., Stiekema, W.J., and Jacobsen, E. (1993). Identification and mapping of three flower colour loci of potato (*S. tuberosum* L.) by RFLP analysis. *Theoretical and Applied Genetics 86:*295-300.

Veilleux, R.E. and Lauer, F.I. (1981). Breeding behaviour of yield components and hollow heart in tetraploid-diploid vs. conventionally derived potato hybrids. *Euphytica 30:*547-561.

Visker, M.H.P.W., Keizer, L.C.P., Van Eck, H.J., Jacobsen, E., Colon, L.T., and Struik, P.C. (2003). Can the QTL for the late blight resistance on potato chromosome 5 be attributed to foliage maturity type? *Theoretical and Applied Genetics 106:*317-325.

Zrenner, R., Salanoubat, M., Willmitzer, L., and Sonnewald, U. (1995). Evidence of the crucial role of sucrose synthase for sink strength using transgenic potato plants (*Solanum tuberosum* L.). *Plant Journal 7:*97-107.

# Chapter 3

# Considerations for Successful Breeding

Jai Gopal

Radiocarbon dating has shown that potato cultivation started at least 8,000 years ago in southeast Peru and northwest Bolivia (Engel, 1970). There are no authentic records to show when potatoes were first the subject of conscious selection by man, but it must have started as soon as potato cultivation was undertaken. It is probable that growers started to exercise choice among the different "types" available to them, and much of the selection required to produce most of the present-day varieties was carried out during the early stages of potato domestication. Potato breeding in the modern sense began in 1807 in England when Knight (1807) made deliberate hybridization between different varieties by artificial pollination. As with other crop species, potato breeding gained impetus with the increase in understanding of the science of heredity and the rediscovery of Mendel's law of inheritance in 1900, initially published in 1865. However, cultivated potato *Solanum tuberosum* proved to be a difficult species for genetical research because of its complex inheritance patterns and it was not until the end of the 1930s that geneticists recognized that it was a tetraploid ($2n = 4x = 48$) that displays tetrasomic inheritance. Furthermore, most traits of economic importance displayed continuous variation for which Mendelian analysis was not possible because distinct classes could not be discerned. As a consequence, potato breeding remained empirical and genetically unsophisticated till now.

The choice of breeding methods to be employed for the improvement of a crop plant depends upon its mode of reproduction and propagation. Although potato reproduces sexually, in commercial cultivation, it is propagated mainly vegetatively through tubers. By this means any genotype can be fixed and all its intra- and inter-locus interactions responsible for phenotypic expression can be maintained, and if perceived desirable, it can be multiplied for commercial cultivation even though initially it may be present as a single plant. As potato is highly heterozygous, in order to maintain its productivity, improved potato varieties are developed by intermating desired parental lines and selecting superior clones from the progeny. The success of any potato-breeding program thus depends on the selection of superior parents and crosses, and selection of superior clones from the progeny. The present chapter concentrates on this aspect including the reproductive biology and other options relevant to developing improved cultivars for propagation through seed tubers. Breeding as well as other considerations for a sexually propagated commercial potato crop have been reviewed separately in Chapter 15 "True Potato Seed."

## *REPRODUCTIVE BIOLOGY*

### *Flowering*

To be able to make the desired cross for any genetic and breeding work, it is essential that the parental genotypes flower and fruit. A photoperiod of 12-14 h and temperatures of 15-20°C favor flower production and berry-setting in potato. A wide genetic variability has been observed for days to flowering, duration of flowering, flowering intensity, berry-setting, etc. (described in Chapter 2 "Genetics of Agrihorticultural Traits"). During the short days of the autumn season when the potato crop is grown in the plains in subtropics, flowering genotypes are too few (Birhman and Kaul, 1989). For further information on flowering and fruiting, the reader is referred to Chapter 15 on "True Potato Seed."

### *Sterility*

Male sterility is a very serious constraint in potato breeding. Though both pollen and ovule sterility can occur, pollen sterility ranging from partial to complete absence of pollen grains is very common in potato (Pushkarnath and Dwivedi, 1961). Almost one-third of the potato cultivars derived from *S. tuberosum* ssp. *tuberosum* do not form berries (Ross, 1986). Male sterility in potato is probably controlled by more than one gene, with partial dominance or by nuclear-cytoplasmic interactions (Grun, 1976; Howard, 1978) and depends partially on environmental factors (Cutter, 1978). Evidence of the importance of cytoplasmic factors (plasmons) in determining male sterility in potato has long been available (Salaman and Lesely, 1922). In cultivated *tuberosum* potato seven types of cytoplasmic factors exist each of which condition, in combination with a specific genotype, a specific type of sterility (Grun, 1970). These cytoplasmic factors (represented by indicated symbols) are: indehiscence ($In^S$), sporads ($Sp^S$), shriveled microspores ($SM^S$), anther style fusion ($Af^S$), thin anthers ($TA^S$), female sterility ($Fm^S$), and deformed flower ($df^S$). Variations, depending on the stage at which development is blocked, have also been observed within a particular type of sterility. A block can occur at any place in the development process and depending on the particular gene-plasmon interaction involved, different kinds of blockage may occur (Grun, 1970, 1976). The prevalence of sterility to such a large extent limits the use of many germplasm lines as parents, but at the same time this is advantageous because male sterile lines do not require emasculation when used as females in controlled pollinations.

### *Crossability*

Crossability in a broad sense can be defined as any natural or artificial fusion of two genetically different cells leading to hybrid progeny. When two parents/species have a normal pollen-pistil relationship, then, after intercrossing, an unimpeded chain of events is initiated, based on a perfect interaction between pollen genes (or gene complexes) and matching pistil genes (or gene complexes). Incomplete matching means one or more missing links in the chain and thus partial or complete failure of hybridization and introgression. Cross-incompatibility can be bilateral (i.e. in both directions) or unilateral (i.e., in only one direction).

Both self- and cross-incompatibility governed by a series of allelomorphs are widely prevalent in potatoes. Self- and cross-incompatibility are closely interrelated. This interrelatedness is based on the evidence that compatible and incompatible interspecific crosses generally obey the following rules (Hanneman, 1999): self-compatible species will cross with other self-compatible species; self-incompatible species will cross with other self-incompatible species; self-incompatible species will cross as females with self-incompatible species, but self-incompatible species as females will not cross with self-compatible species.

One hundred and eighty-one tuber-bearing species of *Solanum* have known ploidy levels (Spooner et al., 2004). Seventy-six percent are diploids, 3 percent exclusively triploids, 12 percent tetraploids, one species exclusively pentaploid, and 7 percent are hexaploids. Most diploid species are obligate outbreeders due to self-incompatibility governed by the classic S locus system (Hosaka and Hanneman, 1998). However, some diploid species from North and Central America are self-compatible, including *S. morreliforme, S. ployadenium,* and *S. verrucosum* (Hawkes, 1990).

Polyploid species are nearly self-compatible (Hawkes, 1990), and there are two classes of polyploids with respect to chromosome pairing. Tetraploid individuals of the cultivated species *S. tuberosum* are an example of a polysomic polyploid in which recombination among all four homologs is possible (Bradshaw and Mackay, 1994). Such species do not exclusively self-fertilize in their natural habitats (Brown, 1993) and maintain high levels of heterozygosity across generations (Hosaka and Hanneman, 1991). In contrast, the polyploid species in North and Central America are disomic polyploids in which only bivalent are typically seen in meiosis, and traits segregate as if the parents are diploids (Everhardt and Rowe, 1974).

Incompatibility barriers can be prezygotic or postzygotic. Prezygotic barriers are expressed between pollination and fertilization as nongermination of pollen on stigma; germination, but no penetration into the stigma; penetration, but inhibition of pollen tube growth to different degrees and at different sites in the style, ovary, or ovulum. Several approaches have been suggested to overcome prezygotic crossability barriers (Hermsen, 1994). For example, selecting matching genotypes within parent species, using a pollen mixture of many male genotypes in each interspecific cross, choosing the right direction of cross, cutting styles and direct pollination of ovules, somatic hybridization etc. Postzygotic barriers are expressed during and after fertilization in ovules, or more specifically in embryo sacs, and also during growth and flowering of plants ($F_1$) produced upon hybridization. These barriers, like the prezygotic ones, may cause partial or complete failure of hybridization and introgression.

### *Endosperm Balance Number*

For normal development of seed, endosperm is indispensable as it nourishes the embryo. Endosperm normally originates from the fusion of two haploid cells (central nucleus) from the female parent and one haploid cell (sperm) from the male parent. However, many interspecific crosses in potato fail, although the species crossed have the same ploidy level and meet the requirement of 2:1 female:male haploid cells for endosperm formation. In order to explain this phenomenon, a hypothetical value endosperm balance number (EBN) has been assigned to various species (Johnston et al., 1980; Hanneman, 1984). According to this hypothesis, for successful interspecific crosses, ratio of female EBN to male EBN in the endosperm is 2:1. This implies that balanced endosperm is obtained when both species have the same EBN. EBNs are independent of ploidy, but have been described as the "effective ploidy" of the parent. EBNs are additive and can be manipulated by meiotic ($2n$ gametes) or mitotic chromosome doubling and by haploidization. That is, if the chromosomes of a $2x$(1EBN) species are synthetically doubled to the tetraploid level, EBN also doubles, resulting in $4x$(2EBN), and a corresponding EBN reduction occurs in the reverse on ploidy reduction. This manipulation can be used to overcome crossability barriers among species of different EBNs. For example, a 2*x*(2EBN) species can only cross with another species that has been designated as 2EBN (i.e., 2*x*[2EBN] or 4*x*[2EBN]). With 2*n* gametes, they can also behave as though they are 4EBN and cross with 4*x*(4EBN) or 6*x*(4EBN) species, thus widening the possibilities for gene transfer. The significance of EBN for introgression breeding is thus in its predictive value for the success of interspecific crosses and of backcrossing, and in its potential to guide the breeder in overcoming barriers to introgression.

EBNs are determined completely empirically relative to other EBNs, and the "number" is meaningful only relative to other numbers based on success with standard test EBN crosses. EBNs have been assigned to most species (see Chapter 1) and can be used to classify species into crossing groups—2*x*(1EBN), 2*x*(2EBN), 4*x*(2EBN), 4*x*(4EBN), and 6*x*(4EBN). EBN has also been used to speculate on major evolutionary trends in tuber-bearing *Solanum* species (Spooner et al., 2004). Seed abortion caused by crosses between EBN incompatible species forms a major biological isolating mechanism in potato. There is a strong association of EBN to the cpDNA clades of Spooner and Sytsma (1992) and Spooner and Castillo (1997).

## *SELECTION OF SUPERIOR PARENTS AND CROSSES*

The choice of parents and crosses for breeding programs constantly puzzles plant breeders. Though, the choice of the right parents is the first important step in the success of a potato breeding program, very little research has been done to effectively predict the performance of the progeny generated from specific parents. Though some combining ability studies have been conducted in recent years, the need is to develop simpler procedures for effective selection of superior parents and crosses.

### Selection of Parents

Generally, regardless of breeding objectives, choice of parent is most often based on its de facto phenotypic expression of the characters of interest. Normally breeders try to pick up pairs of clones/varieties for intermating based on complementary sets of characters with desirable level of expression. Breeders generally try to use improved clones or varieties as parents because these are taken to be of acceptable "horticultural types." However, the choice of parents, based merely on phenotypic expression of desired characters, is often ineffective (Brown and Caligari, 1989; Gopal, 1998; Bradshaw et al., 2000) particularly when only nonadditive gene action is significant for the trait of interest. In such a situation, it is essential to select the parents based on the performance of their progeny. However, where both additive and nonadditive gene actions are important, which is generally the case for most of the characters, general combining ability can be a useful parameter for selecting parents. Though estimating the general combining ability of parents is labor intensive, time-consuming, and tedious due to problems of large-scale sterility and cross-incompatibility in potatoes, there have been many studies on this aspect (reviewed by Bradshaw and Mackay, 1994; Gopal and Gaur, 1997) in the past few years (also refer to Chapter 2). Empirical evidence from previous crossing programs regarding clones or varieties that appear to give a higher proportion of clones within acceptable limits, among their progenies, can also be made use of in future breeding programs.

### Selection of Crosses

It is important to decide in what combinations the selected parents are inter-mated so as to have a good chance of finding superior clones among their progenies. Some considerations in this regard are as follows.

#### Midparent and Midself Values

If parental general combining ability effects are highly correlated with their phenotypes (which approximates their genotypes if accurately assessed), the midparental values should provide a good prediction of the mean performance of the progeny (Bradshaw and Mackay, 1994). Midself values (in analogy to midparent values) measured as mean performance of the self-progenies of parents involved in a cross could be another method of cross prediction. By evaluating self-progenies of *n* parents (one offspring population per parent), a breeder can predict the value of $n(n-1)$ cross combinations. However, this can be used only for self-fertile parents. Brown and Caligari (1989), Gopal (1998), and Bradshaw et al. (2000) have shown that effectiveness of both midparent and midself values for cross prediction has been low to moderate for general impression (a complex character based on all visual characters at harvest including tuber yield, size, color, shape, etc.) and tuber yield, whereas Neele et al. (1991) found these parameters to be quite effective in seed potato harvest (Table 3.1). Midself values were, in general, slightly better than midparent values for cross prediction. This may be because midself values reflect the potential of the parents based on the performance of new genotypic combinations produced after selfing, whereas, midparent values are based on the phenotypic performance of the parents, which may or may not be reflected in their progeny.

*Specific Combining Ability*

The departure of a progeny mean from that expected on the basis of the general combining ability of parents is called specific combining ability. Low predictive values of midparent and midself values suggest that specific combining ability effects are more important than general combining ability effects for characters like tuber yield and its components. In such situations, it is desirable to select the specific cross combinations based on their high specific combining ability for the traits of interest. However, this requires crossing the parents in a specific pattern as for estimating the general combining ability. When this is not possible one can make use of progeny tests.

*Progeny Test*

In recent years a number of efforts have been made to develop progeny tests that can be used to identify superior cross combinations

TABLE 3.1. Correlation coefficients for parameters relevant to identification of superior parents and crosses for some important characters, as observed in some studies.

| Character | Correlation coefficients | | | Reference |
|---|---|---|---|---|
| | GCA vs. parents per se performance | Midparent vs. progeny mean | Midself vs. progeny mean | |
| General impression | 0.46$^{ns}$ | 0.26 | — | Bradshaw et al. (2000) |
| | 0.48$^{ns}$ | 0.41$^{ns}$ | 0.26$^{ns}$ | Brown and Caligari (1989) |
| | 0.39$^{ns}$ | 0.07$^{ns}$ | 0.64 | Gopal (1998)* |
| | — | 0.89 | 0.95 | Neele et al. (1991)* |
| Tuber yield | 0.85 | 0.61 | — | Bradshaw et al. (2000) |
| | 0.90 | 0.55$^{ns}$ | 0.76 | Brown and Caligari (1989) |
| | 0.39$^{ns}$ | 0.31 | 0.47 | Gopal (1998)* |
| | — | 0.90 | 0.96 | Neele et al. (1991)* |
| Mean tuber weight | 0.46$^{ns}$ | 0.40$^{ns}$ | 0.68 | Brown and Caligari (1989) |
| | 0.57 | 0.34 | 0.43 | Gopal (1998)* |
| | — | 0.86 | 0.86 | Neele et al. (1991)* |
| Tuber number | 0.11$^{ns}$ | 0.71 | 0.79 | Brown and Caligari (1989) |
| | 0.17$^{ns}$ | 0.12 | 0.36 | Gopal (1998)* |
| | — | 0.55 | 0.36 | Neele et al. (1991)* |
| Tuber shape | 0.56 | 0.42 | 0.33 | Gopal (1998)* |
| | — | 0.63 | 0.81 | Neele et al. (1991)* |

*Notes:* *highest value as observed in this study; $^{ns}$ = not significant at $p \leq 0.05$, significance level varied due to difference in number of observations in different studies.

from unplanned crosses. The objective is to identify the superior progenies based on a few genotypes per cross (may be 40-50) at the earliest opportunity so that larger populations of the identified crosses are raised to go for individual clone selection in the segregating progenies. The overriding criterion is that progeny means in initial generations must adequately reflect the subsequent performance of the clones when evaluated in advanced generations. Identification of superior crosses based on progeny means in seedling and/or early clonal generations have been found to be effective for a number of characters including resistance to potato cyst nematodes (Phillips, 1981), late blight in foliage (Stewart et al., 1983; Caligari et al., 1984), and in tubers (Wastie et al., 1987), gangrene (Wastie et al., 1990), powdery scab (Wastie, 1991), and even for tuber yield and its components (Brown et al., 1987, 1988; Gopal, 1997; Bradshaw et al., 1998). Besides using progeny means as a guide for selecting superior crosses (Table 3.2), effectiveness of univariate cross prediction method (Jinks and Pooni, 1976) has also been investigated in potato (Caligari and Brown, 1986; Brown and Caligari, 1989; Gopal, 1997). In this method means and within-progeny standard deviation are used to calculate the proportion of clones expected to transgress particular target value. The probabilities can be estimated as the sum of the normal probability integrals corresponding to the value $(x-\mu)/\sigma$ or $(\mu-x)\sigma$ depending on whether the predictions were for values greater than (or equal to) or less than (or equal to) the target value $(x)$. Here, $\mu$ is the progeny mean and $\sigma$ within-progeny standard deviation. Breeders would be mainly interested in knowing whether proportion of clones expected to transgress a particular target value in later generations can be predicted from the observed or predicted proportions transgressing the target value in an earlier generation. Correlation coefficients between predicted and observed proportions of clones exceeding the target values as reported by various workers (Table 3.2) showed that this method has low to moderate efficiency of cross prediction. Gopal (2001b) found that for tuber yield and its components selection of superior crosses based on only progeny mean can be as effective as the one based on both progeny mean and within-progeny standard deviation.

TABLE 3.2. Intergeneration correlation coefficients for parameters relevant to progeny test for some important characters, as observed in some studies.

| Reference | Character | SG vs. FCG | SG vs. SCG | FCG vs. SCG |
|---|---|---|---|---|
| *Correlation coefficients for progeny means between generations* | | | | |
| Bradshaw et al. (1998) | General impression | 0.88 | 0.89 | — |
| Brown et al. (1987) | General impression | 0.71 | 0.69 | 0.83 |
| Brown et al. (1988) | General impression | 0.70 | 0.71 | 0.78 |
| Gopal (1997) | General impression | 0.60 | 0.53 | 0.62 |
| | Tuber yield | 0.56 | 0.34 | 0.50 |
| | Tuber number | 0.56 | 0.34 | 0.50 |
| | Mean tuber weight | 0.26 | 0.21 | 0.49 |
| | Tuber shape | 0.69 | 0.50 | 0.53 |
| | Tuber color | 0.81 | 0.49 | 0.52 |
| *Correlation coefficients between predicted and observed proportions** | | | | |
| Caligari and Brown (1986) | General impression | 0.53 | 0.46 | 0.29 |
| Brown and Caligari (1989) | General impression | — | $0.48^{ns}$ | — |
| | Tuber yield | — | 0.90 | — |
| | Mean tuber weight | — | $0.46^{ns}$ | — |
| | Tuber number | — | $0.11^{ns}$ | — |
| Gopal (1997) | General impression | 0.50 | 0.29 | 0.42 |
| | Tuber yield | 0.45 | 0.27 | 0.33 |
| | Mean tuber weight | 0.24 | $0.07^{ns}$ | 0.27 |
| | Tuber number | 0.28 | 0.28 | 0.44 |

*Notes*: *predicted proportion of clones exceeding the set target value calculated based on progeny mean and within progeny variation and correlated with the observed proportion of clones exceeding the set target value in the next generation; $^{ns}$ = not significant at $p \leq 0.05$, significance level varied due to difference in number of observations in different studies; SG—seedling ($F_1$) generation, FCG—first clonal ($F_1C_1$) generation, SCG—second clonal ($F_1C_2$) generation.

### *Diploid Progenies and Use of* 4x × 2x *Test Crosses*

Neele et al. (1991) used diploid progeny of a tetraploid clone for predicting the breeding value of tetraploid clone, but the results were not encouraging. A possible reason for the poor values of this predictor was that in the diploid progenies many poor growing plants occurred that seemed to be unrelated to the breeding value of the tetraploid parent. Neele et al. (1991) hypothesized that during the production of diploids after pollination of tetraploid with *S. phureja* (see section "Dihaploid and 2*n* gametes" later), recessive deleterious alleles might have been present in a homozygous state. In a tetraploid most such recessive alleles would go unnoticed. Use of diploid progenies is otherwise also not practicable as these are difficult to produce and maintain over subsequent clonal generations.

Neele et al. (1991) also tried 4*x* × 2*x* test crosses as predictors of progeny performance. These provided good estimates for the actual values of the progenies. For many characters the correlation coefficients exceeded 0.8. These were almost of the same value as midparent value in their study (Table 3.1). The high predictive value of these parameters in this study was due to the preponderance of the additive gene action in this population. However, a drawback of 4*x* × 2*x* crosses is the investment in the production of true seeds through 2*n*-FDR pollen producing males.

### *Genetic Divergence and Heterosis*

Heterosis refers to the situation when the progeny surpasses the value of the best parent or the parents' mean for character of interest. Maintaining a high level of heterozygosity is crucial in potato as severe inbreeding depression results from selfing or haploid extraction (De Jong and Rowe, 1971; Mendiburu et al., 1974). Hence, exploitation of positive heterosis is by far the most satisfactory method of potato breeding aimed to enhance tuber yield. To realize heterosis, genetic divergence of parents is generally considered useful. Though studies on genetic divergence in potato have been conducted by many workers, there are few reports about its effectiveness in predicting heterosis. Loiselle et al. (1991) found no relationship between genetic divergence of parents (based on pedigree, agronomic, and molecular data) and tuber yield of progenies. Gopal and Mincoha (1997a) found that genetic divergence based on morphological data is of low to moderate effectiveness in predicting tuber yield and heterosis for tuber yield. In other crops too, where such an association has been observed, yield or hybrid vigor has been related to no more than a moderate degree of divergence. Based on intergroup hybrids, Christopher (1980) and Sanford and Hanneman (1982) suggested the existence of a "heterotic threshold" in the cultivated potato beyond which greater heterozygosity does not result in more vigorous progenies. This appears to be true even for the crosses made on the basis of the genetic divergence of the parents. Bonierbale et al. (1993) reported that value of heterozygosity for yield and its components depends on the genetic background of the parents.

Gopal (1999a) found that the same set of parents may appear to have higher genetic distances under in vitro conditions than under field conditions based on their phenotypic evaluation. But genetic distances estimated under in vitro conditions had no relationship with heterosis under field conditions (Gopal and Minocha, 1997a). Rather, under field conditions too, genetic distance of parents estimated under the conditions similar to those in which crosses were evaluated was more effective in cross prediction than those estimated under different conditions. This reflects the importance of genotype × environment interaction on the use of apparent genetic diversity determined on phenotypic basis of parents for predicting the progeny's performance.

Though all these criteria can be used for identification of superior parents and crosses with varying degrees of effectiveness, the ideal strategy would be to identify the superior parents based on their general combining ability and inter-mating them in all feasible combinations to select a few top combinations based on progeny test. Further, the effectiveness of various genetic parameters in identification of superior parents and crosses vary from material to material, and also by the specific conditions under which these parameters are estimated. Thus, it is important to use them judiciously taking into consideration the extent of their relationship with each other and more importantly with the performance of the progenies generated from parents selected based on these parameters.

## SELECTION OF SUPERIOR CLONES FROM THE PROGENY

### Selection in Early Generations

In the Netherlands, on average one potato variety is developed out of 180,000 seedlings (Neele et al., 1988). Wenzel et al. (1983) calculated that 500,000 to 2,000,000 seedlings must be evaluated in order to select a single new cultivar. This estimate is still consistent with the practices used for most modern breeding programs. These large populations must quickly be reduced to a manageable number before statistically accurate and consistent tests are performed. Major problem in early generations of potato-breeding programs is the limitation in the availability of number of tubers per clone. Assessment is based on nonreplicated plots of just a single plant per genotype in the seedling ($F_1$) generation and a few hills per genotype (3-12) in first ($F_1C_1$) and second clonal ($F_1C_2$) generations. The tubers used for planting various clones also often vary in size. These factors in addition to normal year and location effects make the identification of potentially desirable genotypes in early generations difficult. Clones are selected based on visual assessment of their commercial worth. This appraisal, known as general impression or breeder's preference, involves a complex series of visible traits that may be genetically unrelated. There is a general consensus that selection of individual clones in early generations on the basis of general impression results in a random reduction in number of genotypes (reviewed by Caligari, 1992; Bradshaw and Mackay, 1994). The worst problem appeared to be the inadvertent elimination of superior genotypes when selection was based on overall appearance. In most of the studies aimed at estimating repeatability of clones' performance in early generations (Table 3.3), seedlings were raised in pots in greenhouses. In the plains of subtropics where assured irrigation is available, seedlings can be grown very well in the fields, and this is being followed in all potato-breeding programs in India. Based on studies (Gopal et al., 1992, 1994) conducted under these conditions, the following selection procedure has been suggested for an early and safe elimination of unproductive genotypes for tuber yield and its components, including tuber color, shape, eye depth, etc.

TABLE 3.3. Intergeneration correlation coefficients relevant to early generation clonal selection for some important characters, as observed in some studies.

| Character | Correlation coefficients for clones performance between generations | | | | | | Reference |
|---|---|---|---|---|---|---|---|
| | SG vs. FCG | SG vs. SCG | SG vs. TCG | FCG vs. SCG | FCG vs. TCG | SCG vs. TCG | |
| General impression | — | — | 0.43 | — | 0.35 | 0.55 | Bradshaw et al. (1998) |
| | 0.29 | — | — | — | — | — | Brown et al. (1984)* |
| | 0.48 | 0.36 | — | 0.64 | — | — | Brown et al. (1987) |
| | — | — | — | 0.39 | — | — | Maris, 1966 |
| | — | — | — | 0.60 | — | — | Neele et al. (1988)* |
| Tuber yield | — | — | — | 0.24 | 0.49 | 0.52 | Caligari et al. (1986) |
| | 0.24 | 0.19 | 0.29 | 0.52 | 0.48 | 0.60 | Gopal et al. (1992) |
| | — | — | — | — | — | 0.69 | Love et al. (1997) |
| | — | — | — | 0.63 | — | — | Neele et al. (1988)* |
| | 0.65 | | | | | | Pfeffer (1963)* |
| | — | — | — | 0.31 | — | — | Tai (1975) |
| Mean tuber weight | — | — | — | 0.48 | 0.40 | 0.60 | Caligari et al. (1986) |
| | 0.38 | 0.22 | 0.38 | 0.52 | 0.55 | 0.65 | Gopal et al. (1992) |
| | | | | | | 0.59 | Love et al. (1997 |
| | 0.45 | — | — | — | — | — | Pfeffer (1963)* |
| | — | — | — | 0.23 | — | — | Tai (1975) |
| Tuber number | — | — | — | $0.04^{ns}$ | 0.33 | 0.32 | Caligari et al. (1986) |
| | $0.21^{ns}$ | $0.11^{ns}$ | 0.02 | 0.45 | 0.29 | 0.40 | Gopal et al. (1992) |
| | — | — | — | — | — | 0.63 | Love et al. (1997) |
| | | | | 0.34 | | | Neele et al. (1988)* |
| | 0.29 | — | — | — | — | — | Pfeffer (1963)* |

TABLE 3.3 *(continued)*

| Character | Correlation coefficients for clones performance between generations | | | | | | Reference |
|---|---|---|---|---|---|---|---|
| | SG vs. FCG | SG vs. SCG | SG vs. TCG | FCG vs. SCG | FCG vs. TCG | SCG vs. TCG | |
| Tuber shape | — | — | — | — | — | 0.73 | Love et al. (1997) |
| | — | — | — | 0.74 | — | — | Neele et al. (1988)* |
| | — | — | — | 0.74 | — | — | Maris (1966) |
| | 0.64 | — | — | — | — | — | Pfeffer (1963)* |
| Eye depth | — | — | — | — | — | 0.62 | Love et al. (1997) |
| | — | — | — | 0.61 | — | — | Neele et al. (1988)* |
| | — | — | — | 0.59 | — | — | Maris (1966) |
| | 0.45 | — | — | — | — | — | Pfeffer (1963)* |
| | 0.49 | | | | | | Świezyński (1978) |
| Maturity | — | — | — | 0.52 | — | — | Maris (1966) |
| | — | — | — | 0.51 | — | — | Tai (1975) |

*Notes:* * highest value as observed in this study; [ns] = not significant at $p \leq 0.05$, significance level varied due to difference in number of observations in different studies; SG—seedling ($F_1$) generation, FCG—first clonal ($F_1C_1$) generation, SCG—second clonal ($F_1C_2$) generation, TCG—third clonal ($F_1C_3$) generation.

- Seedlings of poor vigor may be rejected prior to transplanting in the field as they are likely to have low yields in later generations (Figure 3.1).
- Clones with undesirable tuber color, tuber shape, eye depth, and cracking may be rejected from the seedling stage onward, as these characters have a high repeatability over generations.
- Rejection should not be done on the basis of tuber yield, average tuber weight, or number of tubers in the seedling generation.

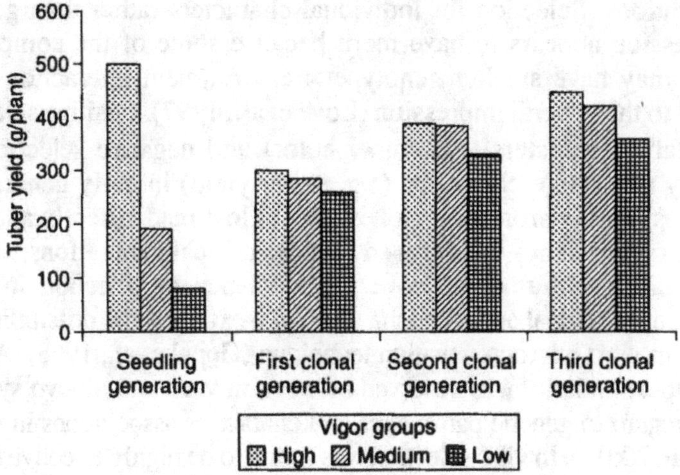


FIGURE 3.1. Mean tuber yield over generations as observed in the clones grouped on the basis of plant vigor in the seedling generation. Though performance of poor vigor seedlings improved in later generations, it was lower than that of high and medium vigor groups. *Source:* Based on Gopal et al. (1992).

- Negative selection (rejection of the poor phenotype) for tuber yield and tuber weight can be initiated from the first clonal generation onward, whereas number of tubers can be considered for the rejection of undesirable types from the second clonal generation onward.
- In the subsequent generation, for simultaneous improvement in tuber yield and its components, average tuber weight should be used as a selection parameter after fixing a standard for the minimum number of tubers required in the selected types.

The low repeatability of tuber yield and its components in early generations (Table 3.3) suggests that the efficiency of visual score (as used in studies reviewed by Bradshaw and Mackay, 1994) perhaps can be improved if the score does not consider tuber yield and related traits as its component characters. Separating the general impression into component traits thus led to the identification of individual traits (e.g., tuber color, tuber shape, eye depth, etc.) selectable in early generations. Selection for individual characters rather than general impression appears to have merit because some of the component traits may have smaller genotype × environment interaction compared to the general impression (Love et al., 1997). Positive selection for stable characters (e.g., tuber color) and negative selection for poorly repeatable characters (e.g., tuber yield) in early generations will insure that promising clones are not lost inadvertently and thus selection efficiency is improved in these critical generations.

In order to further improve the efficiency of selection in early generations, Gopal and Minocha (1997b,c) explored the possibilities of using in vitro micropropagation technique (Gopal et al., 1998). A considerable similarity was observed between in vitro and in vivo systems with regard to genetic parameters and characters associations in potato (Gopal, 2001a). In vitro selection was found to be highly effective for tuber color, number of eyes, and moderately effective for average tuber weight, plant vigor and foliage senescence under specific conditions (Gopal and Minocha, 1997b). They postulated that in vitro selection may be effective even for tuber yield, but need is to simulate and identify the in vitro conditions under which genotypic differences are reflected in phenotypes more realistically as they are expressed in vivo. The problem of fewer tubers in early generations can be overcome by using in vitro propagation to produce microtubers. This can be started after the harvest of seedling generation with the clones selected following the procedure described here. The number of clones required to be multiplied this way can be further reduced if breeding for tuber yield and its components is combined with resistance to diseases such as late blight. Susceptible genotypes can be discarded in the seedling generation itself (Stewart et al., 1983; Caligari et al., 1984). In the crop raised from microtubers, selection can be done for general impression (Gopal and Minocha, 1997c), giving less weightage to tuber yield. The produce from this crop can be used to conduct replicated trials permitting selection for tuber yield, etc. (Gopal et al., 2002). The production of microtubers in sufficient numbers, and the subsequent production of normal seed tubers from them, will result in earlier large-scale trials thereby substantially reducing the period taken to develop a variety. The suggested selection procedure based on only in vivo propagation and also combined with in vitro propagation is illustrated in Figure 3.2.

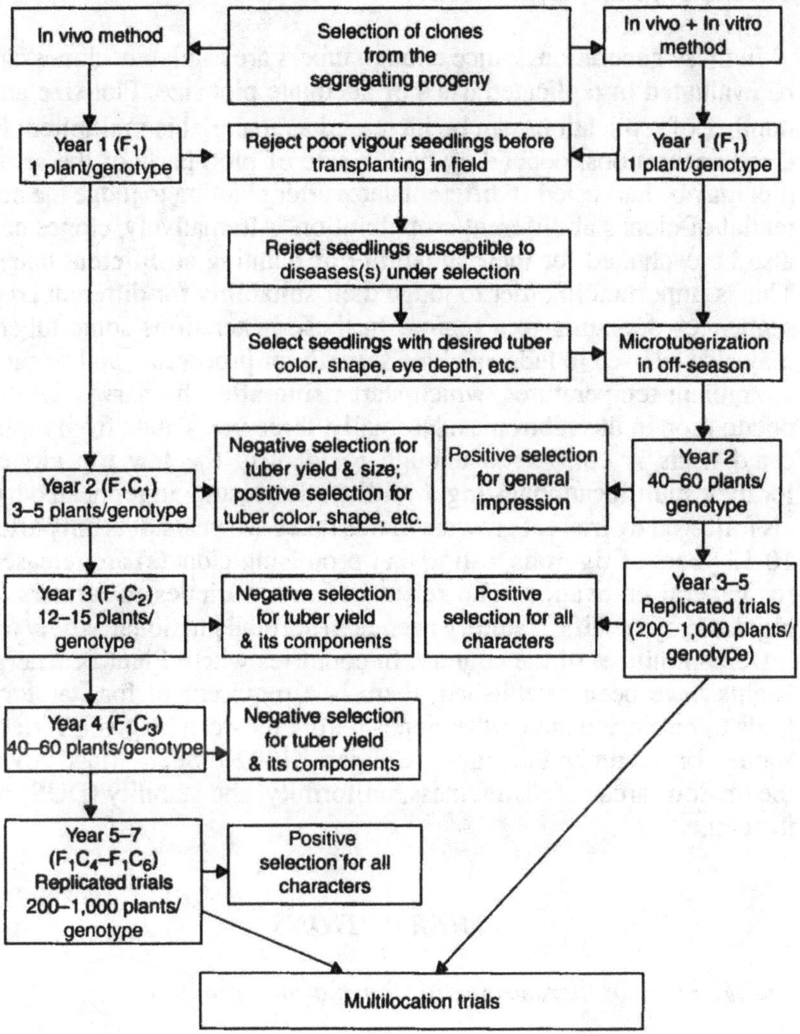


FIGURE 3.2. Stepwise procedure for selection of superior clones for agronomic characters in potato breeding indicating the time advantage of combining in vitro and in vivo methods.

### *Selection in Later Generations*

In these generations, since enough tubers are available, clones can be evaluated in replicated trials of adequate plot size. Plot size and number of replications can be increased as material is multiplied. In these generations, depending on the size of plot, parts of the same plot may be harvested at different dates after planting to judge the potential of clones at different crop duration. Alternatively, clones can also be evaluated for their suitability of planting at different dates. This is important in order to judge their suitability for different crop sequences operating in a region. In these generations some tubers may also be used to judge a clone's worth for processing and storage at ambient temperatures, which start rising after the harvest of the potato crop in the subtropics. Normally, three years' data from replicated trials is considered enough to identify the few top clones for their multilocation testing. Multilocation testing in replicated trials followed by trials at growers' fields takes 4-6 years. It is only after 10-12 years of rigorous testing that promising clone(s) are released for general cultivation. The release of new varieties to farmers is regulated by specific statutory agencies (normally national variety release committee) of the country. In countries where Plant Breeders' Rights have been established, there is a requirement for statutory trials to be carried out on the clones before they can be given variety status for commercialization (Caligari, 1992). Such trials cover the obvious areas of distinctness, uniformity, and stability (DUS) of the clone.

## *OTHER OPTIONS*

### *Andigena, Neotuberosum, and Phureja/Stenotomum*

*Solanum tuberosum* ssp. *andigena* potato is short day-adapted and is an important evolutionary source of *Solanum tuberosum* ssp. *tuberosum* potato, which is long day-adapted. Since *tuberosum* descended from a very small number of introductions of *andigena* potato, and the late blight epidemic of the 1840s greatly reduced the amount of variability present in *tuberosum* potatoes, it is known to have a narrow genetic base (Hawkes, 1967; Grun, 1990). Little improvement is observed in tuber yield among the recombinants produced by crossing among *tuberosum* parental clones (Howard, 1963; Simmonds, 1976). To overcome these problems attempts were made in Europe and North America (where the crop is grown under temperate long days) to use *andigena* in potato-breeding programs. But the short-day requirement for tuberization of *andigena* resulted in poor performance of *tuberosum* × *andigena* progenies (Howard, 1963). However, the use of *andigena* preselected for long days, in crosses with *tuberosum* led to progenies yielding higher than *tuberosum* × *tuberosum* progenies (Glendinning, 1969; Cubillos and Plaisted, 1976). In the 1970s, the name *neo-tuberosum* was proposed for advanced *andigena* material adapted to long days (Simmonds, 1976). Likewise, during the period 1962-1979, Caroll (1982) employed a mass selection method to produce a population of Group Phureja/Stenotomum adapted to long-day north European conditions. Direct hybridization of members of this improved diploid population with tetraploid *tuberosum* cultivars via unreduced pollen grains ($4x \times 2x$ crosses) resulted in tetraploid hybrids, some of which were superior to standard tetraploid cultivars in both total and marketable yield (Caroll and De Maine, 1989). Similar work has been done in North America (Plaisted and Hoopes, 1989).

In the plains of subtropics, potato is grown under short days during winter. Under these conditions, *andigena* adapted to short days can be used advantageously for exploiting heterosis. The progenies from such crosses, however, are generally late maturing. To overcome this drawback and avoid the undesirable tuber characters in *tuberosum* × *andigena* populations, there is a need to select early bulking/maturing *andigena* clones with desired tuber characters like uniform white or red skin with shallow eyes. Considerable variability exists for such characters and heritability values for maturity and tuber size are known to be moderate to high under short-day conditions (Gopal, 1999b).

### *Other Primitive Cultivated and Wild Species*

The potato is outstanding in having the richest genetic resources among the cultivated plant species. Wild potato genetic resources have proved useful in breeding programs for disease resistances, environmental stresses, and other agronomic traits of interest (Ross, 1986; Hawkes, 1990; Spooner et al., 1991; Ochoa, 1999; Jansky, 2000). Species in North and Central America are particularly rich in such traits (Spooner et al., 2004). For example, many of the wild species have high levels of resistance against *Phytophthora infestans* (Mont.) de Bary (Bamberg et al., 1994), which causes late blight, the most damaging fungal disease of potato. *Solanum stoloniferum* has long been recognized as a source of resistance to potato virus Y (Pavek and Corsini, 2001). Nematode diseases are particularly difficult to combat without genetic resistance, such as that found in species like *S. bulbocastanum* and *S. hougasii* (Brown et al., 1991). In addition to protecting the crop from diseases, pests, and stresses, potato breeders also must maintain high quality standards. Brown et al. (1999) discovered resistance to nonenzymatic tuber browning in *S. hjertingii* and incorporated it into cultivated gene pool through breeding. Use of wild potato species, however, can lead to increased glycoalkaloid contents in the breeding populations. Glycoalkaloids in high concentration are thought to impart protection against pathogens and pests (Dale and Mackay, 1994; Maga, 1994). But they also convey a bitter taste and in very high concentrations a burning sensation and various symptoms associated with food poisoning (Maga, 1994). The widely accepted safety limit for glycoalkaloids levels in tubers is 0.2 $mg{\cdot}g^{-1}$ fresh weight (for more information on quality improvement the reader is referred to Chapter 5).

According to Ross (1986) nearly all traits found in wild species can also be detected in primitive cultivars. Thus, to avoid the transfer of excessively large number of undesirable characters and the requirement for many years of backcrossing to eliminate them, searches for newly required genes should proceed in the sequence: ssp. *tuberosum* cvs. and breeding parents, ssp. *andigena* and other cultivated species, and finally wild species. Nevertheless, primitive and wild spp. are indispensable for broadening the genetic base of the cultivated potatoes.

As majority of wild *Solanum* spp. are diploid and they do not cross directly with tetraploid *tuberosum*, special efforts are required to transfer useful genes from wild to cultivated potatoes. Ross (1986) has given a vivid account of transferring some useful characters from primitive and wild species to the cultivated potato. Prebreeding of wild species to eliminate the undesired characters by recurrent selection of desired clones from sib-mated/open-pollinated populations, can facilitate the use of wild species in potato variety improvement programs.

### *Dihaploids and 2n Gametes*

In potato, the haploid of a tetraploid is called a dihaploid and, the haploid of a diploid is referred to as a monohaploid. Dihaploids of potato were first reported by Hougas and Peloquin (1957) who used $4x \times 2x$ crosses between S. *tuberosum* × *S. phureja* (pollinator) for parthenogenetic development of dihaploids (Hougas and Peloquin, 1958). Anther culture has also been used to produce dihaploids in potatoes, but the application of this approach is limited due to the lack of response of many genotypes to anther culture. Dihaploids of *tuberosum* cross readily with many diploid wild species and thus provide the opportunity to introgress useful traits from alien sources to cultivated background. Dihaploids with disomic rather than tetrasomic inheritance facilitate the genetics and breeding of potatoes. Smaller populations are required to detect recessive genes in $2x$ crosses than in $4x$ intercrossings (Peloquin et al., 1990).

The final goal of potato breeding at the diploid level is to develop $2x$ clones with specific attributes, good breeding value for agronomic traits and $2n$ gametes. Production of $2n$ (unreduced) gametes by diploid-dihaploid hybrids is important to revert back to tetraploid level through $4x \times 2x$ or $2x \times 4x$ (unilateral) or $2x \times 2x$ (bilateral) polyploidization. The natural production of $2n$ gametes in many tuber-bearing *Solanum* species is well established (Watanabe and Peloquin, 1991), although their frequency may vary greatly within and between species owing to genetic, environmental, and physiological factors. Such $2n$ gametes may originate in different ways, but genetically the basic distinction is between $2n$ FDR (first division restitution) gametes (reduction of $2n$ to $n$ chromosomes followed by restitution to $2n$ through incomplete first division) and $2n$ SDR (second division restitution) gametes (reduction of $2n$ to $n$ chromosomes followed by restitution to $2n$ through incomplete second division).

The mechanisms involved in the origin of FDR and SDR gametes and their genetic consequences have been reviewed by Tai (1994). It can be calculated that FDR gametes introgress 80 percent of the intact parental genotype to the hybrid progeny and SDR gametes 40 percent. This high percentage of intact gene transfer through FDR gametes is especially important when the parent plant carries many desirable genes for qualitative as well as quantitative characters. The analytical breeding scheme of Chase (1963) based on dihaploids recommends the use of 2*n* gametes for production of tetraploids with maximal heterozygosity.

### *Biotechnology*

Potato is highly amenable to tissue culture and has been used widely for the development of biotechnology techniques. These techniques are now becoming handy in the potato variety improvement programs. The potential of in vitro micropropagation in improving the efficiency of selection in early generations and for an early conduct of replicated trials has been described earlier. The other techniques worth exploitation in potato-breeding programs are as follows: somaclonal variations (Larkin and Scowcroft, 1981) for creating useful variability, somatic hybridization (Möllers and Wenzel, 1992) for overcoming barriers to normal crossing, molecular markers (Bonierbale et al., 1990; Melchinger, 1990) for marker-assisted selection, and genetic transformation (Huisman et al., 1992) for transfer of single, specific genes from any source to the existing varieties. The potential of biotechnology in potato improvement has been reviewed by Professor Gerhard Wenzel in Chapter 4. Use of biotechnology in potato improvement is already a reality, but these techniques are likely to remain as adjuncts (Caligari, 1992) to the more classical methods of breeding, which will themselves continue to evolve in efficiency as the underlying mechanisms and properties of genetic systems become more clearly known.

Through this recourse I have tried to present the theoretical and practical considerations for successful potato breeding. These revolve around the basic requirements of making crosses between pairs of parents, generation of genetic variation and phenotypic selection over a number of vegetative generations. Since the middle of the nineteenth century, many cultivars have been produced worldwide through conventional breeding. For example, in 1998, the Scottish Agricultural Science Agency in Edinburgh had in their reference collection, 831 potato varieties that had been grown in Scotland, of which 80 were still considered commercial (SASA, 1998). In India, 35 potato varieties have been released since 1958 after the establishment of the Central Potato Research Institute, Shimla in 1949. The world catalogue of potato varieties (Hils and Pieterse, 2005) lists nearly 4,000 cultivars from 100 countries. Farmers and end users, however, will judge these achievements of conventional breeding by how well currently available cultivars meet their needs. It appears that they had not benefited as much as they should have from the new cultivars given the continuing popularity of many older ones such as Russet Burbank. This was released in 1876 and still accounted for 38 percent of the certified seed acreage in the United States (Douches and Jastrzebski, 1993). Similarly, in India, more than 37 percent of the breeder seed produced in 2001-2002 (CPRI, 2003) was of only three potato cultivars, namely, Kufri Chandramukhi, Kufri Jyoti and Kufri Sindhuri, released way back in 1968. It is easy to find similar examples in other countries and this must be a cause for concern and something that needs to be rectified. Therefore, potato breeders must strive hard to develop efficient breeding methods as outlined in this chapter. They must also consider how effective they have been in utilizing the biodiversity available to them in the tuber-bearing *Solanum* species. Plaisted and Hoopes (1989) reported the repeated appearance of a few ancestors in the pedigrees of American varieties and a Chilean Tuberosum cultivar "Rough Purple Chili" introduced into the United States in 1851 was found in the pedigrees of almost all modern American varieties leading to their narrow genetic base. European (Hawkes, 1979; Ross, 1986) as well as Indian (Gopal and Oyama, 2005) potato varieties too have been reported to have a narrow genetic base. Therefore, a breeding methodology that could exploit unadapted genetic resources more efficiently is required. Besides helping in the development of substantially superior genotypes, this would widen the genetic base of cultivated potatoes, and hence provide some protection against the unexpected outbreaks of pests.

For additional reviews on potato breeding, the reader is referred to Ross (1986), Mackay (1987), Caligari (1992), Tarn et al. (1992), Douches and Jastrzebski (1993), and Bradshaw and Mackay (1994).

## REFERENCES


Bamberg, J.B., Martin, M.W., and Schartner, J.J. (1994). *Elite Selections of Tuber-bearing Solanum Species Germplasm.* Madison: University of Wisconsin Press.

Birhman, R.K. and Kaul, M.L.H. (1989). Flower production, male sterility and berry setting in *andigena* potato. *Theoretical and Applied Genetics 78:*884-888.

Bradshaw, J.E., Dale, M.F.B., Swan, G.E.L., Todd, D., and Wilson, R.N. (1998). Early-generation selection between and within pair crosses in a potato *(Solanum tuberosum* subsp. *tuberosum)* breeding programme. *Theoretical and Applied Genetics 97:*1331-1339.

Bradshaw, J.E. and Mackay, G.R. (1994). Breeding strategies for clonally propagated potatoes. In Bradshaw, J.E. and Mackay, G.R. (Eds.), *Potato Genetics* (pp. 467-498).Wallingford, U.K.: CAB International.

Bradshaw, J.E., Todd, D., and Wilson, R.N. (2000). Use of tuber progeny tests for genetical studies as part of a potato *(Solanum tuberosum* subsp. *tuberosum)* breeding programme. *Theoretical and Applied Genetics 100:*772-781.

Bonierbale, M.W., Ganal, M.W., and Tanksley, S.D. (1990). Applications of restriction fragment length polymorphisms and genetic mapping in potato breeding and molecular genetics. In Vayda, M.E. and Park, W.D. (Eds.), *The Molecular and Cellular Biology of the Potato* (pp. 13-24). Wallingford, U.K.: CAB International.

Bonierbale, M.W., Plaisted, R.L., and Tanksley, S.D. (1993). A test of the maximum heterozygosity hypothesis using molecular markers in tetraploid potatoes. *Theoretical and Applied Genetics 86:*481-491.

Brown, C.R. (1993). Outcrossing rate in cultivated autotetraploid potato. *American Potato Journal 68:*445-452.

Brown, C.R., McNabay, M., and Dean, B.B. (1999). Genetic characterization of reduced melanin formation in tuber tissue of *Solanum hjertingii* and hybrids with cultivated diploids. *American Journal of Potato Research 76:*37-43.

Brown, C.R., Mojtahedi, H., and Santo, G.S. (1991). Resistance to Columbia root-knot nematode in *Solanum* ssp. and in hybrids of *S. hjertingii* and hybrids with tetraploid cultivated potato. *American Potato Journal 68:*445-452.

Brown, J. and Caligari, P.D.S. (1989). Cross prediction in a potato breeding programme by evaluation of parental material. *Theoretical and Applied Genetics 77:*246-252.

Brown, J., Caligari, P.D.S., Dale, M.F.B., Swan, G.E.L., and Mackay, G.R. (1988). The use of cross prediction methods in a practical potato breeding programme. *Theoretical and Applied Genetics 76:*33-38.

Brown, J., Caligari, P.D.S., and Mackay, G.R. (1987). The repeatability of progeny means in the early generations of a potato breeding programme. *Annals of Applied Biology 110:*357-363.

Brown, J., Caligari, P.D.S., Mackay, G.R., and Swan, G.E.L. (1984). The efficiency of seedling selection by visual preference in a potato breeding programme. *Journal of Agriculture Science, UK 103:*339-346.

Caligari, P.D.S. (1992). Breeding new varieties. In Harris, P.M. (Ed.), *The Potato Crop,* Second edition (pp. 334-372). London, U.K.: Chapman and Hall.

Caligari, P.D.S. and Brown, J. (1986). The use of univariate cross prediction methods in the breeding of a clonally reproduced crop *(Solanum tuberosum). Heredity 57:*395-401.

Caligari, P.D.S., Brown, J., and Abbott, R.J. (1986). Selection for yield and yield components in the early generations of a potato breeding programme. *Theoretical and Applied Genetics 73:*218-222.

Caligari, P.D.S., Mackay, G.R., Stewart, H.E., and Wastie, R.L. (1984). A seedling progeny test for resistance to potato foliage late blight *(Phytophthora infestans* (Mont.) de Bary). *Potato Research 27:*43-50.

Caroll, C.P. (1982). A mass-selection method for the acclimatization and improvement of edible diploid potatoes in the United Kingdom. *Journal of Agricultral Science, UK 99:*631-640.

Caroll, C.P. and De Maine, M.J. (1989). The agronomic value of tetraploid F1 hybrids between potatoes of group Tuberosum and group Phureja/Stenotomum. *Potato Research 32:*447-456.

Chase, S.C. (1963). Analytical breeding of *Solanum tuberosum. Canadian Journal of Genetics and Cytology 5:*359-363.

Christopher, S.J. (1980). Utilizing *Solanum* germplasm to increase yields of the potato. I. Population improvement. II. Reciprocal differences. III. Heterosis. *Dissertation Abstracts. International (B) 41:*1577.

CPRI (2003), *Annual Report,* Shimla, India: Central Potato Research Institute.

Cubillos, A.G. and Plaisted, R.L. (1976). Heterosis for yield in hybrids between *S. tuberosum* ssp. *tuberosum* and *tuberosum* ssp. *andigena. American Potato Journal 53:*143-150.

Cutter, E.G. (1978). Structure and development of potato plant. In Harris, P.M. (Ed.), *The Potato Crop* (pp. 70-147). London, U.K.: Chapman and Hall.

Dale, M.F.B. and Mackay, G.R. (1994). Inheritance of table and processing quality. In Bradshaw, J.E. and Mackay, G.R. (Eds.), *Potato Genetics.* Wallingford, U.K.: CAB International.

De Jong, H. and Rowe, P.R. (1971). Inbreeding in cultivated diploid potatoes. *Potato Research 14:*74-83.

Douches, D.S. and Jastrzebski, K. (1993). Potato. In Kalloo, G. and Bergh, B.O. (Eds.), *Genetic Improvement of Vegetable Crops* (pp. 605-644). Oxford, U.K.: Pergamon Press.

Engel, F.A. (1970). Exploration of the Chilea-Canyon, Peru. *Current Anthropology 11:*55-58.

Everhardt, E.R. and Rowe, P.R. (1974). Disomic inheritance of anthocyanins and flavonol glycosides in the tetraploid tuber-bearing species *Solanum stoloniferum. American Potato Journal 51:*287-294.

Glendinning, D.R. (1969) The performance of progenies obtained by crossing Groups Andigena and Tuberosum of *Solanum tuberosum. European Potato Journal 12:*13-19.

Gopal, J. (1997). Progeny selection for agronomic characters in early generations of a potato breeding programme. *Theoretical and Applied Genetics 95:*307-311.

———. (1998). Identification of superior parents and crosses in potato breeding programmes. *Theoretical and Applied Genetics 96:*287-293.

———. (1999a). In vitro versus in vivo genetic divergence in potato. *Theoretical and Applied Genetics 98:*299-304.

———. (1999b). Genetic parameters and character association for clonal selection in potato breeding programmes. *Agronomie 19:*531-539.

———. (2001a). In vitro and in vivo genetic parameters and character associations in potato. *Euphytica 118:*145-151.

———. (2001b). Between and within families variation and family selection in potato breeding programmes. *Journal Genetics & Breeding 55:*313-318.

Gopal, J. and Gaur, P.C. (1997). Potato genetic resources in India. *Indian Journal of Plant Genetic Resource 10:*205-215.

Gopal, J., Gaur, P.C., and Rana, M.S. (1992). Early generation selection for agronomic characters in a potato breeding programme. *Theoretical and Applied Genetics 84:*709-713.

———. (1994). Heritability, and intra- and inter-generation associations between tuber yield and its components in potato (*Solanum tuberosum* L.). *Plant Breeding 112:*80-83.

Gopal, J. and Minocha, J.L. (1997a). Genetic divergence for cross prediction in potato. *Euphytica 97:*269-275.

———. (1997b). Effectiveness of in vitro selection for agronomic characters in potato. *Euphytica 103:*67-74.

———. (1997c). Effectiveness of selection at microtuber crop level in potato. *Plant Breeding 116:*293-295.

Gopal, J., Minocha, J.L., and Dhaliwal, H.S. (1998). Microtuberization in potato (*Solanum tuberosum* L.). *Plant Cell Reports 17:*794-798.

Gopal, J. and Oyama, K. (2005). Genetic base of Indian potato selections as revealed by pedigree analysis. *Euphytica 142:*23-31.

Gopal, J., Raj Kumar, and Kang, G.S. (2002). Minituber crop and its effectiveness in selection for agronomic characters in potato breeding programmes. *Potato Research 45:*145-151.

Grun, P. (1970). Cytoplasmic sterilities that separate the cultivated potato from its putative diploid anscestors. *Evolution 24:*750-758.

———. (1976). *Cytoplasmic Genetics and Evolution.* NY: Columbia University Press.

———. (1990). The evolution of cultivated potatoes. In Bretting, P.K. (Ed.), *New Perspectives on the Origin and Evolution of New World Domesticated Plants. Economic Botany 44* (3 Suppl.):39-55.

Hanneman, R.E. (1984). Assignment of endosperm balance number (EBN) to tuber-bearing *Solanum* species. *Botanical Review 41:*362-390.

———. (1999). The reproductive biology of the potato and its implication for breeding. *Potato Research 42:*283-312.

Hawkes, J.G. (1967). The history of the potato. *Journal Royal Horticultral Society 92:*288-302.

———. (1979). Genetic poverty of the potato in Europe. *Proceedings Conference; Broadening Genetic base of Crops* (pp. 19-27). Wageningen, The Netherlands.

———. (1990). *The Potato: Evolution, Biodiversity and Genetic Resources.* Oxford, U.K.: Belhaven Press.

Hermsen, J.G.T. (1994). Introgression of genes from wild species, including molecular and cellular approaches. In Bradshaw, J.E. and Mackay, G.R. (Eds.), *Potato Genetics* (pp. 515-538).Wallingford, U.K.: CAB International.

Hils, U. and Pieterse, L. (2005). *World Catalogue of Potato Varieties.* Bergen/ Dumme, Germany: Buchedition Agrimedia Gmb H.

Hosaka, K. and Hanneman, R.E., Jr. (1991). Seed protein variation within accessions of wild and cultivated potato species and inbred *Solanum chacoense. Potato Research 34:*419-428.

———. (1998). Genetics of self-compatability in a self-incompatible wild diploid potato species *Solanum chacoense.* I. Detection of an S locus inhibitor (Sli) gene. *Euphytica 99:*191-197.

Hougas, R.W. and Peloquin, S.J. (1957). A haploid plant of potato variety Katahdin. *Nature 180:*1209-1210.

———. (1958). The potential of potato haploids in breeding and genetic research. *American Potato Journal 35:*701-707.

Howard, H.W. (1963). Some potato breeding problems. *Report of the Plant Breeding Institute* 1961-1962 (pp. 5-21) Cambridge, U.K.

———. (1978). The production of new varieties. In Harris P.M. (Ed.), *The Potato Crop* (pp. 607-644). London, U.K.: Chapman and Hall.

Huisman, M.J., Cornelissen, B.J.C., and Jongedijk, E. (1992). Transgenic potato plants resistant to viruses. *Euphytica 63:*187-197.

Jansky, S.H. (2000). Breeding for disease resistance in potato. *Plant Breeding Review 19:*69-155.

Jinks, J.L. and Pooni, H.S. (1976). Predicting the properties of recombinant lines derived by single seed descent. *Heredity 45:*305-312.

Johnston, S.A., de Nijs, A.P.M., Peloquin, S.J., and Hanneman, R.E. (1980). The significance of genetic balance to endosperm development in interspecific crosses. *Theoretical and Applied Genetics 57:*5-9.

Knight, T.A. (1807). On raising of new and early varieties of the potato *(Solanum tuberosum). Transactions Horticultural Society, London 1:*57-59.

Larkin, P.J. and Scowcroft, W.R. (1981). Somaclonal variation- a novel source of variability from cell cultures for plant improvement. *Theoretical and Applied Genetics 60:*197-214.

Loiselle, F., Tai, G.C.C., and Christie, B.R. (1991). Pedigree, agronomic and molecular divergence of parents in relation to progeny performance in potato. *Potato Research 34:*305-316.

Love, S.L, Werner, B.K., and Pavek, J.J. (1997). Selection for individual traits in the early generations of a potato breeding program dedicated to producing cultivars with tubers having long shape and russet skin. *American Potato Journal 74:*199-213.

Maga, J.A. (1994). Potato flavor. *Food Reviews International 10:*1-48.

Maris, B. (1966). The modifiability of characters important in potato breeding. *Euphytica 15:*18-31.

Mackay, G.R. (1987). Screening and breeding for better potato cultivars. In Abbott, A.J. and Atkin, R.K. (Eds.), *Improving Vegetatively Propagted Crops* (pp. 181-196). London, U.K.: Academic Press.

Melchinger, A.E. (1990). Use of molecular markers in breeding for oligogenic disease resistance. *Plant Breeding 104:*1-19.

Mendiburu, A.O., Peloquin, S.J., and Mok, D.W.S. (1974). Potato breeding with haploids and 2n gametes. In Kasha, K. (Ed.), *Haploids in Higher Plants* (pp. 249-258). Ontario, Canada: University of Guleph.

Möllers, C. and Wenzel, G. (1992). Somatic hybridization of dihaploid potato protoplasts as a tool for potato breeding. *Botanica Acta 105:*133-139.

Neele, A.E.F., Barten, J.H.M., and Louwes, K.M. (1988). Effects of plot size and selection intensity on efficiency of selection in the first clonal generation of potato. *Euphytica S:*27-35.

Neele, A.E.F., Nab, H.J., and Louwes, K.M. (1991). Identification of superior parents in a potato breeding programme. *Theoretical and Applied Genetics 82:*264-272.

Ochoa, C. (1999). *Las Papas de Sudamerica: Peru (Parte I).* Lima, Peru: International Potato Center.

Pavek, J.J. and Corsini, D.L. (2001). Utilization of potato genetic resources in variety development. *American Journal of Potato Research 78:*433-441.

Peloquin, S.J., Werner, J.E., and Yerk, G.L. (1990). The use of potato haploids in genetics and breeding. In Gupta, P.K., and Tsuchiya, T. (Eds.), *Chromosome Engineering in Plants, Part B* (pp. 79-92). Barking, Essex, England: Elsevier.

Pfeffer, C. (1963). Vergleichende Untersuchungen über Auslesemöglichkeiten von im Freiland und in Töpfen kultivierten Kartoffelsämlingen. *Zuchter 33:*6-11.

Phillips, M.S. (1981). A method of assessing potato seedling progenies for resistance to the white potato cyst nematodes. *Potato Research 24:*101-103.

Plaisted, R.L. and Hoopes, R.W. (1989). The past record and future prospects for the use of exotic potato germplasm. *American Potato Journal 66:*603-627.

Pushkarnath and Dwivedi, R.S. (1961). Potato floral biology series. II. Pollen and pollination. *Indian Potato Journal 3:*76-85.

Ross, H. (1986). *Potato Breeding-Problems and Perspectives.* Berlin, Germany: Springer-Verlag.

Salaman, R.N. and Lesley, J.W. (1922). Genetic studies in potatoes: Sterility. *Journal Agricultural Sciences, Cambridge 12:*31-39.

Sanford, J.C. and Hanneman, R.E. Jr. (1982). A possible heterotic threshold in the potato and its implications for breeding. *Theoretical and Applied Genetics 61:*151-159.

SASA. (1998). *Guide to Potato Plots 1998.* Edinburgh, U.K.: Scottish Agricultural Science Agency.

Simmonds, N.W. (1976). Neotuberosum and the genetic base in potato breeding. *Agricultural Research Council Research Review 2:*9-11.

Spooner, D.M., Bamberg, J.B., Hjerting, J.P., and Gomez, J. (1991). Mexico, 1998 potato germplasm collecting expedition and utility of the Mexican potato species. *American Potato Journal 68:*29-43.

Spooner, D.M. and Castillo, R.T. (1997). Reexamination of series relationships of South American wild potatoes (Solanaceae: *Solanum* sect. *Petota*): Evidence from chloroplast DNA restriction site variation. *American Journal of Botany 84:*671-685.

Spooner, D.M. and Sytsma, K.J. (1992). Reexamination of series relationships of Mexican and Central American wild potatoes *(Solanum* sect. *Petota)*: Evidence from chloroplast DNA restriction site variation. *Systematic Botany 17:*432-448.

Spooner, D.M., van den Berg, R.G., Rodriguez, A., Bamberg, J., Hijmans, R.J., and Cabrera, S.I.L. (2004). *Wild Potatoes* (Solanaceae, *S* section *Petota) of North and Central America, Systematic Botany Monographs* (Vol. 68), The American Society of Plant Taxonomists.

Stewart, H.E., Taylor, K., and Wastie, R.L. (1983). Resistance to late blight in foliage *(Phytophthora infestans)* of potatoes assessed as true seedlings and as adult plants in the glasshouse. *Potato Research 26:*363-366.

Świeżyński, K.M. (1978). Selection of individual tubers in potato breeding. *Theoretical and Applied Genetics 53:*71-80.

Tai, G.C.C. (1975). Effectiveness of visual selection for early clonal generation seedlings of potato. *Crop Science 15:*15-18.

———. (1994). Use of 2n gametes. In Bradshaw, J.E. and Mackay, G.R. (Eds.), *Potato genetics* (pp. 109-132). Wallingford, U.K.: CAB International.

Tarn, T.R., Tai, G.C.C., De Jong, H., Murphy, A.M., and Seabrook, J.E.A. (1992). Breeding potatoes for long-day, temperate climates. In Janick, J. (Ed.), *Plant Breeding Reviews* 9 (pp. 217-332). NY: John Wiley & Sons.

Watanabe, K. and Peloquin, S.J. (1991). The occurrence and frequency of 2n pollen in 2x, 4x, and 6x wild, tuber bearing, *Solanum* species from Mexico, and Central and South America. *Theoretical and Applied Genetics 82:*621-626.

Wastie, R.L. (1991). Resistance to powdery scab of seedling progenies *of Solanum tuberosum. Potato Research 34:*249-252.

Wastie, R.L., Caligari, P.D.S., Stewart, H.E., and Mackay, G.R. (1987). A glasshouse progeny test for resistance to tuber blight. *Potato Research 30:*533-538.

Wastie, R.L., Mackay, G.R., Caligari, P.D.S., and Stewart, H.E. (1990). A glasshouse progeny test for resistance to gangrene *(Phoma foveata). Potato Research 33:*131-133.

Wenzel, G., Bapat, V.A., and Uhrig, H. (1983). New strategy to tackle breeding problems of potato. In Sen, S.K. and Giles, K.L. (Eds.), *Plant Cell Culture in Crop Improvement* (pp. 337-349). NY: Plenum Press.

# Chapter 4

# Biotechnology in Potato Improvement

Gerhard Wenzel

With the onset of biotechnology-based plant breeding, potato became an outstanding example for rapid in vitro propagation, use of haploids, protoplast regeneration and fusion, genomics, gene cloning, gene transfer, and now proteomics. Due to its tetraploid nature and high levels of self-incompatibility, potato improvement is not an easy task with classical genetic tools. Clonal nature and good in vitro growth of potato is advantageous for successful application of biotechnology. Potato is a pioneer under the aspect of genomics, marker development, and marker-assisted selection. It is also a forerunner in genetic engineering; its transformation is not too difficult and the number of cloned and transferred genes is increasing.

Plant breeding relies on two areas: (1) maintenance of genetic resources and (2) breeding of new cultivars. The variability created by the evolution and enlarged by directed recombination has to be preserved to address new breeding challenges, some of which might not even be formulated today. Consequently, keeping a broad gene pool is as important for breeding success as powerful selection systems during the development of superior cultivars. This chapter reviews the role of biotechnology both in management of genetic resources and breeding improved potato cultivars.

## BIOTECHNOLOGY FOR CONSERVATION AND CHARACTERIZATION OF GERMPLASM

Potato with a large primary as well as secondary and tertiary gene pool is in a fortunate situation as far as availability of genetic variability is concerned. However, in the vegetatively propagated outbreeder potato, maintenance of genotypes is not easy, since clones are often lost due to biotic and abiotic stresses. Consequently, it would be a great gain (a) to keep valuable material under conditions where infections by pathogens are minimized, (b) to make sure that a broad phylogenetic spectrum is represented and (c) to recognize duplicates. Biotechnology offers following advances to reach these objectives:

- tissue culture, allowing in vitro conservation of collections protected from negative influences of environments, diseases, and pests;
- the use of fingerprint techniques, helping to detect duplicates and to reveal descents based on polymorphism at the DNA level.

### *In Vitro Culture for Genotype Maintenance*

#### *Rapid Propagation*

In vitro living collections are established from a pathogen-free mother clone. Meristem culture helps in virus-free regenerants (also see Chapter 11 "Viruses and Viroids"). By culturing the starting material at 30-35°C, the percentage of healthy plants can be further increased. Mother cultures so established are used for subsequent rapid propagation. Rapid propagation is done on MS (Murashige and Skoog, 1962) or slightly modified MS medium. During rapid propagation, there is interplay between the genotype, the tissue source, the culture medium, and the culture conditions. But neither the culture medium nor the culture conditions (about 22-26°C, 3,000-4,000 Lux for 16 h) are very critical. Potato breeders use tens of thousands of rapidly propagated plants per year. Presently 38.4 Mio plants are rapidly propagated in Germany (more than 200 Mio plants in the world) from which 1.5 percent are potatoes (Preil, 2003).

Potato plantlets can be stored at temperatures between 4°C and 10°C. Depending on the genotype, one transfer between six months and two years is sufficient for maintenance (Keller, 2003). Callus formation, however, must be prevented in order to keep the genotype stable. Conservation at normal propagation temperatures (25°C) on slow-growth medium offers an alternative strategy for normal growth of in vitro plantlets (Gopal et al., 2002).

Protoplasts offer in theory a much faster propagation of a given genotype since a single potato leaf may contain several hundred thousand cells, each of which may grow into a functional plant. Although numerous protocols exist today for an efficient protoplast regeneration of potato through a callus phase, this technique is not used routinely for rapid propagation. One reason is its rather laborious and sophisticated nature; the other is the appearance of protoclonal variation. Thus, a crucial prerequisite of rapid propagation—genome stability—is not fulfilled. By shortening the callus phase this variability could be reduced but not prevented.

### *Mini- and Microtubers*

Since storage and transport of in vitro shoots is not always easy, mini- and microtubers are used. Minitubers are produced from the in vitro propagated plantlets after transfer to soil. Microtubers develop in vitro on media with 8 percent sucrose under short day (6 - 8 h), or complete darkness (Gopal et al., 1998; Alix et al., 2001). A problem for using such microtubers is still the insufficient dormancy, small size, and the relatively low number of microtubers per single-node microcutting. Efforts to increase the dormancy period of microtubers using abscisic acid in the microtuber induction medium did not yield expected results (Gopal et al., 2004). For the sake of curiosity Kordyum et al. (1997) showed that minitubers even developed in microgravity from shoots grown for eight days on board of the MIR orbital space station exhibited apparently normal tuber characteristics.

### *Cryoconservation*

Potatoes can be stored in liquid nitrogen, to avoid the laborious transfer process in micropropagated collections, and to make sure that no undesired somaclonal variation influences a valuable genotype. In the German potato collection, for example, 970 potato clones (status September 2003) are cryoconserved (Keller, 2003). The central problem is the formation of callus during regeneration vulnerable to somaclonal variation besides a limited survival rate. To avoid any callus growth, scientists attempted to cryoconserve embryos or embryoids and to regenerate plantlets directly via embryoids. Along this line, Hirai and Sakai (1999) succeeded in encapsulation of in vitro grown meristems. They found that encapsulated meristems when osmoprotected and dehydrated at 0°C prior to plunging into liquid nitrogen produced more shoots without intermediate callus formation. Somatic embryogenesis was achieved from excised 3-4 week old in vitro shoot tip meristems of the German cultivar Tomensa on an embryo formation medium supplemented with 0.1 mg·$l^{-1}$ GA3 and 0.05 mg·$l^{-1}$ zeatin (Fiegert et al., 2000). Embryoids appeared about 30 days after transfer. The costs of cryoconserved plantlet are rather high: 1-2 US$/plant (Keller, 2003).

### *Artificial Seeds*

These are encapsulated embryoids or other regenerable tissues that can be sown and grown into a plantlet. Embryoids might be suitable for the production of artificial seeds with or without a cryoconservation step. A prerequisite to use such a system under applied conditions is mass induction of embryoids and a simple technique for encapsulation. Seabrook and Douglass (2001) achieved an efficient method to produce somatic embryos on explants of internodes. However, the germination rate and the storage survival are still critical. Embryoids from internodes could be encapsulated and they grew sufficiently (Fiegert et al., 2000). Trials with encapsulated potato shoot tips showed 93 percent germination on soil in the greenhouse and up to 25 percent germination in the field. Nondesiccated artificial seeds showed about 60 percent viability after 12 months storage at 4°C, desiccated seeds were not viable (Nyende et al., 2002). Presently, the process is not reproducible and cheap enough to allow direct application in a commercial propagation system.

### *Biotechnological Tools for Characterization of Genetic Resources*

Hitherto the most successful markers for characterization of germplasm were morphological. But since these are few and affected by environment, discovery of biochemical and molecular markers proved advantageous in supplementing the morphological markers. However, biochemical markers normally isozymes are also few and could be used only for specific tasks, for example differentiation of homo- and heterocaryotic somatic hybrids (Möllers and Wenzel, 1992). In the tetraploid potato four alleles exist per locus, and most markers show only the presence or absence but not the quantity of an allele.

#### *Molecular Markers*

The most powerful markers are molecular ones since they are unlimited in number, easy to generate, independent from developmental stages, and without epigenetic effects. The first system developed was the restriction fragment length polymorphism (RFLP, Botstein et al., 1980). RFLPs are detected upon hybridization of single-copy DNA probes to restriction enzyme-hydrolyzed, agarose gel electrophoresis-separated, and nylon membrane-bound genomic DNA. Marker alleles are identified by size differences of the restriction fragments to which the probes hybridize and are correlated to similarly segregating phenotypic traits of the mapping population (Gupta et al., 2002). However, RFLP analysis is a time-consuming and labor-intensive task and, therefore, not suitable for the rapid genotyping of large segregating populations. This is where random amplified polymorphic DNA (RAPD, Williams et al., 1990) finds its use. Addition of a 10-base oligomer to restricted DNA results in RAPDs that can be directly visualized after PCR amplification in conventional agarose gel electrophoresis. This has striking advantages such as low costs, the small amount of DNA needed (10-100 ng), and the use of universal primers. However, the DNA profiles are often nonreproducible. Today, amplified fragment length polymorphism (AFLP, Vos et al., 1995) is the most efficient technique used for detecting polymorphisms. It is also based on PCR amplification of a set of restriction fragments under high stringency conditions from a pool of DNA fragments. This is generated through digestion with a pair of restriction endonucleases, one of them being a four-base cutter, and the other being a six-base cutter. To allow for PCR tagging of restriction fragments, oligonucleotide adapters are ligated to the ends of the DNA fragments. The number of DNA fragments to be amplified is decreased by adding 2-3 selective nucleotides to the core sequence of the primers. This method generates many bands, but not too many to create difficulties in scoring. PCR-based methods offer the striking advantage of being rapid and requiring only tiny bits of plant material for DNA extraction. A technique used quite often for potatoes is simple sequence repeats (SSRs) or microsatellites. These are tandem repeats of short nucleotide sequence motifs made visible through hybridization (Tautz and Renz, 1984). SSR loci are extremely variable in the number of repeat units and can be easily typed. Since they are evenly spaced throughout the genome as simple PCR markers, they represent an ideal resource for a marker class. For potato a number of SSR markers are publicly available (Milbourne et al., 1998).

The primary focus for the development of the next generation of genetic markers is on single nucleotide polymorphism (SNP) (Reiter, 2001). An SNP is a single nucleotide position in a defined DNA stretch at which there is variation between different clones. SNPs are of particular interest for potato improvement, since they (1) represent the most frequent variations in the genome, thus, offering the potential to find informative markers for a distinct genomic region and (2) can be simply treated as di-allelic markers and, therefore, make them amenable to automated high-throughput genotyping. The most advanced SNP procedures are no longer gel based and therefore readily automated, allowing cost-effective screening of large populations.

The most successful application of molecular markers is thus the comparison of complex DNA polymorphisms produced by the RAPD, RFLP, or by AFLP techniques. The marker bands are comparable to fingerprints. Fingerprint techniques leading to information on genetic distances of wild potato species were used to trace back the origin of the cultivated tetraploid potato on the basis of the DNA similarity of the nuclear genomes (Debener et al., 1990). The results fitted perfectly to all morphological descriptors.

Cytoplasmic genomics have also been used for characterization (Hosaka, 1995; Bryan et al., 1999). This backed the information developed on the basis of the nuclear DNA. Frei et al. (1998) characterized different mitochondrial types in dihaploid and tetraploid clones. Five distinct classes of mitochondrial DNA were separated. Plastid (pt)DNA is less variable and has predominantly two pt-types: *tuberosum* (T)-type and wild (W)-type chloroplasts in a 1:1 ratio. Provan et al. (1999) emphasized the need to increase levels of non-T-type cytoplasms in the cultivated potato gene pool, since most promising new traits are coming from wild species and primitive cultivars carrying W-type plastids.

### *Genetic Distances*

Kim et al. (1998) estimated by AFLPs the genetic variation and relationship in American potato cultivars. Isenegger et al. (2001) screened cultivars of Australia, McGregor, Greyling et al. (2000) those of South Africa, and Bäumler et al. (2003) the German ones. Figure 4.1 shows the genetic distances of 69 German cultivars. Several clusters, particularly the groups 1, 2+3, 4+5, and 6 could be separated. The different groups often reflect the descents of one dominant clone or cultivar. Clusters 3 and 4, for example, can be traced back to introgression of two clones from the Max Planck Institute in the late 1960s carrying extreme resistance to PVX and PVY (Braun, 2002). A selection of 75 tetraploid cultivars from different regions of the world was probed with 6 AFLP primer combinations. By principal coordinate analysis of the AFLP patterns, different cultivars could be distributed to clearly separated pools (Figure 4.2). This will help in parent selection to look for highest amounts of genetic distances and thus for heterozygosity and heterotic yield parameters.

### *Comparison of Marker Systems*

The molecular marker types are numerous and being followed with such a speed that new and more efficient systems become available before the latest marker system could be optimized and

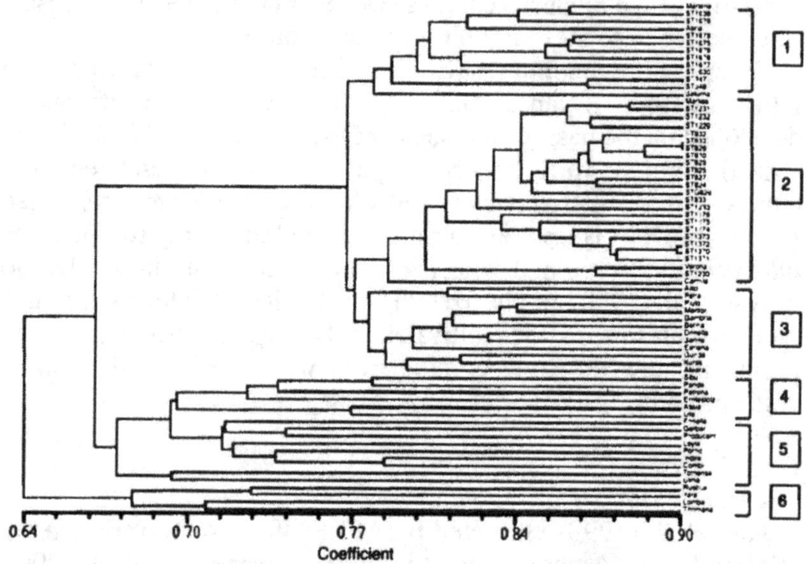


FIGURE 4.1. Dendrogram of 69 German tetraploid potato cultivars or breeding lines probed with 6 AFLP primer combinations (Braun and Wenzel, 2005).

transferred to application. A drawback of RFLP is, for example, the high costs for a single test resulting in limited genotype numbers that could be screened at a reasonable price. This underlines the need for a decision on which marker system shall be used for which aim. Comparisons and evaluations on their applicability, reproducibility and costs are well documented (Virk et al., 2000). McGregor, Lambert et al. (2000) developed the following genotype indices based on the relative number of bands generated by primer or primer pair to indicate the usefulness of the different systems in potato: AFLP = 1; multilocus SSR = 0.8; RAPD = 0.5; single locus SSR = 0.4. The number of bands is, however, not equivalent to the number of polymorphic or informative bands. SNP markers will gain importance when whole genome sequences of potato become known.

All marker techniques help to characterize potato gene pools; they are thus an important tool for parent selection. Whether they should be also used in seed boards for legal variety characterization is under

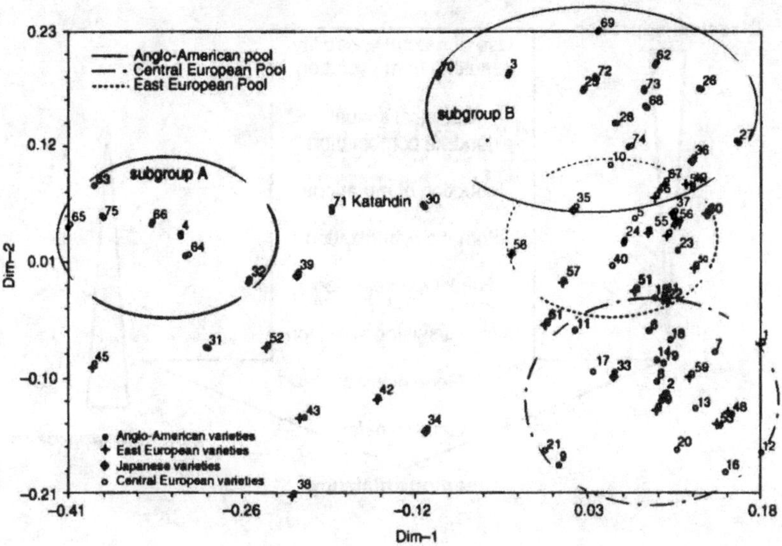


FIGURE 4.2. Principal coordinate analysis of 75 cultivars from different locations of the world using AFLPs produced by 6 primer combinations. See Plate 4.2 in the color gallery.

discussion. As long as most markers are only correlated to traits, often located in introns rather than in exons or functional genes, marker profile variation is no proof for a deviation in a registered character. Thus, varieties may easily express 5 percent variation in their AFLP profile during successive years without any morphological change. Despite this variability, Bonnel (2003) proposes to use 50-300 AFLP markers for variety identification and quality control.

## *BREEDING NEW CULTIVARS USING BIOTECHNOLOGY*

Biotechnology offers new and more efficient breeding strategies than the classical approaches. Both processes are complementing each other as Figure 4.3 shows. The classical process uses less sophisticated rather cheap methods whereas the molecular counterpart

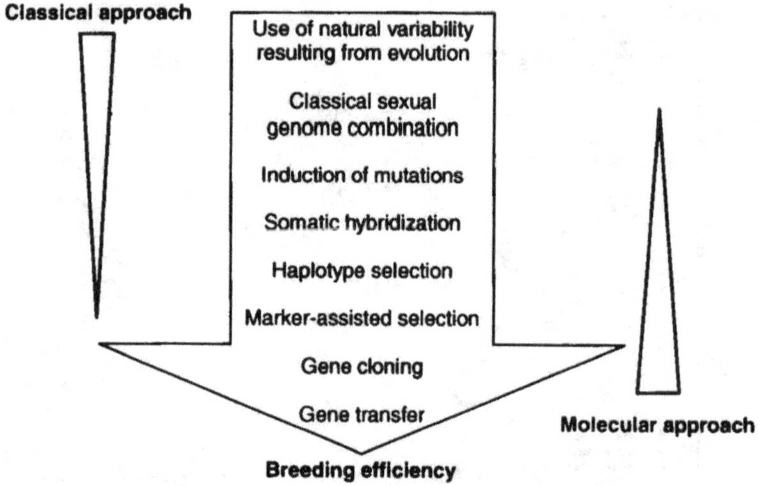


FIGURE 4.3. Considerations for using biotechnology in potato breeding.

has the advantage of a better specificity or allows working with larger populations efficiently. Mutagenesis and cell culture approaches bridge both techniques. The following opportunities exist to speed up and increase efficiency of potato breeding:

- Mono- and/or dihaploids: this makes segregation easier.
- Genes transfer: performed either via symmetric and asymmetric fusion, or by DNA transfer.
- Genomics: allowing a better description of the genome offering the chance of marker-assisted selection and finally gene cloning.
- High-throughput techniques: allowing screening of large populations for desired characters.

These techniques can also be combined, for example, genome analysis on dihaploids or identification of somatic hybrids using molecular markers.

### *Production and Use of Haploids*

Tetrasomic inheritance in cultivated potato *(Solanum tuberosum)* makes its breeding a difficult task. It is thus desirable to reduce the ploidy level as with these reductions the probability for finding homozygous traits is increased. If an important character is controlled by one recessive allele the homozygous situation in a dihaploid appears in a ratio 1:4 whereas in tetraploids full homozygosity is achieved in a ratio 1:36. For five alleles the ratios are 1:1,024 in dihaploids compared to 1:60,466,176 in tetraploids (Kuckuck et al., 1991).

Adequate percentages of dihaploid potatoes ($2n = 2x = 24$) can be produced parthenogenetically by crossing $4x$ with specific $2x$ *S. phureja* clones (Hougas and Peloquin, 1957). This technique worked well also for the reduction of the ploidy level from $2x$ to monohaploids ($2n = 1x = 12$; van Breukelen et al., 1975). The parthenogenetic reduction from $4x$ to $2x$ is very efficient with up to 46 percent dihaploids calculated on the bases of fertilized flowers (Wenzel and Foroughi-Wehr, 1984). For the reduction from $2x$ to $1x$, the parthenogenetic process of haploid production is rather laborious because the triploid block that causes lethality of the normal hybrids does not work and the few monohaploids have to be selected among many diploid seeds.

Anther culture is another technique useful for ploidy reduction, particularly for production of monohaploids. The production of androgenetic monohaploids in potatoes can be summarized as follows: Anthers from dihaploid clones grafted on tomato root stocks are kept for 3 days at 6°C before excision. Surface sterilized buds are placed in 25 ml Erlenmeyer flasks and shaken at 26°C with 120 strokes/min in half strength, slightly modified Murashige and Skoog (1962) medium, supplemented with 0.1 $mg \cdot l^{-1}$ indolylacetic acid, 0.05 percent activated charcoal, and 6 percent sucrose (Sopory et al., 1978). The majority of the regenerated plants proved to be dihaploid, with rare monohaploid exceptions. Such monohaploids were used for somatic hybridization resulting in vigorous somatic hybrids. Johnson et al. (2001) used more than 100 androgenetic monohaploids for somatic hybridization and found 31 monohaploid genotypes, most of them suitable for hybrid production.

A central parameter in increasing regeneration success in microspore culture is the selection of clones with a high response to anther culture and to transfer this ability to progenies by sexual crosses. Wenzel and Uhrig (1981) demonstrated that the character regeneration capacity is heritable and can be transferred, by combination breeding, to clones with low response. The Mendelian segregation of this trait implicated nuclear origin, but an additional influence of the plasmone during regeneration processes was apparent. These results were confirmed by Sonnino et al. (1989) who combined a dihaploid clone with regeneration capacity with a root-knot nematode resistant clone. Among $F_1$ progenies tested, a wide range of androgenetic capability was found. The authors conclude that the character regeneration capacity is controlled by more than one major gene, that influences the induction of embryoid formation or its frequency. Additional minor genes are involved in this process.

There is no difference whether the ploidy level reduction is performed by partheno- or androgenesis. Due to the higher number of microspores, the chance to find a desired genotype is greater through androgenesis. De Vicente and Tanksley (1991) found in tomato significantly less recombination for male gametes. Wenzel and Uhrig (1981) produced androgenetic doubled monohaploid clones possessing PVX, PVY, PLRV, and *Globodera pallida* resistance in the homozygous condition. These experiments include monogenically inherited traits (PVX/Y resistance) and polygenic ones (PLRV, *G. pallida* resistance), demonstrating that recombination works sufficiently.

Wenzel et al. (1979) proposed the combination of haploidy, protoplast fusion, and the classical breeding steps for combining several traits of potato in a stepwise fashion. First the monohaploid ploidy level is reached via two successive haploidization steps. Homozygous dihaploids are subsequently obtained by doubling of 1*x* clones. In these clones additive traits are homozygous and will be inherited during a sexual combination to heterozygous interdihaploids. These interdihaploids are subjected to intensive sexual combination and selection to give superior heterozygous interdihaploids. The final step is the somatic fusion of these hybrids to combine several characters and to reach, at the same time, the tetraploid level expressing the highest amount of heterozygosity. About 15 years after this scheme had been proposed, first analytical synthetic clones combining different resistances were tested in the field (Thach et al., 1993). Today, the use of dihaploids and somatic fusion of these became routine in private companies; monohaploids are, however, of basic interest only.

### *Somatic Cell Fusion*

Sexual combination is always accompanied by meiosis and recombination, and thus its specific success is unpredictable. Fusion of protoplasts allows to combine asexually—without meiotic segregation—two complex nuclear genomes and to combine plasmones as well. Fusion can be induced chemically by the polycation polyethyleneglycol or via electrofusion. Presently, the electrofusion technique is applied in most laboratories with a lamellar fusion chamber at a voltage of 100-200 $V \cdot cm^{-1}$ and a frequency of 1 MHz. After 30-40 sec the majority of the protoplasts are aligned; the fusion is obtained by applying a pulse of 2,000 $V \cdot cm^{-1}$ for 20-40 µsec (Möllers and Wenzel, 1992). Hybrid identification is predominantly performed by molecular probes that allow the use of undifferentiated calli so that waiting for morphogenesis is not necessary. For DNA probing repetitive as well as single copy probes can be used (Frei et al., 1998).

### *Symmetric Fusion*

Today, symmetric protoplast fusion of *S. tuberosum* clones is used in applied breeding programs (Hofferbert and Wenzel, 1997; Schwarzfischer et al., 2002). In such experiments, Möllers et al. (1994) could demonstrate high flower production, a pollen fertility of 60-90 percent, and a substantial tuber yield increase in somatic hybrids compared to mean tuber yield of the parents. There were, however, large variations between different combinations. Different combining abilities were suggested as an explanation for the observed variation.

The fusion of primitive cultivars or wild species with *S. tuberosum* concentrates on the integration of characters not available in a heritable form in the primary gene pool of *S. tuberosum* (Table 4.1). Interspecific protoplast fusion also helps to introduce novel

TABLE 4.1. Selected characters combined via somatic fusion.

| Parents | Characters of interest | Reference |
|---|---|---|
| *Symmetric fusion* | | |
| *S. tuberosum* (+) *S. tuberosum* | PVX/PVY | Thach et al. (1993) |
| S. tuberosum (+) S. tuberosum | Nematode resistance | Möllers and Wenzel (1992) |
| *S. tuberosum* (+) *S. brevidens* | PLRV resistance | Austin et al. (1985) |
| *S. tuberosum* (+) *S. brevidens* | PVX resistance | Gibson et al. (1988) |
| *S. tuberosum* (+) *S. brevidens* | PVY resistance | Gibson et al. (1988) |
| *S. tuberosum* (+) *S. brevidens* | *Erwinia* resistance | McGrath et al. (2002) |
| *S. tuberosum* (+) *S. bulbocastanum* | *Phytophthora* resistance | Helgeson et al. (1998) |
| *S. tuberosum* (+) *S. bulbocastanum* | *Meloidogyne* resistance | Mojtahedi et al. (1995) |
| *S. tuberosum* (+) *S. commersonii* | *Erwinia* resistance | Carputo et al. (2000) |
| *S. tuberosum* (+) *S. commersonii* | Frost tolerance | Cardi et al. (1993) |
| *S. tuberosum* (+) *S. tarijense* | *Erwinia* resistance | Carputo et al. (2000) |
| *S. tuberosum* (+) *S. verrucosum* | PLRV resistance | Carrasco et al. (2000) |
| *Asymmetric fusion* | | |
| *S. tuberosum* (+) *S. commersonii* | *Verticillium* resistance | Bastia et al. (2000) |
| *S. tuberosum* (+) *S. circaeifolium* | Chilling tolerance | Oberwalder et al. (2000) |
| *S. tuberosum* (+) *Lycopersicon esculentum* | | Melchers et al. (1992) |

germplasm from sexually incompatible species or sterile clones in order to broaden the genetic base of parents. Austin et al. (1988) and McGrath et al. (2002) demonstrated the presence of *Erwinia* and *Phytophthora* resistance in hybrids between *S. tuberosum* and *S. brevidens.*

The genotypes influence the fusion process. Some fuse easily while others never result in hybrids. Potato multifusion can help to identify reasons for this different behavior (Frei, 1994). In all fusion events, in addition to the nuclear fusion, organelle segregation and recombination takes place and has a substantial influence on fusion success. Fusion provides the opportunity to optimize not only the genetic composition of the nuclear genome but also that of the cytoplasm and the interaction of both.

### *Asymmetric Fusion*

Asymmetric hybrids with unequal contributions of the cytoplasmic and nuclear genetic material of each parent are sometimes of greater interest than symmetric fusions. The asymmetry may go so far that one parent delivers the nucleus and the other one the cytoplasm only forming a cybrid. The problem of combining two nuclear genomes with the probability to incorporate undesired genes and the consequence of doubling the ploidy level could thus be overcome. In the simplest procedure, protoplasts of a donor species (normally a distantly related genotype) are irradiated, destroying the nuclear DNA before fusion with the recipient protoplast (normally a cultivar) where the metabolism of the cytoplasm is blocked by rhodamin-6-G or iodoacetamide (Sidorov et al., 2000). As a result, the nuclear genome of the cultivar is combined with the plastome of a distantly related species. Since under these conditions nuclear-cytoplasmic incompatibilities appear, de novo CMS systems can be created. In asymmetric hybrids, for example between *S. tuberosum* and *Lycopersicon esculentum* the regenerants are not stable. During propagation the phenotype changes, probably due to the loss of chromosomes of either parent. Melchers et al. (1992) achieved finally very asymmetric hybrids that were in phenotype tomatoes but expressed the potato genes for chilling resistance.

### *Fused Products*

Clones regenerated after protoplast fusion of two parents may have the same fused nuclear genome but varying cytoplasm compositions. Chloroplasts from two potato parents are normally not mixed in one cell after fusion. By using chloroplast-specific DNA-probes, Möllers and Wenzel (1992) demonstrated that the chloroplasts segregate in a 1:1 ratio soon after fusion. The genes of the mitochondrial genome might, however, be recombined. Molecular strategies for selection are useful when the agronomically important characters are not coded just by one but by several genes. McGrath et al. (1994) could demonstrate in the second backcross generation from a somatic hybrid of *S. tuberosum* and *S. brevidens* that a series of new linkage groups were formed that could be detected by RFLP and RAPD markers. They detected, for example one clone with 48 chromosomes carrying a single *S. brevidens* marker, suggesting transfer of just this *S. brevidens* DNA piece into *S. tuberosum.* Such hybridization techniques come very close to gene transfer approaches, but final trait expression appears only in a nonpredictable way.

### Nuclear Cytoplasmic Interaction

Mitochondria and chloroplasts represent a source of genetic diversity as they contain DNA and the machinery to express this information as RNA and, after RNA editing and tRNA import, by their own translation system, as proteins. Somatic cell fusion circumvents the normal uniparental inheritance of the plasmone and permits complete combination of cytoplasms. Cytoplasms differ in potato, Lössl et al. (1999) revealed five distinct mitochondrial (mt-DNA) types ($\alpha$, $\beta$, $\gamma$, $\delta$, $\epsilon$) associated with three chloroplast (pt-DNA) types T, W, and S. The interaction of these compartments influences not only several yield traits, but also the success of somatic hybrid regeneration. To judge the importance of the interactions, Lössl et al. (2000) combined distinct parental cytoplasms in symmetric tetraploid somatic potato hybrids. This allowed, in the presence of nearly isogenic nuclear genomes, to estimate the contribution of mt- and pt-DNA to the genomes. Evaluations of the correlation between the cytoplasmic types showed that an interaction existed between starch content and mt-pt combinations. In general, high-field performance was associated with those cytoplasmic configurations that appeared at a high frequency after completion of the segregation process in vitro. Identification of clones with an optimized organelle composition is advantageous for early selection and such clones regenerate in vitro preferentially (Frei et al., 2003). The effect was, however, not large enough in large-scale field trials to produce significantly more starch and yield or superior morphological characteristics (Braun, 2002).

### Directed Gene Transfer

In addition to the techniques already discussed, the transfer of cloned genes allows improvement of a superior cultivar in just one or a few missing traits. Prerequisites are

- availability of the trait to be transferred as cloned DNA;
- availability of a powerful transfer system;
- availability of a reliable regeneration system.

For potato, the last bullet point is guaranteed by single-cell regeneration. Transfer of monogenic characters is not a problem and even the transfer of more than two genes is becoming routine. Presently an unsolved question is how to address complex traits. Since most traits of interest are not coded by a single gene but rather by oligo- or polygenes, different strategies have to be applied for complex characters. Here, marker-assisted selection is allowing faster success than the transfer of isolated or even monogenic genes.

*Directed Alterations by Nuclear Transformations*

Microbombardment and vector-assisted transfer via the *Agrobacterium* system are the two transfer systems of choice for directed gene transfer. The results by Romano et al. (2001) indicate that an efficient particle bombardment protocol is available for potato internodal segments. More general is the use of *Agrobacterium* where large fragments can be transferred (Shibata and Liu, 2000).

Several characters have been transferred today using microbombardment and *Agrobacterium* systems (Table 4.2). Of particular interest are genes responsible for the carbohydrate metabolism and for disease resistances. First isolations and functional analysis were performed in the model plant *Arabidopsis* followed by tobacco and tomato. It was striking that *R*-genes for a wide range of pathogens code for structurally similar proteins, a similarity making a high amount of conservation of the signal transduction chain visible (Wenzel, 1998). From tobacco and tomato it was a small step to potato, and now the functional analysis of resistance genes from potato *(R1; Gpa; Rx)* showed that they all belong to the same superfamily of plant genes for pathogen resistance sharing a nucleotide-binding-site and a leucine rich repeat domain: the NBS-LRR type (Ballvora et al., 2002). Although these monogenic resistance genes can easily be transferred via classical breeding, understanding the function of *R*-genes is of considerable importance and useful for the development of chemical protectants.

A number of genes transferred and expressed in potatoes are of foreign origin, often from *Arabidopsis.* Thiele et al. (1999) achieved an overexpression of phytochrome B by the responsible *Arabidopsis* gene in potato. It resulted in an increased yield under greenhouse

TABLE 4.2. Selection of traits expressed in potato after gene transfer.

| Trait | Reference |
|---|---|
| PVX resistance | Hoekema et al. (1989) |
| PVY resistance | Kaniewski et al.(1990) |
| PLRV resistance | Kawchuk et al. (1991) |
| Bt tolerance | Donegan et al. (1996) |
| Herbicide tolerance | De Block (1988) |
| Inulin production | Hellwege et al. (2000) |
| Increased level of phytochrome B | Thiele et al. (1999) |
| Palatinose production | Börnke et al. (2002) |
| Zeaxanthin production | Lübeck et al. (2002) |
| *R*, resistance gene against *Phytophthora* | Ballvora et al. (2002) |
| Pectate lyase | Wegener (2001) |
| *Erwinia* resistance | Düring et al. (1993) |
| Starch synthase | Visser et al. (1991) |
| Branching enzyme | Kossmann et al. (1991) |

conditions; however, these results could not be verified under field conditions (Braun et al., 2002; Schittenhelm et al., 2002). Very stable are the alterations of the carbohydrate metabolism, as potatoes producing inulin, or palatinose are being successfully grown for several years under field conditions (Table 4.2). The increased production of zeaxanthin in potatoes by blocking the pathway to violaxanthin with antisense or cosuppression (Lübeck et al., 2002) could be verified in the field (Schenkel and Freising, unpublished).

### *Directed Alterations by Plasmone Transformations*

Plasmone transformation allows an even more specific approach: adding one or a few genes to a plastome or chondriome resulting in transgene organelles. Due to their better genetic stability plastids are the major object in transferring organelle genes. The transformation of pt-genomes has a number of advantages compared to the nuclear genome. Plastid transformation usually results in a high level of transgene expression partly attributed to the presence of 50-60 chloroplasts per leaf cell and 60-100 chromosomes per plastid, compared to the usual two copies of each of many chromosomes in the nucleus. The chloroplasts of potato encode 120-150 genes, many of which are grouped in operons. In contrast, the nuclear chromosome contains thousands of genes arranged separately and transcribed individually. As the vast majority of interesting agronomic traits are multigenic, their introduction into the nuclear genome has to be done one gene at a time with unpredictable expression level. The introduction of even short biosynthetic chains is laborious and time-consuming. The principal method to introduce DNA into *Solanum* chloroplasts is biolistic bombardment (Sidorov et al., 2000) or the use of polyethylene glycol (Koop et al., 1996).

## *Molecular Markers in Selection*

Molecular marker-based approaches are opening up the field of genomics dealing with the genes/enzymes directly. Figure 4.4 summarizes the correlations of molecular approaches and its possible application in potato breeding. The increasing amount of information about DNA documented in molecular marker collections and dense gene maps, together with an excellent bioinformation system, allows the development of new strategies for creation of superior cultivars. Common principles are uncovered, for example in defense reactions

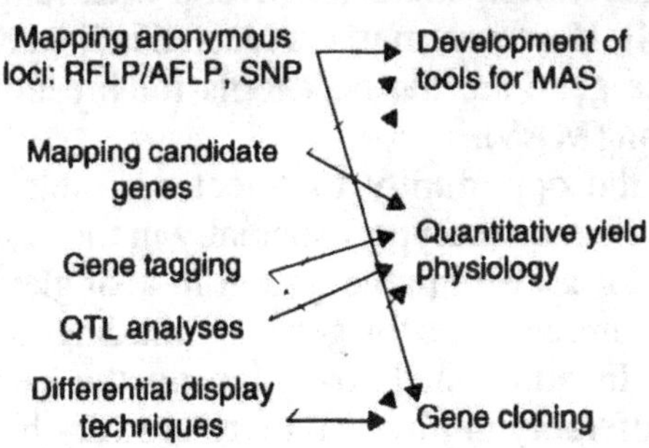


FIGURE 4.4. Correlations of different molecular processes.

to biotic and abiotic stresses, offering strategies like the candidate gene approach (Lübberstedt et al., 2002). Molecular markers are a powerful tool for (1) higher efficiency in parental selection allowing a choice of better combining and thus heterotic parents and (2) pyramiding of different traits to result in more complex characters. Additionally, they are presently the most efficient system to uncover a rare but desired genotype in a large segregating population.

### *Marker Correlation to Traits*

Most mapping and marker isolation has been done at the diploid level for monogenic traits, since in the tetraploid potato the dosage effect by doubling the ploidy level might have a striking influence (Luo et al., 2001). A huge number of agronomically important genes have been correlated to molecular markers, but only some of them are gene-sequence derived and only these do really cosegregate with the target gene. Markers that have been elaborated and validated and that are used in current potato breeding are listed in Table 4.3.

Of utmost interest are alleles responsible for resistance to diseases and pests. Understanding the genetic basis of pathogen resistance in potato has been broadened tremendously by DNA-marker technologies. Resistance gene analogs (RGAs) are important molecular markers for disease and pest resistance as these often have more general diagnostic value. RGAs are primarily located close to resistance genes and are often members of multigene families. In potato, hot spots with numerous resistance loci exist and such areas are often associated with the relevant characters over a wide range of genotypes (Gebhardt, 2004). However, markers located in the gene and not only correlated to the gene are the best tools for breeding and marker-assisted selection (MAS).

MAS offers the opportunity to select desirable lines based on genotype rather than phenotype, especially in the case of combining different genes for a given pathosystem in a single genotype (gene pyramiding). A prerequisite for gene pyramiding is that characters are not allelic. In addition, knowledge on the gene distances in molecular or preferably of physical maps are very helpful.

Most characters relevant for the tuber crop show continuous phenotypic variation because they are controlled quantitatively by

TABLE 4.3. Important mapped DNA markers.

| Trait | Gene | Chromosome | Reference |
|---|---|---|---|
| *Resistance to* | | | |
| Potato virus Y | $Ry_{adg}$ | XI | Sorri et al. (1999) |
| | $Ny_{tbr}$ | IV | Celebi et al. (2002) |
| | $Ry_{sto}$ | | Brigneti et al. (1997) |
| Potato virus X | *Rx1; Rx2* | XII; V | Bendahamane et al. (2000) |
| | *Na* | XI | Gebhardt and Valkonen (2001) |
| | *Nb* | V | Marano et al. (2002) |
| | *Nx* | IX | Gebhardt et al. (2001) |
| Potato virus S | *Ns* | VIII | Marczewski et al. (2002) |
| Potato leaf roll virus | *PLRV* QTL | XI | Marczewski et al. (2001) |
| *Globodera rostochiensis* | *Gro* | III; VII; X; XI | Barone et al. (1990) |
| | *H1* | V | Pineda et al. (1992) |
| *Globodera pallida* | *Gpa* QTL | IV; V; IX; XII | Bryan et al. (2002) |
| | *Gpa2* | XII | Van der Voort et al. (1997) |
| *Meliodogyne chitwoodi* | $R_{mc1}$ | XI | Brown et al. (1996) |
| *Phytophthora infestans* | *RB* | X | Song et al. (2003) |
| | *R1* | V | Ballvora et al. (2002) |
| | *R2* | IV | Gebhardt et al. (2001) |
| | *R3; R6; R7* | XI | El-Kharbotly et al. (1994; 1996) |
| | $R_{blc}$ | VIII | Gebhardt and Valkonen (2001) |
| | QTL | | Meyer et al. (1998); Oberhagemann et al. (1999); Trognitz et al. (2002) |
| | QTL | V | Visker et al. (2003) |
| *Synchytrium endobioticum* | *Sen1* | XI | Hehl et al. (1999) |
| *Erwinia carotovora* | QTL | | Zimnoch-Guzowska et al. (2000) |
| *Quality and yield traits* | | | |
| Tuber starch; tuber yield | QTL | I-XII | Schäfer-Pregel et al. (1998) |

TABLE 4.3 *(continued)*

| Trait | Gene | Chromosome | Reference |
|---|---|---|---|
| Tuberization | QTL | | Van den Berg et al. (1996) |
| Tuber dormancy | QTL | | Fryre et al. (1994) |
| Cold sweetening | QTL | V; IX | Menendez et al. (2002) |
| Skin color | QTL | X | Gebhardt et al. (2001) |
| Tuber shape | | X | Van Eck et al. (1994) |
| Tuber flesh color | QTL | III | Bonierbale et al. (1988) |
| Chip color | QTL | | Douches and Fryre (1994) |

multiple genes (QTL) and by environmental factors. The impact of factors like temperature or day length on gene expression is multifaceted and will require much more understanding of global gene functions. A new promising approach is "association mapping," that is based on linkage disequilibrium (Remington et al., 2001). Though labor intensive, this approach has been used to identify candidate gene QTL loci for tuber starch, yield, and chipping quality (Table 4.3). The molecular haplotypes of known candidate genes were analyzed in breeding material related by descent (Li et al., 2002). Several strategies for using QTL are being tested.

### *Genetic Maps*

In potato, the heterozygosity of the parents allows the construction of two linkage maps in a single $F_1$ mapping population, based on meiotic recombination in the female parent and in the male parent (Ritter et al., 1990). Many genes have been introgressed into potato cultivars and breeding clones from other tuber-bearing *Solanum* species by sexual and somatic hybridization. Such clones are used now for constructing molecular maps with different sets of RFLP, RAPD, AFLP, and SSR markers. Molecular maps are now available for diploid potatoes (Table 4.4) reflecting the localization of genes for qualitative and quantitative characters. A linkage map covering the 48 chromosomes of tetraploid potato based on AFLP markers was published by Meyer et al. (1998). Dong et al. (2000) anchored a cytogenetic map to

TABLE 4.4. Selected molecular maps of diploid/dihaploid *Solanum*.

| Parental species | Marker type | No. of marker loci | Reference |
|---|---|---|---|
| *S. tuberosum*<br>*S. phureja*<br>*S. chacoense* | RFLP | ~130 | Bonierbale et al. (1988) |
| *S. tuberosum* | RFLP, SSR | ~450 | Gebhardt et al. (2001) |
| *S. tuberosum*<br>*S. phureja* | RFLP, AFLP, cDNA-AFLP | ~1,200 | Brugmans et al. (2002) |
| *S.tuberosum*<br>*S. spegazinii* | RFLP | ~445 | Gebhardt et al. (2003) |
| *S. tuberosum* | AFLP | ~10,000 | Van der Voort et al. (1997); http://potatotdbase.dpw.wau.nl/UHDdata.html |

the molecular maps by fluorescence in situ hybridization (FISH) and a physical map is under construction (van Eck et al., 2003).

The molecular maps revealed that genes for tuber starch content, sugar content, and yield are located on chromosome V. This genomic region also harbors QTLs for plant maturity and vigor (Oberhagemann et al., 1999). Furthermore, some genes functional in carbohydrate metabolism or transport are closely linked to QTL for tuber traits (Menendez et al., 2002). Such clusters make map-based cloning easier, since sequencing strategies can concentrate on particular DNA fragments.

For the development of a physical map of potato, bacterial artificial chromosome (BAC) libraries are necessary. For *S. bulbocastanum* a library exists consisting of nearly 24,000 clones with an average size of 155 kb, representing approximately 3.7 times of the diploid potato genome (Song et al., 2000). For *S. tuberosum* 70,000 BACs are under development with an average insert size of >120 kb covering the genome 8-10 times (van Eck et al., 2003).

Another successful gene cloning approach uses transposable elements for gene tagging. An additional strategy is based on differential display techniques directly coupled to gene functions (Ros et al., 2004). For map-based cloning ultra dense maps with distances of less than 0.1 cm (1,000 kb) between trait and marker have to be available, furthermore, BAC libraries covering the whole genome

2-4 times, and some good luck is needed. Since these are very heavy prerequisites, cloning of genes where the gene products are unknown is a most difficult task. Consequently, most genes cloned today were isolated starting from proteins and enzymes and their amino acid sequence.

*Gene Tagging*

The analysis of the phenotypes of mutants deficient in the functional gene allows the assignment of a specific function to a certain gene. The most efficient method to generate populations saturated with loss-of-function mutations is insertional mutagenesis. In plants, insertional elements with the ability to integrate randomly within chromosomes are transposons such as *Ac/Ds* or *En/Spm* (Martienssen, 1998). Insertion mutants that destroy individual genes are an important tool in the functional analyses of genes. The first gene of a higher plant cloned without knowing the gene product was the *cf* gene responsible for resistance against *Cladosporium* in tomato. It was cloned using the gene tagging technique (Hammond-Kosack et al., 1994). Potato is not very amenable to the deployment of transposon tagging for gene cloning and gene identification. But, van Enckevort et al. (2000) described a selection method for direct transposon tagging for a gene of interest in a heterozygous background. Diploid potato plants with a *Ds* transposon linked to the desired gene of interest (the *Phytophthora infestans R1* locus) in a heterozygous state were used for the development of a transposon tagging strategy. After crossing to a diploid *Ac* transposon-containing genotype, *R1Ds/r-; Ac/-* seedlings were selected that showed active *Ds* transposition as displayed by DNA blot hybridization. Protoplast isolation and the use of the hygromycin gene as a cell-specific selection marker of *Ds* excision enabled the direct selection of *Ds* excision sectors in these highly chimeric seedlings. This somatic selection of *Ds* transpositions and the regeneration through protoplasts resulted in the development of a large population of almost 2,000 hygromycin resistant plants. Southern blot analysis confirmed the insertion of *Ds* at independent positions in the genome. Every selected plant displayed independent *Ds* excisions and reinsertions due to the expression of the *Ac* transposase. The population with the desired *R1* gene in a heterozygous state is directly useful for searching for transposon tagged *R1* mutants.

### *High-Throughput DNA Techniques*

Presently, application of biotechnology relies to a great extent on using available techniques rather than having a programmed approach solving problems in a logical way. The knowledge of physiological pathways and cascades is still too limited to effectively manipulate specific enzymes. The main activities today are characterized by efforts to construct comprehensive genomic maps. In the next phase, research activities will focus on the utilization of these maps for gene cloning and elucidating the function of the responsible genes. A systematic molecular evaluation of the complete genetic information of potato and the resulting cellular activities, such as transcription and protein expression will be essential for future developments. In order to obtain this information, a hierarchic approach has been undertaken, namely genome sequencing, transcriptional profiling, and proteome analysis. The number of genes, probably about 20,000-40,000 and the even higher number of descending enzymes is too large to be handled. High-throughput techniques together with powerful bioinformatics allow overcoming this bottleneck.

In potatoes, like in other eukaryotes, biological processes such as growth, adaptation to physiological shifts, or defense against stress, are mediated by distinct programs of differential gene expression. The identification of the subset of genes differentially expressed under certain growth conditions will facilitate breeding of potatoes that can better cope with pathogens or biotic stress. An experimental approach to gain insight into gene function is the analysis of the gene expression patterns at a genomewide scale, ideally of the total mRNA (transcript) complement expressed by the genome (analysis of the transcriptome). Clues of the biological function of the genes can be obtained from their spatial and temporal expression patterns. In this way, cDNAs can be very helpful to analyze complex genomes. Sequencing of expressed sequence tags (EST), serial analysis of gene expression (SAGE), and oligonucleotide chips or DNA microarrays are powerful tools for mRNA expression monitoring. ESTs from different tissues provide preliminary information on expression patterns for the more abundant transcripts. Using robotics, thousands of sequences can be generated within a short time. EST databases turned out to be excellent resources for identifying genes by sequence comparison, for discovering new genes, and for assigning exons in genomic sequences. They have also provided molecular markers for genetic and physical mapping and clones for expression analyses. The technology receiving major attention is the DNA chip or microarray technology, allowing the analysis of large numbers of genes in parallel (Hauser et al., 1998). For expression profiling, the arrays are hybridized to total mRNA pools that have been converted to cDNA in the presence of labeled nucleotides (Brown and Botstein, 1999). By measuring the signal intensity at each position on the array, the identity and quantity of the components in the labeled mixture can be determined.

A drawback of microarray hybridization is that it preferably detects mRNA species at middle to high abundance. To identify low abundance transcripts an enrichment procedure has to be applied. Diatchenko et al. (1996) equalized the representation of differently expressed genes with suppression subtractive hybridization (SSH). Ros et al. (2004) used SSH in combination with cDNA array hybridization to search for genes that are induced during the infection of potato with *Phytophthora infestans.* Leaflets of the tolerant cultivar Bettina and the susceptible cultivar Indira were inoculated with *P. infestans.* The infection of the potato leaves was quantified by real-time quantitative PCR. Using infected and control tissue, two cDNA libraries were prepared, highly enriched for *P. infestans* induced genes. For the SSH process the tester was prepared from a mixture of RNA from leaflets incubated with *P. infestans* for 16 h and for 72 h, and the driver from a mixture of RNA from control leaflets. The second cDNA library was produced from the susceptible cultivar Indira. From these libraries more than 500 clones were randomly picked and sequenced. Of the nearly 300 unigenes found, 182 clones were selected for further analysis of the differential gene expression by cDNA array hybridization. The two subtractive cDNA libraries utilized proved to be highly enriched for *P. infestans* induced genes. Only 3 percent of 182 analyzed cDNAs did not exhibit differential expression patterns. This allowed access to a source of *P. infestans* induced genes, such as pathogenesis-related protein 1, lipid saturase, peroxidase, chitinase, glucanase, with increased activity and others like proteinase inhibitor, leucin aminopeptidase, or RNA binding protein having reduced activity. Differences in defense responses of the two cultivars could be clearly determined.

Quantitative resistance is achieved when several genes interact. Interestingly, the majority of genes derived from the cDNA library of the tolerant cultivar encoded general metabolic proteins, whereas most of the genes derived from the cDNA library of the susceptible cultivar encoded stress- and/or defense-related genes. The genes that have the highest induction levels in Bettina are mainly involved in the energy metabolism or post translational modifications (Ros et al., 2004). It appears that defense reactions need energy; taking this energy from the plant growth metabolism, the result will be a smaller plant. Since resistant potato cultivars are not less yielding, a breeder always selects intuitively for both: increased resistance and increased growth metabolism.

### *Proteomics*

The term proteome refers to the total protein complement expressed by the genome and generally means the high-throughput systematic separation, identification, and characterization of proteins (Wilkins et al., 1997). A major tool for proteomics is in silico biology making comprehensive use of the data derived from the genome sequencing efforts. However, the information that can be deduced from the genome data about their encoded proteins is rather limited. Techniques are still needed for the parallel analyses of thousands of proteins and one has to await further developments in high-throughput, highly automated systems to be able to study total protein complements in a crop like potato. Fortunately, nature is rather conservative, as evidenced by the high similarity of gene sequences. Closely related plant genomes are composed of similar and collinear arranged genomic building blocks, whereas their intergenic DNA contents vary. Due to this synteny, it will not be necessary to start gene isolation and function analysis always right from the beginning: heterologous probes will allow the quick identification of homologous genes. The Solanaceae is a family expressing a high level of synteny (Grube et al., 2000), and even good correlations to the model plant *Arabidopsis* could be shown (Gebhardt et al., 2003). Thus, potato is again in a fortunate situation.

Though many promising biotechnological improvements have been made, results documented in vitro or in the greenhouse were not confirmed under field conditions. In conclusion, it appears that for potato, as for all other crops, biotechnology alone will not improve a cultivar. Biotechnology needs the classical field testing.

Finally, following are the key strategies fitting breeders' all dreams:

- secure germplasm conservation;
- handling a crop plant at different ploidy levels via sexual and clonal propagation;
- reproducible characterization of monogenic and polygenic traits in large diploid and polyploid populations;
- programmed combination of such traits by symmetric and asymmetric somatic hybridization;
- transfer of single traits via gene transfer to the nucleus or plastids;
- knowledge directed to construction of metabolic pathways resulting in plants superior in quality, resistance, and yield stability.

I predict it will be the potato where these strategies will be developed and applied. Although enough has been already achieved through conventional and biotechnological tools, even the most superior clone produced will not remain superior. The dynamics of evolution will make breeding of new cultivars necessary and demand the steady attention of scientists.

## REFERENCES

Alix, M.J., Savides, S., Blake, J., Herrmann, R., and Hornung, R. (2001). Effects of illumination source, culture ventilation and sucrose on potato *(Solanum tuberosum)* microtuber production under short days. *Annals of Applied Biology 139:*175-187.

Austin, S., Baer, M.A., and Helgeson, J.P. (1985). Transfer of resistance to potato leaf roll virus from *Solanum brevidens* into *Solanum tuberosum* by somatic fusion. *Plant Science 39:*75-82.

Austin, S., Lojkowska, E., Ehlenfeldt, M.K., Kelman, A., and Helgeson, J.P. (1988). Fertile interspecific somatic hybrids of *Solanum:* A novel source of resistance to *Erwinia* soft rot. *Phytopathology 78:*1216-1220.

Ballvora, A., Ercolano, M.R., Weiss, J., Meksem, K., Bormann, C.A., Oberhagemann, P., Salamini, F., and Gebhardt, C. (2002). The *R1* gene for potato resistance to late blight *(Phytophthora infestans)* belongs to the leucine zipper/NBS/LRR class of plant resistance genes. *The Plant Journal 30:*361-371.

Barone, A., Ritter, E., Schachtschnabel, U., Debener, T., Salamini, F., and Gebhardt, C. (1990). Localization by restriction fragment length polymorphism in potato of a major gene conferring resistance to the potato cyst nematode *Globodera rostochiensis. Molecular and General Genetics 224:*177-182.

Bastia, T., Carotenuto, N., Basile, B., Zoina, A., and Cardi, T. (2000). Induction of novel organelle DNA variation and transfer of resistance to frost and *Verticillium* wilt in *Solanum tuberosum* through somatic hybridization with 1 EBN *S. commersonii. Euphytica 116:*1-10.

Bäumler, M., Braun, A., Meisel, B., Schenkel, W., Schullehner, K., Thümmler, F., and Wenzel, G. (2003). Characterization of genetic diversity in 2*x* and 4*x* clones by means of AFLP, SSR markers or differentially induced genes. In Bradshaw, J.E., van Eck, H., and Valkonen, J. (Eds.), *Breeding and Adaptation of Potatoes* (p. 17). Oulu, Finland: *EAPR-EUCARPIA* University of Helsinki.

Bendahmane, A., Querci, M., Kanyuka, M., and Baulcombe, D.C. (2000). *Agrobacterium* transient expression system as a tool for the isolation of disease resistance genes: Application to the *Rx2* locus in potato. *The Plant Journal 21:*73-81.

Bonierbale, M., Plaisted, R.L., and Tanksley, S.D. (1988). RFLP maps based on a common set of clones reveal modes of chromosomal evolution in potato and tomato. *Genetics 120:*1095-103.

Bonnel, E. (2003). DNA markers: Tools for the variety registration, intellectual property and quality seed management. In Bradshaw, J.E., van Eck, J.E., and Valkonen, J. (Eds.), *Breeding and Adaptation of Potatoes* (p. 27). Oulu, Finland: *EAPR-EUCARPIA* University of Helsinki.

Börnke, F.M., Hajirezaei, M., and Sonnewald, U. (2002). Potato tubers as bioreactors for palatinose production. *Journal Biotechnology 96:*119-124.

Botstein, D., White, R.L., Skolnick, M., and Davis, R.W. (1980). Construction of a genetic linkage map in man using restriction fragment length polymorphisms. *American Journal of Human Genetics 32:*314-331.

Braun, A. (2002). Einfluß von Cytoplasma und Heterozygotie auf die Merkmalsvariabilität in Kreuzungsnachkommenschaften von *Solanum tuberosum* L. *Dissertation,* TU München.

Braun, A., Gatz, C., Frei, U., and Wenzel, G. (2002). Field experiments with transgenic potato clones. In Wenzel, G. and Wulfert, I. (Eds.), *Potatoes Today and Tomorrow* (p. 210). Königswinter, Germany: WPR.

Braun, A., Schullehner, K., and Wenzel, G. (2005). Molecular analysis of genetic variation in potato (*Solanum tuberosum* L.) II. International cultivar spectrum. *Potato Research 47:*93-99.

Braun, A. and Wenzel, G. (2005). Molecular analysis of genetic variation in potato (*Solanum tuberosum* L.) I. German cultivars and advanced clones. *Potato Research 47:*81-92.

Brigneti, G., Garcia-Mas, J., and Baulcombe, D.C. (1997). Molecular mapping of the potato virus Y resistance gene $Ry_{sto}$ in potato. *Theoretical and Applied Genetics 94:*198-203.

Brown, C.R., Yang, C.P., Mojtahedi, H., and Santo, G.S. (1996). RFLP analysis of resistance to Columbia root-knot nematode derived from *Solanum bulbocastanum* in a $BC_2$ population. *Theoretical and Applied Genetics 92:*572-576.

Brown, P.O. and Botstein, D. (1999). Exploring the new world of the genome with DNA microarrays.*Nature Genetics 21 (Suppl.):*33-37.

Brugmans, B., del Carmen, A.F., Bachem, C.W.B., van Os, H., van Eck, H.J., and Visser, R.G.F. (2002). A novel method for the construction of genome wide transcriptome maps. *The Plant Journal 31:*211-222.

Bryan, G.J., McNicoll, J., Ramssay, G., and Meyer, R.C. (1999). Polymorphic simple sequence repeat markers in chloroplast genomes of Solanaceous plants. *Theoretical and Applied Genetics 99:*859-867.

Bryan, G.J., van Eck, H.J., Milbourne, D., Isidore, E., McLean, K., McNicoll, J., Linton, S., Tierney, I., Purvis, A., van Os, H. et al. (2002). Application of UHD genetic linkage map for BAC landing and contig initiation in a region of linkage group V. In Wenzel, G. and Wulfert, I. (Eds.), *Potatoes Today and Tomorrow* (p. 128). Königswinter, Germany: WPR.

Cardi, T., Puite, K.J., Ramulu, K.S., D'Ambreosio, F., and Fruscicante, L. (1993). Production of somatic hybrids between frost tolerant *Solanum commersonii* and *S. tuberosum:* Protoplast fusion, regeneration and isozyme analysis. *American Potato Journal 70:*753-764.

Carputo, D., Basile, B., Cardi, B., and Frusciante, L. (2000). *Erwinia* resistance in backcross progenies of *Solanum tuberosum* x *S. tarijense* and *S. tuberosum* (+) *S. commersonii* hybrids. *Potato Research 43:*135-142.

Carrasco, A., Ruiz de Galarreta, J.I., Rico, A., and Ritter, E. (2000). Transfer of PLRV resistance from *Solanum verrucosum* to potato (*S. tuberosum* L.) by protoplast electrofusion. *Potato Research 43:*31-42.

Celebi, T.F., Slack, S.A., and Jahn, M.M. (2002). A new gene, $Ny_{tbr}$, for hypersensitivity to Potato virus Y from *Solanum tuberosum* maps to chromosome IV. *Theoretical and Applied Genetics 104:*669-674.

Debener, U., Salamini, F., and Gebhardt, C. (1990). Phylogeny of wild and cultivated *Solanum* species based on nuclear restriction fragment length polymorphisms (RFLPs). *Theoretical and Applied Genetics 79:*360-368.

De Block, M. (1988): Genotype independent leaf disk transformation of potato (*Solanum tuberosum* L.) using *Agrobacterium tumefaciens. Theoretical and Applied Genetics 76:*767-774.

De Vincente, M.C. and Tanksley, S.D. (1991). Genome-wide reduction in recombination of backcross progeny derived from male versus female gametes in an interspecific cross of tomato. *Theoretical and Applied Genetics 83:*173-178.

Diatchenko, I., Lau, Y.F., Campbell, A.P., Chenchik, A., Moqadam, F., Hiuang, B., Lukyanov, S., Lukyanov, K., Gurskaya, N., Sverdlov, E.D., and Siebert, P.D. (1996). Suppression subtractive hybridisation: A method for generating differentially regulated tissue specific cDNA probes and libraries. *Proceedings National Academy of Sciences USA 93:*14863-14868.

Donegan, K.K., Schaller, D.L., Stone, J.K., Ganio, L.M., Reed, G., Hamm, P.B., and Seidler, R.J. (1996). Microbial populations, fungal species diversity and plant pathogen levels in field plots of potato plants expressing the *Bacillus thuringiensis* var. *tenebrionis* endotoxin. *Transgenic Research 5:*25-35.

Dong, F., Song, J., Naess, S.K., Helgeson, J.P., Gebhardt, C., and Jiang, J. (2000). Development and applications of a set of chromosome-specific cytogenetic DNA markers in potato. *Theoretical and Applied Genetics 101:*1001-1007.

Douches, D.S. and Freyre, R. (1994). Identification of genetic factors influencing chip colour in diploid potato (*Solanum* spp.). *American Potato Journal 71:*581-590.

Düring, K., Porsch, P., Fladung, M., and Lörz, H. (1993). Transgenic potato plants resistant to the phytopathogenic bacterium *Erwinia carotovora. Plant Journal 3:*587-598.

El-Kharbotly, A., Leonhards-Schippers, C., Huigen, D.J., Jacobsen, E., Pereira, A., Stiekema, W.J., Salamini, F., and Gebhardt, C. (1994). Segregation analysis and RFLP mapping of the *R1* and *R3* alleles conferring race specific resistance to *Phytophthora infestans* in progeny of dihaploid potato parents. *Molecular and General Genetics 242:*749-754.

El-Kharbotly, A., Palomino-Sanchez, C., Salamini, F., Jacobsen, E., and Gebhardt, C. (1996). *R6* and *R7* alleles of potato conferring race-specific resistance to *Phytophthora infestans* (Mont.) de Bary identified genetic loci clustering with the R3 locus on chromosome XI. *Theoretical and Applied Genetics 92:*880-884.

Fiegert, A.K., Mix-Wagner, G., and Vorlop, K.D. (2000). Regeneration of *Solanum tuberosum* L. cv. Tomensa: Induction of somatic embryogenesis in liquid culture for the production of "artificial seeds." *Landbauforschung Völkenrode 50:*3-4.

Frei, U. (1994). Multifusions. *Potato Research 37:*338.

Frei, U., Peiretti, E.G., and Wenzel, G. (2003). Significance of cytoplasmic DNA in plant breeding programs. *Plant Breeding Review 23:*175-210.

Frei, U., Stattmann, M., Lössl, A., and Wenzel, G. (1998). Aspects of fusion combining ability of dihaploid *Solanum tuberosum* L.: Influence of the cytoplasm. *Potato Research 41:*155-162.

Freyre, R., Warnke, S., Sosinski, B., and Douches, D.S. (1994). Quantitative trait locus analysis of tuber dormancy in diploid potato *(Solanum* spp.). *Theoretical and Applied Genetics 89:*474-480.

Gebhardt, C. (2004). Potato genetics: Molecular maps and more. In Lörz, H. and Wenzel, G. (Eds.), *Molecular Marker Systems in Plant Breeding and Crop Improvement* (pp. 215-227). Berlin, Germany: Springer.

Gebhardt, C., Ritter, E., and Salamini, F. (2001). RFLP map of the potato. In: Phillips, R.L. and Vasil, I.K. (Eds.), *DNA-Based Markers in Plants, Advances in Cellular and Molecular Biology of Plants,* (*Vol. 6,* pp. 319-336). Dordrecht, the Netherlands: Kluwer Academic Publishers.

Gebhardt, C. and Valkonen, J.P.T. (2001). Organization of genes controlling disease resistance in the potato genome. *Annual Review Phytopathology 39:* 79-102.

Gebhardt, C., Walkemeier, B., Henselewski, H., Barakat, A., Delseny, M., and Stüber, K. (2003). Comparative mapping between potato *(Solanum tuberosum)* and *Arabidopsis thaliana* reveals structurally conserved domains and ancient duplications in the potato genome. *The Plant Journal 34:*529-541.

Gibson, R.W., Jones, M.G.K., and Fish, N. (1988). Resistance to potato leaf roll virus and potato virus Y in somatic hybrids between *Solanum tuberosum* and *S. brevidens. Theoretical and Applied Genetics 76:*113-117.

Gopal, J., Chamail, A., and Sarkar, D. (2002). Slow-growth in vitro conservation of potato germplasm at normal propagation temperature. *Potato Research 45:* 203-213.

———. (2004). In vitro production of microtubers for conservtion of potato germplasm: Effect of genotype, abscisic acid and sucrose. In vitro *Cellular and Development Biology-Plant 40:*485-490.

Gopal, J., Minocha, J.L., and Dhaliwal, H.S. (1998). Microtuberization in potato (*Solanum tuberosum* L.). *Plant Cell Reports 17:*794-798.

Grube, R.C., Radwanski, E.R., and Jahn, M. (2000). Comparative genetics of disease resistance within the Solanaceae. *Genetics 155:*873-887.

Gupta, P.K., Varshney, R.K., and Prasad, M. (2002). Molecular markers: Principles and methodology. In Jain, S.M., Brar, D.S., and Ahloowalia, B.S. (Eds.), *Molecular Techniques in Crop Improvement* (pp. 9-54). Dordrecht, the Netherlands: Kluwer Academic Publishers.

Hammond-Kosack, K.E., Jones, D.A., and Jones, J.D.G. (1994). Identification of two genes required in tomato for full Cf-9-dependent resistance to *Cladosporium fulvum. Plant Cell 6:*361-374.

Hauser, N.C., Vingron, M., Scheideler, M., Krems, B., Hellmuth, K., Entian, K.D., and Hoheisel, J.D. (1998). Transcriptional profiling on all open reading frames of *Saccharomyces cerevisiae. Yeast 14:*1209-1221.

Hehl, R., Faurie, E., Hesselbach, J., Salamini, F., Whitham, S., Baker, B., and Gebhardt, C. (1999). TMV resistance gene *N* homologues are linked to *Synchytrium endobioticum* resistance in potato. *Theoretical and Applied Genetics 98:*379-386.

Helgeson, J.P., Pohlman, J.D., Austin, S., Haberlach, G.T., Wielgus, S.M., Ronis, D., Zambolim, L., Tooley, P., McGrath, J.M., James, R.V., and Stevenson, W.R. (1998). Somatic hybrids between *Solanum bulbocastanum* and potato: A new source of resistance to late blight. *Theoretical and Applied Genetics 96:*738-742.

Hellwege, E.M.A., Czapla, A., Jahnke, A., Willmitzer, L., and Heyer, A.G. (2000). Transgenic potato *(Solanum tuberosum)* tubers synthesize the full spectrum of inulin molecules occurring in globe artichoke *(Canara scolymus)* roots. *Proceedings National Academy of Sciences (USA) 97:*8699-8704.

Hirai, D. and Sakai, A. (1999). Cryopreservation of in vitro-grown meristems of potato (*Solanum tuberosum* L.) by encapsulation-vitrification. *Potato Research 42:*153-160.

Hoekema, A., Huisman, M.J., Molendyk, L., van der Elsen, P.J.M., and Cornelissen, B.J.C. (1989). The genetic engineering of two commercial potato cultivars for resistance to potato virus X. *Biotechnology 7:*273-278.

Hofferbert, H.R. and Wenzel, G. (1997). Alternative Hybridzüchtung—somatische Hybridisation bei der Kartoffel. *Berichte Arbeitstagung Gumpenstein 48:*111-114.

Hosaka, K. (1995). Successive domestication and evolution of the Andean potatoes as revealed by chloroplast DNA restriction endonuclease analysis. *Theoretical and Applied Genetics 90:*356-363.

Hougas, R.W. and Peloquin, S.J. (1957). A haploid plant of the potato variety Katahdin. *Nature 180:*1202-1210.

Isenegger, D.A., Taylor, P.W.J., Ford, R., Franz, P., McGregor, G.R., and Hutchinson, J.F. (2001). DNA fingerprinting and genetic relationship of potato cultivars (*Solanum tuberosum* L.) commercially grown in Australia. *Australian Journal of Agricultural Research 52:*911-918.

Johnson, A.A.T., Piovano, S.M., Ravichandran, V., and Veilleux, R.E. (2001). Selection of monohaploids for protoplast fusion and generation of intermonoploid somatic hybrids of potato. *American Journal of Potato Research 78:*19-29.

Kaniewski, W., Lawson, C., Sammons, D., Haley, L., Haret, J., Delannay, X., and Tumer, N.E. (1990). Engineering resistance to mixed infection into a commercial potato cultivar: Resistance to potato virus X and Y in transgenic Russet Burbank. *Biotechnology 8:*127-143.

Kawchuk, L.M., Martin, R.R., and McPherson, J. (1991). Sense and antisense RNA-mediated resistance in potato leaf roll virus in Russet Burbank potato plants. *Molecular Plant–Microbe Interactions 4:*247-253.

Keller, J.E.R. (2003). In-vitro-Kultur zur Erhaltung pflanzengenetischer Ressourcen. ADVK Meeting (pp. 51-52). Geisenheim, Germany.

Kim, J.H., Joung, H., Kim, H.Y., and Lim, Y.P. (1998). Estimation of genetic variation and relationship in potato (*Solanum tuberosum* L.) cultivars using AFLP markers. *American Journal of Potato Research 75:*107-112.

Koop, U., Steinmüller, K., Wagner, H., Roessler, C., Eibl, C., and Sacher, L. (1996). Integration of foreign sequences into the tobacco plastome via polyethylene glycol-mediated protoplast transformation. *Planta 199:*193-201.

Kordyum, E, Baranenko, V., Nedukha, E., and Samoilov, V. (1997). Development of potato *(Solanum tuberosum)* minitubers in microgravity. *Plant Cell Physiology 38:*1111-1117.

Kossmann J., Visser, R., Müller-Röber, B., Willmitzer, L., and Sonnewald, U. (1991). Cloning and expression analysis of potato cDNA that encodes branching enzyme: Evidence for co-expression of starch biosynthetic genes. *Molecular and General Genetics 230:*39-45.

Kuckuck, H., Kobabe, G., and Wenzel, G. (1991). *Fundamentals of Plant Breeding.* Berlin, Germany: Springer Verlag.

Li, L., Strahwald, J., Hofferbert, H.R., Lübeck, H.R., Tacke, E., Junghans, H., Salamini, F., and Gebhardt, C. (2002). Association mapping based on candidate genes for tuber traits in tetraploid potato. In Wenzel, G. and Wulfert, I. (Eds.), *Potatoes Today and Tomorrow* (p. 262). Königswinter, Germany: WPR.

Lössl, A., Adler, N., Horn, R., Frei, U., and Wenzel, G. (1999). Chondriome-type characterization of potato: mt α, β, γ, δ, ε and novel plastid-mitochondrial configurations in somatic hybrids. *Theoretical and Applied Genetics 99:*1-10.

Lössl, A., Götz, M., Braun, A., and Wenzel, G. (2000). Molecular markers for cytoplasm in potato: Male sterility and contribution of different plastid-mitochondrial configurations to starch production. *Euphytica 116:*221-230.

Lübberstedt, T., Mohler, V., and Wenzel, G. (2002). Function of genetic material: Genes involved in quantitative and qualitative resistance. *Progress in Botany 63:*80-105.

Lübeck, J., Römer, S., Adomat, C., and Sandmann, G. (2002). Genetic engineering of potato tuber carotenoid composition. In Wenzel, G. and Wulfert, I. (Eds.), *Potatoes Today and Tomorrow* (p. 92). Königswinter, Germany: WPR.

Luo, Z.W., Hackett, C.A., Bradshaw, J.E., McNichol, J.W., and Milbourne, D. (2001). Construction of a genetic linkage map in tetraploid species using molecular markers. *Genetics 157:*1369-1385.

Marano, M.R., Malcuit, I., de Jong, W., Baulcombe, D.C., and de Jong, W. (2002). High resolution genetic map of *Nb*, a gene that confers hypersensitive resistance to potato virus X in *Solanum tuberosum. Theoretical and Applied Genetics 105:*192-200.

Marczewski, W., Flis, B., Syller, J., Schäfer-Pregl, R., and Gebhardt, C. (2001). A major QTL for resistance to potato leafroll virus (PLRV) is located in a resistance hotspot on potato chromosome XI and is tightly linked to *N*-gene-like markers. *Molecular Plant–Microbe Interactions 14:*1420-1425.

Marczewski, W., Hennig, J., and Gebhardt, C. (2002). The potato virus S gene *Ns* maps to potato chromosome VIII. *Theoretical and Applied Genetics 105:*564-567.

Martienssen, R.A. (1998). Functional genomics: Probing plant gene function and expression with transposons. *Proceedings National Academy of Sciences (USA) 95:*2021-2026.

McGrath, J.M., Wielgus, S.M., Uchytil, T.F., Kim, L.H., Haberlach, G.T., Williams, C.E., and Helgeson, J.P. (1994). Recombination of *Solanum brevidens* chromosomes in the second backcross generation from a somatic hybrid with *S. tuberosum. Theoretical and Applied Genetics 88:*917-924.

McGrath, J.M., Williams, C.E., Haberlach, G.T., Wielgus, S.M., Uchytil, T.F., and Helgeson, J.P. (2002). Introgression and stabilization of *Erwinia* tuber soft rot resistance into potato after somatic hybridisation of *Solanum tuberosum* and *S. brevidens. American Journal of Potato Research 79:*19-24.

McGregor, C.E., Greyling, M.M., and Warnich, L. (2000). The use of simple sequence repeats (SSRs) to identify commercially important potato (*Solanum tuberosum* L.) cultivars in South Africa. *South African Journal of Plant Soil 17:*177-179.

McGregor, C.E., Lambert, C.A., Greyling, M.M., Louw, J.H., and Warnich, L. (2000). A comparative assessment of DNA fingerprinting techniques (RAPD, ISSR, AFLP and SSR) in tetraploid potato (*Solanum tuberosum* L.) germplasm. *Euphytica 113:*135-144.

Melchers, G., Mohri, Y., Watanabe, K., Wakabayashi, S., and Harada, K. (1992). One-step generation of cytoplasmic male sterility by fusion of mitochondrial-inactivated tomato protoplasts with nuclear-inactivated *Solanum* protoplasts. *Proceedings of National Academic Sciences (USA) 89:*6832-6836.

Menendez, C.M., Ritter, E., Schäfer-Pregl, R., Walkemeier, B., Kalde, A., Salamini, F., and Gebhardt, C. (2002). Cold-sweetening in diploid potato. Mapping QTL and candidate genes. *Genetics 162:*1423-1434.

Meyer, R.C., Milbourne, D., Hackett, C.A., Bradshaw, J.E., McNichol, J.W., and Waugh, R. (1998). Linkage analysis in tetraploid potato and association of markers with quantitative resistance to late blight *(Phytophthora infestans). Molecular and General Genetics 259:*150-160.

Milbourne, D., Meyer, R.C., Collins, A.J., Ramsay, L.D., Gebhardt, C., and Waugh, R. (1998). Isolation, characterisation and mapping of simple sequence repeat loci in potato. *Molecular and General Genetics 259:*233-245.

Mojtahedi, H.A., Brown, C.R., and Santo, G.S. (1995). Characterization of resistance in a somatic hybrid of *Solanum bulbocastanum* and *S. tuberosum* to *Meloidogyne chitwoodii. Journal of Nematology 27:*86-93.

Möllers, C., Frei, U., and Wenzel, G. (1994). Field evaluation of tetraploid somatic hybrids of potato. *Theoretical and Applied Genetics 88:*147-152.

Möllers, C. and Wenzel, G. (1992). Somatic hybridization of dihaploid potato protoplasts as a tool for potato breeding. *Botanica Acta 105:*133-139.

Murashige, T. and Skoog, F. (1962). A revised medium for rapid growth and bioassay with tobacco tissue culture. *Physiologia Plantarum 15:*473-497.

Nyende, A.B., Mix-Wagner, G., and Schittenhelm, S. (2002). Plant regeneration from synthetic seeds of potato. In Wenzel, G. and Wulfert, I. (Eds.), *Potatoes Today and Tomorrow* (p. 272). Königswinter, Germany: WPR.

Oberhagemann, P., Chatot-Balandras, C., Bonnel, E., Schäfer-Pregl, R., Wegener, D., Palomino, C., Salamini, F., and Gebhardt, C. (1999). A genetic analysis of quantitative resistance to late blight in potato: Towards marker assisted selection. *Molecular Breeding 5:*399-415.

Oberwalder, B., Schilde-Rentschler, L., Loffelhardt-Ruoss, B., and Ninnemann, H. (2000). Differences between hybids of *Solanum tuberosum* L. and *Solanum circaeifolium* Bitt. obtained from symmetric and asymmetric fusion experiments. *Potato Research 43:*71-82.

Pineda, O., Bonierbale, M.W., Plaisted, R.L., Brodie, B.B., and Tanksley, S.D. (1992). Identification of RFLP markers linked to the *H1* gene conferring resistance to the potato cyst nematode *(Globodera rostochiensis). Genome 36:*152-156.

Preil, W. (2003). In vitro Vermehrung gartenbaulicher Kulturpflanzen in Europa—aktueller Stand und Trends ADVK Meeting (pp. 51-52). Geisenheim, Germany.

Provan, J., Powell, W., Dewar, H., Bryan, G., Machray, G.C., and Waugh, R. (1999). An extreme cytoplacmic bottleneck in the modern European cultivated potato (*Solanum tuberosum*) is not reflected in decreased levels of nuclear diversity. *Proceedings of the Royal Society (London) B 266:*633-639.

Reiter, R. (2001). PCR-based marker systems. In Phillips, R.L. and Vasil, I.K. (Eds.), *DNA-Based Markers in Plants, Advances in Cellular and Molecular Biology of Plants (Vol. 6,* pp. 9-29). Dordrecht, the Netherlands: Kluwer Academic Publishers.

Remington, D.L., Thornsberry, J.M., Matsuoka, Y., Wilson, L.M., Whitt, S.R., Doebley, J., Kresvick, S., Goodman, M.M., and Buckler, E.S. (2001). Structure of linkage disequilibrium and phenotypic associations in the maize genome. *Proceedings of the National Academy of Sciences (USA) 98:*11479-11484.

Ritter, E., Gebhardt, C., and Salamini, F. (1990). Estimation of recombination frequencies and construction of RFLP linkage maps in plants from crosses between heterozygous parents. *Genetics 125:*645-654.

Romano, A., Raemakers, K., Visser, R., and Mooibrock, H. (2001). Transformation of potato *(Solanum tuberosum)* using particle bombardment. *Plant Cell Reports 20:*198-204.

Ros, B., Thümmler, F., and Wenzel, G. (2004). Analysis of differentially expressed genes in a susceptible and moderately resistant potato cultivar upon *Phytophthora infestans* infection. *Molecular Plant Pathology 5:*191-201.

Schäfer-Pregl, R., Ritter, E., Concilio, L., Hesselbach, J., Lovatti, L., Walkemeier, B., Thelen, H., Salamini, F., and Gebhardt, C. (1998). Analysis of quantitative trait loci (QTL) and quantitative trait alleles (QTA) for potato tuber yield and starch content. *Theoretical and Applied Genetics 97:*834-846.

Schittenhelm, S., Menge-Hartmann, U., and Oldenburg, E. (2002). Growth analysis of transgenic potato overexpressing *Arabidopsis thaliana* phytochrome B. In Wenzel, G. and Wulfert, I. (Eds.), *Potatoes Today and Tomorrow* (p. 289). Königswinter, Germany: WPR.

Schwarzfischer, A., Song, Y., Scholz, H., Schwarzfischer, J., and Hepting, L. (2002). Haplopidiezüchtung, Protoplastenfusion und Entwicklung von genetischen Markern zur gezielten Sortenentwicklung bei Kartoffeln. *Vorträge für Pflanzenzüchtung 54:*123-130.

Seabrook, J.E.A. and Douglass, L.K. (2001). Somatic embryogenesis on various potato tissues from range of genotypes and ploidy levels. *Plant Cell Reports 20:*175-182.

Shibata, D., and Liu, Y.G. (2000). Technical focus *Agrobacterium*-mediated plant transformation with large DNA fragments. *Trend in Plant Science 5:*354-357.

Sidorov, V.A., Kasten, D., Pang, S.Z., Hajdukiewicz, P., Staub, J.M., and Nehra, N.S. (2000). Stable chloroplast transformation in potato: Use of fluorescent protein as a plastid marker. *The Plant Journal 19:*209-216.

Song, J., Bradeen, J.M., Naess, S.K., Raasch, J.A., Wielgus, S.M., Haberlach, G.T., Liu, J., Kuang, H., Austin-Phillips, S., Buell, C.R. et al. (2003). Gene RB cloned from *Solanum bulbocastanum* confers broad spectrum resistance to potato late blight. *Proceedings of the National Academy of Sciences (USA) 100:*9128-9133.

Song, J., Dong, F., and Jiang, J. (2000). Construction of a bacterial artificial chromosome (BAC) library for potato molecular cytogenetics research. *Genome 43:*199-204.

Sonnino, A., Tanaka, S., Iwanaga, M., and Schilde-Rentschler, L. (1989). Genetic control of embryo formation in anther culture of diploid potatoes. *Plant Cell Reports 8:*105-107.

Sopory, S.K., Jacobsen, E., and Wenzel, G. (1978). Production of monohaploid embryoids and plantlets in cultured anthers of *Solanum tuberosum. Plant Science Letters 12:*47-54.

Sorri, V.A., Watanabe, K.N., and Valkonen, J.P.T. (1999). Predicted kinase-3a motif of a resistance gene-like fragment as a unique marker for PVY resistance. *Theoretical and Applied Genetics 99:*164-170.

Tautz, D. and Renz, M. (1984). Simple sequences are ubiquitous repetitive components of eukaryotic genomes. *Nucleic Acids Research 12:*4127-4138.

Thach, N.Q., Frei, U., and Wenzel, G. (1993). Somatic fusion for combining virus resistance in *Solanum tuberosum* L. *Theoretical and Applied Genetics 85:*863-867.

Thiele, A., Herold, M., Lenk, I., Quail, P.H., and Gatz, C. (1999). Heterologous expression of *Arabidopsis* Phytochrome B in transgenic potato influences photosynthetic performance and tuber development. *Plant Physiology 120:*73-82.

Trognitz, F.C., Manosalva, P.M., Nino-Liu, D.O., Herrera, M.R., Ghislain, M., Trognitz, B.R., and Nelson, R.J. (2002). Plant defense genes associated with quantitative resistance to potato late blight in *Solanum phureja* x *S. tuberosum* hybrids. *Molecular Plant–Microbe Interactions 15:*587-597.

van Breukelen, E.W., Ramanna, M.S., and Hermsen, J.G.T. (1975). Monohaploids ($n = x = 12$) from autotetraploids *Solanum tuberosum* ($2x = 4x = 48$) through two successsive cycles of female parthenogenesis. *Euphytica 24:*567-574.

van den Berg, J.H., Ewing, E.E., Plaisted, R.L., McMurry, S., and Bonierbale, M.W. (1996). QTL analysis of potato tuberization. *Theoretical and Applied Genetics 93:*307-316.

van der Voort, J., Wolters, P., Folkertsma, R., Hutten, R., van Zandvoort, P., Vinke, H., Kanyuka, K., Bendahmane, A., Jacobsen, E., Janssen, R. et al. (1997). Mapping of the cyst nematode resistance locus *Gpa2* in potato using a strategy based on co-migrating AFLP markers. *Theoretical and Applied Genetics 95:*874-880.

van Eck, H.J., de Boer, J., Brugmans, B., Borm, T., Bakker, E., Bakker, J., and Visser, R. (2003). Development of a genome-wide physical map of potato. In Bradshaw, J.E., van Eck, H., and Valkonen, J. (Eds.), *Breeding and Adaptation of Potatoes* (p. 29). Oulu, Finland: EAPR/EUCARPIA.

van Eck, H.J., Jacobs, J.M.E., Stam, P., Ton, J., Stiekema, W.J., and Jacobsen, E. (1994). Multiple alleles for tuber shape in diploid potato detected by qualitative and quantitative genetic analysis using RFLPs. *Genetics 137:*303-309.

Van Enckevort, L.J.G., Bergervoet, J.E.M., Stiekema, W.J., Pereira, A., and Jacobsen, E. (2000). Selection of independent Ds transposon insertions in somatic tissue of potato by protoplast regeneration. *Theoretical and Applied Genetics 101:* 503-510.

Virk, P.S., Newbury, H.J., Jackson, M.T., and Ford-Lloyd, B.V. (2000). Are mapped markers more useful for assessing genetic diversity? *Theoretical and Applied Genetics 100:*607-613.

Visker, M.H.P.W., Keizer, L.C.P., van Eck, H.J., Jacobsen, E., Colon, L.T., and Struik, P.C. (2003). Can the QTL for late blight resistance on potato chromosome 5 be attributed to foliage maturity type? *Theoretical and Applied Genetics 106:*317-325.

Visser, R.G.F., Stolte, A., and Jacobsen, E. (1991) Expression of a chimeric granule-bound starch synthase-GUS gene in transgenic potato plants. *Plant Molecular Biology 17:*691-699.

Vos, P., Hogers, R., Bleeker, R., Reijans, M., van de Lee, T., Hornes, M., Frijters, A., Pot, V., Peleman, J., Kupier, M. et al. (1995). AFLP: A new technique for DNA fingerprinting. *Nucleic Acids Research 23:*4407-4414.

Wegener, C.B. (2001). Transgenic potatoes expressing an *Erwinia* pectate lyase gene—results of a 4-year field experiment. *Potato Research 44:*401-410.

Wenzel, G. (1998). Function of genetic material responsible for disease resistance in plants. *Progress in Botany 59:*80-107.

Wenzel, G. and Foroughi-Wehr, B. (1984). Production and use of isogenic lines. In Vasil, I.K. and Thorpe, T.A. (Eds.), *Plant Cell and Tissue Culture* (pp. 153-172). Dordrecht, The Netherlands: Kluwer Publishers.

Wenzel, G., Schieder, O., Przewozny, T., Sopory, S.K., and Melchers, G. (1979). Comparison of single cell culture derived *Solanum tuberosum* L. plants and a model for their application in breeding programs. *Theoretical and Applied Genetics 55:*49-55.

Wenzel, G. and Uhrig, H. (1981). Breeding for nematode and virus resistance in potato via anther culture. *Theoretical and Applied Genetics 80:*359-365.

Wilkins, M.R., Williams, K.L., Appel, R.D., and Hochstrasser, D.F. (1997). Proteome Research: New Frontiers in Functional Genomics. Berlin, Germany: Springer.

Williams, J.G.K., Kubelik, A.R.K., Livak, J.L., Rafalski, J.A., and Tingey, S.V. (1990). DNA polymorphisms amplified by random primers are useful genetic markers. *Nucleic Acids Research 18:*6531-6535.

Zimnoch-Guzowska, E., Marczewski, W., Lebecka, R., Flis, B., Schäfer-Pregl, R., Salamini, F., and Gebhardt, C. (2000). QTL analysis of new sources of resistance to *Erwinia carotovora* ssp. *atroseptica* in potato done by AFLP, RFLP and resistance-gene-like markers. *Crop Science 40:*1156-1167.

# Chapter 5

# Quality Improvement

T. R. Tarn
G. C. C. Tai
Qiang Liu

Potatoes are consumed in a great variety of forms either fresh as table stock or converted into processed products. Fresh consumption as table stock includes three main forms of cooking the raw potatoes: boiling, baking, and microwaving. Industrial processing transforms the raw potatoes into a variety of processed products. French fries and potato chips represent the two major processed potato products in the food industry.

In Chapter 14, "Processing and Value Addition," of this book, Xiu-Qing Li and co-authors have described the food value of potato and the industrial technology in use for potato processing and improving the properties of various potato products. In this chapter, we concentrate on genetic manipulation of potato for improving its food value and characters important to processing.

## *BREEDING OBJECTIVES FOR QUALITY IMPROVEMENT*

The breeding objectives for quality improvement vary according to the forms of food consumption. Fresh-consumed potatoes require an attractive tuber shape with shallow eyes and good skin finish. Skin color, flesh color, and texture requirements differ in different regions. The cooked products must be free of enzymatic browning and after-cooking darkening. Boiled potatoes should have firm texture and be free of sloughing. Baked potatoes generally have a more mealy texture. While there is rarely selection for specific flavors, absence of off-flavors is important. Texture, generally correlated with dry matter content and measured by specific gravity, is the single most important trait determining the general cooking quality of the potatoes.

Production of frozen French fries is the largest sector of the processing industry. The key requirements for this are large size tubers with blocky, uniform, oblong to long shape and good storability. Shallow eyes and smooth skin of raw potatoes are required to minimize losses in peeling and slicing. Most importantly, potatoes with high dry matter and low reducing sugars with minimal variation through the tuber tissue are essential to produce French fries with low oil content, uniform texture, and light golden color.

Production of potato chips requires cultivars with round to oval tubers of medium size (ca. 75 mm diameter). Again, high dry matter with uniform distribution throughout the tuber flesh is essential to minimize oil absorption and prevent blistering while frying. Potatoes used for processing are expected to be cold stored at 3-5°C for as long as seven months. Development of cultivars resistant to cold temperature sweetening represents an important objective for many breeding programs (Thill and Peloquin, 1995; Love et al., 1998). Maintenance of reducing sugars around 0.1 percent (on fresh weight basis) throughout the storage period results in light golden color chips (Storey and Davis, 1992).

There are a number of traits that do not directly affect the qualities of cooked or processed products, but ensure production of raw potatoes that meet the physiological conditions for cooking/processing purposes and nutritional and health standards for food uses. These include sensitivity in responding to management practices and environmental conditions so as to resist formation of defective tubers (greening and sunburn, knobs, growth cracks, cuts and bruises, hollow heart, brown center, disease lesions, etc.), response of tuber dormancy and sprouting to temperature regimes in storage over a long period, nutritional traits (proteins, vitamins, minerals), and glycoalkaloids (see Tarn et al., 1992; Dale and Mackay, 1994). These traits, while subjects for specific genetic studies, are each screened according to a predetermined threshold at a later stage in the breeding program, but not as targeted ones in the mating designs.

Potatoes are also consumed as various types of products such as canned and prepeeled potatoes, granules and flakes from dehydrated mashed potatoes, noodles with potato flour, etc. Again, these foods are not specifically targeted in a breeding program.

## *GENETIC IMPROVEMENT OF QUALITY TRAITS*

The strategies of selection for breeding lines with superior qualities depend a great deal on the knowledge of genetic mechanisms and technologies of measuring the various traits. The genetics of quality traits have most recently been reviewed by Tarn et al. (1992), Dale and Mackay (1994), and Kerlan and Ellissèche (2000). Most culinary and processing quality traits are metric in nature; hence breeding methods are based on the principles of quantitative genetics. Research is also underway on mapping of the genome and identification of gene loci for specific quality traits.

### *Texture*

Texture is a key component of cooked potato quality, typically assessed as mouth feel by sensory evaluation panels and frequently used to group cultivars by quality. It is a complex trait determined by structural and biochemical properties of tuber tissue, with some environmental influence. However, genotype is considered to play a significant role. These broader aspects of texture have been reviewed by Storey and Davies (1992).

The work of Killick (1977), which identified only small differences in texture attributable to specific combining ability among progenies from a half-diallel with six parents, remains one of the few genetic studies of this trait. A useful step in identifying the components of texture is the use of principal component analysis. Van Marle et al.

(1997) identified the descriptors mealy/crumbly, waxy, and firm as the major properties determining the texture, but concluded that dry matter remains the most practical property to determine texture with respect to mealiness. Dale and Mackay (1994) noted that more research on the components of this complex trait is required before the genetics of this trait is analyzed.

### *Dry Matter*

Dry matter is important for processing and cooking quality. Specific gravity is closely related to dry matter and is extensively used as an estimate of quality (Storey and Davies, 1992). Killick (1977) reported the importance of specific combining ability for specific gravity in tetraploid lines, whereas Tai (1976) and Henninger et al. (2000) found heritability for tuber dry matter in tetraploid potatoes to be moderate to high. Studying two categories of tuber size, 64-83 mm and > 83 mm, Henninger et al. (2000) reported broad sense heritabilities for specific gravity as 0.92 (95% confidence intervals 0.85, 0.97) and 0.77 (0.54, 0.91), respectively.

Specific gravity has been the focus of a series of studies by Haynes based on a long-day adapted hybrid diploid population derived from *Solanum tuberosum* Group Phureja and Group Stenotomum. Haynes et al. (1989) grew parents and populations in several high temperature environments in North Carolina and obtained 24 estimates of narrow sense heritability. Sixteen of these were unbiased with parents and offspring grown at different locations and different years, and the mean narrow sense heritability was 0.28 with a range of −0.07 to 0.77. The remaining three groups of estimates were biased and in these heritabilities were 0.57 (range 0.51-0.63), 0.38 (range 0.29-0.46), and 0.41 (range 0.40-0.42). In a further study with this material, Haynes et al. (1995) investigated the broad sense heritability and stability of specific gravity in the cooler climate of the northeast United States. In this study they obtained a broad sense heritability of 0.66, with a 95 percent confidence interval of 0.50-0.78. They found a majority of the genotypes to be stable with respect to specific gravity. They also concluded that in this population the additive and nonadditive genetic variances were of similar magnitude. A third study with this material involved two cycles of recurrent selection in which Haynes (2001) estimated narrow sense heritabilities as 0.37 ± 0.25 and 0.43 ± 0.27 for the first and second selection cycles, respectively. Gaur et al. (1983) reported that both additive and nonadditive gene actions were equally important for tuber dry matter in Tuberosum potatoes, whereas in Andigena potatoes the nonadditive component was found to be more important by Kumar and Gopal (2003).

Freyre and Douches (1994) proposed a model for marker-assisted selection for specific gravity. Using data across several environments and involving nine loci their model could explain 37 to 57 percent of the variation in specific gravity. Schäfer-Pregl et al. (1998) identified 18 putative QTLs for tuber starch content and a few of these were highly stable. High dry matter (specific gravity) has been successfully transferred from diploid germplasm to tetraploid progeny using 2*n* pollen producing diploids. Wannamaker and Collins (1992) successfully transferred high dry matter to the progeny of 7 tetraploid breeding lines and 15 diploids from Haynes' Phureja × Stenotomum hybrid population. Sterrett et al. (2003) were also successful with a similar set of crosses in transmitting both high and stable specific gravity to tetraploid progeny.

A low amount of variation in specific gravity within and between tubers of a given cultivar is required by processors to obtain products with consistent texture. Love and Pavek (1991) reported that selection for high specific gravity could be achieved without a corresponding increase in variability. Werner et al. (1998), in a study of the relationship between mean specific gravity and standard deviation, found differences among the cultivars, but concluded that there was not sufficient evidence to show that higher specific gravity was associated with greater variability.

### *Sugars and Cold-Induced Sweetening*

Sugars are simple carbohydrates. There are three major types of sugars in potatoes: the disaccharide sucrose and two monosaccharides—D-glucose and D-fructose. Sucrose is the nonreducing sugar and the monosaccharides are hexose-reducing sugars. Sugars are synthesized in the leaves during growth, transported to the tubers, and converted into starch through enzymatic reactions. During tuber bulking, increase in the tuber size is accompanied by a gradual reduction of sugar content and an increase in starch and other polysaccharides. The reducing sugars are at their lowest levels at tuber maturity. Tuber sugar content continues to change in storage after harvest. Reducing sugar content of 0.1 percent is ideal for processing and unacceptable when over 0.33 percent (see Dale and Mackay, 1994; also Chapter 13 "Storage").

The sugar content of tubers is a genotypic characteristic, that is, it varies from cultivar to cultivar (Dale and Mackay, 1994; Tai and Coleman, 1999). Genotypes by environment interactions, however, are of considerable magnitude (Tai and Coleman, 1999). The environmental and management stresses experienced by the tubers, aside from long-term storage conditions, include those induced by short periods of chilling and warming temperatures, moisture, and handling during storage (Sowokinos et al., 1987). Senescent sweetening of potatoes is caused by extended storage at temperatures above 10°C. The loss of color resulting from this process is irreversible. Senescent sweetening most likely occurs when the tubers have short dormancy and thus sprout sooner.

The distribution of sugars varies between tissues and positions within the tuber. Heat or moisture stress during tuber development may cause the so-called sugar end development (Eldredge et al., 1996; Sowokinos et al., 2000). During stress, the sugars are transferred to the stem end to keep it from drying out. This creates a visible difference in stem end color after cooking.

Sowokinos (1978) pointed out that the level and availability of sucrose at harvest is the critical factor affecting the initial rate of reducing sugar formation and thus the quality of potatoes following storage. A sucrose rating value (SR = mg sucrose$\cdot g^{-1}$ tuber FW) was proposed by Sowokinos to measure the potential influence of harvest sucrose content on processing maturity and storage life of tubers. Nelson and Sowokinos (1983) indicated that sucrose levels measured during early tuber development, when differences in sucrose concentration are greatest among cultivars, may be used to predict their relative chipping quality following long-term storage. Herrman et al. (1996) observed high negative correlations between chip quality and low reducing sugars for tubers stored in the cold (4°C) over six months. Copp et al. (2000) reported that tuber respiration rate provided a more reliable method to monitor or predict chip color than contents of either sucrose or reducing sugars during long-term storage.

Reconditioning is a process in which potatoes stored at low temperatures are subjected to 20-30°C and 85-90 percent RH conditions for up to six weeks. Acceleration of the rate of respiration in a reconditioning environment decreases the level of reducing sugars in tubers (Talburt et al., 1987). Cultivars differ in their ability to respond to reconditioning (Walsh, 1995).

Studies have been made on inheritance (Douches and Freyre, 1994; Thill and Peloquin, 1994) and biochemical pathways of cold chipping (Sowokinos, 2001b), and introgression of cold chipping quality into cultivars through interspecific crosses and ploidy manipulation (Ortiz, 1998; Hayes and Thill, 2002a,b). Thill and Peloquin (1994) reported that three key loci were involved in the process of cold sweetening in potatoes. According to Lynch et al. (2003) two independent loci control the response of fry color to low temperature storage. Tetraploid *S. tuberosum* clones with low-temperature sugar stability were reported to exist (Brown et al., 1990; Mackay et al., 1990). The heritabilities of sugar contents were reported to be moderate to high (Tai, 1976; Pereira et al., 1994; Xiong et al., 2002).

Genetic modification of carbohydrate metabolism and transport in developing and stored potatoes may lead to the control of the regulation process of cold sweetening (Sowokinos, 2001a). Carbohydrate metabolism in developing and stored potatoes, especially cold-induced sweetening in stored potatoes, has been extensively investigated. Sowokinos (2001b) presented a scheme for the five related carbohydrate pathways in potato tubers: (a) starch synthesis, (b) starch breakdown, (c) glycolysis, (d) hexogenesis, and (e) mitochondrial respiration. The hexogenic pathway is implicated as the major cold-labile step leading to the formation of reducing sugars in potato tubers. Among enzymes involved in hexogenic reactions, UDPase (UDP-Glucose Pyrophosphorylase) catalyzes the first step committed to the sweetening pathway (Sowokinos, 2001a,b). UGPase is a single-copy gene and two alleles, *UgpA* and *UgpB*, have been described in potatoes. The allelic ratio *UgpA:UgpB* affects the degree of cold resistance to sweetening in tetraploid cultivars. The tetraploid potatoes contain four different alleles for the gene. Tetraploid cultivars with a degree of cold resistance to sweetening showed a ratio of the alleles *Ugp*A:*Ugp*B of 4:0 or 3:1 in favor of *Ugp*A, whereas the cold sensitive cultivars exhibited 1:3 or 0:4 ratio in favor of *Ugp*B. Five isozymes of UGPase have been identified in potatoes (UGP1 to UGP5). Most of the cold resistant cultivars contain UGP4 and UGP5, which are the only isozymes with fast-moving kinetic properties. Other molecular, biochemical, and compositional properties are also involved in enhancing the resistance to cold sweetening, and quantitative genes may be responsible for their inheritance.

Gene markers located near quantitative gene loci (QTL) responsible for the inheritance of economic traits can be used to enhance the efficiency of response to selection for resistance to cold sweetening in the breeding program. Douches and Freyre (1994) reported six QTLs for chip color in an interspecific cross of diploid potatoes. Menéndez et al. (2002) from their experiments on two $F_1$ populations of diploid potato to map QTL for cold sweetening reported a total of 24 QTLs for sugar content with 1-3 QTLs per chromosome. QTLs each explaining more than 10 percent of the variability for reducing sugars were located on six linkage groups.

The sugar content in tubers can be determined by three procedures: (1) colorimetric—Anthrone (Sowokinos, 1978) and Dinitrophenol (Ross, 1975) methods have been used to measure sucrose and glucose contents, respectively; (2) chromatographic—a High Performance Liquid Chromatographic (HPLC) method can be used to precisely determine content of all three sugars; and (3) enzymatic—the Yellow Spring Instrument (YSI) employs an immobilized glucose oxidase on a silver membrane that oxidizes glucose to generate electric current. Sucrose is determined after conversion into glucose by invertase. Various methods for the fast measurement of blood glucose have been introduced for research and industrial use in potatoes. These are: Clinitest (Otazu and Secore, 1980), Reflocheck glucose test strips (Louwes and Neele, 1987), and Acu-Check III pocket instrument (Coleman et al., 1993). Misener et al. (1995) developed a small desktop electronic Sugarmeter that uses a blood glucose-testing strip for measuring glucose content. The sugarmeter requires a drop of potato juice and less than 30 sec to measure the sugar content of a sample without limiting the range of readings like the blood glucose instrument.

### Starch

The quality and quantity of starch in potato tuber can be altered by genetic and other factors important for photosynthetic carbon fixation, sucrose transport to the developing tuber, and starch synthesis in the tuber. A molecular map of genes and their allelic variants with a functional role in carbohydrate metabolism and transport has been constructed by Chen et al. (2001).

Potato starch has two components, that is, amylose and amylopectin. The amount of amylose and amylopectin in potato starch varies with cultivars and growth environment. In some mutants, potato starch granules contain nearly 100 percent amylopectin.

Starches with different ratios of amylose to amylopectin offer different functional properties. For example, starch with a high level of amylose results in higher retrogradation after processing than starch containing a high level of amylopectin. Starch retrogradation (see Chapter 14 "Processing and Value Addition") plays a very important role in the quality of starchy food, such as freeze-thaw stability in frozen foods. It leads to syneresis, a separation of the starch gel and water phases. Stabilization of the starch structure is normally achieved by chemical modification to prevent these changes from occurring. High amylose starches are widely used as thickeners, gelling agents, and films because of their rapid setting properties. Since high amylose starches can hamper the penetration of oil, they are used in snack foods or other foods to decrease the fat intake by the consumer. High amylopectin starch shows increased susceptibilities to amylolysis than high amylose starch. The amylopectin starch improves the uniformity, stability, and texture of some food systems. Starch structure, especially amylopectin, is influenced by starch branching and debranching enzymes. A higher degree of branching with more short chains was found in amylose-free potato starch (Kortstee et al., 1998). This starch is extremely freeze-thaw stable and shows no syneresis even after five freeze-thaw cycles (Zheng and Sosulski, 1998).

By changing the levels and activities of starch synthases and starch branching enzymes through genetic modification, potato starches with different ratios of amylose to amylopectin could be obtained. One granule-bound starch synthase (GBSS I), three isoforms of soluble starch synthase, and two isoforms of starch branching enzyme have already been identified in potato. An amylose-free starch with short-chain amylopectin can be produced by simultaneous antisense down-regulation of three starch synthase genes. Antisense inhibition of GBSS I led to amylose-free potato starch (Smith, 1999). However, attempts to increase the amylose level in potato by inhibiting amylopectin synthesis have been much less successful.

The introduction of bacterial ADPGPP (ADP glucose pyrophosphorylase) genes increases the level of starch in potato (Stark et al., 1992). ADP glucose is a common substrate for starch synthase enzymes involved in the synthesis of amylose and amylopectin through addition of the glucosyl moiety of ADP glucose to the nonreducing end of a preexisting starch molecule.

Potato starch granules contain 0.01-0.6 percent (w/w) phosphorus. Although phosphorus is present at very low levels in starch granules, yet it plays a significant role in physicochemical properties of the starch. Potato starch is widely used in manufacture of paper. Phosphorus is in part responsible for the high swelling power and stable-paste properties of potato starch. Because of the charged nature of phosphate monoesters, electrostatic repulsion between molecules increases and the properties of starch gelatinization and pasting change. Different potato cultivars and soil fertilization regimes for cropping result in different phosphorus contents in potato starch (Liu, 1997). By controlling the enzymes responsible for the incorporation of phosphate groups into starch, it is possible to obtain potato starch that has various phosphorus contents, hence the desired functionality for different applications.

## *Nutritional Components*

### *Proteins*

Tuber nitrogen content is composed of soluble and insoluble protein fractions and nonprotein N that has organic and inorganic components. The organic N fraction includes free amino acids and amides (glutamine and asparagine) (Dale and Mackay, 1994). Potato protein is relatively high in lysine and low in the sulfur-containing amino acids (methionine, cystine) compared to cereals (Woolfe, 1987). Protein concentration is higher in the cortex, including the peel, than in the tuber pith. However, because of its large volume, the pith contains about 71 percent of the total tuber protein (Munshi and Mondy, 1989).

Reported cultivar differences in tuber protein content range from 3.5 to 23 percent on dry weight basis (Dale and Mackay, 1994). Sotelo et al. (1998) studied four Mexican potato species and found crude protein content to range between 8.50 and 11.27 percent compared to 10.00 percent in cultivar Alpha. However, three species gave a protein N percentage ranging from 51.76 to 59.03, compared to 89.28 percent in *S. ehrenbergii* and 95.0 percent in Alpha.

The major potato tuber protein is patatin, which accounts for 40 percent of the protein content and is generally considered to be a storage protein. Because of its structural similarities with pollen allergens, patatin is considered to be the most important allergen in raw potatoes (Koppelman et al., 2002). Patatin has been considered to be a heat-labile allergen. Aggregation of patatin occurs on heating, especially in the presence of other potato proteins, resulting in up to 110-fold decrease in allergenicity.

Several approaches to engineering changes in tuber protein content have been reported. Methionine content was increased by 2.66 times in tubers from mature plants following selection of Russet Burbank protoplasts on a medium containing the amino acid analogue ethionine (Langille et al., 1998). A tuber methionine increase of up to 30-fold was obtained in *S. tuberosum* cv Désirée expressing a threonine synthase antisense construct (Zeh et al., 2001). Lines with low to medium inhibition showed marginal changes of phenotype whereas the line showing the strongest inhibition had severe growth retardation and tuber yield reduction.

A 35-45 percent increase in total protein and an increase in most essential amino acids was achieved through the introduction of the amaranth seed albumin gene (AmA1) from *Amaranthus hypochondriacus* into a late blight resistant diploid genotype, A16 (Chakraborty et al., 2000). In lines with higher levels of gene expression tuber number was increased over 2-fold and yield up to 3.5-fold. Protein content ranged from 14.57 to 16.58 mg·g$^{-1}$ tuber compared to 11.15 for the A16 clone.

### *Vitamins*

Ascorbic acid (vitamin C) is the major vitamin present in potato. Munshi and Mondy (1989) have shown levels to be significantly higher in tuber pith than in the cortex. Sinden et al. (1978) found significant differences for ascorbic acid content between 98 Tuberosum genotypes. They also reported differences between 10 wild species. Losses in storage were highly correlated with initial levels. They estimated narrow sense heritability to be 0.45 and suggested that ascorbic acid levels could be enhanced by breeding and selection. Inheritance of ascorbic acid was considered to be polygenic.

Love et al. (2004) evaluated 75 clones from 12 North American breeding programs and found that ascorbic acid content was distributed continuously between 11.5 and 29.8 mg·100 g$^{-1}$. Mean values were similar for 11 of the breeding programs (the twelfth was represented only by two high content clones), which was considered not to be a surprise since there was no selection for the trait. A study with a subsample of this material showed variation in expression of stability. The observed high variability with a high genotypic component and a high (0.71) broad sense heritability suggested that rapid breeding gains could be expected. Low G × E interaction suggested that a single location would be adequate while AMMI analysis suggested years to be more important than locations in explaining variability. Davies et al. (2002) reported a 24-chromosome Phureja-haploid Tuberosum hybrid that retained twofold higher ascorbic acid content after storage at 5°C. A survey of progeny of this clone showed significant family and ploidy effects on tuber ascorbic acid levels during cold storage consistent with transmission of higher ascorbic acid by FDR 2*n*-pollen. These results were considered encouraging for the further study of 2*n*-pollen and wild species germplasm to breed for increased ascorbic acid levels. Whereas genetic evidence indicates potential for progress, location, cultivation methods, and storage also influence ascorbic acid levels (Storey and Davies, 1992).

*Antioxidants*

Several constituents of potato tubers, such as polyphenols (including anthocyanins), carotenoids, and organic acids, have antioxidant activity and can be significant sources of antioxidants in human nutrition (Al-Saikhan et al., 1995). Friedman (1997) has provided a comprehensive review of the chemistry, biochemistry, and dietary role of potato polyphenols.

In white-fleshed potatoes, chlorogenic acid constitutes up to 90 percent of the total tuber phenolic content, and most of the phenols are distributed between the cortex and the skin (Friedman, 1997). Anthocyanins are of particular interest for their red and blue pigmentation. While they occur most commonly in tuber skin, flesh with anthocyanins in various distribution patterns, rings, arcs and radiating stars, are also observed, particularly in native cultivars of the Andes (Brown et al., 2003). Distribution of anthocyanins in tuber flesh is controlled by a single dominant gene, *Pf*, closely linked with gene *I* and dependent on the pigment production genes *P* and *R* (De Jong, 1987, also see Chapter 2, "Genetics of Agrihorticultural Traits"). Brown et al. (2003) reported a multigenic control for the degree of flesh pigmentation. Between clones a fivefold difference in total anthocyanin levels has been found. In red-fleshed clones anthocyanin values ranged from 6.9 to 35.0 $mg \cdot 100\ g^{-1}$ FW, whereas in purple-fleshed clones they ranged from 5.5 to 17.1 $mg \cdot 100\ g^{-1}$ FW. Antioxidant measurements had high values of 25.0 for Oxygen Radical Absorbance Capacity and 12.5 for Ferrous Radical Absorbance Plasma. The authors suggested that potential exists to increase anthocyanin and antioxidant levels beyond those found in their study.

The carotenoids violaxanthin and lutein are the major determinants of yellow flesh pigmentation in tetraploid (Iwanzik et al., 1983, Lu et al., 2001, Nesterenko and Sink, 2003) and diploid (Lu et al., 2001) cultivated potatoes. Total carotenoid levels range between 343 $\mu g \cdot 100\ g^{-1}$ FW in tetraploid cultivars (Iwanzik et al., 1983) to 1,435 $\mu g \cdot 100\ g^{-1}$ FW in diploid potatoes (Lu et al., 2001), with the highest concentration close to the skin and lowest in the flesh (Iwanzik et al., 1983). Total carotenoid levels are at a maximum early in tuber development and decrease as dry weight increases. During storage total carotenoid levels decrease only slightly while there is a shift in relative amounts of individual carotenoids (Morris et al., 2004). The intensity of yellow flesh correlates better with total carotenoid content than with individual carotenoids, a relationship that was first considered to be linear (Iwanzik et al., 1983) but later proved to be exponential (Lu et al., 2001). An inverse relationship has been reported between tuber weight and yellow-flesh intensity (Haynes et al., 1994). Haynes et al. (1996) reported that clone × environment interaction was significant for yellow flesh but too small to be of practical importance. Broad sense heritability was found to be 0.93 (Haynes et al., 1996) and narrow sense heritability to be 0.99 (Haynes, 2000), which is consistent with control of yellow flesh by the single dominant gene *Y*.

Zeaxanthin is the principal carotenoid leading to orange flesh (Nesterenko and Sink, 2003). Its content was up to 2,055 $\mu g \cdot 100\ g^{1}$ FW in diploid populations derived from *S. phureja* and *S. stenotomum* (Brown et al., 1993). It is controlled by the allele *Or* at the *Y* locus which is dominant over both *Y* and *y* (Brown et al., 1993). Römer et al. (2002) successfully increased the amount of zeaxanthin and total carotenoids by down-regulation of the zeaxanthin epoxidase gene in two cultivars tested.

### Flavor

Flavor of raw, boiled, baked, and various processed potato products has been reviewed by Maga (1994). There is considerable literature on compounds that contribute to flavor. Included are amino acids and sugars that can be precursors of volatiles produced during heating, nucleotides that can act as flavor potentiators, various organic acids, and lipids that through oxidation reactions have the potential to produce off-flavors (Maga, 1994).

There are few studies comparing the variation of volatile flavor components among cultivars. In breeding programs flavor is normally evaluated by organoleptic tests by sensory evaluation panels usually conducted to eliminate off-flavors rather than to select for specific potato flavors. Since flavor is an individual preference, it has been suggested that selection for a characteristic potato flavor would be undesirable (Dale and Mackay, 1994; Kerlan and Ellissèche, 2000).

Jensen et al. (1999) found significant differences among four cultivars in the content of potato off-flavor (POF) compounds. They suggested that breeders could contribute to reduced development of POF by breeding cultivars with low levels of enzyme activity and unsaturated fatty acids. Sensory profiles and instrumental aroma analysis of boiled potato flavor were studied by Ulrich et al. (2000). These authors identified seven character impact compounds and found three cultivars to have similar basic patterns for intensive odor notes, but noted clear differences in substances associated with off-flavor peaks. Identification of a few essential compounds and simplified methods of analysis are required before instrumental sensory analysis can be applied to breeding.

Oruna-Concha et al. (2001) found quantitative and qualitative differences in the volatile composition of skin and flesh of baked potatoes of four cultivars. Aroma values for key volatiles confirmed the effect of cultivar and location on flavor. Duckham et al. (2001) extended this research to a wider range of cultivars, again showing significant differences among cultivars. In a comparison of four cultivars baked conventionally and by microwave, Maga and Twomey (1977) found differences in the aroma and flavor not only between cultivars but also between baking methods. Oruna-Concha et al. (2002) identified cultivar-specific volatile compound profiles for eight cultivars following microwave baking.

### *Glycoalkaloids*

Glycoalkaloids, nitrogen-containing steroidal glycosides, are found throughout potato plants (also see Chapter 14 "Processing and

Value Addition"). The highest levels are in flowers, unripe berries, young leaves, and sprouts, and they are considered to confer a degree of protection against various potato pathogens and pests. In tubers the glycoalkaloids are generally at much lower levels and are concentrated in the skin (Wünsch and Munzert, 1994), and in cultivars they are lower than in most wild species (Van Gelder et al., 1988). There are a number of steroidal glycosides, but those of most common concern are the solanidine glycosides α-chaconine and the somewhat less toxic α-solanine.

Glycoalkaloids impart a bitter taste and can be toxic above threshold levels (Smith et al., 1996; Friedman and McDonald, 1997). This has led to an official guideline in many countries that glycoalkaloid levels in tubers of commercial cultivars should be less than 200 $mg \cdot kg^{-1}$ FW (Friedman and McDonald, 1997). However, Parnell et al. (1984) recommend a glycoalkaloid level of less than 60 $mg \cdot kg^{-1}$ as a guideline for breeders submitting new cultivars for testing in the United Kingdom. Owing to the influences of postharvest conditions on glycoalkaloid levels, determining the value for a cultivar solely by an analysis at harvest is not considered adequate to protect consumer safety. Cultivars should have both an initial low glycoalkaloid concentration and a low rate of increase in glycoalkaloids in the environments that induce changes (Griffiths et al., 1997; Percival, 1999). There are two reports of cultivars with high glycoalkaloid levels being withdrawn from cultivation. The case of withdrawal of cultivar Lenape from production in the United States is well known (Sinden and Webb, 1972) and, more recently, Magnum Bonum was removed from the National List in Sweden following reports of glycoalkaloid levels as high as 665 $mg \cdot kg^{-1}$ fresh weight (Hellenäs et al., 1995). Several authors have reviewed the health safety aspects of glycoalkaloids, and discussed the implications for breeding and cultivar testing (Smith et al., 1996; Valkonen et al., 1996; Friedman and McDonald, 1997).

The amount of total glycoalkaloids and the proportions of α-solanine and α-chaconine vary with cultivar and are primarily determined by genetic factors. Tuber glycoalkaloid levels are also influenced by maturity, tuber size, and other physiological factors, as well as crop management, tuber damage, temperature, light exposure, and storage history (Dale and Mackay, 1994; Valkonen et al., 1996; Friedman and McDonald, 1997). Cultivar × environment interaction is important for many of these factors. For breeders it is particularly noteworthy that the greatest increases in glycoalkaloid levels occur at 4°C and the least at 10°C, with significant differences among cultivars (Griffiths et al., 1997). Tubers transferred to colder conditions after initial storage at 10°C accumulated glycoalkaloids at a lower rate than when directly placed in cold storage following harvest. This rapid accumulation of glycoalkaloids at low storage temperatures was confirmed in a second study (Griffiths et al., 1998). Nitithamyong et al. (1999) found temperature to have a larger effect on glycoalkaloid levels than light intensity, day length, carbon dioxide concentration, or humidity.

Wünsch and Munzert (1994) found no significant change in glycolakloid levels of five cultivars over six months of storage. Griffiths et al. (1994) too found no significant effect of three months storage on glycoalkaloid content of the stored tubers. However, increases in alkaloid levels with increase in length of storage were reported by Percival et al. (1993) and Griffiths et al. (1997).

Exposure of tubers to light results in increased amounts of chlorophyll as well as increased levels of glycoalkaloids. However, there has been disagreement over the relationship between chlorophyll and glycoalkaloid accumulation. Griffiths et al. (1994) found no relationship between photo-induced increases in glycoalkaloids and chlorophyll. Edwards et al. (1998) reported that chlorophyll synthesis inhibitors that reduced greening had no effect on alkaloids. On the other hand, Percival (1999) found a strong relationship between glycoalkaloid and chlorophyll accumulation in response to light. Dale et al. (1997) also found a relationship between synthesis of glycoalkaloids and chlorophyll while studying the impact of gamma radiation on tubers.

Johns and Alonso (1990) studied 12 species of potato that were either reputed progenitors of the cultivated potato or that have been in cultivation themselves. The Mexican species *S. cardiophyllum* and *S. ehrenbergii* had glycoalkaloid levels $< 6$ mg·100 $g^{-1}$ FW, whereas levels ranged from 23 to 164 mg·100 $g^{-1}$ FW in accessions of other species. Van Gelder et al. (1988) surveyed glycoalkaloids in tubers of 14 *Solanum* species grown in a greenhouse and levels varied between

123 and 7,348 mg·kg fresh weight. Cultivars grown under such conditions had glycoalkaloid levels two to three times those of field-grown tubers. In addition to the solanidine glycosides (α-solanine and α-chaconine), these authors also reported the presence of demissidine and tomatidine glycosides in *S. acaule,* and solasodine glycosides in *S. berthaultii* and *S. vernei.* They also found unidentified steroidal type alkaloids in many wild species. Van Gelder and Scheffer (1991) reported that *S. vernei* had high levels of steroidal alkaloids with levels influenced by the growing conditions. Hybrids of *S. vernei* with a diploid *S. tuberosum* clone had lower levels of solanidine and solasodine compared to the *S. vernei* parent. However, even after repeated backcrossing, species derived cultivars still had potentially toxic concentrations of alkaloids. Mattheij et al. (1992) reported that somatic hybrids of *S. circaeifolium* and 2*x S. tuberosum,* contained the novel glycoside demissidine in addition to the solanidine glycoside from the *S. tuberosum* parent and tomatidine and tomatidenol glycosides from the *S. circaeifolium* parent. Louwes et al. (1992) had reported similar results for sexual hybrids between these species. Also working with somatic hybrids, Kozukue et al. (1999) found that tuber cortex of 11 of 12 hybrids between tetraploid *S. tuberosum* and dihaploid *S. acaule* possessed the glycoalkaloids from both fusion parents: α-chaconine and α-solanine from *S. tuberosum* and α-tomatine and demissidine from *S. acaule.* The lack of one alkaloid in one clone was attributed to somaclonal variation. Glycoalkaloid levels in interspecific hybrids were mostly intermediate between those of the parents. *S. chacoense,* a wild species that has been used extensively in breeding for resistance to insect pests, has high glycoalkaloid content. It is the parent of the cultivar Lenape that was withdrawn because of its high glycoalkaloids levels (Sinden and Webb, 1972).

Sanford et al. (1995) reported the glycoalkaloid content in hybrids and backcross populations of *S. tuberosum* and *S. chacoense.* The glycoalkaloid levels of parent genotypes from *S. chacoense* and *S. tuberosum* were 245 and 4 mg·100 $g^{-1}$ FW, respectively. The $F_2$ and $F_4$ population mean values of glycoalkaloids were 39 and 35 mg·100 $g^{-1}$ FW, respectively. Levels were reduced further in backcrosses of $F_2$ genotypes to *S. tuberosum,* and $BC_1$ and $BC_2$

populations had mean total glyoalkaloid levels of 15 and 9 mg·100 $g^{-1}$ FW, respectively. There was considerable variation in levels in all four populations so that a small proportion of selections in the $F_2$ and $F_4$ populations, and a majority of the selections in the two backcross populations, had values below 20 mg·100$g^{-1}$ FW.

Moehs et al. (1997) cloned the gene for solanidine UDP-glucose glucosyltransferase, an enzyme that catalyzes the synthesis of γ-chaconine. This construct has been used as antisense to produce transformed Lenape and Désirée lines with significantly lower glycoalkaloid levels (McCue et al., 2003). Using a different approach, Arnquist et al. (2003) modified the composition of sterols, which occur in the alkaloid synthetic pathway, by overexpressing an inserted soybean sterol methyltransferase. The resulting transformed potatoes were phenotypically similar to the control cultivar, but with reduced levels of free cholesterol and glycoalkaloid in leaves and tubers.

Yencho et al. (1998) and Ronning et al. (1999) undertook QTL mapping of the compounds involved in the synthetic pathways of solanidine and solasidine, and of leptinidine, respectively. Their results showed that a major locus determining solanidine accumulation in potato is on chromosome 1, or that this chromosome region harbors two or more genes determining accumulation of steroidal aglycones. Though this work was conducted on foliar alkaloids, a moderate correlation has been reported between foliar and tuber alkaloids (Sanford et al., 1996) and gene(s) controlling alkaloid production in leaves and tubers appear to be related (Yencho et al., 1998).

### *Enzymatic Browning*

Enzymatic browning occurs at the surface of cut and peeled tubers and blackspot bruising occurs as a result of damage at harvest or moving potatoes in or out of storage (McGarry et al., 1996). Both of these are the result of the same enzymatic reaction: the oxidation of phenolic substrates by polyphenol oxidase (PPO) to quinines that spontaneously polymerize brown, black, or gray pigments commonly known as melanins. The phenolic substrates, principally tyrosine and chlorogenic acid in potato tubers (Sabba and

Dean, 1994), are localized in the cell vacuole whereas the PPO is localized in plastids. Tissue damage leads to the loss of this compartmentalization of substrate and enzyme and results in browning reaction (Thygesen et al., 1995).

It is generally agreed that blackspot susceptibility is related to the amount of free tyrosine. Corsini et al. (1992) reported a correlation coefficient of 0.90 between blackspot susceptibility and free tyrosine levels in 12 cultivars, but recognized that the degree of susceptibility was influenced by additional undefined factors. Similar results were obtained by Sabba and Dean (1994). Pavek et al. (1993) estimated the heritability of blackspot bruise reaction from a half diallel experiment using an abrasive-peel method for bruising evaluation. Narrow sense heritability was $h^2 = 0.85$. Analysis showed the predominance of additive genetic variance, with 59 percent of the variance attributed to GCA and 9 percent to SCA. No significant reciprocal differences were detected.

Thygesen et al. (1995) isolated from potato tubers five distinct cDNA clones encoding for PPO, and concluded that there is a small multigene family with each gene having a different expression pattern. Bachem et al. (1994) developed a series of antisense PPO constructs and produced transformed clones inhibiting PPO production and enzymic browning. Coetzer et al. (2001) inserted into a very bruise susceptible cultivar, Russet Burbank, sense and antisense constructs of tomato PPO cDNA and an antisense construct of potato PPO cDNA. Transformed lines with tomato constructs produced tubers with significantly reduced levels of browning; greater reduction was with the antisense construct. The lines transformed with the potato PPO construct had no significant reduction of browning. These results were explained on the basis of varying expressions of genes within the PPO gene family and the interactions between them. Taking a different approach, Love et al. (1993) exposed tuber eye-pieces of the cultivar Lemhi Russet to gamma irradiation and screened the resulting lines for increased blackspot bruise resistance. A total of 8 of 5,692 lines had significantly higher bruise resistance, but the level of resistance was not high enough to be commercially valuable. Taylor et al. (1993) examined protoplast derived somaclones of the cultivar Crystal for bruise resistance and found 20 of 33 clones to be more resistant than the original cultivar.

Enzymatic browning has been found to be absent or reduced in tubers of several accessions of wild species (Woodwards and Jackson, 1985). The lack of browning in the Mexican tetraploid, 2EBN species *S. hjertingii* was quantified colorimetrically by Gubb et al. (1989) and no PPO enzyme activity was found in these by Sim et al. (1997). However, the latter authors did identify PPO mRNA and protein and considered it to be structurally altered or lacking a cofactor. Brown et al. (1999) measured PPO activity and melanin formation in four crosses between *S. hjertingii* and cultivated diploids and the parents. In all cases the mean performance of the progeny was closest to the *S. hjertingii* parent, demonstrating nearly complete dominance or overdominance. Culley et al. (2002) crossed *S. hjertingii* with a cultivated diploid to produce triploid hybrids, doubled these to produce hexaploids and crossed a hexaploid with blackspot bruise susceptible Tuberosum cultivars. Tubers of the resulting progeny showed very little browning indicating the potential of this approach for developing blackspot bruise resistant cultivars.

### *After-Cooking Darkening*

After-cooking darkening is the result of a nonenzymatic oxidation that occurs when boiled, baked, fried, or dehydrated potatoes are exposed to air. A ferrous-chlorogenic acid complex is formed during cooking and on exposure to air oxidizes to form a bluish-gray ferri-dichlorogenic compound (Wang-Prusk and Nowak, 2004). Darkening is greater in the cortex than in the pith, greater at the stem end than at the bud end, and increases during storage. The severity of darkening is determined by the ratio of chlorogenic acid to citric acid. Chlorogenic acid levels have been shown to vary with each cultivar whereas citric acid levels are more sensitive to environmental factors (Hughes and Evans, 1967). Iron is considered to be uniformly distributed in the tuber and not to be a determinant in darkening (Wang-Pruski and Nowak, 2004).

There is limited information on the genetics of this trait. Dalianis et al. (1966), studying four groups of material, found significant general combining ability for after-cooking darkening, thereby confirming the value of using parents that showed little or no darkening. Significant additive gene effects for after cooking darkening were also reported by Pika et al. (1984, cited in Dale and Mackay, 1994). Killick (1977) found specific combining ability to be more important than general combining ability. Pika et al. (1984) reported that broadsense heritability ranged between 0.60 and 0.68, and that narrow sense heritability was between 0.33 and 0.63. Success in predicting cross-performance for after-cooking darkening was noted by Caligari (1992) based on the ranking of predicted and observed values for six progenies. Veerman (2001), found that the largest component (32%-40%) of the total variance in after-cooking darkening was attributable to clones. Wang-Pruski et al. (2003) evaluated 48 diploid clones for after-cooking darkening over two years and two locations and found the character to be quite stable.

Through the above account we have tried to demonstrate the progress made in the understanding of the genetics of processing and nutritional traits including proteins, vitamins, and antioxidants, leading to the possibilities for improving the nutritional value of this major world crop. However, the increased knowledge of the genetics of quality has not yet resulted in improvement of cultivars for the production of frozen French fries, as the old cultivar Russet Burbank continues to dominate this sector of potato production. For the French fry market the need is to look for a higher level of expression of uniform, smooth, blocky, oblong to long tubers, with uniform high dry matter within and between tubers and uniform low levels of sugars. On the other hand, considerable improvements have been made in the development of chip cultivars with low sugars even after cold storage (Douches et al., 1996; Love et al., 1998). Knowledge of sugar chemistry and its genetic control should lead to continuing improvements in chip cultivars and to the introduction of the low temperature processing trait into French fry cultivars.

Technologies are now available to better study the flavor with the potential to take the potato beyond being a bland tasting staple food. The germplasm pool of the Andean cultivars offers a rich diversity of textures and flavors, little known outside this region. Progress made at the molecular level is helping the identification of genes and the manipulation of synthetic pathways determining the quality of potatoes. Complex nutritional and quality traits are particularly well suited to the development and use of marker-assisted selection. It is now also conceivable, through the use of antisense inserts, to eliminate tuber glycoalkaloid production and hence overcome health risks resulting from the consumption of potatoes with high glycoalkaloid levels. Potential exists to make considerable further improvements in potato quality, including nutritional quality.

## REFERENCES

Al-Saikhan, M.S., Howard, L.R., and Miller, J.C., Jr. (1995). Antioxidant activity and total phenolics in different genotypes of potato (*Solanum tuberosum* L.). *Journal of Food Science 60:*341-343.

Arnquist, L., Dutta, P.C., Johnsson, L., and Sitbon, F. (2003). Reduction of cholesterol and glycoalkaloid levels in transgenic potato plants by overexpression of a type 1 sterol methyltransferase cDNA. *Plant Physiology 131:*1-8.

Bachem, C.W.B., Speckmann, G.-J., van der Linde, P.C.G., Verheggen, F.T.M., Hunt, M.D., Steffens, J.C., and Zabeau, M. (1994). Antisense expression of polyphenol oxidase genes inhibits enzymatic browning in potato tubers. *Bio/Technology 12:*1101-1105.

Brown, C.R., Edwards, C.G., Yang, C.-P., and Dean, B.B. (1993). Orange flesh trait in potato: Inheritance and carotenoid content. *Journal of the American Society of Horticultural Science 118:*145-150.

Brown, C.R., McNabnay, M., and Dean, B. (1999). Genetic characterization of reduced melanin formation in tuber tissue of *Solanum hjertingii* and hybrids with cultivated diploids. *American Journal of Potato Research 76:*37-43.

Brown, C.R., Wrolstad, R., Durst, R., Yang, C.-P., and Clevidence, B. (2003). Breeding studies in potatoes containing high concentrations of anthocyanins. *Journal of the American Society of Horticultural Science 80:*241-250.

Brown, J., Mackay, G.R., Bain, H., Griffith, D.W., and Allison, M.J. (1990). The processing potential of tubers of the cultivated potato, *Solanum tuberosum* L. after storage at low temperatures. 2. Sugar concentration. *Potato Research 33:*219-227.

Caligari, P.D.S. (1992). Breeding new varieties. In Harris, P.M. (Ed.), *The Potato Crop,* Second edition (pp. 334-372). London, U.K.: Chapman & Hall.

Chakraborty, S., Chakraborty, N., and Datta, A. (2000). Increased nutritive value of transgenic potato by expressing a nonallergenic seed albumin gene from *Amaranthus hypochondriacus. Proceedings of the National Academy of Sciences (USA) 97:*3724-3729.

Chen, X., Salamini, F., and Genhardt, C. (2001). A potato molecular-function map for carbohydrate metabolism and transport. *Theoretical and Applied Genetics 102:*284-295.

Coetzer, C., Corsini, D., Love, S., Pavek, J., and Turner, N. (2001). Control of enzymatic browning in potato *(Solanum tuberosum* L.) by sense and antisense RNA from tomato polyphenol oxidase. *Journal of Agricultural and Food Chemistry 49:*652-657.

Coleman, W.K., Tai, G.C.C., Clayton, S., Howie, M., and Pireirea, A. (1993). A portable monitor for the rapid assessment of processing quality of stored potato tubers. *American Potato Journal 70:*909-923.

Copp, L.J., Blenkinsop, R.W., Yada, R.Y., and Marangoni, A.G. (2000). The relationship between respiration and chip color during long-term storage of potato tubers. *American Journal of Potato Research 77:*279-287.

Corsini, D.L., Pavek, J.J., and Dean, B. (1992). Differences in free and protein-bound tyrosine among potato genotypes and their relationship to internal blackspot resistance. *American Potato Journal 69:*423-435.

Culley, D.E., Dean, B.B., and Brown, C.R. (2002). Introgression of the low browning trait from the wild Mexican species *Solanum hjertingii* into cultivated potato *(S. tuberosum* L.). *Euphytica 125:*293-303.

Dale, M.F.B., Griffiths, D.W., Bain, H., and Goodman, B.A. (1997). The effect of gamma irradiation on glycoalkaloid and chlorophyll synthesis in seven potato cultivars. *Journal of the Science of Food and Agriculture 75:*141-147.

Dale, M.F.B. and Mackay, G.R. (1994). Inheritance of table and processing quality. In Bradshaw, J.E. and Mackay, G.R. (Eds.), *Potato Genetics* (pp. 285-315). Wallingford, Oxon: CAB International.

Dalianis, C.D., Plaisted, R.L., and Peterson, L.C. (1966). Selection for freedom from after cooking darkening in a potato breeding program. *American Potato Journal 43:*207-215.

Davies, C.S., Ottman, M.J., and Peloquin, S.J. (2002). Can germplasm resources be used to increase the ascorbic acid content of stored potatoes? *American Journal of Potato Research 79:*295-299.

De Jong, H. (1987). Inheritance of pigmented tuber flesh in cultivated diploid potatoes. *American Potato Journal 64:*337-343.

Douches, D.S. and Freyre, R. (1994). Identification of genetic factors influencing chip color in diploid potato (*Solanum* spp.). *American Potato Journal 71:*581-590.

Douches, D.S., Maas, D., Jastrzebski, K., and Chase, R.W. (1996). Assessment of potato breeding progress in the USA over the last century. *Crop Science 36:*1544-1552.

Duckham, S.C., Dodson, A.T., Bakker, J., and Ames, J.M. (2001). Volatile flavour components of baked potato flesh. A comparison of eleven potato cultivars. *Nahrung/Food 45:*317-323.

Edwards, E.J., Saint, R.E., and Cobb, A.H. (1998). Is there a link between greening and light-enhanced glycoalkaloid accumulation in potato *(Solanum tuberosum* L.) tubers? *Journal of the Science of Food and Agriculture 76:*327-333.

Eldredge, E.P., Holmes, Z.A., Mosley, A.R., Shock, C.C., and Stieber, T.D. (1996). Effects of transitory water stress on potato tuber stem-end reducing sugar and fry color. *American Potato Journal 73:*517-529.

Freyre, R. and Douches, D.M. (1994). Development of a model for marker-assisted selection of specific gravity in diploid potato across environments. *Crop Science 34:*1361-1368.

Friedman, M. (1997). Chemistry, biochemistry, and dietary role of potato polyphenols. A review. *Journal of Agricultural and Food Chemistry 45:*1523-1540.

Friedman, M. and McDonald, G.M. (1997). Potato Glycoalkaloids: Chemistry, analysis, safety, and plant physiology. *Critical Reviews in Plant Science 16:*55-132.

Gaur, P.C., Gopal, J., and Rana, M.S. (1983). Combining ability for yield, its components and tuber dry matter in potato. *Indian Journal of Agricultural Sciences 53:*876-879.

Griffiths, D.W., Bain, H., and Dale, M.F.B. (1997). The effect of low-temperature storage on the glycoalkaloid content of potato *(Solanum tuberosum)* tubers. *Journal of the Science of Food and Agriculture 74:*301-307.

———. (1998). Effect of storage temperature on potato (*Solanum tuberosum* L.) tuber glycoalkaloid content and the subsequent accumulation of glycoalkaloids and chlorophyll in response to light. *Journal of Agriculture and Food Chemistry 46:*5262-5268.

Griffiths, D.W., Dale, M.F.B., and Bain, H. (1994). The effect of cultivar, maturity and storage on photo-induced changes in the total glycoalkaloid and chlorophyll contents of potatoes *(Solanum tuberosum). Plant Science 98:*103-109.

Gubb, I.R., Hughes, J.C., Jackson, M.T., and Callow, J.A. (1989). The lack of enzymic browning in the wild potato species *Solanum hjertingii* compared with commercial *Solanum tuberosum* varieties. *Annals of Applied Biology 114:*579-586.

Hayes, R.J. and Thill, C.A. (2002a). Selection for potato genotypes from diverse progenies that combine 4°C chipping with acceptable yields, specific gravity and tuber appearance. *Crop Science 42:*1343-1349.

———. (2002b). Introgression of cold (4°C) chipping from 2x (2 endosperm balance number) potato species into 4x(4EBN) cultivated potato using sexual polyploidization. *American Journal of Potato Research 79:*421-431.

Haynes, K.G. (2000). Inheritance of yellow-flesh intensity in diploid potatoes. *Journal of the American Society of Horticultural Science 125:*63-65.

———. (2001). Variance components for yield and specific gravity in a diploid potato population after two cycles of recurrent selection. *American Journal of Potato Research 78:*69-75.

Haynes, K.G., Haynes, F.L., and Henderson, W.R. (1989). Heritability of specific gravity of diploid potato under high temperature growing conditions. *Crop Science 29:*622-625.

Haynes, K.G., Potts, W.E., Chittams, J.L., and Fleck, D.L. (1994). Determining yellow-flesh intensity in potatoes. *Journal of the American Society of Horticultural Science 119:*1057-1059.

Haynes, K.G., Sieczka, J.B., Henninger, M.R., and Fleck, D.L. (1996). Clone x environment interactions for yellow-flesh intensity in tetraploid potatoes. *Journal of the American Society of Horticultural Science 121:*175-177.

Haynes, K.G., Wilson, D.R., and Kang, M.S. (1995). Genotype x environment interactions for specific gravity in diploid potatoes. *Crop Science 35:*977-981.

Hellenäs, K.-E., Branzell, C., Johnsson, H., and Slanina, P. (1995). High levels of glycoalkaloids in the established Swedish potato variety Magnum Bonum. *Journal of the Science of Food and Agriculture 68:*249-255.

Henninger, M.R., Sterrett, S.B., and Haynes, K.G. (2000). Broad-sense heritability and stability of internal heat necrosis and specific gravity in tetraploid potatoes. *Crop Science 40:*977-984.

Herrman, T.J., Love, S.L., Shafii, B., and Dwelle, R.B. (1996). Chipping performance of three processing potato cultivars during long-term storage at two temperature regimes. *American Potato Journal 73:*411-425.

Hughes, J.C. and Evans, J.L. (1967). Studies on after-cooking blackening in potatoes. IV. Field experiments. *European Potato Journal 10:*16-36.

Iwanzik, W., Tevini, M., Stute, R., and Hilbert, R. (1983). Carotinoidgehalt und—zusammensetzung verschiedener deutscher Kartoffelsorten und deren Bedeutung für die Fleischfarbe der Knolle. *Potato Research 26:*149-162.

Jensen, K., Petersen, M.A., Poll, L., and Brockhoff, P.B. (1999). Influence of variety and growing location on the development of off-flavor in precooked vacuum-packed potatoes. *Journal of Agricultural and Food Chemistry 47:*1145-1149.

Johns, T. and Alonso, J.G. (1990). Glycoalkaloid change during the domestication of the potato, *Solanum* Section *Petota. Euphytica 50:*203-210.

Kerlan, M.-C. and Ellissèche, D. (2000). Tuber quality: Objectives in breeding research. *Plant Breeding and Seed Science 44:*21-36.

Killick, R.J. (1977). Genetic analysis of several traits in potatoes by means of a diallel cross. *Annals of Applied Biology 86:*279-289.

Koppelman, S.J., van Koningsveld, G.A., Knulst, A.C., Gruppen, H., Pigmans, I.G.A.J., and de Jongh, H.H.J. (2002). Effect of heat-induced aggregation on the IgE binding of patatin (Sol t 1) is dominated by other potato proteins. *Journal of Agricultural and Food Chemistry 50:*1562-1568.

Kortstee, A.J., Suurs, L.C.J.M., Vermeesch, A.M.G., Keetels, C.J.A.M., Jacobsen, E., and Visser, R.G.F. (1998). The influence of an increased degree of branching on the physico-chemical properties of starch from genetically modified potato. *Carbohydrate Polymers 37:*173-184.

Kozukue, N., Misoo, S., Yamada, T., Kamijima, O., and Friedman, M. (1999). Inheritance of morphological characters and glycoalkaloids in potatoes of somatic hybrids between dihaploid *Solanum acaule* and tetraploid *Solanum tuberosum. Journal of Agricultural and Food Chemistry 47:*4478-4483.

Kumar, R. and Gopal, J. (2003). Combining ability of Andigena accessions for yield components and tuber dry matter in third clonal generation. *Journal of Indian Potato Association 30:*3-4.

Langille, A.R., Lan, Y., and Gustine, D.L. (1998). Seeking improved nutritional properties for the potato: Ethionine-resistant protoclones. *American Journal of Potato Research 75:*201-205.

Liu, Q. (1997). *Characterization of Physico-chemical Properties of Starch from Various Potatoes and Other Sources.* Ph.D. thesis, Laval University, Quebec, Canada.

Louwes, K.M., Hoekstra, R., and Mattheij, W.M. (1992). Interspecific hybridization between the cultivated potato *Solanum tuberosum* subspecies *tuberosum* L. and the wild species *S. circaeifolium* Bitter exhibiting resistance to *Phytophthora infestans* (Mont.) de Bary and *Globodera pallida* (Stone) Behrens. 2. Sexual hybrids. *Theoretical and Applied Genetics 84:*362-370.

Louwes, K.M. and Neele, A.E.F. (1987). Selection for chip quality and specific gravity of potato clones: Possibility for early generation selection. *Potato Research 30:*241-251.

Love, S.L. and Pavek, J.J. (1991). Relationship of clonal mean to the uniformity and stability of tuber specific gravity in potatoes. *American Potato Journal 68:*543-550.

Love, S.L., Pavek, J.J., Thompson-Johns, A., and Bohl, W. (1998). Breeding progress for potato chip quality in North American cultivars. *American Journal of Potato Research 75:*27-36.

Love, S.L., Salaiz, T., Mosley, A.R., and Thornton, R.E. (2004). Stability of expression and concentration of ascorbic acid in North American potato germplasm. *HortScience 39:*156-160.

Love, S.L., Thompson-Johns, A., and Baker, T. (1993). Mutation breeding for resistance to blackspot bruise and low temperature sweetening in the potato cultivar Lemhi Russet. *Euphytica 70:*69-74.

Lu, W., Haynes, K., Wiley, E., and Clevidence, B. (2001). Carotenoid content and color in diploid potatoes. *Journal of the American Society of Horticultural Science 126:*722-726.

Lynch, D.R., Kawchuk, L.M., Yada, R., and Armstrong, J.D. (2003). Inheritance of the response of fry color to low temperature storage. *American Journal of Potato Research 80:*341-344.

Mackay, G.R., Brown, J., and Torrance, C.J.W. (1990). The processing potential of tubers of the cultivated potato, *Solanum tuberosum* L. after storage at low temperatures. 1. Fry color. *Potato Research 33:*211-218.

Maga, J.A. (1994). Potato flavour. *Food Reviews International 10:*1-48.

Maga, J.A. and Twomey, J.A. (1977). Sensory comparison of four potato varieties baked conventionally and by microwaves. *Journal of Food Science 42:*541-542.

Mattheij, W.M., Eijlander, R., de Koning, J.R.A., and Louwes, K.M. (1992). Interspecific hybridization between the cultivated potato *Solanum tuberosum* subspecies *tuberosum* L. and the wild species *S. circaeifolium* Bitter exhibiting resistance to *Phytophthora infestans* (Mont.) de Bary and *Globodera pallida* (Stone) Behrens. 1. Somatic hybrids. *Theoretical and Applied Genetics 83:*459-466.

McCue, K.F., Allen, P.V., Rockhold, D.R., Maccree, M.M., Belknap, W.R., Shepherd, L.V.T., Davies, H., Joyce, P., Corsini, D.L., and Moehs, C.P. (2003). Reduction of total steroidal glycoalkaloids in potato tubers using antisense constructs of a gene encoding a solanidine glucosyl transferase. *Acta-Horticulturae 619:*77-86.

McGarry, A., Hole, C.C., Drew, R.L.K., and Parsons, N. (1996). Internal damage in potato tubers: A critical review. *Postharvest Biology and Technology 8:*239-258.

Menéndez, C.M., Ritter, E., Schäfer-Pregl, R., Walkemeier, B., Kalde, A., Salamini, F., and Gebhardt, C. (2002). Cold sweetening in diploid potato: Mapping quantitative trait loci and candidate genes. *Genetics 162:*1423-1434.

Misener, G.C., Gerber, W.A., Tai, G.C.C., and Embleton, E.J. (1995). Measurement of glucose concentrations of potato extract using a blood glucose test strip. *Canadian Agricultural Engineering 38:*59-62.

Moehs C.P., Allen, P.V., Friedman, M., and Bellknap, W.R. (1997). Cloning and expression of solanidine UDP-glucose glucosyltransferase from potato. *The Plant Journal 11:*227-236.

Morris, W.L., Ducreux, L., Griffiths, D.W., Stewart, D., Davies, H.V., and Taylor, M.A. (2004). Carotenogenesis during tuber development and storage in potato. *Journal of Experimental Botany 55:*975-982.

Munsrhi, C.B. and Mondy, N.I. (1989). Ascorbic acid and protein content of potatoes in relation to tuber anatomy. *Journal of Food Science 54:*220-221.

Nelson, D.C. and Sowokinos, J.R. (1983). Yield and relationships among tuber size, sucrose and chip color in six potato cultivars on various harvest dates. *American Potato Journal 60:*949-958.

Nesterenko, S. and Sink, K.C. (2003). Carotenoid profiles of potato breeding lines and selected cultivars. *HortScience 38:*1173-1177.

Nitithamyong, A., Vonelbe, J.H., Wheeler, R.M., and Tibbitts, T.W. (1999). Glycoalkaloids in potato tubers grown under controlled environments. *American Journal of Potato Research 76:*337-343.

Ortiz, R. (1998). Potato breeding via ploidy manipulations. *Plant Breeding Reviews 16:*15-86.

Oruna-Concha, M.J., Bakker, J., and Ames, J.M. (2002). Comparison of the volatile components of eight cultivars of potato after microwave baking. *Lebensmittel-Wissenschaft und -Technologie 35:*80-86.

Oruna-Concha, M.J., Duckham, S.C., and Ames, J.M. (2001). Comparison of volatile compounds isolated from the skin and flesh of potato cultivars after baking. *Journal of Agricultural and Food Chemistry 49:*2414-2421.

Otazu, V. and Secor, G.A. (1980). Clinitest: A simple technique for estimation of reducing sugar content of potatoes. *American Potato Journal 57:*15-20.

Parnell, A., Bhuva, V.S., and Bintcliffe, E.J.B. (1984). The glyocalkaloid content of potato varieties. *Journal of the National Institute of Agricultural Botany 16:*535-541.

Pavek, J.J., Brown, C.R., Martin, M.W., and Corsini, D.L. (1993). Inheritance of blackspot bruise resistance in potato. *American Potato Journal 70:*43-48.

Percival, G. (1999). The influence of light upon glycoalkaloid and chlorophyll accumulation in potato tubers *(Solanum tuberosum* L.). *Plant Science 145:*99-107.

Percival, G.C., Harrison, J.A.C., and Dixon, G.R. (1993). The influence of temperature on light enhanced glycoalkaloid synthesis in potato. *Annals of Applied Biology 123:*141-153.

Pereira, A. da S., Tai, G.C.C., Yada, R.Y., Coffin, R.H., and Sousa-Machado, V. (1994). Potential for improvement by selection for reducing sugar content after cold storage for three potato populations. *Theoretical and Applied Genetics 88:*678-684.

Pika, N.A., Tarasenko, V.A., and Mitsko, V.N. (1984). Combining ability of potato varieties and hybrids for tuber flesh blackening after cooking. *Selektsiia-i-Semenorodstro,* USSR *7:*16-17.

Römer, S., Lübeck, J., Kauder, F., Steiger, S., Adomat, C., and Sandmann, G. (2002). Genetic engineering of a zeaxanthin-rich potato by antisense inactivation and co-suppression of carotenoid epoxidation. *Metabolic Engineering 4:*263-272.

Ronning, C.M., Stommel, J.R., Kowalski, S.P., Sanford, L.L., Kobayashi, R.S., and Pineada, O. (1999). Identification of molecular markers associated with leptine production in a population of *Solanum chacoense* Bitter. *Theoretical and Applied Genetics 98:*39-46.

Ross, A.G. (1975). Dinitrophenol method for reducing sugars. In Talburt, W.F., and Smith, O. (Eds.), *Potato processing* Third edition (pp. 696-697). Westport, CT: The Avi Publishing Co., Inc.

Sabba, R.P. and Dean, B.B. (1994). Sources of tyrosine in genotypes of *Solanum tuberosum* L. differing in capacity to produce melanin pigments. *Journal of the American Society of Horticultural Science 119:*770-774.

Sanford, L.L., Deahl, K.L., Sinden, S.L., and Kobayashi, R.S. (1995). Glycoalkaloid content in tubers of hybrid and backcross populations from a *Solanum tuberosum* x *S. chacoense* cross. *American Potato Journal 72:*261-271.

Sanford, L.L., Kobayashi, R.S., Deahl, K.L., and Sinden, S.L. (1996). Segregation of leptines and other glycoalkaloids in *Solanum tuberosum* (4x) x *S. chacoense* (4x) crosses. *American Potato Journal 73:*21-33.

Schäfer-Pregl, R., Ritter, E., Concilio, L., Hesselbach, J., Lovatti, L., Walkemeier, B., Thelen, H., Salamini, F., and Genhardt, C. (1998). Analysis of quantitative trait loci (QTLs) and quantitative trait alleles (QTAs) for potato tuber yield and starch content. *Theoretical and Applied Genetics 97:*834-846.

Sim, S.K., Ohmann, S.M., and Tong, C.B.S. (1997). Comparison of polyphenol oxidase in the tubers of *Solanum tuberosum* and the non-browning tubers of *S. hjertingii. American Potato Journal 74:*1-13.

Sinden, S.L. and Webb, R.E. (1972). The effect of variety and location on the glycoalkaloid content of potatoes. *American Potato Journal 49:*334-338.

Sinden, S.L., Webb, R.E., and Sandford, L.L. (1978). Genetic potential for increasing ascorbic acid content in potatoes. *American Potato Journal 55:*394-395.

Smith, A.M. (1999). Making starch. *Current Opinion in Plant Biology 2:*223-229.

Smith, D.B., Roddick, J.G., and Jones, J.L. (1996). Potato glycoalkaloids: Some unanswered questions. *Trends in Food Science and Technology 7:*126-131.

Sotelo, A., Contreras, E., Sousa, H., and Hernandez, V. (1998). Nutrient composition and toxic factor content of four wild species of Mexican potato. *Journal of Agricultural and Food Chemistry 46:*1355-1358.

Sowokinos, J.R. (1978). Relationship of harvest sucrose content to processing maturity and storage life of potatoes. *American Potato Journal 55:*333-344.

———. (2001a). Pyrophosphorylase in *Solanum tuberosum* L.: Allelic and isozyme patterns of UDP-Glucose pyrophosphorylase as a marker for cold-sweetening resistance in potatoes. *American Journal of Potato Research 78:*57-64.

———. (2001b). Biochemical and molecular control of cold-induced sweetening in potatoes. *American Journal of Potato Research 78:*221-236.

Sowokinos, J.R., Orr, P.H., Knoper, J.A., and Varns, J.L. (1987). Influence of potato storage and handling stress on sugars, chip quality and integrity of the starch (amyloplast) membrane. *American Potato Journal 64:*213-226.

Sowokinos, J.R., Shock, C.C., Stieber, T.D., and Eldredge, E.P. (2000). Compositional and enzymatic changes associated with the sugar-end defect in Russet Burbank potatoes. *American Journal of Potato Research 77:*47-56.

Stark, D.M., Timmerman, K.P., Barry, G.I., Preiss, J.P., and Kishore, G.M. (1992). Regulation of the amount of starch in tissues by ADP-glucose pyrophosphorylase. *Science 258:*287-292.

Sterrett, S.B., Henninger, M.R., Yencho, G.C., Lu, W., Vinyard, B.T., and Haynes, K.G. (2003). Stability of internal heat necrosis and specific gravity in tetraploid x diploid potatoes. *Crop Science 43:*790-796.

Storey, R.M.J. and Davies, H.V. (1992). Tuber quality. In Harris, P.M. (Ed.), *The Potato Crop,* Second edition (pp. 507-569). London, U.K.: Chapman & Hall.

Tai, G.C.C. (1976). Estimation of general and specific combining abilities in potato. *Canadian Journal of Genetics and Cytology 18:*463-470.

Tai, G.C.C. and Coleman, W.K. (1999). Genotype x environment interaction of potato chip color. *Canadian Journal of Plant Science 79:*433-438.

Talburt, W.F., Schwimmer, S., and Burr, H.K. (1987). Structure and chemical composition of the potato tuber. In Talburt, W.F. and Smith, O. (Eds.), *Potato Processing,* Fourth edition (pp. 11-46). NY: Van Nostrand Reinhold Co.

Tarn, T.R., Tai, G.C.C., De Jong, H., Murphy, A.M., and Seabrook, J.E.A. (1992). Breeding potatoes for long day temperate climates. *Plant Breeding Reviews 9:*217-332.

Taylor, R.J., Secor, G.A., Ruby, C.L., and Orr, P.H. (1993). Tuber yield, soft rot resistance, bruising resistance and processing quality in a population of potato (cv. Crystal) somaclones. *American Potato Journal 70:*117-130.

Thill, C.A. and Peloquin, S.J. (1994). Inheritance of potato chip color at the 24-chromosome level. *American Potato Journal 71:*629-646.

———. (1995). A breeding method for accelerated development of cold chipping clones in potato. *Euphytica 84:*73-80.

Thygesen, P.W., Dry, I.B., and Robinson, S.P. (1995). Polyphenol oxidase in potato. *Plant Physiology 109:*525-531.

Ulrich, D., Hoberg, E., Neugebauer, W., Tiemann, H., and Darsow, U. (2000). Investigation of the boiled potato flavour by human sensory and instrumental methods. *American Journal of Potato Research 77:*1111-1117.

Valkonen, J.P.T., Keskitalo, M., and Vasara, T. (1996). Potato glycoalkaloids: A burden or a blessing? *Critical Reviews in Plant Sciences 15:*1-20.

Van Gelder, W.M.J. and Scheffer, J.J.C. (1991). Transmission of steroidal glycoalkaloids from *Solanum vernei* to the cultivated potato. *Phytochemistry 30:*165-168.

Van Gelder, W.M.J., Vinke, J.H., and Scheffer, J.J.C. (1988). Steroidal alkaloids in tubers and leaves of *Solanum* species used in potato breeding. *Euphytica* (Suppl.):147-158.

Van Marle, J.T., van der Vuurst de Vries, R., Wilkinson, E.C., and Yuksel, D. (1997). Sensory evaluation of the texture of steam-cooked table potatoes. *Potato Research 40:*79-90.

Veerman, A. (2001). *Variatie in Knolkwaliteit Tussen en Binnen Partijen van Consumptieaar Dappelrassen.* Ph.D. Dissertation, Wageningen University, Wageningen, The Netherlands.

Walsh, J.R. (1995). Utilizing the stored crop. *American Potato Journal 72:*481-492.

Wang-Pruski, G., Astatkie, T., De Jong, H., and Leclerc, Y. (2003). Genetic and environmental interactions affecting potato after-cooking darkening. *Acta Horticulturae 619:*45-52.

Wang-Pruski, G. and Nowak, J. (2004). Potato after-cooking darkening. *American Journal of Potato Research 81:*7-16.

Wannamaker, M.J. and Collins, W.W. (1992). Transfer of high dry matter from 2x to 4x level in potato. *American Potato Journal 69:*613.

Werner, B.K., Loe, S.L., and Shafii, B. (1998). Comparison of seasonal development of tuber specific gravity among seven potato clones. *American Journal of Potato Research 75:*121-127.

Woodwards, L. and Jackson, M.T. (1985). The lack of enzymic browning in wild potato species, series *Longipedicellata,* and their crossability with *Solanum tuberosum. Zeitschrift für Pflanzenzüchtung 94:*278-287.

Woolfe, J.A. (1987). *The Potato in the Human Diet*. Cambridge, UK: Cambridge University Press.

Wünsch, A. and Munzert, W. (1994). Einfluss von Lagerung und Sorte auf die Verteilun der Glykoaloide in der Kartoffelknolle. *Potato Research 37:*3-10.

Xiong, X., Tai, G.C.C., and Seabrook, J.E.A. (2002). Effectiveness of selection for quality traits during the early stage in the potato breeding program. *Plant Breeding 121:*441-444.

Yencho, G.C., Kowalski, S.P., Kobayashi, R.S., Sinden, S.L., Bonierbale, M.W., and Deahl, K.L. (1998). QTL mapping of foliar glycoalkaloid aglycones in Solanum tuberosum x *S. berthaultii* potato progenies: Quantitative variation and plant secondary metabolism. *Theoretical and Applied Genetics 97:*563-574.

Zeh, M., Casazza, A.P., Kreft, O., Roessner, U., Bieberich, K., Willmitzer, L., Hoefgen, R., and Hesse, H. (2001). Antisense inhibition of threonine synthase leads to high methionine content in transgenic potato plants. *Plant Physiology 127*:792-802.

Zheng, G.H. and Sosulski, F.H. (1998). Determination of water separation from cooked starch and flour pastes after refrigeration and freeze-thaw. *Journal of Food Science 63*:134-139.

# Chapter 6

# Ecophysiology and Agronomic Management

P. M. Govindakrishnan
A. J. Haverkort

Plant growth occurs in phases and each phase has specific ecophysiological requirements. However, agricultural crops are often grown in a range of environments not necessarily fully suited to a particular crop, exposing one or the other phase of its growth to varying degrees of stress. Agronomic practices are the means to provide the best possible environment to a crop so that the rate of development of each phase is optimized for maximum exploitation of the genotypic potential in terms of economic yield.

Potato is grown in a wide range of environments ranging from summer season in temperate regions and tropical highlands to the winter season in the tropical lowlands. High-economic yield with minimum adverse impact on the environment are the main considerations to be borne in mind while developing cultural practices for diverse environments. In this chapter, we discuss the various growth phases of the crop and the ecophysiological factors affecting them, and the manipulation of the agronomic practices for optimum resource use efficiency. Examples of how these principles have been embodied into decision support systems are also presented.

## GROWTH PHASES

Four growth phases have been recognized in most tuber crops including potato. These are (1) establishment phase, (2) growth of autotrophic organs—predominantly leaf, (3) overlapping period of leaf growth and induction of storage organs, and (4) the period of predominantly storage organ growth (Kawakami, 1978). These growth phases broadly correspond to the growth phases (numbered 0-3) identified by Kooman and Haverkort (1995). The resource requirements in the different phases vary and depending upon the prevalent conditions, the onset, duration, and the growth accrued in the different growth phases is determined. The major ecophysiological factors affecting the growth and development in each of these phases are discussed below.

### *Establishment Phase*

Quick establishment is desirable especially when the duration of the growing season is short so that the plants become autotrophic early and there is sufficient time for the other growth phases to exploit the available resources. Normally, nondormant tubers from the previous season are used as seed in potato. The extent of sprout growth in the seed tubers varies depending upon its physiological age. Physiological age is measured by the accumulated day degrees above a base temperature (normally 4°C) from the end of dormancy till the end of storage (Allen and O'Brien, 1986) and is a method of manipulating the growth and development of the crop. The process of advancing sprout growth is called presprouting and is achieved by exposing the seed tubers before planting to elevated temperatures for different durations. The stage at which the seed tubers are exposed to higher temperatures (rather than the total accumulated day degree) determines the effectiveness of temperature for presprouting and later exposures are known to be more effective (Jenkins et al., 1993).

Physiologically young tubers take relatively longer time for emergence than the well-sprouted ones. Planting of sprouted tubers besides leading to quick emergence also results in a uniform crop with more stems/hill. The sprouts need to be stout and robust to avoid their damage during planting. This is achieved by exposing the seed tubers to diffused light. Natural white light is reported to be the best for inhibiting sprout elongation (Ogilvy and Roberts, 1992).

Size of the seed tubers is another important factor affecting quick establishment of the crop. Initially after planting, the mother tuber is the main source of energy for the growing sprouts. Therefore, bigger seed tubers supply more energy to the growing sprouts promoting faster emergence and early vigorous crop. However, there is need to optimize the seed size and spacing because increasing the seed size affects the number of stems/hill, the size distribution in the produce and also has economic implications since increasing the seed size increases the seed requirement per unit area.

Temperature, especially soil temperature also affects the establishment phase. The optimum temperature for emergence has been reported to be between 22°C and 24°C (Sale, 1979), and up to this temperature, the emergence rate increases linearly with increase in temperature. Temperature below 20°C delays sprouts growth (Firman et al., 1992). Temperatures higher than the optimum also reduce sprout growth (Midmore, 1984). In such cases, treatments that reduce soil temperature (mulching, irrigation, etc.) are beneficial for quick emergence and early canopy cover. The slower emergence rate at higher than optimum temperature is reported to be due to subapical necrosis arresting sprout growth (McGee et al., 1986). Besides temperature, sufficient soil moisture is also required to ensure quick root establishment and shoot emergence. Salinity is a problem in many tropical and subtropical regions affecting emergence. Yadav et al. (2001) found emergence to be delayed by 15 days at salinity level of 10 $dS \cdot m^{-1}$.

### *Autotrophic Organs Growth Phase*

As soon as the plants become autotrophic, haulm growth predominates. It has mainly two components, namely stems and leaves, which together affect the duration and rate of haulm growth. The rate and duration of haulm growth affect the amount of incident radiation intercepted by the plants, which consequently determines the yield. Thus, ground-cover duration is positively correlated with yield (van der Zaag and Demagante, 1987) because yield is a product of intercepted radiation and radiation/light use efficiency. Early establishment of complete canopy cover as well as its continuance for a longer period ensures greater interception of radiation leading to higher yield.

The production of stems is affected mainly by the age and size of the seed tubers and temperature. Stems from small tubers have relatively slow rate of initial growth and are also thinner (Lommen, 1994) whereas aged seeds give rise to stems with a larger proportion of branched shoots (Gillison et al., 1987). The relation between stem elongation and temperature is almost linear up to 35°C (Manrique, 1990) and is stimulated by high day and low night temperatures while high night temperature promotes branching (Moreno, 1985). Leaf growth is also affected by physiological age as well as size of the seed tubers, day length, temperature, moisture, and nutritional status. The leaf area attained is a function of the rate of leaf appearance, their number, and individual leaf extension rate. Older seed tubers result in the early production of more leaves, higher leaf area, and leaf dry weight, but this advantage is seen only during the early growth period (Knowles and Botar, 1991). Early establishment of peak leaf area is also observed when large-sized seed tubers are used (Wiersema, 1989).

Leaf growth is affected both by high as well as low temperatures. Low temperatures reduce the overall growth rate of the plants and consequently production of new leaves and the rate of their expansion are slow. The rate of leaf appearance is linearly correlated with temperature in 9-25°C range and no further increase occur beyond 25°C (Kirk and Marshall, 1992), whereas 25°C was found to be the optimum for leaf expansion also (Benoit et al., 1983).

At very high temperature, leaves do not expand fully thereby reducing light interception. Prange et al. (1990) observed that though the leaf area was less under warm conditions, the dry weight of the leaves was not affected indicating that the leaf expansion was reduced. Therefore, genotypes grown in hot conditions have lower specific leaf area (Midmore and Prange, 1991). In potato models, leaf expansion has been considered to increase linearly up to 24°C, thereafter decreasing linearly to 0 at 35°C (Kooman and Haverkort, 1995).

Long days also increase leaf area (Lemaga and Caesar, 1990) whereas drought reduces both rate of canopy expansion as well as maximum leaf area index (Jefferies and Mackerron, 1993). The effect of moisture and nutrition are discussed separately.

### *Phase of Overlapping Period of Leaf Growth and Induction of Storage Organ*

In this phase the haulms continue to grow but tubers are also initiated, therefore, partitioning of photosynthates into tubers starts during this period. The extent of haulm growth by the time tuber initiation occurs is an important factor affecting the yield. Early tuberization, which is desirable for obtaining high yield at early harvest, occurs when physiologically old seed tubers are used. Temperature is another critical factor affecting tuber initiation. Low temperature, especially at night is required for tuber initiation and night temperatures above 21°C have been reported to hamper tuber initiation (Pushkarnath, 1976). From studies on exposing different plant parts to differential temperatures, Slater (1968) showed that tuber initiation occurred earlier when shoots were exposed to low night temperature while it was advanced further when both shoot and root were exposed to low night temperature. No tuberization was observed in potato clones at 40/30°C (day/night temperatures) and also in soils maintained at ca. 32°C (Reynolds and Ewing, 1986), indicating that the potential for induction of the plants to tuberize is affected principally by air temperature but the expression of the signal can be blocked by high soil temperature. Response to temperature is, however, modified by day length and light intensity (Bodlaendar, 1963); long days and low light intensity promote haulm growth but delays tuber initiation.

### *Phase of Tuber Growth*

The rate of bulking and its duration affect the amount of photosynthates accumulated in the tubers. Initially the tuber bulking is slow but soon it attains a constant rate during which yield increases linearly. However, with the onset of foliage senescence the bulking rate decreases. The rate of translocation of photosynthate into storage organs is determined mainly by temperature. Plants growing under warm conditions are taller with higher stem to leaf and lower tuber to stem dry matter ratios (Ben Khedhar and Ewing, 1985). At 28°C only 50 percent of assimilates produced are translocated into the tubers, whereas at 18°C more than two-thirds of the assimilates were translocated (Randeni and Caesar, 1986). Thus, at higher temperatures, tops get priority over the tubers for assimilates leading to an increased haulm growth and reduced tuber yield. Pushkarnath (1976) reported night temperature between 12°C and 18°C to be ideal for high yield.

## *MODELING*

A successfully raised crop should use optimum amount of resources without any adverse effect on the environment. With increase in cost of inputs, decreasing returns on investment and increased awareness toward environmental degradation, this is an increasingly important but difficult task. Agronomists have always been in search of tools (models) to aid decision making for raising a crop successfully. As more and more information on the ecophysiology of the crop is available, this is leading to better models that would help in fine-tuning the cultural requirements of the crop for different situations quickly, efficiently, and at a lesser cost.

When the crop is supplied with adequate nutrients and protected from pests and diseases, the yield limit is determined by the environmental factors. The maximum yield, which can be obtained under a given set of environmental conditions, other factors being nonlimiting, is called potential productivity. Knowledge of the gap between the potential yield and the yield actually obtained reflects the technological gap that can be bridged by the development of appropriate agrotechniques (Van der Zaag and Horton, 1983). Knowledge of the potential productivity may also be useful in choosing production sites and manipulating agronomic practices. Estimate of the potential yield can be made by the use of models where environmental factors are used as inputs. Basically, the technique for development of systems' dynamic models consists of demarcating the system boundary. The various stocks in the system within the boundary are defined and the flows that feed or drain the stock are then identified.

The initial values of the stocks are estimated followed by defining the logic that determines the magnitude of each flow.

### *Potato Models*

Many models have been developed incorporating different variables in different ways and ranging in complexity from simple empirical models to detailed simulation models. Anonymous (1978) and Doorenbos and Kassam (1979) developed simple models that can be used to estimate the potential yield of crops in a given area. Both these methods use the gross assimilation of carbon dioxide by a standard crop (de Wit, 1965) for the incident photosynthetically active radiation in a given latitude. Standard crop is defined as a crop growing at 20°C and intercepting all the incident radiation. Corrections are then made for the temperature prevailing during the growing period, development of leaf area that affects the amount of photosynthetically active radiation intercepted and harvest index to derive the net yield from the total biomass. The major difference between the two methods lies in calculating the correction factors. Doorenbos and Kassam (1979) used seasonal mean temperature to calculate the photosynthesis and respiration whereas the other method utilizes the mean day temperature to calculate photosynthesis and the mean 24 h temperature to calculate respiration, which in turn is obtained according to the formula of McCree (1974). In the case of Doorenbos and Kassam's method, respiration is assumed to be 40 percent at temperatures below 20°C and 50 percent at higher temperatures. These two models are general-purpose models used for estimating the final yield. Versteeg and van Keulen (1986) simplified the calculations further and developed a set of graphs and tables to determine gross photosynthesis and respiration.

Van der Zaag (1984) developed a model specifically for potato to determine the potential production at a given location. In this model, the dry matter production of the standard crop is obtained at monthly intervals as per de Wit (1965). The dry matter production of the standard crop is then adjusted to the actual growing conditions by multiplying with a factor 'K' (a function of maximum and minimum temperatures) to account for respiration and actual prevailing temperatures. The product is then multiplied by ground cover values to account for the reduction in intercepted radiation by the crop as compared to a standard crop. Thus, the improvement in this method is the weightage given to crop development over time and the method of accounting for respiration. This model has been used to predict the potential yield (Kunkel and Campbell, 1987) as well as to explain the difference in yielding ability of different potato cultivars (van der Zaag and Doorenbos, 1987). The advantage of these models is that they are simple and require a few easily obtainable inputs, namely maximum and minimum temperature, sunshine hours or evaporation or irradiation, canopy cover or leaf area index, etc.

Ng and Loomis (1984) and Ingram and McCloud (1984) developed detailed simulation models for potato. These models take into account various factors affecting the crop at different levels of hierarchy and simulate the various physiological processes of the plant like photosynthesis, respiration, etc. and consequently the growth and development of the crop. These models are useful in understanding the growth and development of the crop under diverse situations but they require many inputs.

Another approach based on the Light Interception and its Utilization (LINTUL) concept has also been used in the development of models for potato. In this case, the yield is considered to be a function of the intercepted radiation (radiation use efficiency). Mackerron and Waister (1985) used this approach to develop a model that recognized three different stages of crop growth namely planting to emergence, emergence to canopy closure, and canopy closure to haulm destruction. Different weather factors or their combinations for each stage determined the progress of the crop from one phase to the next. The development of canopy during each phase is simulated to estimate the intercepted radiation during each phase. The sum of the radiation intercepted during each phase is obtained and multiplied by 1.84 to obtain the total dry matter since the conversion efficiency of intercepted radiation into dry matter is assumed to be 1.84 $g \cdot MJ^{-1}$. The tuber yield is derived from the total dry matter based on the harvest index. This method has been used to estimate the potential yield in the Columbia Basin, USA (Kunkel and Campbell, 1987) as well as in the subtropics (Prihar et al., 1995). Haverkort and Harris (1987) opined that the conversion of intercepted radiation into dry matter is not a constant but is a function of temperature, and developed another simple model using intercepted radiation and a net conversion coefficient dependent on temperature. According to them the temperature dependence of the conversion coefficient of intercepted radiation into dry matter is important for its applicability in the low altitude tropics. Spitters and Schapendonk (1990) used the LINTUL approach to assess genotypes for drought tolerance whereas van Keulen and Stol (1995) used this approach for agroecological zonation of potato.

## *AGROECOLOGICAL ZONING*

The delineation of geographical area into suitability classes based on ecophysiological factors is called Agroecological Zoning. It helps to make strategic decisions regarding where and how to grow potatoes, and what kind of yields may be expected. Ideotyping shows how genotypes perform under such conditions and what breeding should aim at. van Keulen and Stol (1995) used the potato crop growth model (Spitters, 1987) LINTUL for developing a map of agroecological zones of potato in the world. In this model, they used a constant value of light use efficiency to estimate dry matter production. The effect of day length on dry matter production and partitioning was also not included. Haverkort et al. (2004) developed a decision support system "POTATO-ZONING" for determining where potatoes can yield the best. They followed the same approach as that of van Keulen and Stol (1995) but tried to remove the lacunae of their model. The assumptions of this model are as follows:

- A potato crop grows only between average daily temperatures of 5°C and 21°C. When it is cooler there is risk of frost, when it is warmer the tuber growth stops.
- The model is initiated by defining the plant density and the initial light interception by the foliage at emergence and the initial relative growth rate of that foliage. The fraction of intercepted ($F_{int}$) photosynthetically active radiation (PAR between 400 and 700 nm wavelength) is calculated from subsequent foliar increase.

The model assumes that a two-week period is required for emergence of the crop when planted at suitable temperatures.

- The length of the growth cycle of the varieties match exactly with that of the available length of the growing season, resulting in a harvest index of 0.80 at crop senescence but is reduced at higher temperatures to reach 0 at 28°C.
- Daily growth *(W)* is calculated as $W = \text{LUE} \times F_{int}$ PAR, with LUE = 2.9 $g \cdot MJ^{-1}$ (Haverkort, 1990). Light use efficiency (LUE) was made dependent on temperature and the base value of LUE is multiplied by a reduction factor TF, whose value depends on the daily mean temperature as shown in Figure 6.1.
- Roots are expected to grow 60 cm deep and infiltration of precipitation, percolation, soil surface evaporation, and crop transpiration are calculated depending on soil type and crop development.
- At an LAI (leaf area index) of 4, transpiration is maximal. Where the actual/potential transpiration ratio is smaller than 1 the crop accumulates drought stress linearly affecting LUE and when its cumulative sum exceeds 10 it linearly decreases light interception assuming that leaves are shed.

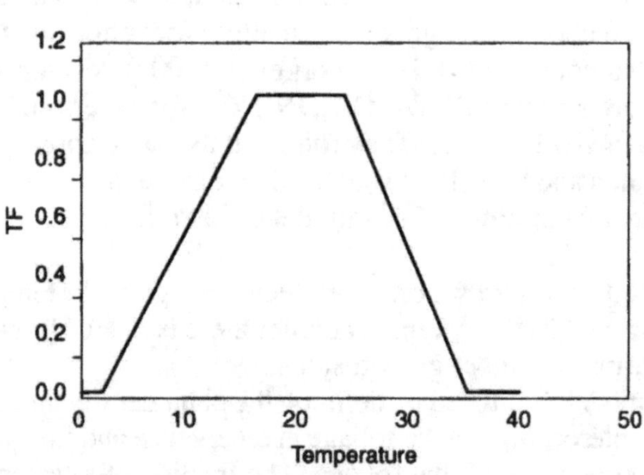


FIGURE 6.1. Reduction factor TF versus daily mean temperature. This factor accounts for the effect of temperature on LUE (light use efficiency). *Source:* Kooman and Haverkort (1995).

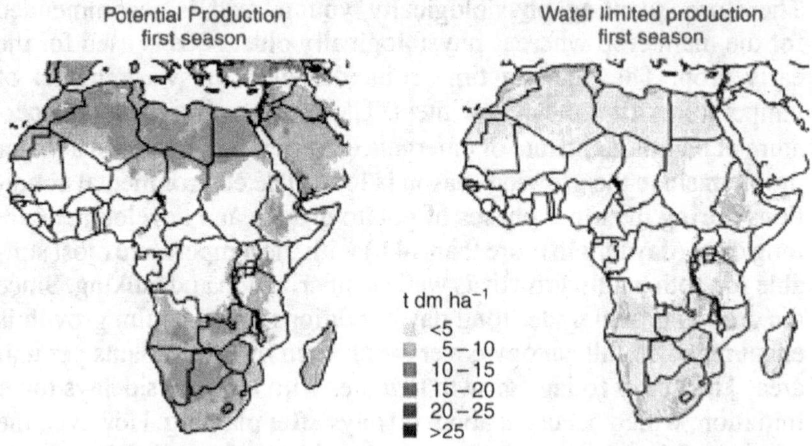


FIGURE 6.2. Calculated potential (left) and water limited yields (right) for Africa. *Source:* Haverkort et al., 2004.

The zoning model needs meteorological data (daily minimum, maximum temperatures, evaporation, and precipitation) and soil maps showing water-holding capacity. Figure 6.2 shows the results of the model runs for Africa under assumed irrigated and rain-fed conditions in which areas where potato production is possible, as well as potential yield of the place, are delineated. Such an analysis is useful not only for agronomists but also for potato-processing companies planning production sites, or for governmental bodies planning irrigation schemes.

## *POTATO GROWING ENVIRONMENTS AND THE ROLE OF AGROTECHNIQUES*

Potato growing environments have been categorized into four zones namely the temperate zone, the highland zone, the lowland tropical, and subtropical zone and the Mediterranean zone (Guenthner, 2001). Major potato area is in the temperate zone (Europe and North America) followed by the tropical and subtropical zone (Asia).

In the temperate zone, planting is done in spring and harvested in autumn, thus, the crop duration is fairly long (4-5 months).

Therefore, planting physiologically young seed is recommended for the main crop whereas physiologically older seed is used for the early crop. The planting time coincides with the warming up of temperatures after the cold winters. Therefore, owing to low temperature at planting, the rate of emergence is slow, but this is not a major problem since the growing season is long. The environmental conditions during the other phases of potato growth and development are long days (day length more than 14 h) with the temperature most suitable for both haulm growth as well as tuberization and bulking. Since the crop is grown under long day conditions, good haulm growth is encouraged so full canopy cover is achieved by fewer plants per unit area. The heavy foliage growth together with long days delays tuber initiation, which occurs at about 60 days after planting. However, the big size of the source as well as the prevalence of congenial temperatures is conducive to high bulking rate and hence high yields are obtained in this zone. The crop here is generally rain fed with supplemental irrigation in some places.

The lowland tropics and subtropics represent the other extreme of environmental conditions where potato cultivation is successful due to the fine-tuning of agrotechniques to mitigate the adverse impact of environmental factors. Here, the crop is raised during the cool season, by planting in autumn in areas where the winters are severe and frosting is regular and is harvested before the frost occurs. Thus the growing season is quite short (about 100-120 days). Since planting is done during relatively warmer conditions and the temperature cools down as the season progresses, scheduling the planting time is a critical decision. It is to be ensured that the sprout and haulm growth does not get unduly affected by the prevalent high temperature and also that the tuberization stage coincides with the fall in temperature to that required for good tuberization. Since day length as well as temperatures decrease as the potato season progresses, agrotechniques aimed at quick emergence, early establishment of complete canopy cover with early tuber initiation and high bulking rate have been developed. These include planting well-sprouted medium-sized seed tubers at 10-12 cm depth with ample supply of nutrients and water, including 1-2 irrigations during emergence phase. These practices ensure quick emergence (10-12 days after planting) and early establishment of full canopy cover (about 50–60 days after planting). Short days and the fall in temperature (to below 20°C) soon after emergence ensure early tuber initiation (about 25-30 days after planting) and rapid bulking. Thus, high yields (40-50 $t{\cdot}ha^{-1}$) are obtained in 100-120 days as in the Indo-gangetic plains. In areas where the winters are milder and frosts are infrequent, planting is usually done in winter so that the crop emergence, growth, and development proceeds commensurate with temperature and day length increase. Between these extreme situations, there is a whole range of environmental conditions in which potato is cultivated.

Land preparation, planting, nutrient management, water management, and weed management are the major cultural practices that have a bearing on the growth and yield as well as profitability of the crop. In the sections following, the effect of these cultural practices on the growth and development of the crop is discussed.

### *Land Preparation*

The tillage requirements of potato vis-à-vis other crops particularly cereals differ as in potato the planting material is the tuber that is much bigger compared to seeds of other crops. The depth of planting is critical since it affects emergence rate and growth. The manipulation of planting depth is possible only if the soil is well prepared. Potato planting, therefore, requires a well-prepared field. Moreover, the root system of potato requires well-prepared soil for easy penetration to deeper layers. A well-prepared soil is also necessary for having enough air space and moisture for the respiration and growth of both roots and tubers. The underground stolons and tubers also need to be kept covered by soil to avoid exposure to sunlight and this also requires a finely prepared soil. These requirements make land preparation very important. Blake et al. (1962) studied the tillage requirements of potato and reported that it consists of an initial loosening of the soil that is necessary for breaking the plough pans in some situations to facilitate proper movement of water and deep penetration of roots. This is achieved through subsoiling and it has been found to increase yield under acute drought (Ross, 1986). Miller and Martin (1987) also reported subsoiling to promote deep rooting and avoid water stress on sandy soils.

A secondary preplanting tillage is done to remove weeds and make the soil friable for easy planting and making the ridges. Nelson and Giles (1989) found 2-3 cultivations sufficient for effective weed control. After bringing the soil to proper tilth, ridging is invariably done when potatoes are manually planted, while ridging and planting are done simultaneously in the case of mechanical planting. The third category of tillage is the postemergence tillage, which is done primarily to control weeds. However, when herbicides are used, this tillage operation is not necessary (Blake et al., 1962). Generally, in the tropical/subtropical plains where the growing season is short, planting is done on small ridges to reduce the time taken for emergence and the full ridge is made up after manual/mechanical weeding at the time of stolon formation. Remaking of ridges is also often necessary when surface irrigation method is adopted because irrigation erodes the ridges and has to be remade to ensure adequate coverage of stolons and growing tubers.

## *Planting*

### *Seed Size and Spacing*

Seed accounts for almost half to one-third of the cost of potato cultivation especially in tropical and subtropical countries. Hence seed tubers need to be used judiciously. Since the range of variation in seed tuber sizes is very high in potato, use of large tubers would increase the seed rate tremendously whereas small tubers without proper adjustment of the spacing would lead to yield loss. Bremner and El Saeed (1963) reviewed experiments on seed size and spacing and opined that tuber yield is invariably greater from large than from small tubers of the same population. The curve for increase in yield with increase in seed size depended upon variety. Varieties that produced more stems were less responsive to increase in seed size than those that produced fewer stems. This review indicated that the yield difference between plants from large and small seeds were due to a difference in durations of tuber bulking rather than rate of bulking due to early tuber initiation in plants from large-size tubers.

The unit of population in potato is the number of main stems aboveground (Wurr, 1974). The number of eyes from which stems develop depends upon the tuber weight and the minimum weight per stem (Birecki and Roztropowicz, 1963). The weight of tuber per eye is higher in big tubers but the weight per stem is about the same in all tuber sizes. Therefore, bigger tubers produce more stems per plant than smaller ones, and the more the number of stems, the greater is the assimilation area per plant and consequently the higher the yield. Wurr (1974) reported the relationship between stem number and yield to be of the form of $y = a + bx = cx^{1/2}$. This would mean that there is an optimum stem density for obtaining maximum yield, the value of which would vary with location, variety, etc. Allen and Wurr (1992) analyzed the effect of plant density and opined that increasing the seed size or decreasing within-row spacing or both results in more stems. The optimum stem density for ware yield is lower than that for total yield. At equal within-row spacing, large seed tubers produce greater total yield than small tubers until the spacing is reached at which the small seed tubers achieve the optimum density. This within-row spacing for small seed tubers is much closer than that at which large seed tubers achieve the same stem density. They also inferred that when seeds of different sizes are grown at their optimum stem densities they would give similar total yields. However, in actual practice use of stem number for decision making is difficult since farmers can decide only on the basis of seed weight or riddle size.

### *Advisory System for Seed Rate*

At the Scottish Crop Research Institute a decision support system called 'The MAPP' (The Management Advisory Package for Potatoes) has been developed (MacKerron et al., 2004) to determine the correct seed size and spacing to be adopted under varying cultural practices and price situations. With the help of this advisory system, a user can examine the consequences of choice of cultivar, seed size, seed rate, planting date, depth, etc. on the crop, given a number of constraints such as the soil type, the likely weather, possibilities for irrigation, preferred sizes of tubers to be produced, and their prices (market specification).

MAPP uses models of tuber multiplication and size grade distribution to recommend the proper seed size. Once the user has selected a

TABLE 6.1. The influence of market specification on the seed rate giving maximum return.

| Valuable size fraction | Value of size fraction | | | Seed rate giving max. margin (t·ha) | Cost of seed (£·ha⁻¹) | Max. margin (£) |
|---|---|---|---|---|---|---|
| | 30-40 mm | 40-50 mm | 50-100 mm | | | |
| Small | £120 | £60 | £25 | 5.6 | £980 | £1,451 |
| Level | £100 | £110 | £120 | 4.8 | £840 | £5,485 |
| Big | £25 | £60 | £120 | 3.2 | £560 | £4,625 |

*Source:* MacKerron et al. (2004).
*Notes:* cultivar: Premiere, size of seed = 650 tubers · 50 kg $^{-1}$; price of seed = £175 · t $^{-1}$; anticipated yield = 55 t · ha $^{-1}$; Value of 10-30 mm = £0.

seed rate, MAPP provides an estimate of the likely daughter tuber population for a given cultivar, size of seed, and plant spacing.

Output of MAPP illustrates how readily the user can follow the likely outcome of each choice that is made. The market specification—the prices to be paid for particular sizes of tuber—is particularly important in determining the return from a crop. The example presented in Table 6.1 shows that for a given size and price of seed and the anticipated yield from it, the seed rate that leads to the maximum return over the cost of the seed changes under varying price structure for the harvested crop.

MAPP can also help the grower to accommodate changes in the market that occur during the growing season by changing the dates of haulm destruction and harvest. The seed rate for the greatest margin over its cost changes nonlinearly with seed price (Table 6.2) and also leads to a nonlinear change in the expected number of daughter tubers.

The seed rate for maximum return over the cost of the seed is strongly dependent upon the size of the seed (Table 6.3), which leads to differing expectations of the number/grades of daughter tubers. Where a seed lot comprises a wide range of sizes, MAPP provides the means to quantify the advantages of split grading the seed and the best seed rates for each resulting fraction. Table 6.3 shows the contrast between the seed rates for the whole seed lot and its small and

TABLE 6.2. The influence of the price of seed on the seed rate giving maximum return and on the associated population of daughter tubers.

| Price of seed (£) | Seed rate giving max. margin ($t{\cdot}ha^{-1}$) | Cost of seed ($£{\cdot}ha^{-1}$) | Max. margin (£) | No. of daughter tubers $\cdot$ $m^{2}$ |
|---|---|---|---|---|
| £80 | 7.0 | £560 | £6,028 | 79 |
| £175 | 4.8 | £840 | £5,485 | 66 |
| £250 | 4.0 | £1,000 | £5,155 | 59 |

*Source:* MacKerron et al. (2004).
*Notes:* cultivar: Premiere, size of seed = 650 tubers/ 50 kg; anticipated yield = 55 t.ha $^{1}$; values of size fractions in daughter crop as in 'Level' in Table 6.1 .

TABLE 6.3. The influence of the size of the seed on the seed rate giving maximum return and on the associated seed spacing, seed cost, and population of daughter tubers.

| Size of seed | Spacing (cm) | Seed rate giving max. margin (t/ha) | Cost of seed (£/ha) | Max. margin (£) | No. of daughter tubers / $m^{2}$ |
|---|---|---|---|---|---|
| 650·50 kg $^{1}$ (35-55 mm) | 18 | 4.8 | £840 | £5,485 | 66 |
| 1,205·50 kg $^{1}$ (35-40 mm) | 12 | 4.0 | £700 | £5,750 | 71 |
| 433·50 $kg^{-1}$ (50-55 mm) | 24 | 5.3 | £927 | £5,329 | 63 |

*Source:* MacKerron et al. (2004).
*Notes:* cultivar: Premiere; price of seed = £175·t $^{1}$; anticipated yield = 55 t·ha $^{1}$; values of size fractions in daughter crop as in 'Level' in Table 6.1 ; ridge spacing = 90 cm.

large fractions. Obviously, seed rates would also differ with the choice of cultivar to be grown.

## *Time and Method of Planting*

The time of planting has to be so adjusted that the environmental factors are conducive to emergence. Moreover, the environmental requirements of the other growth phases that follow should also be met once the plants emerge. For example, in the Indo-gangetic plains, where the autumn season is the main potato season, the time of planting is dictated by temperature and it is so adjusted that the maximum temperature at the time of planting is in the range of 30-32°C. It is also ensured that the minimum temperature falls to less than 20°C within 25-30 days after planting (the time of tuber initiation).

Planting is normally done on ridges since it leads to more uniform plants and yield increase by 6-10 percent in light soils and more than 10 percent in heavy soils (Zanker, 1986). Ridging is also desirable when surface irrigation method is adopted since irrigation in furrows allows better control over amount and movement of water in the furrows. Planting in ridges is also advantageous for better soil coverage of emerging stolons and the growing tubers and if this is ensured in flat beds also, both planting methods are equally good. Dambroth and Bramm (1984) did not find any significant advantage in tuber yield, percent of saleable tubers, or in response to Nitrogen (N) by planting in ridges over level planting

Depth of planting is critical when the moisture and temperature conditions are not favorable. For example when the soil moisture is less, deep planting is advocated so that the seed tubers lie in the region of relatively more moisture. Similarly when the temperatures at the time of planting are higher, as in the early autumn crop in the Indo-gangetic plains, deep planting is advocated whereas shallow planting is recommended when the temperatures at the time of planting are relatively cool and it is necessary to hasten emergence.

### *Nutrient Management*

In potato crop, the mineral nutrient uptake ranges from 2.28 to 3.57 kg N, 0.04 to 0.12 kg P, and 3.7 to 5.41 kg K per ton fresh tuber weight (Gunasena, 1968). It also takes up high quantities of micronutrients. The ranges of different micronutrients in tuber flesh on dry weight basis of 19 Indian potato cultivars/hybrids were estimated to range between 10 and 18 ppm Zn, 21 and 53 ppm Fe, 2.0 and 4.3 ppm Cu, and 5 and 31 ppm Mn (Trehan and Sharma, 1996). Corresponding ranges in tuber peels were 14-23 ppm Zn, 452-1,021 ppm Fe, 5-17 ppm Cu, and 18-86 ppm Mn. Therefore, potato needs a heavy dressing of fertilizers and micronutrients to meet the demand. Moreover, the use of pesticides and herbicides is also fairly high in potato, making it a high chemical energy input intensive crop. The need, however, is to use the various inputs judiciously.

*Nitrogen*

It is the first limiting nutrient for potato production affecting almost all phases of the growth and development of the crop. Published literature on the effect of fertilizer application on emergence is scanty. Trehan et al. (1998) did not find any effect of N on the emergence. As regards root growth, Pushkarnath (1976) reported that N application led to higher root dry weight especially at early stages of crop growth.

The major influence of N is, however, on canopy development. It increases leaf area index and relative growth rate (Perumal and Sahota, 1986). This effect is more in the early part of the season (Haverkort and Rutavisire, 1986), thus, N application helps in early attainment of complete canopy cover. N also delays senescence (Santeliz and Ewing, 1981). The early attainment of canopy cover and delayed senescence leads to increased interception of solar radiation (Martin, 1995).

N also exerts profound influence on tuber initiation and its bulking rate. High N early in the season delays tuberization (Biemond and Vos, 1992) but due to its effect of delaying senescence, it results in prolonged bulking. Therefore, the rate of N application has to be adjusted in accordance with the available growing period in such a way that early canopy cover is established but at the same time tuber initiation is not unduly delayed. Sufficient canopy cover early in the season is also required for establishing high bulking rate since bulking rate is a function of the leaf area index at tuber initiation. Senescence should also be delayed as much as possible so that maximum use of the available growing period is made for tuber bulking leading to economic yields. Therefore, the timing and placement of N application is important, more so for early than the late cultivars (Maidl, 1995), as in the early crop yield needs to be maximized within the limited growing period.

Soil fertility status, variety, source, time, and method of N application determine the actual quantity of N to be applied, the number of splits, the quantity to be applied at each split, etc. These aspects are required to be worked out depending on the local situations.

### *Soil Fertility Status*

Soil organic matter and $NO_3$ N content influences the quantity of fertilizer N to be added. In the case of soil organic matter, it is not only the quantity but also its mineralization rate (influenced by temperature and soil moisture) that determines its effectiveness to meet the N requirement of the crop. Sharma and Sharma (1987) reported that under Indian conditions soil nitrate N and organic carbon status accounted for 59 and 25 percent, respectively, of the variation in yield. Soils with high organic matter (Sharma et al., 1988a) and nitrate N status (Stieberitz et al., 1986) require less N fertilizer application. No fertilizer N was needed when soil nitrate N status was more than 250 $kg \cdot ha^{-1}$ (Stieberitz et al., 1986). Since both organic matter and nitrate N status of the soil determine the fertilizer N to be applied, the methods that take into account both the sources for arriving at the N fertilizer dose to be applied are better. Thus, the Dutch system (Neeteson, 1989) that considers soil type and recently applied organic manure for determining the required N dose was found to be good in temperate situations. In tropical countries, soil organic matter content is generally very low, hence contribution of nutrients by this source is low especially in a short duration crop like potato. Soil test procedures have been developed to determine the soil mineral N available to the crop. The alkaline permanganate extract method (Subbiah and Asija, 1956) is widely used to denote the available N status of soils especially in India. Relations between N content in the permanganate extract, yield target, and fertilizer N requirement have been derived. However, even for similar soil test values the response of the crop varies under different situations due to the effect of crop factors. Therefore, soil tests need to be calibrated to the crop response depending upon the local situations. Velayutham (1979) and Adinarayana (1987) reported that fertilizer N requirement can be calculated from the equations $FN = 0.88\ T - 0.33\ SN$ and $FN = 1.9\ T - 0.98\ SN$ for variety Kufri Chandramukhi in Grey Brown and Mollisol soils, respectively. The variables FN, T, and SN in the equations represent the N to be applied through fertilizer in $kg \cdot ha^{-1}$, targeted yield in q·ha$^{-1}$ and N in soil estimated using alkaline permanganate extractant, respectively. This represents a simple method to relate the fertilizer N requirement to the soil N status and the yield target. However, such equations have to be worked out for every situation, which makes it expensive and time-consuming.

*Variety*

The maturity class of the variety, its yield potential, as well as the inherent nature of the variety to produce tubers of different grades affect its response to applied N. Long duration varieties are more responsive to high rates of N than short duration varieties. Similarly varieties with high-yield potential require more N than those with low potential.

*Time and Method of Application*

Sufficient quantity of N has to be applied at the time of planting to ensure rapid development of foliage without delaying tuber initiation. Continued N supply is required for delaying the senescence and ensuring high bulking rate for a longer period. This is achieved by splitting the N dose. The quantity to be applied at each stage and the number of splits are required to be worked out for each location. Vos (1999) reported that when the N dose was applied in two splits, that is at planting and at 60 days after emergence under temperate long-day conditions, it did not affect dry matter yield at maturity but improved the utilization of N. Under short-day tropical conditions the split dose is to be applied earlier due to the short duration of the crop. The second dose is generally applied around the time of stolon formation or tuber initiation.

Response to N application is also affected by the method of application. Placement of fertilizer near the seed tubers is advocated in case of ammonium sulfate and calcium ammonium nitrate, which contain ammonical N. It is also advocated for rain-fed conditions so that in the event of moisture stress its availability to the growing roots is not hampered. In many studies high yield and N uptake were reported when N fertilizer was banded (Westerman and Sojka, 1996). Fertilizers containing nitrate form of N can, however, be broadcasted since they are mobile, and thus are easily available to the growing roots along with the soil moisture.

The N fertilizer use efficiency of potato ranges around 50 percent in many places in the tropical and subtropical countries where it is a short-duration crop and the winters are not severe, that is temperature and radiation are not the constraints. However, in these regions more than one crop is grown in a year in the same field. The nutrients left over by potato can be utilized by other crops grown in sequence after potato, thus increasing the fertilizer use efficiency of the system as a whole. Studies in the northern Indo-gangetic plains have shown that wheat grown in sequence after potato utilizes about 13.5 percent of the N applied to the preceding potato crop (Sharma et al., 1988a).

### *Decision Support Systems for Nitrogen Application*

Fertilizer is a costly input. It also adversely affects the environment when used indiscreetly. Normally the economic optimum dose of N for a location is worked out from field trials by fitting a regression equation to the yield at different doses of N. However, farmers, especially in less developed countries, apply less than the economic optimum dose due to fiscal constraints. It is therefore desirable to know the N dose required for obtaining a targeted yield for a given variety under a given set of conditions. Alternatively, it may be useful to know whether the present yield levels can be obtained through lesser N dose by planting a more efficient variety. Normally, the economic optimum is derived from the quadratic regression equation, which usually defines the yield response to different N levels. Govindakrishnan et al. (1999) suggested an improvement and proposed that N dose for getting any targeted yield can be calculated from the formula: N dose $= \{-b \pm \sqrt{[b^2 - 4c(a - F)]}\}/2c$, where '*a*', '*b*', and '*c*' are the coefficients of the quadratic equation and '*F*' is the targeted yield. Using this method they showed that the quantity of N fertilizer to be applied to cultivar Kufri Ashoka was much less compared to that to Kufri Chandramukhi to get the same targeted yield. Such differences in N fertilizer requirement for obtaining a targeted yield were observed in other varieties also. Though such differences may be because of many physiological traits associated with the variety, still this equation can be used to select efficient genotypes through statistical comparison. These methods are in vogue in less developed countries where large-scale soil and plant analysis facilities are lacking, and fertilizer recommendations are to be based on field trials only. Moreover, owing to the short growing season, the adjustment of N dose during the growing season is difficult.

In many developed countries farmers have access to the agri-consultancy firms, which assist them to optimize the amount of N to be applied. Here only about 50-70 percent of the total N requirement is applied at planting and the rest in one or two top dressings after one or two months of emergence. By the time of top dressing the amount of N mineralized from organic matter and a fair estimate of final yield is known and these are used to determine the N to be top-dressed. A decision on the amount of N to be applied is based on the following balance (Chambenoit et al., 2004):

N fertilization = N absorbed by the crop + soil mineral N not available to the crop − soil mineral N available at the start of the cycle − N resulting from mineralization of soil organic matter

This concept has been developed by the Institut National de Recherche Agronomique (the National Agricultural Research Institute of France, INRA) for all the common annual crops, with the help of a decision support system called "Azobil" (Machet et al., 1990). The Azobil method defines the potato crop's critical curve for nitrogen (Duchenne et al. 1997). The curve describes the greatest possible production of total dry matter (haulms + tubers) with the lowest possible nitrogen content (Figure 6.3).

The method next defines the nitrogen nutrition index (NNI) as the ratio of the measured N concentration to the critical N concentration ($N_{ct}$).

Nitrogen nutrition index, NNI = %N measured / %Nct

This NNI characterizes the crop's N nutritional status. On the critical curve, the value of the NNI is 1 indicating that the crop has reached the optimum state of nitrogen nutrition for biomass production. Below the curve, NNI is less than 1, which indicates that

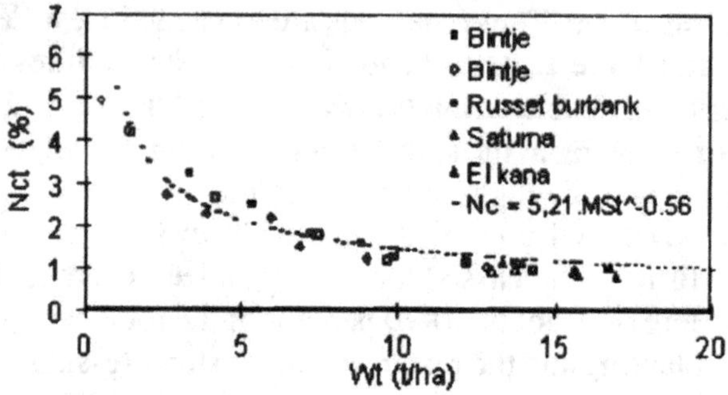


FIGURE 6.3. The potato crop's critical curve. *Source:* After Duchenne et al. (1997). Nct is the critical crop nitrogen concentration.

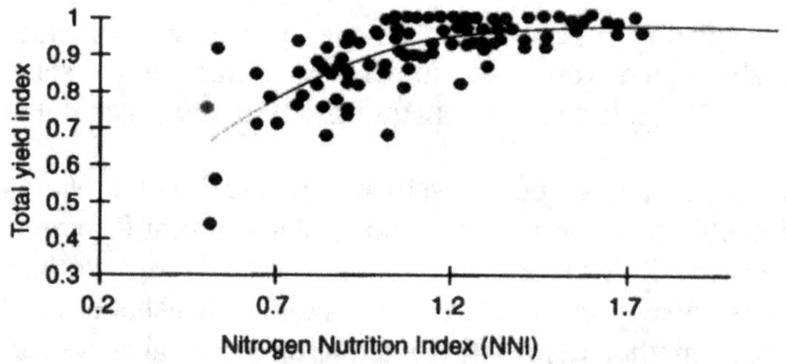


FIGURE 6.4. The nitrogen nutritional level needed for maximizing tuber yield. *Source:* After Chambenoit et al. (2004).

N limits growth. Above the curve, the value of NNI is greater than 1 indicating, "luxury consumption."

Figure 6.4 shows the relationship between NNI measured just before haulm destruction and the yield measured at harvest from the results of 24 trials carried out between 1991 and 2000 with different varieties on different soil types in France.

The decision support system needs to know the expected requirement and uptake of N by the crop. Therefore, for each production area it calculates the expected yield using Monteith's concepts (1972, 1977). This simulated growth is indexed (maximum yield not limited

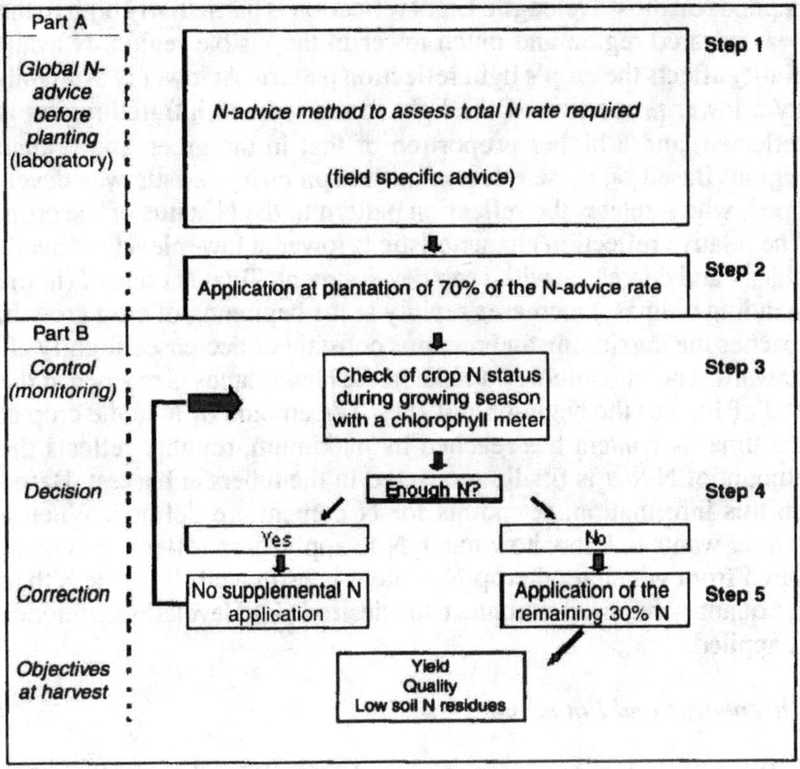


FIGURE 6.5. Steps in the strategy for management of nitrogen fertilization of the potato crop including monitoring of leaf chlorophyll concentration. *Source:* After Goffart and Olivier (2004).

by N = 1) and from Figure 6.4 the corresponding N requirement of the crop is derived.

The soil needs to be sampled before planting to know how much nitrate and organic matter is present in the soil. The latter is used to estimate mineralization during the growing season. In Belgium, the Azobil method has been further improved (Figure 6.5) by including leaf chlorophyll concentration in the decision support system.

In the Netherlands, another method was developed (Booij and Uenk, 2004) based on the relation between the crop N status and canopy light reflection in certain wave bands. The crop's light reflection depends on the wavelength. Light reflection is particularly high in the near infrared region and much lower in the visible region. N availability affects the crop's light reflection pattern. At lower N availability a lower proportion of the light in the red and infrared region is reflected, and a higher proportion of that in the green and orange region. Based on these effects a reflection characteristic was developed, which relates the reflection pattern to the N status of the crop. The relative reflection characteristic is lower at lower levels of available N and decreases with crop development. Total N content (in the standing biomass) increases rapidly at the beginning of crop growth, reaches the maximum, and remains constant or decreases slightly afterward. The maximum N uptake in the Netherlands is reached at the end of June or the beginning of July. The amount of N in the crop at the time its content has reached its maximum, roughly reflects the amount of N that is finally recovered in the tubers at harvest. Based on this information, set points for N content are defined. When a farmer wants to know how much N to apply, crop reflection is measured from which total crop N content is estimated. If it is less than the quantity required to achieve the desired yield level, the difference is applied.

### *Phosphorus and Potassium*

Potato is a heavy feeder of phosphorus (P) and potassium (K). Their application increases the root growth (De, 1960, Pushkarnath, 1976) as well as plant height. Pushkarnath (1976) reported 20 percent increase in plant height in alluvial soils while in acidic hill soils, where P fixation is high, an increase of up to 60 percent has been reported (Sharma et al., 1981). Both P and K nutrition affects the leaf area development but the pattern varies. P application increases the leaf area throughout the growth period whereas K application increases leaf area only at later stages (Reddy et al., 1986). Jenkins and Ali (1999) reported rapid expansion of foliage due to P application leading to early attainment of maximum canopy cover and higher yield at early harvests. In addition to the growth characters, P and K nutrition also affects the yield directly through its effect on either tuber number or their size or both. Sharma and Arora (1987) reported an increase in both tuber number and weight due to P and K application. The proportion of tubers in different grades is affected differently by P and K nutrition. P decreased the number of tubers in the large and medium grade and increased the proportion of small-sized tubers. K application on the other hand increased the number of large and medium-sized tubers and decreased small-sized tubers (Grewal and Trehan, 1993).

K also has profound influence on the quality and stress resistance of the potato crop. It increases the contents of total and reducing sugars while tuber dry matter decreases, the decrease is more with muriate of potash (MOP) than when sulfate of potash (SOP) is applied. Application of SOP produces tubers with higher dry matter, starch, and vitamin C content. Application of K through MOP decreases enzymatic discoloration and phenol content thereby reducing the browning of potato chips. Tubers with higher K content show better skin curing and hardening, and lesser rottage (Grewal and Trehan, 1993).

A strong interaction exists between frost injury and K nutrition of potato in K deficient soils. Crop grown on soils supplied with sufficient quantity of K is less liable to be affected by frost (Figure 6.6). High K concentration in leaf is said to help in lowering the freezing point of cell sap thereby helping the plants to escape frost injury. In general, highly frost-susceptible varieties are also highly responsive to K application (Grewal and Singh, 1980).

Potassium also helps the plants to achieve water economy. Its application although does not economize on water needs of the crop, it increases water use efficiency in terms of tuber yield$\cdot$mm$^{-1}$ water. This effect is more in rain-fed crop where drought stress often occurs during plant emergence and tuber initiation stage (Grewal and Trehan, 1993).

Adequate K application also imparts resistance to late blight. It has also been reported to reduce incidence of black scurf as well as many other diseases and pests in potato (Perrenoud, 1977). Variety, soil fertility, season, source of nutrients, time, and method of application are some of the factors affecting the P and K requirement of potato.

### *Variety*

Differential response of potato varieties to both P and K application has been reported (Pushkarnath, 1976). These differences have

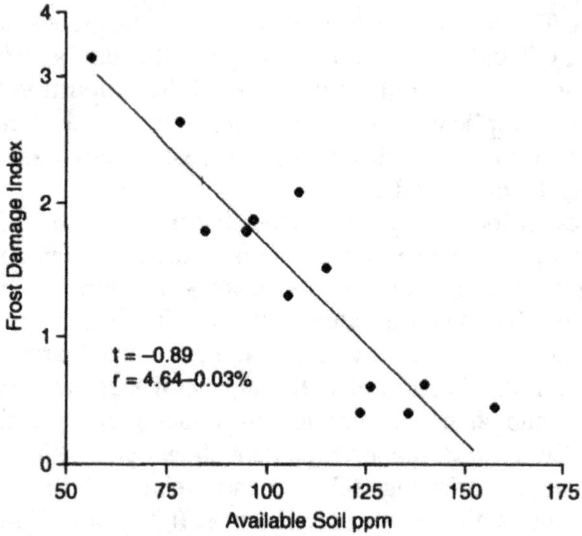


FIGURE 6.6. Relation between available soil K status and frost damage. *Source:* After Grewal and Singh (1980).

been attributed to differences in the root volume and to a strong interaction of the yield potential and the proportion of large-sized tubers in the produce. Short duration, rapid bulking varieties producing large-sized tubers show better response to K application than long-duration varieties yielding a large proportion of small-sized tubers (Grewal and Trehan, 1993). The difference in cation exchange capacity of the roots is also responsible for the differential behavior of the varieties (Grewal and Sharma, 1978).

### *Soil Fertility*

The nutrient supplying capacity of the soil affects the response to both P and K. The response varied greatly not only between different soil groups but also within the same soil group. The critical nutrient content of soils below which response to P and K application is likely to be obtained has been estimated (Table 6.4). The critical limits varied with season, soil type, and also the extractant used to determine the P and K content and have to be worked out for each situation.

TABLE 6.4. Critical limits of available phosphorus and potassium in soils.

| Soil group | Critical limits | Season | Method of estimation |
|---|---|---|---|
| Acidic Brown hill | 16 ppm P | Summer | Olsen's method |
| Acidic Brown hill | 35 ppm P | Summer | Bray $P_1$ method |
| Acidic Brown hill | 120 ppm K | Summer | $NH_4OAC$ method |
| Acidic Brown hill | 65 ppm K | Summer | Morgan's method |
| Alluvial Soils | 10 ppm P | Autumn | Olsen's method |
| Alluvial Soils | 13 ppm P | Spring | Olsen's method |
| Alluvial Soils | 105 ppm K | Autumn | $NH_4OAC$ method |
| Alluvial Soils | 112 ppm K | Autumn | Morgan's method |

*Source:* After Grewal and Trehan (1993).

### *Season*

The season influences P and K uptake through its effect on nutrient availability as well as yield potential. In a cool season, response to P application was higher than in a warmer season (Grewal and Trehan, 1993). The critical limits of available P were also higher in a cool season than in a warmer season. In the case of K, response was higher in a season with higher yield potential and in a season with ample moisture in the case of a rain-fed crop.

### *Time of Application*

Invariably both P and K are applied at the time of planting since they are required early in the life of the crop (Sahota and Govinda-krishnan, 1984). K, however, can be applied in splits also (Trehan and Grewal, 1991).

### *Source*

Water-soluble P fertilizers are preferred for potato due to its shallow root system. However, in acidic soils water-insoluble P fertilizers (rock phosphates) can also be used in combination with water-soluble P source in a suitable ratio that is required to be worked out for each location (Pushkarnath, 1976; Sahota and Sharma, 1988). Regarding K sources, sulfate of potash has an edge over muriate of potash (Sharma et al., 1976). Scheonite, however, is only as effective as muriate of potash (Grewal and Sharma, 1978).

### *Method of Application*

The availability of P is a problem both in acidic and alkaline soils. In acidic soils Fe and Al ions fix P whereas in alkaline soils Ca and Mg ions fix it. Therefore, it is generally recommended to be placed near the seed tubers (Grewal and Sud, 1990). Availability of K is, however, not a big problem (Grewal and Trehan, 1993) because the different forms of K, namely solution K, exchangeable K, and the fixed K are in dynamic equilibrium and also because K is mobile. Its application in furrows is, however, recommended to minimize leaching losses. Foliar application of K through 2 percent KCl is also advocated in some long-duration varieties (Grewal and Trehan, 1993).

### *Micronutrients*

In recent years, deficiency of micronutrients has frequently been observed especially in tropical countries due to increase in cropping intensity, increased use of high-analysis fertilizers, higher micronutrient requirements of high-yielding varieties, and less use of organic manures (Trehan and Sharma, 1999). Zinc is generally the most deficient micronutrient in potato-growing soils, followed by iron, copper, and manganese. Cultivar, soil types and their nutrient status, organic manures, crop residues, cropping pattern, cultural practices, and nutrient interactions influence the response to micronutrients (Trehan and Grewal, 1983; Trehan and Sharma, 1999; Singh et al., 1994).

Rapid bulking, short duration, and high-yielding cultivars such as Kufri Chandramukhi and Kufri Alankar showed higher response to zinc as compared to Kufri Badshah, Kufri Chamatkar, Kufri Sindhuri, and Kufri Jyoti. This trend, however, was not observed for response to other micronutrients. Kufri Badshah and JH-222 gave higher response (21-34 $q \cdot ha^{-1}$) to iron than Kufri Bahar, Kufri Chandramukhi, Kufri Jyoti, and Kufri Sindhuri (Trehan and Sharma,

1999). Kufri Badshah was also more responsive to boron than Kufri Chandramukhi, Kufri Bahar, JH-222, Kufri Jyoti, and Kufri Sindhuri, whereas Kufri Jyoti gave higher response to manganese as compared to other cultivars (Trehan and Sharma, 1999).

Long-term adoption of intensive crop rotations decreased micronutrient status of soils but crop residue incorporation improved it. Singh et al. (1994) reported decrease in micronutrient status in paddy-potato-sunflower rotation when only NPK fertilizers were applied but when residues of the crops were incorporated then the micronutrient status improved. Incubation studies showed that residues of sunflower, paddy, and maize increased zinc availability up to 50, 20, and 10 percent, respectively after 50 days of incubation (Trehan and Sharma, 1999).

Cultural practices like land shaping for gravity irrigation of crops exposes the subsoil that is generally deficient or poor in available micronutrients (Trehan and Sharma, 1999). Hence, micronutrient deficiency has been noticed as a limiting factor in potato production on recently leveled lands. The interaction of macro and micronutrients also affects their availability and the response of potato to a particular micronutrient. High levels of either phosphorus or zinc create imbalance between phosphorus and zinc concentration in leaves, which adversely affects the tuber yield.

The response of potato to the micronutrients is also dependent on available NPK level. Different micronutrients (zinc, iron and manganese) increased the response of potato to NPK fertilizers by 3-19 kg of tubers for every kilogram of NPK applied indicating better utilization in the presence of micronutrients (Trehan and Sharma, 1999).

### *Environmental Factors Causing Micronutrient Deficiencies*

Trehan and Sharma (1999) reviewed the requirement of micronutrients for potato and opined that soil properties such as pH, redox potential, organic matter content, and environmental factors, such as soil water content, temperature, and light influence micronutrient deficiency or toxicity symptoms. For example, the severity of iron chlorosis is increased both by a combination of high soil moisture and cool temperature, and by hot and dry summers. They reported that in the northern Indo-gangetic plains, potato crop exhibits iron deficiency symptoms in December/January when the temperatures are low. Manganese availability is lowest in cold soils, but increases rapidly as the soil warms up and passes through several wetting and drying cycles. Copper availability is a problem in organic soils. Zinc deficiency symptoms normally are more severe when plants are young and temperatures are relatively low. Boron deficiency is more prevalent during a dry summer, particularly if it follows a wet winter or spring. Environmental factors apparently have a relatively small direct role on the incidence of molybdenum deficiency except that its deficiency is more pronounced in high rainfall areas.

*Methods of Ameliorating Micronutrient Deficiency*

Soil application of a micronutrient carrier is the most widely used method. This is particularly suitable when a relatively large quantity has to be applied to compensate the high fixation capacity or quick reversion of the nutrient elements to unavailable forms. Foliar application of micronutrients is also widely adopted especially in situations where it can be combined with other plant growth/protection chemicals. The third approach consists of treating seed tubers with micronutrient compound/solution.

Correction of zinc and copper deficiencies may be made through inorganic fertilizers, for example zinc sulfate and copper sulphate or chelates. Spray application of zinc was superior to soil application and seed treatment when soil was marginal in available zinc. In highly deficient soil, seed treatment with zinc salts was the best. The dip treatment of seed tubers in 2 percent zinc oxide prior to planting is recommended to meet zinc needs of potato crop. Application of farmyard manure at 10 tons·ha$^{-1}$ is also able to meet zinc requirements of potato crop (Trehan and Sharma, 1999).

Correction of iron deficiency in the field is best accomplished by iron sprays. Trehan and Sharma (1999) reported that foliar sprays with acidified solution of 0.3 percent ferrous sulphate or with 0.1 percent iron citrate reduced iron chlorosis and significantly increased potato tuber yield of cv. Kufri Jawahar.

### *Other Effects of Micronutrients*

Apart from influencing yield, micronutrients can also help to mitigate insect pests and disease damage. Zinc sulfate per se has some fungicidal value (Trehan and Sharma, 1999). It also increased the fungicidal efficiency of tribasic copper sulfate, copper oxychloride, and Bordeaux mixture. Likewise, boric acid treatment of potato tubers controls common scab and black scurf (Somani, 1986). Black Scurf incidence was reduced from 97 to 12 percent by dipping the mother seed tubers in 0.05 percent $ZnSO_4$ + 1 percent acetic acid solution for 15 min before planting (Somani, 1986). Leaf spots observed in the early planted crop in Indo-gangetic plains were controlled by foliar sprays of 0.2 percent $MnSO_4$ (Trehan and Sharma, 1999). Spray application of boron also reduced leaf spot intensity significantly. Foliar sprays of micronutrients (Zn, Cu, Zn + Cu + Mo) proved effective in reducing late blight by 65.4 to 92.6 percent and increasing tuber yield by 36.3 to 45.9 percent. In blight epidemic years the micronutrient mixture (Zn+Cu+Mo+Mn) and mancozeb sprays were significantly effective in controlling late blight by 42.5 and 31.2 percent, respectively, and increased tuber yield, though, this treatment was less effective than systemic fungicide (95.7%) metalaxyl + mancozeb (Trehan and Sharma, 1999).

Micronutrients also affect the nutritional quality of potato. Zinc, copper, manganese, boron and molybdenum application increased ascorbic acid content of tubers (Das and Jena, 1973; Trehan and Sharma, 1999). Zinc fertilization reduced the content of tyrosine, orthodihydroxy, and total phenols but its high dose enhanced reducing sugars, sucrose, and total sugars in tubers (Marwaha, 1989). Application of $ZnSO_4$ up to 20 $kg \cdot ha^{-1}$ in zinc deficient alluvial soils improved processing quality as it increased starch content and reduced phenolic compounds without an increase in sugars (Marwaha, 1989). Molybdenum and boron application have also been reported to increase starch content of tubers (Das and Jena, 1973; Dwivedi and Dwivedi, 1992).

Micronutrients also affect storage properties. Application of boron and molybdenum to the potato crop reduced the percentage of rot in all grades of tubers under laboratory conditions at 126 days after harvest (Dass and Jena, 1973). Application of molybdenum and zinc to potato crop decreased weight loss of potato from 20 to 18 and 19 percent, respectively, during the storage period of 16 weeks at room temperature (Trehan and Sharma, 1999). The relationship calculated between percent loss in weight and nutrient concentrations in potato tubers of nine genotypes indicated that the loss in tuber weight was negatively correlated with copper ($r = -0.463$) and potassium ($r = -0.420$) concentration in tuber peel. Thus, genotypes having higher concentration of copper and potassium in tuber peel are expected to be better keepers. This is corroborated by the fact that Phulwa, an old indigenous Indian potato variety, which has good keeping quality even in the country stores, is very rich in micronutrients (Trehan and Sharma, 1999).

### *Water Management*

Water productivity of potato has been reported to range between 54 and 120 $kg \cdot ha \cdot mm^{-1}$ (Wright and Stark, 1990). This water should be available to the crop at low soil water tension because potato is very sensitive to drought stress. Many reasons have been attributed to its high sensitivity to drought stress. Leaf water potential of potato is lower than that of other crops and its recovery is also slower (Ackerson et al., 1977). Photosynthesis also decreases at a relatively higher leaf water potential (Munns and Pearson, 1974) and so is the case with stomatal closure (Rijtjema and Abou Khaled, 1975). Moreover, the root system in potato is invariably confined to the topsoil layer only. Though the rooting depth of potato is reported to range from 59 to 140 cm (Stalham and Allen, 2001), most studies suggest that it is confined to shallow depths due to various soil and moisture limitations. Ovaa and de Smet (1984) observed it to be 50 cm while Gregory and Simmonds (1992) estimated that top 30 cm of the soil had one-third of the total dry weight of the roots but accounted for 82 percent of the root volume.

The response to water deficit depends upon the stage of the crop as well as the severity of water deficit. Sufficient soil moisture at planting ensures quick emergence (Cavagnero et al., 1971) whereas stress during the period of emergence and canopy development reduces the number, size, and longevity of the leaves (Jefferies, 1989; Walworth and Carling, 2002), which reduces the radiation intercepted by the crop. Jefferies and MacKerron (1993) reported that yield is related to the leaf area duration and that drought stress reduces it. Drought also affects the tuber development through its effect on the number and size of tubers (Mac Kerron and Jefferies, 1986; Belanger et al., 2001). The bulking rate is reduced due to decrease in leaf area, and a reduction to the extent of 40 percent has been reported (Belanger et al., 2001). Yield reduction due to drought in later stages is caused by the decrease in proportion of large tubers (MacKerron and Jefferies, 1988).

Drought stress also affects the quality of produce; the specific gravity gets reduced and the proportion of tubers with dark end increases (Stark and McCann, 1992). Short periods of water stress lead to knobby or dumb bell-shaped tubers as well as cracking in some varieties. According to Jefferies and MacKerron (1987) tuber-cracking is either due to rewetting after prolonged dry periods or due to occurrence of wet conditions and rapid tuber growth.

### *Irrigation*

When, how, and how much to irrigate are the common decisions to be taken regarding irrigation. Bailey (2000) described the principles of irrigation scheduling that can be used when water availability is not a constraint. After heavy rain or irrigation, the soil is saturated for a few hours. Surplus water soon starts to drain downward, wetting the layers below leaving the layers above at field capacity. It is convenient to consider this water in terms of the suction required to remove some of it. At field capacity, water removal requires suction (soil moisture tension) somewhere between −5 and −10 kPa. As crop removes water from the soil, the soil arrives at a soil moisture deficit (SMD). The SMD at field capacity is zero. If a crop extracts 50 mm of water from a soil, and there has been no rainfall or irrigation to replenish it, the soil has an SMD of 50 mm. If, during the following four days, the crop extracts a further 10 mm of water, the SMD is then 60 mm. If rainfall or irrigation then adds 20 mm of water back to the soil, the SMD becomes 40 mm. As this process occurs, and water is removed from the soil, the water remaining is that which is held at a greater tension, that is it requires a greater suction to remove it. At field capacity, or when the SMD is low, water supply is unrestricted and crop water uptake is at a maximum, the rate being determined by weather conditions. As the SMD increases, the soil water tension increases, water becomes more difficult to extract, and plants respond by closing stomata and thereby reducing water loss to the atmosphere. This is an important stage, and the SMD at which it occurs is known as the critical SMD, because closing the stomata will result in less intake of carbon dioxide, and therefore lower rates of photosynthesis and potentially lower yields. This stage is referred to as the refill point. Irrigation scheduling aims at maintaining soil moisture somewhere between the refill point and field capacity. Therefore when there is ample supply of water, irrigation at or before critical SMD needs to be resorted to. The critical soil moisture status depends upon soil texture and nature of the crop. Typical values for critical SMD in different textured soils (Table 6.5) have been reported for potato (Bailey, 2000).

It is also to be ensured that the amount of water applied does not exceed the quantity required to return to field capacity because irrigating beyond field capacity will only initiate drainage, reduce irrigation efficiency, and possibly give rise to problems associated with poor soil aeration except under special situations like saline soils.

TABLE 6.5. Thresholds of soil water potential (tension) and crop responses.

| **Scale of water potential as it reflects the need for irrigation** | |
|---|---|
| 0 KPa | Saturated soil. If prolonged, anaerobic conditions likely. |
| 1-10 KPa | Surplus water in the soil. If prolonged, the soil is not draining. Conditions favorable for disease development. |
| 10-20 KPa | Plenty of water and air in the soil. Do not irrigate. |
| 20-40 KPa | Adequate water in the soil. Prepare to irrigate as the readings approach the upper limit. |
| 40-60 KPa | Sufficient water in fine soils if the crop has a well-developed root system. In medium soils consider irrigation. In sandy soils, irrigate. |
| 60-80 KPa | Readily available water is almost all used except on heavy soils. Wilting imminent. |
| 80-100 KPa | Do not let the soil get this dry. On rewetting there may be problems with tuber growth cracking. Tensiometer no longer reliable. |

*Source:* De Neve et al. (2000).

When the water resources are limited, it is necessary to schedule irrigation according to priority so that stages at which water stress causes high-yield loss get priority over other stages. This is especially important where surface irrigation is adopted because in this system of irrigation even if the actual water requirement is less, a minimum quantity (normally 50 cm) has to be compulsorily applied due to practical problems. One of the methods that have been tested to determine the water stress susceptibility of different crops is the stress day index (SDI) approach. The SDI is the product of the crop susceptibility factor (CS) and the stress days (SD) and indicates the actual crop stress that is responsible for the reduction of potential yield of the crop (Hiller et al., 1972). The CS factor indicates the susceptibility of the crop stage to a given water deficit whereas SD indicates the stress (extent and duration) that a particular crop stage can tolerate. Using this concept to schedule irrigation a saving in water to the extent of 100-150 mm has been reported in the Indo-gangetic plains (Hukkeri, 1982). Another approach is to irrigate only alternate furrows at each irrigation so that every furrow is irrigated only half the total number of times. Although this has been found to economize water by 38 percent, there is a corresponding yield loss of 10.7 percent (Bhan and Dhama, 1982).

As regards method of irrigation, it can be done through sprinklers, trickle (drip), and surface (furrow) methods. Sprinkler and drip systems are efficient in regulating the amount of water to be applied and conveyance losses are also minimized, but they are expensive to establish and maintain. Therefore, surface methods are common especially in developing countries.

### *Weed Management*

Weeds compete with potato for nutrients, light, space, as well as moisture thus aggravating their stresses especially when in short supply. In addition, they retard the growth of the crop through their allelopathic effect. They also hamper smooth harvest of the crop leading to tuber damage and increase in the cost of harvesting. Studies on nutrient uptake by weeds in short-duration potato crop in the Indo-gangetic plains showed that 21, 7, and 19 $kg \cdot ha^{-1}$ of N, P, and K, respectively, were taken up by the weeds (Govindakrishnan and

Singh, 1985). In the northwestern hills of India where the crop duration is longer, the uptake of N, P, and K by the weeds to the extent of 43, 8, and 49 kg·ha$^{-1}$, respectively, have been reported (Nankar and Singh, 1982). Thus, weeds result in a great degree of competition to the potato crop.

The stage of weed control is important for efficient weed management. The period during which weeds cause greatest damage is called the critical period. In the Indo-gangetic plains where plant emergence occurs within 10-12 days of planting in the autumn season, the first 25 days after planting were found to be critical (Jaiswal, 1992). In the highlands where emergence is slow and the autotrophic growth sets in later, the critical period would also be later. Similarly, in the temperate zone also where the onset of autotrophic growth phase is late, Nieto et al. (1968) found that weed-free period of three weeks between six and nine weeks after planting is necessary for high yield.

As regards method of weed control, mechanical weed control is well suited to potato crop especially when full soil cover of the seed tubers is not done at planting to hasten emergence. An intercultural operation—earthing up, is therefore necessary to cover the emerging stolons and tubers. This cultural operation necessitates loosening of the soil, and during this operation weeds can be easily removed. Weed control through cultural means is desirable since this would decrease the dependence on herbicides. Conley et al. (2001) reported cultivation and hilling to be possible alternatives to herbicide use. However, its effectiveness is determined by many factors. The mechanical methods are generally effective if weather is favorable (Irla, 1995), otherwise weed growth is heavy. Wesenberg and Wulf (1995) reported mechanical weed control to be affected by weather, risk of injury due to postemergence cultivation, and late weed flushes. Hilling and one cultivation reduced weed biomass by 98-99 percent when weed intensity is moderate, whereas at high weed densities only 30-61 percent of weed biomass was removed (Eberlein et al., 1997). Further, the mechanical methods are feasible only early in the season as development of canopy prevents cultivation after about four weeks. Connell et al. (1999) therefore, advocated combined use of herbicides and mechanical weed control.

Mulching and growing cover crops have also been tried for weed control. In the Indo-gangetic plains, mulching combined with earthing up gave up to 82 percent weed control in addition to increasing the tuber yield (Jaiswal and Lal, 1996). Mulching also helped conserve soil moisture in addition to controlling weeds, in a dry season (Sharma, 1994). Cui et al. (2000) also reported that mulching effectively controlled weeds, but its effect diminished as the canopy developed (Cui et al., 2002). Interseeding of cover crops supplemented with post emergence herbicides also decreased herbicide requirement (Rajalahti et al., 1999).

*Decision Support in Weed Control*

MLHD (Minimum Lethal Herbicide Dose) is a decision support system for herbicide use in integrated weed control. Ketel and Lotz (1997, 1998) derived a formula for calculating the MLHD for photosynthesis-inhibiting herbicides metribuzin and fenmedifam for controlling the weeds *Chenopodium album* and *Polygonum persicaria.*

$$MLHD = FW_{leaf} \times T \times MW_{herb} \times R \times 10^{-4}$$

where $FW_{leaf}$ = mean fresh weight of the leaves per surface unit (g), substituted later by developmental stage of weeds; $T$ = target size, composed of the number of target sites (chloroplasts) per gram of leaf fresh weight and the number of herbicide molecules needed to inactivate one target. $MW_{herb}$ is the molecular weight of the herbicide. $R$ is a ratio parameter and $10^{-4}$ is a scale constant.

Standard values for $T$ and $R$ were estimated with regression analysis. Parameters for weed-sensitivity class, for correction of weather effects, and formulation have also been introduced.

The MLHD decision support concept basically consists of three steps (Kempenaar and van den Boogaard, 2004):

1. Advice on a minimum lethal herbicide dose is obtained on the basis of field-specific information on the crop, weeds (species and stages), and weather.
2. Shortly after herbicide application (two days or more), a sensing technique is used to assess whether the dose will be

effective. The technique may consist of using either a chlorophyll fluorescence meter or a meter based on the photosystem I (PS1) technique that measures light-induced changes in absorbance at around 820 nm. This wavelength is specific for reactions occurring in PS1. If herbicides block electron transport from photosystem II to photosystem I, reactions in photosystem I will be affected.

3. A decision is made whether additional herbicide application is needed based on the readings of the measurements under step 2, and crop and weed information. If readings are below a certain level, an additional treatment is recommended.

## *MULTIPLE CROPPING*

The tropical countries are endowed with plenty of sunshine. This coupled with moderate temperatures during winter makes cultivation of more than one crop possible wherever water is not a limitation. Therefore, during the initial stages of development of agriculture in these regions, emphasis was laid on increasing the irrigated area and raising the cropping intensity. The important crops during the rainy season in these countries are rice, jute, bajra (Pearl millet), sugarcane, and groundnut whereas during the winter season wheat, potato, sorghum, and vegetables are the main crops. Among these, rice in the rainy season and wheat during the winter season have emerged as the most important cropping systems.

In the tropical plains, especially in regions where potato is a short-duration crop (80-120 days), cultural practices requiring high inputs have been adopted for getting high yield. Salient potato-production practices in these countries include mechanization for faster cultural operations, presprouting and use of medium-sized tubers to hasten emergence and early growth, frequent irrigation for quick emergence as well as to mitigate the effect of higher temperature at planting, and high fertilizer doses to ensure early establishment of full crop canopy.

The widespread adoption of rice-wheat cropping system (the major staple crops in the region) has led to many sustainability issues cropping up. Some of them are depletion of groundwater, declining soil fertility and factor productivity, increase in the weed intensity etc. Therefore, diversification of rice—wheat cropping system is called for and development of appropriate cropping systems is one the important ways of improving the input use efficiency of the system as a whole. In other cropping systems also, inclusion of potato is desirable since the crops would complement each other and the efficiency of input use of the system as a whole would be improved. Effective utilization of residual nutrients applied to potato by the other crops in cropping systems has been widely studied. Wheat following potato utilized 13.5 percent of the fertilizer N applied to potato (Sharma et al., 1988a) and 6.5 percent of the P applied to potato (Sharma et al., 1988b). In another study Satnarayana and Arora, (1984) reported that onion following potato fertilized with 150 Kg $N \cdot ha^{-1}$ gave yield equivalent to that obtained by applying 50 kg $N \cdot ha^{-1}$ directly to it. Weed control efficiency was also improved in potato-based cropping systems. Govindakrishnan and Singh (1985) reported that application of Metribuzin and Oxyfluorfen for weed control in potato provided effective weed control in the succeeding wheat crop also.

In conclusion, it can be inferred that a good understanding of the ecophysiological requirements of the different growth phases of the potato crop has been reached. This has led to its cultivation in a wide range of environments and development of agrotechniques for these areas. The quantification of the ecophysiological requirements has also enabled modeling of its growth and development, which would give further fillip to the sustainable development of the potato crop. Lastly, to ensure the delivery of the scientific knowledge to the actual users in a manner suitable for its adoption, decision support systems on various aspects of potato cultivation have been developed.

Potato production is sustainable if the three P's requirements are met, that is, People should benefit from healthy nourishment and socially acceptable production methods, there must be Profit in it for all links of the chain, and the Planet should not suffer. This calls for integrated management of pests, diseases, crops, and farms increasingly making use of DSS-based models and sensing techniques. There is an increased need for decision support systems to make the fullest use of available knowledge in order to make the best use of less or fewer resources. A second reason is that in large farms the farmers benefit more from assistance with their decisions as the cost per unit area would be low. Third, and maybe most important, improved access to knowledge as such and to knowledge aggregated in models, and the availability of sensing techniques and of information and communication technology can improve the services given by extension agents. Lastly, governments are rapidly reducing their direct support of the primary production sector of society, especially the farming community. Many private consultancy firms are being formed and they will serve farmers best if tools are available for decision support. These four reasons are likely to become more important in the coming years. With these developments it is hoped that potato industry will continue to flourish, arresting the declining trend of its growth in Europe and maintaining its increasing trend in Asia, the two major potato growing continents.

## REFERENCES

Ackerson, R.C., Krieg, D.R., Miller, T.D., and Stevens, R.G. (1977). Water relations and physiological activity of potatoes. *Journal American Society of Horticultural Science 102:*572-575.

Adinarayana, V. (1987). Requirement of nitrogen, phosphorus and potassium for targeted yield of potato (*Solanum tuberosum* L.) variety Kufri Chandramukhi in Nainital tarai region. *Andhra Agricultural Journal 34:*45-52.

Allen, E.J. and O'Brien, P.J. (1986). The practical significance of accumulated day-degrees as a measure of physiological age of seed potato tubers. *Field Crops Research 14:*141-151.

Allen, E.J. and Wurr, D.C.E. (1992). Plant density. In Harris, P.M. (Ed.), *The Potato Crop* (pp. 292–333). London: Chapman & Hall, .

Anonymous (1978). *Report on the Agroecological Zones Project Volume 1. Methodology and results for South and Central America.* World Soil Resources Report 48/3 FAO, Rome.

Bailey, R.J. (2000). Practical use of soil water measurement in potato production. In Haverkort A.J. and MacKerron, D.K.L. (Eds.), *Management of Nitrogen and Water in Potato Production* (pp. 206-218). Wageningen, The Netherlands: Wageningen Press.

Belanger, G., Walsh, J.R., Richards, J.E., Milburn, P.H., and Ziadi, N. (2001). Tuber growth and biomass partitioning of two potato cultivars grown under different N fertilization rates with and without irrigation. *American Journal of Potato Research 78:*109-117.

Ben Khedher, M. and Ewing, E.E. (1985). Growth analysis of eleven potato cultivars grown in greenhouse under long photoperiods with and without heat stress. *American Potato Journal 62:*537-554.

Benoit, G.R., Stanley, C.D., Grant, W.J., and Torrey, D.B. (1983). Potato top growth as influenced by temperatures. *American Potato Journal 60:* 489-502.

Bhan, S. and Dhama, C.S. (1982). Effect of frequency and method of irrigation on potato. *Indian Journal of Agronomy 27:*227-230.

Biemond, H. and Vos J. (1992). Effects of nitrogen on the development and growth of the potato plant. 2. The partitioning of dry matter, nitrogen and nitrate. *Annals of Botany 70:*37-45.

Birecki, M. and Roztropowicz, S.T. (1963). Studies of seed potato size and productivity. *European Potato Journal 6:*1-13.

Blake, G.R., French, G.W., and Nylund, R.E. (1962). Seed bed preparation and cultivation studies on potatoes. *American Potato Journal 39:*227-234.

Bodlaender, K.B.A. (1963). Influence of temperature, radiation and photoperiod on development and yield. In Ivins, J.D. and Milthorpe F.L. (Eds.), *The Growth of the Potato* (pp. 199-210). London: Butterworths.

Booij, R. and Uenk, D. (2004). Crop reflection based DSS for supplemental nitrogen dressings in potato production. In MacKerron, D.K.L. and Haverkort, A.J. (Eds.), *Decision Support Systems in Potato Production: Bringing Models to Practice* (pp. 43-49). Wageningen, The Netherlands: Academic Publishers.

Bremner, P.M. and El Saeed, A.K. (1963). The significance of seed size and spacing. In Ivins, J.D. and Milthorpe, F.L. (Eds.). *The Growth of the Potato* (pp. 267-280). London: Butterworths.

Cavagnero, J.B., de Lis, B.R., and Tizio, R.M. (1971). Drought hardening of the potato plant as an after effect of soil drought conditions at planting. *Potato Research 14:*181-192.

Chambenoit, C., Laurent, F.Machet, J.M., and Boizard, H. (2004). Development of a decision support system for nitrogen management on potatoes. In , MacKerron D.K.L. and Haverkort, A.J. (Eds.), *Decision Support Systems in Potato Production: Bringing Models to Practice* (pp. 50-63). Wageningen, The Netherlands: Academic Publishers.

Conley, S.P., Binning, L.K., and Connell, T.R. (2001). Effect of cultivar, row spacing, and weed management on weed biomass, potato yield, and net crop value. *American Journal of Potato Research 78:*31-37.

Connell, T.R., Binning, L.K., and Schmitt, W.G. (1999). A canopy development model for potatoes. *American Journal of Potato Research 76:*153-159.

Cui, R.X., Lee, B.W., and Lee, H.L. (2000). Growth and yield of potato as affected by paper, oil-treated paper and urea-coated paper mulching in spring season culture. *Korean Journal of Crop Science 45:*216-219.

Cui, R.X., Lee, S.H., and Lee, B.W. (2002). Effect of recycled paper mulch on yield of fall-grown potato. *Korean Journal of Crop Science 47:*68-73.

Dambroth, M. and Bramm, A. (1984). Is ridging necessary in potato cultivation? *Kartoffelbau 35:*12-14.

Das, R.C. and Jena, M.K. (1973). Studies on the effect of soil application of molybdenum, boron and paper mill sludge on the post harvest qualities of potato tubers (*Solanum tuberosum* L.) *Madras Agricultural Journal 60:*1026–1029.

De, R. (1960). Effect of potassium on yield, mineral composition and carbohydrate content of potato varieties. *Indian Potato Journal 2:*70-77.

De Neve, S., MacKerron, D.K.L., and Igras, J. (2000). Measurement techniques for soil and plant water status. In Haverkort, A.J. and MacKerron, D.K.L. (Eds.), *Management of Nitrogen and Water in Potato Production* (pp. 188-205). Wageningen, The Netherlands: Academic Publishers.

de Wit, C.T. (1965). Photosynthesis of leaf canopies. *Agricultural Research Report 663.* Wageningen, The Netherlands.

Doorenbos, J. and Kassam, A.H. (1979). Yield response to water. *FAO Irrigation and Drainage Paper* No 33, FAO, Rome.

Duchenne, T., Machet, J.M., and Martin, M.(1997). Diagnosis of a potato nitrogen status. In Lemaire, G. (Ed.), *Diagnosis of the Nitrogen Status in Crops* (pp. 119-130). Berlin, Germany: Springer.

Dwivedi, G.K. and Dwivedi, M.(1992). Efficacy of different models of application of copper, zinc and boron to potato. *Annals of Agricultural Research 13:*1-6.

Eberlein, C.Y., Patterson, P.E., Guttieri, M.I.J., and Stark, J.C. (1997). Efficacy and economics of cultivation for weed control in potato (*Solanum tuberosum*). *Weed Technology 11:*257-264.

Firman, D.M., O'Brien, P.J., and Allen, E.J. (1992). Predicting the emergence of potato sprouts. *Journal of Agricultural Science, Cambridge 118:* 55-61.

Gillison, T.C., Jenkins, P.D., and Hayes, J.D. (1987). Some factors affecting the expression of the physiological age of potato seed tubers. *Journal of Agricultural Science, Cambridge 108:*437-451.

Goffart, J.P. and Olivier, M. (2004). Management of N-fertilisation of the potato crop using total N-advice software and in-season chlorophyll-meter measurements. In MacKerron, D.K.L. and Haverkort, A.J. (Eds.), *Decision Support Systems in Potato Production: Bringing Models to Practice* (pp. 64-79). Wageningen, The Netherlands: Academic Publishers.

Govindakrishnan, P.M., Sharma, R.C., Khan, I.A., and Sharma, H.C. (1999). A technique for selecting potato genotypes with low nitrogen requirement. *Indian Journal of Agricultural Science 69:*350-354.

Govindakrishnan, P.M. and Singh, J.P. (1985). Weed control in potato-wheat sequence in north-western plains of India. *Pesticides 19:*34-35.

Gregory, P.J. and Simmonds, L.P. (1992). Water relations and growth of potatoes. c.f. Mineral nutrition. In Harris, P.M. (Ed.), *The Potato Crop* (pp. 215-246). London: Chapman & Hall.

Grewal, J.S. and Sharma, R.C. (1978). Response of potato to potassium fertilizers in India. *Proceedings of the International Symposium on Potassium in Soils and Crops* (pp. 309-326). New Delhi, India.

Grewal, J.S. and Singh, S.N. (1980). Effect of potassium nutrition on frost damage and yield of potato plants on alluvial soils of the Punjab. *Plant and Soil 57:*105-110.

Grewal, J.S. and Sud, K.C. (1990). Potato responses to phosphorus in Northern India. *Fertiliser News 25:*25-30.

Grewal J.S. and Trehan, S.P. (1993). Phosphorus and potassium nutrition of potato. In Chadha, K.L. and Grewal, J.S. (Eds.), *Advances in Horticulture Vol 7 – Potato* (pp. 261-297). New Delhi, India: Malhotra Publishing House .

Guenthner, J.F. (2001) *The International Potato Industry.* Cambridge, UK: Woodhead Publishing Ltd.

Gunasena, H.P.M. (1968). *Studies on the Growth of the Potato with Particular Reference to the Efficient Use of Nitrogen and Potassium.* Ph.D. Thesis. Reading University, UK.

Haverkort, A.J. (1990). Ecology of potato cropping systems in relation to latitude and altitude. *Agricultural Systems 32:*251-272.

Haverkort, A.J. and Harris, P.M. (1987). A model for potato growth and yield under tropical highland conditions. *Agricultural and forest Meteorology 39:*271-282.

Haverkort, A.J. and Rutavisire, C. (1986). Fertilizer utilization under tropical conditions. 2. Effect of applying nitrogen, phosphorus and potassium on the relationship between intercepted radiation and yield of potatoes in central Africa. *Potato Research 29:*357-365.

Haverkort A.J., Verhagen, A., Grashoff, C., and Uithol, P.W.J. (2004). POTATO-ZONING: A decision support system on expanding the potato industry through agro-ecological zoning using the LINTUL simulation approach. In MacKerron, D.K.L. and Haverkort, A.J. (Eds.), *Decision Support Systems in Potato Production: Bringing Models to Practice* (pp. 24-40). Wageningen, The Netherlands: Academic Publishers.

Hiler, E.A., van Bavel, C.H.M., Hossain, M.M., and Jordan, W.R. (1972). Sensitivity of Southern Peas to plant water deficit at three growth stages. *Agronomy Journal 64:*60-64.

Hukkeri, S.B. (1982). Irrigation scheduling for higher water use efficiency in potato production. In Nagaich, B.B. (Ed.), *Potato in Developing Countries, Proceedings of the International Seminar on Approaches towards Increasing the Potato Production in Developing Countries* (pp. 144-148). Jalandhar (Punjab): India, Central Potato Research Station.

Ingram, K.T. and McCloud, D.E. (1984). Simulation of potato growth and development. *Crop Science 24:*21-27.

Irla, E. (1995). Tending technique and mechanical weed control in potatoes: Environment friendly tending methods are successful. Berichte, Switzerland: FAT No. 462, pp. 1-7.

Jaiswal, V.P. (1992). Crop weed competition studies in potato. *Journal of Indian Potato Association. 18:*131-134.

Jaiswal, V.P. and Lal, S.S. (1996). Efficacy of cultural and chemical weed-control methods in potato (*Solanum tuberosum*). *Indian Journal of Agronomy 41:*454-456.

Jefferies, R.A. (1989). Water-stress and leaf growth in field-grown crops of potato (*Solanum tuberosum L.*). *Journal of Experimental Botany 40:*1375-1381.

Jefferies, R.A. and MacKerron, D.K.L. (1987). Observations on the incidence of tuber growth cracking in relation to weather patterns. *Potato Research 30:*613-623.

———. (1993). Responses of potato genotypes to drought. II. Leaf area index, growth and yield. *Annals of Applied Biology 122:*105-112.

Jenkins, P.D. and Ali, H. (1999). Growth of potato cultivars in response to application of phosphate fertilizer. *Annals of Applied Biology 135:*431-438.

Jenkins, P.D, Gillison, T.C., and Al-Saidi, A.S. (1993). Temperature accumulation and physiological ageing of seed potato tubers. *Annals of Applied Biology 122:*345-356.

Kawakami, K. (1978). Physiology of yield of underground storage organs. In Gupta, U.S. (Ed.), *Crop Physiology* (pp. 269-309). New Delhi, India: Oxford and IBH.

Kempenaar, C. and van den Boogaard, R. (2004). MLHD, a decision support system for rational use of herbicides: Developments in potatoes. In MacKerron, D.K.L. and Haverkort, A.J. (Eds.), *Decision Support Systems in Potato Production: Bringing Models to Practice* (pp. 183-192). Wageningen, The Netherlands: Academic Publishers.

Ketel, D.H. and Lotz, L.A.P. (1997). A new method for application of minimum-lethal herbicide dose rates. *Proceedings of the Tenth EWRS Symposium 1997* (p. 150). Poznan, Poland.

———. (1998). Influence of allocation and detoxification of metribuzin in *Chenopodium album* on the reliability of prediction of the minimum lethal herbicide dose rate. *Weed Research 38:*267-274.

Kirk, W.W. and Marshall, B. (1992). The influence of temperature on leaf development and growth in potatoes in controlled environments. *Annals of Applied Biology 120:*511-525.

Knowles, N.R. and Botar, G.I. (1991). Modelling the effects of potato seed-tuber age on plant establishment. *Canadian Journal of Plant Science 71:*1219-1232.

Kooman, P.L. and Haverkort, A.J. (1995). Modelling development and growth of the potato crop influenced by temperature and daylength: LINTUL-POTATO. In Haverkort, A.J. and MacKerron, D.K.L. (Eds.), *Potato Ecology and Modelling of Crops Under Conditions Limiting Growth* (pp. 41-60). London: Kluwer Academic Publishers.

Kunkel, R. and Campbell, G.S. (1987). Maximum potential potato yield in the Columbia Basin, USA. Model and measured values. *American Potato Journal 64:*355-366.

Lemaga, B. and Caesar, K. (1990). Relationships between numbers of main stems and yield components of potato (*Solanum tuberosum* L. cv. Erntestolz) as influenced by different day lengths. *Potato Research 33:*257-267.

Lommen, W.J.M. (1994). Effect of weight of potato mini tubers on sprout growth, emergence and plant characteristics at emergence. *Potato Research 37:*315-322.

Machet, J.M., Dubrulle, P., and Louis, P. (1990). AZOBIL: A computer program for fertiliser N recommendation based on a predictive balance sheet method (p. 21-22). *Proceedings of First Congress of the European Society of Agronomy.*

MacKerron, D.K.L and Jefferies, R.A. (1986). The influence of early soil moisture stress on tuber numbers in potato. *Potato Research 29*:299-312.

———. (1988). The distributions of tuber sizes in droughted and differing cultivars. *Potato Research 31*:269-278.

MacKerron, D.K.L., Marshall, B., and McNicol, J.W. (2004). MAPP, the management advisory package for potatoes. In MacKerron, D.K.L. and Haverkort, A.J. (Eds.), *Decision Support Systems in Potato Production: Bringing Models to Practice* (pp. 115-137). Wageningen, The Netherlands: Academic Publishers.

Mackerron, D.K.L. and Waister, P.D. (1985). A simple model of potato growth and yield. Part I Model development and sensitivity analysis. *Agricultural and Forest Meteorology 34*:241-252.

Maidl, F.X. (1995). Considerations on targeted nitrogen application. *Kartoftelbau 46*:56-58.

Manrique, L.A. (1990). Growth and yield of potato grown in the green house during summer and winter in Hawaii. *Communications in Soil Science and Plant Analysis 212*:237-249.

Martin, R.J. (1995). The effect of nitrogen fertilizer on the recovery of nitrogen by a potato crop. *Proceedings of the Annual Conference Agronomy Society of New Zealand 25*:97-104.

Marwaha, R.S. (1989). Changes in chemical constituents of potato tuber with the application of zinc. *Proceedings National Academy Science (India)* 59:229-233.

McCree, K.J. (1974). Equations for the rate of dark respiration of white clover and grain sorghum as functions of dry weight, photosynthetic rate and temperature. *Crop Science 14*:509-514.

McGee, E., Jarvis, M.C., and Duncan, H.J. (1986). The relationship between temperature and sprout growth in stored seed potatoes. *Potato Research 29*:521-524.

Midmore, D.J. (1984). Potato (*Solanum* sp.) in the hot tropics I. Soil temperature effects on emergence, plant development and yield. *Field Crops Research 8*:255-271.

Midmore, D.J. and Prange, R.K. (1991). Sources of heat tolerance amongst potato cultivars, breeding lines and *Solanum* species. *Euphytica 55*:235-245.

Miller, D.E. and Martin, M.W. (1987). The effect of irrigation regime and subsoiling on yield and quality of three potato cultivars. *American Potato Journal 64*:17-25.

Monteith, J.L. (1972). Solar radiation and productivity in tropical ecosystems. *Journal of Applied Ecology* 9:747-766.

———. (1977). Climate and efficiency of crop production in Britain. *Philosophical Transactions, Royal Society (London) 281*:277-294.

Moreno, U. (1985). Environmental effects on growth and development of potato plants. In Li, P.H. (Ed.), *Potato Physiology* (pp. 481-501) London, UK: Academic Press Inc.

Munns, R. and Pearson, C.J. (1974). Effects of water deficit on translocation of carbohydrate in *Solanum tuberosum. Australian Journal of Plant Physiology 1*:529-537.

Nankar, J.T. and Singh, M. (1982). Nutrient removal and uptake as influenced by weed management in potato in Shimla hills. In Nagaich, B.B. (Ed.), *Proceedings of the International Seminar on Approaches towards increasing the Potato Production in Developing Countries* (pp. 144-148). Jalandhar (Punjab), India, Central Potato Research Station,.

Neeteson, J.J. (1989). Evaluation of the performance of three advisory methods for nitrogen fertilization of sugar beet and potatoes. *Netherlands Journal of Agricultural Science 37:*143-155.

Nelson, D.C. and Giles, J.F. (1989). Weed management in two potato (*Solanum tuberosum*) cultivars using tillage and pendimethalin. *Weed Science 37:*228-232.

Ng, E. and Loomis, R.S. (1984). *Simulation of Growth and Yield of the Potato Crop.* Simulation Monographs, Wageningen, The Netherlands: PUDOC.

Nieto, J.H., Brondo, M.A., and Gonzalez, J.T. (1968). Critical periods of crop growth cycle for competition from weeds. *Pest Articles and News Summaries 14:*159-166.

Ogilvy, S.E. and Roberts, A.J. (1992). Colour and duration of light for chitting potatoes. *Aspects of Applied Biology 33:*51-58.

Ovaa, I and de Smet, L.A.H. (1984). Root growth in relation to soil profile and tillage system. c.f. Water relations and growth of potatoes. In Harris, P.M. (Ed.), *The Potato Crop* (pp. 214-246). London: Chapman & Hall.

Perenoud, S. (1977). *Potassium and Plant Health.* IPI Research Topic Series 3, Switzerland, International Potash Institute.

Perumal, N.K. and Sahota, T.S. (1986). Investigations on growth and tuberization of potato at different planting dates and nitrogen levels. *International Journal of Tropical Agriculture 4:*63-72.

Prange, R.K., McRae, K.B., Midmore, D.J., and Deng, R. (1990). Reduction in - potato growth at high temperature. *Role of Photosynthesis and Dark Respiration. American Potato Journal 67:*357-370.

Prihar, S.S., Arora, V.K., Singh, G., and Singh, R. (1995). Estimating potato tuber yield in a sub tropical environment with simple radiation based models. *Experimental Agriculture 31:*65-73.

Pushkarnath (1976). *Potato in Sub-Tropics.* New Delhi, India: Orient Longman.

Rajalahti, R.M., Bellinder, R.R., and Hoffmann, M.P. (1999). Time of hilling and inter seedling affects weed control and potato yield. *Weed Science 47:*215-225.

Randeni, G. and Caesar, K. (1986). Effect of soil temperature on the carbohydrate status in the potato plant (*Solanum tuberosum* L.). *Journal of Agronomy and Crop Science 156:*217-224.

Reddy, L.V., Arora, P.N., and Sharma, R.P. (1986). Effect of different doses of phosphorus and potassium on the growth and tuber yield of potato varieties. *Indian Journal of Agricultural Science 56:*497-502.

Reynolds, M.P. and Ewing, E.E. (1986). Effects of air and soil temperatures on potato tuberization. *American Potato Journal 63:*451-452.

Rijtema, P.E. and Abou-Khaled, A. (1975). Crop water use in the Arab Republic of Egypt. *FAO Report.* RNEA Cairo.

Ross, C.W. (1986). The effect of subsoiling and irrigation on potato production. *Soil and Tillage Research 7*:315-325.

Sahota, T.S. and Govindakrishnan, P.M. (1984). Effect of split application of phosphorus to potato in acidic hill soils of Shillong. *Journal of the Indian Potato Association 11*:126-130.

Sahota, T.S. and Sharma, R.C. (1988). Comparative performance of Mussoorie rock phosphate, single superphosphate and their combinations in potato. *Journal of Agricultural Science, Cambridge 107*:155-159.

Sale, P.J.M. (1979). Growth of potatoes (*Solanum tuberosum* L) to the small tuber stage as related to soil temperature. *Australian Journal of Agricultural Research 30*:667-675.

Santeliz, G. and Ewing, E.E. (1981). Effects of nitrogen fertilization on growth and development of potatoes. *American Potato Journal 58*:517-518.

Satyanarayana, V. and Arora, P.N. (1984) Effect of fertilizers on potato and their residual effect on succeeding onion. *Indian Journal of Agronomy 29*:309-311.

Sharma, J.P. (1994). Production potential and economics of potato under different schedules of irrigation, levels of potassium and methods of weed control. *Journal of Potassium Research 10*:278-281.

Sharma, R.C., Govindakrishnan, P.M., Singh, R.P., and Sharma, H.C. (1988a). Effect of organic manures on economy and utilisation of N by potato and succeeding wheat. *Indian Journal of Agricultural Science 58*:362-397.

———. (1988b). Effect of green manures and fodder crops on crop yield and phosphorus needs of potato – wheat systems in the north – western plains of India. *Experimental Agriculture 24*:97-103.

Sharma, R.C. and Sharma, H.C. (1987). Determination of fertilizer requirements of potato on the basis of soil tests in the north-western hills of India. *Journal of Agricultural Science, Cambridge 109*:171-175.

Sharma, R.C., Singh, M., and Sud, K.C. (1976). Relative responses of the potato to three sources of potassium. *Indian Journal of Agronomy 21*:341-348.

Sharma, U.C. and Arora, B.R. (1987). Effect of nitrogen, phosphorus and potassium application on yield of potato tubers. *Journal of Agricultural Science, Cambridge 108*:321-328.

Singh, J.P., Jaiswal, V.P., and Trehan, S.P. (1994). Nutrients interactions of two intensive potato based cropping systems in north western plains of India. In Shekhawat, G.S., Paul Khurana, S.M., Pandey, S.K., and Chandla, V.K. (Eds.), *Potato – Present and Future* (pp. 110). Shimla, India: Indian Potato Association.

Slater, J.W. (1968). The effect of night temperature on tuber initiation of the potato. *European Potato Journal 11*:14-22.

Somani, A.K. (1986). Non hazardous chemical control of black scurf of potato. *Indian Journal of Agricultural Sciences 56*:366-369.

Spitters, C.J.T. (1987). An analysis of variation in yield among potato cultivars in terms of light utilization and dry matter partitioning. *Acta Horticulturae 214*:71-84.

Spitters, C.J.T. and Schapendonk, A.H.C.M. (1990). Evaluation of breeding strategies for drought tolerance in potato by means of crop growth simulation. *Plant and Soil 123:*193-203.

Stalham, M.A. and Allen, E.J. (2001). Effect of variety, irrigation regime and planting date on depth, rate, duration and density of root growth in the potato *(Solanum tuberosum)* crop. *Journal of Agricultural Science, Cambridge 137:*251-270.

Stark, J.C. and McCann, I.R. (1992). Optimal allocation of limited water supplies for Russet Burbank potatoes. *American Potato Journal 69:*413-421.

Stieberitz, J., Kopp, R., and Lanfermann, M. (1986). Studies on the consideration of inorganic soil nitrogen in the measurement of nitrogen fertilizer application in potato cultivation. *Archive fur Acker und Pflanzenbau und Bodenkunde 30:*623-630.

Subbiah, B.V. and Asija, G.L. (1956). A rapid procedure for the estimation of available nitrogen in soils. *Current Science 25:*259-260.

Trehan, S.P. and Grewal, J.S. (1983). Available micronutrient status of alluvial and hill soils as influenced by intensive cropping and manuring. *Journal of the Indian Society of Soil Science 31:*343-61.

———. (1991). Effect of time and level of potassium application on tuber yield and K composition of plant tissue and tubers of two potato cultivars. *Journal of Indian Potato Association 18:*115-121.

Trehan, S.P. and Sharma, R.C. (1996). Mineral nutrient content in peels and flesh of tubers of some potato genotypes. *Journal of Indian Potato Association 23:*139-143.

———. (1999). Micronutrient requirements of potato. Technical Bulletin No 23. Shimla, India: Central Potato Research Institute.

Trehan, S.P., Sharma, U.C., and Grewal, J.S. (1983). Response of potato to foliar spray of Alpha Napthalene acetic acid and zinc sulphate in different soils. *Indian Journal of Ecology 10:*150-153.

Trehan, S.P., Singh, J.P., and Sharma, R.C. (1998). Nitrogen management through different sources in potato-wheat-maize cropping system in north-western plains of India. *Journal of Indian Potato Association 25:*103-108.

Van der Zaag, D.E. (1984). Reliability and significance of a simple method of estimating the potential yield of the potato crop. *Potato Research 27:*51-73.

Van der Zaag, D.E. and Doorenbos, J.H. (1987). An attempt to explain the differences in yielding ability of potato cultivars based on differences in cumulative light interception, utilization efficiency of foliage and harvest index. *Potato Research 30:*551-568.

Van der Zaag, D.E. and Horton, D. (1983). Potato production and utilization in world perspective with special reference to the tropics and sub tropics. In Hooker, W.J. (Ed.), *Research for the Potato in the Year 2000* (pp. 44-58). Lima, Peru: CIP.

Van der Zaag, P. and Demagante, A.L. (1987). Potato (*Solanum* spp.) in an isohyperthermic environment. I Agronomic management. *Field Crops Research 17:*199-217.

Van Keulen, H. and Stol, W. (1995). Agroecological zonation for potato production. In Haverkort, A.J. and MacKerron, D.K.L. (Eds.), *Potato Ecology and Modelling of Crops under Conditions Limiting Growth* (pp. 357-372). The Netherlands: Kluwer Academic Publishers.

Velayutham, M. (1979). Fertiliser recommendations based on targeted yield concept. *Fertiliser News 24:*12-20.

Versteeg, M.N. and van Keulen, H. (1986). Potential crop production prediction by some simple calculation methods as compared with computer simulations. *Agricultural Systems 19:*249-272.

Vos, J. (1999). Split nitrogen application in potato: Effects on accumulation of nitrogen and dry matter in the crop and on the soil nitrogen budget. *Journal of Agricultural Science, Cambridge 133:*263-274.

Walworth, J.L. and Carling, D.E. (2002). Tuber initiation and development in irrigated and non-irrigated potatoes. *American Journal of Potato Research 79:*387-395.

Wesenberg, M. and Wulf, B. (1995). Mechanical weed management. *Kartoffelbau 46:*100-103.

Westerman, D.T. and Sojka, R.E. (1996). Tillage and nitrogen placement effects on nutrient uptake by potato. *Soil Science Society of America Journal 60:*1448-1453.

Wiersema, S.G. (1989). Comparative performance of three small seed tuber sizes and standard size seed tubers planted at similar stem densities. *Potato Research 32:*81-89.

Wright, J.L. and Stark, J.C. (1990). Potato. In Stewart, B.A. and Nielson, D.R. (Eds.), *Irrigation of Agricultural Crops. Agronomy Monograph No 30* (pp. 859-888). USA: ASA-CSSA-SSSA.

Wurr, D.C.E. (1974). Some effects of seed size and spacing on the yield and grading of the main crop potato varieties I Final yield and its relationship to plant population. *Journal of Agricultural Science, Cambridge* 82:37-45.

Yadav, A.C., Sharma, S.K., Kapoor, Ashok, Singh, Avtar, and Mangal, J.L. (2001). Response of potato to sodic water irrigation with and without amendments. *Haryana Journal of Horticultural Sciences 31:*129-132.

Zanker, J. (1986). Pre-ridging of potatoes—a procedure with complex effects. *Tagungsbericnt Akademie-der-Landwirtschaftswissenschaften-der-Deutschen-Demokratischen-Republik 250:*37-41.

# Chapter 7

# Abiotic Stresses

Kazuto Iwama
Junichi Yamaguchi

The highest average yield of potato (*Solanum tuberosum* L.) is obtained in regions with moderate climate, such as the northern part of the United States and the northwestern parts of Europe. The longer days and the moderate temperatures of the temperate regions facilitate longer periods of photosynthesis, efficient translocation of assimilates from the leaves to the tubers, and low transpiration loss. MacKerron et al. (1982) reported that one of the high-yielding cultivars, Maris Piper, had the fresh tuber yield of 107 $t \cdot ha^{-1}$ in the United Kingdom. Isoda et al. (1988) reported the fresh tuber yield of 135 $t \cdot ha^{-1}$ in the northern Japan. However, these high tuber yields were achieved with plentiful irrigation or rainfall and using modern cultivation methods. In the United States, tuber yields in the irrigated fields in the state of Washington have been rising at 0.9 $t \cdot ha^{-1}$ per year since 1950 and reached almost 70 $t \cdot ha^{-1}$ by 1990, whereas the yield in the rain-fed fields in Maine has been stagnant during the period and was less than 30 $t \cdot ha^{-1}$ in 1990 (Walker et al., 1999). In tropical and semitropical climates, potato is generally exposed to high day and night temperatures combined with either high humidity or comparatively dry atmosphere. Though during the past 50 years increase in potato area has been mainly in the subtropical lowlands (Hijmans, 2001), the yield levels in these regions are much lower than those in the temperate regions, mainly due to heat stress (Van der Zaag and Demagante, 1987; Malik et al., 1992).

The predicted increase in temperature due to anticipated global climate change (Houghton et al., 2001) may result in severe drought and heat stress. Abiotic stresses would thus pose a great challenge to sustained growth in world potato production. This chapter describes various aspects of potato growth under abiotic stresses, especially focusing on drought and high temperature.

## *PHYSIOBIOCHEMICAL BASIS OF TUBER FORMATION AND BULKING*

Tuber formation starts with the swelling of the first internode behind the apical bud of the stolon (Cutter, 1992). Internode swelling is due to the radial expansion resulting from cell division or enlargement. The environmental and physiological factors affecting tuber formation have long been investigated (reviewed by Cutter, 1992; Struik and Ewing, 1995 and also in Chapter 6, "Ecophysiology and Agronomic Management" of this book). In principle, short day length and low temperature (especially at night) promote tuberization leading to early tuber initiation and more number of tubers. Some growth substances are generated in the plant in relation to these environmental factors. Gibberellic acid (GA) generally inhibits tuber formation. Cytokinines and abscisic acid (ABA) promote tuber formation. Kraus (1978) and Krauss and Marschner (1982) showed that the ratio of GA/ABA influenced the growth of the potato plant: a relatively high ratio promoted shoot growth and inhibited tuber growth whereas a relatively low ratio limited stem growth and promoted tuber growth.

High GA level was found to reduce the activity of ADPG-pyrophosphorylase in the tuber, slowing or stopping tuber growth (Mares et al., 1981). This may explain the reported association of tuber formation with high concentrations of soluble sugars in the stolon tips (Slater, 1968). The higher level of sucrose required in the media for in vitro tuber formation (Gopal et al., 1998) also supports this. Jasmonic acid and its related compounds (tuberonic acid and its glycosides) have also been reported to stimulate tuber-induction (Koda et al., 1988) due to the expansion of cells (Takahashi et al., 1994), but their effect varied with genotype (Jackson and Willmizer, 1994). Much work has been done to elucidate the mode of action of growth regulators leading to tuber-induction, but experimental results are unequivocal (Cutter, 1992; Gopal et al., 2004).

Within a few days of tuber initiation, tuber becomes about 1 cm in diameter (Kawakami et al., 2003). Thereafter, the weight and the volume of tuber increase almost linearly, which is called tuber bulking (Cutter, 1992). Despite the absence of synchrony in tuber initiation within the plant, in most genotypes the formation of new tubers almost stops with flower initiation. The duration and the rate of tuber bulking also vary with cultivars and environmental conditions. Generally speaking, the duration of tuber bulking is longer and the rate is lower in late-maturing genotypes than in early-maturing ones, and in long days than in short days (Ewing, 1981). On the other hand, shoots and roots continue to grow longer and become greater in late-maturing genotypes (Iwama et al., 1981) and in long days (Iwama, 1988), indicating that requirements for shoot and root growth are opposite to those required for early tuberization and fast tuber bulking. Root and shoot growths are strongly affected by partitioning of dry matter to tubers during an early phase of tuber bulking (Iwama et al., 1995).

## *DROUGHT STRESS*

Although water is essential for plant growth in all species, potato is very sensitive to drought. Much work has been done on this aspect and periodically reviewed (Harris, 1978; Loon, 1981; Burton, 1989; Gregory and Simmonds, 1992; Jefferies, 1995; Bowen, 2003). However, the progress in the improvement for drought-tolerance is still limited. The following sections deal with aspects relevant to understanding and managing drought stress.

### *Transpiration*

Burton (1989) estimated the potential transpiration in potato based on the theoretical water emission from leaf stomata due to inevitable carbon dioxide absorption. Climatic conditions, such as solar radiation, air temperature and humidity, and wind speed largely affect water vapor pressure deficit between leaf and air. Roughly 200 g water would inevitably be lost by transpiration for every gram of dry matter incorporated in the plant in moist temperate climate regions, whereas some 300 g of water is lost under irrigated hot arid conditions. If the potential tuber fresh yield of 100 $t \cdot ha^{-1}$ is assumed, an accompanying average water usage would be about 5,600 t in temperate regions, and about 8,400 t in hot arid regions, which are equivalent to 560 and 840 mm of rain plus irrigation, respectively.

### *Leaf Growth, Photosynthesis, and Dry Matter Production*

When water absorption by roots cannot meet leaf transpiration, the plant has some water stress as indicated by a decrease of water potential (WP, negative value) and/or of relative water content. The first symptom of water stress is the restriction of leaf growth as a result of the decrease of cell expansion and division (Hsiao, 1973). In potato, Gandar and Tanner (1976) found that, in pot experiments, when leaf WP was higher (wetter) than –0.5 MPa, leaves elongated during day and night, at almost equal rates. When leaf WP was lower than this, there was little leaf expansion. Under field conditions, although most leaf elongation occurred at night in both stressed and unstressed treatments, the enlargement of leaves also appeared to cease at leaf WP lower than –0.5 MPa. They suggested that a relatively short period of water shortage could reduce leaf enlargement and result in a substantial reduction in total leaf growth in potato. Jefferies (1989) measured the growth of individual leaves under field conditions. He revealed that potato leaf growth in the drought treatment was reduced as compared to that in the irrigated treatment when leaf WP of midday was less than –0.28 MPa and ceased when it was –1.1 MPa. In addition, drought resulted in a reduction in both rate and duration of leaf growth, number of leaves and branches, and stem height (Deblonde and Ledent, 2001).

Leaf growth restriction results in reduced stomatal conductance, and thus reduced transpiration and photosynthesis (Hsiao, 1973). Stomatal conductance in potato decreased at relatively higher leaf

WP of –0.4 to –0.6 MPa in comparison with many other species, for example –1.1 to –2.5 MPa in wheat, and from –0.8 to – 1.2 MPa in beans, lucerns, and soybeans (Gregory and Simmonds, 1992). It may be noted, however, that the relationship between stomatal conductance and leaf WP is inevitably complicated, because stomatal conductance responds to many environmental factors, such as temperature, irradiance, and humidity (Vos and Groenwold, 1989a). In addition, even on the same day in a well-watered field, stomatal conductance showed a reduction in the afternoon (Dwelle et al., 1981). In maize, when a part of root system was in dry soil and the other part was in highly wet soil, the stomata closed, even though the leaf WP did not change (Blackman and Davies, 1986).

Closure of stomata inevitably reduces photosynthesis as there is a reduction in carbon dioxide intake through stomata (Burton, 1989). Vos and Groenwold (1989a) showed that the response of photosynthetic rate to stomatal conductance was curvy-linear. Stomatal conductance declined earlier, more rapidly, and to a greater extent than photosynthetic rate in the stressed plants. The ratio of photosynthetic rate to stomatal conductance increased with the decrease (more negative) of soil WP. The critical level of soil WP for the change in the ratio was between –0.025 and –0.05 MPa.

Schapendonk et al. (1989) reported that water stress reduced photosynthesis initially as a consequence of stomatal closure. But with the progress of water stress, it directly inhibited the photosynthetic system. On the other hand, Jefferies (1994) could not find a significant effect of drought on the functioning of Photosystem II (a part of photosynthetic system) in the field-grown potato. These contradictory results may be partly explained by the difference in the stress level.

Reduced leaf growth due to drought adversely affects the amount of solar radiation captured by the plant canopy, resulting in reduced dry matter production and tuber yield (Wolfe et al., 1983; Jefferies, 1995). Based on the results for two years and two irrigation treatments with several cultivars, Jefferies and MacKerron (1987) concluded that the differences between treatments in total dry matter were attributed to differences in the amount of solar radiation intercepted rather than to differences in the photosynthesis and respiration of these cultivars. In another experiment, Jefferies and MacKerron (1994) measured carbon isotope discrimination for 19 cultivars and reported that total dry matter production in irrigated and drought conditions was determined largely by processes other than carbon assimilation rate of individual leaves. However, Trebejo and Midmore (1990) observed that the efficiency of conversion of intercepted radiation to dry matter (radiation-use efficiency, RUE) was reduced by drought, especially in the hot season in Peru. Demagante et al. (1996) also found in Philippines that reduction in total DW was attributed not only to low intercepted radiation due to the restricted canopy size and accelerated senescence of the canopy, but also due to small RUE, particularly for genotypes with late tuber formation and low harvest index. Jefferies and MacKerron (1989) also reported some decrease of RUE when the soil water deficit (difference of evapo-transpiration to supply of water by precipitation or irrigation) became greater than 47 mm in the United Kingdom.

From these results we can hypothesize that the reduction in dry matter accumulation in drought-affected potato is attributed primarily to lower interception of radiation as a result of the reduced size of canopy especially in cool climate regions. But, in hot dry regions, where the soil dries up more or the transpiration demand is severe, a decrease of RUE also affects dry matter production (Midmore, 1992). In maize, which is more drought-tolerant than potato, Earl and Davis (2003) reported that in two years when the effect of water stress on grain yield was drastically different, reduction in intercepted radiation was responsible only for a very minor yield loss. The yield loss was rather attributable mainly to reduced RUE and reduced harvest index. Therefore, the reduction of intercepted radiation in potato even under mild water-stress condition may be a reflection of higher sensitivity of its leaf growth to water stress.

### *Tuber Growth and Yield*

Tuber growth and yield are greatly restricted by drought under all climatic conditions ranging from moist temperate climate (Iwama et al., 1999), in which the potential loss of water from the crop canopy is less, to a hot and arid climate (Belanger et al., 2000; Ferreira and Carr, 2002), in which the loss of water from the crop canopy is high. Harris (1978) showed a linear decline of fresh tuber yield with an increase in soil-water deficit from 0 to 100 mm. The critical soil WP level below which yield started to decline, was –20 kPa to –60 kPa, which was almost equal or near to soil-field capacity in the densely rooted soil layer (Shimshi and Susnoschi, 1985; Lynch and Tai, 1989). Therefore, Burton (1981) emphasized that, if there is no rainfall, soil WP in silty soil is supposed to reach to the critical level in less than a day in a hot and semiarid region and in three days even in a moist temperate region.

The effect of drought on marketable tuber yield is more complicated than that on total tuber yield. Marketable tuber yield is determined not only by total tuber yield but also by the proportion of tubers above the marketable size, which is dependent upon the balance between total tuber yield and number of tubers (Harris, 1978; Burton, 1989). The number of tubers is reduced if water stress occurs very early in the growing season, either at the time of or before tuber initiation (Shock et al., 1992). MacKerron and Jefferies (1986) found a positive correlation between number of tubers and number of days when soil WP was below –25 kPa before and during tuber initiation. Deblonde and Ledent (2001), however, indicated that the effect of drought on number of tubers in early cultivars could be explained by the conditions in the early part of the season, whereas in the case of medium-maturing cultivars, besides the effect of the foreseason, the intensity of water shortage in the middle of the season also influenced final tuber number. In addition, the effect of drought on number of tubers at harvest varies with cultivar (Levy, 1983), year, and planting date (Miller and Martin, 1985). As mean tuber size depends on total tuber yield and number of tubers, it also varies depending on the time, intensity, and the duration of drought. Mean tuber size is larger if drought occurs right at the tuber initiation stage, but smaller if drought occurs after this stage (Shock et al., 1998; Fabeiro et al., 2001). Therefore, the effect of drought on marketable tuber yield is not parallel to that on total tuber yield, and the ratio between them varies depending on years, locations, and cultivars (Shock et al., 1992; Waddell et al., 1999; Deblonde and Ledent, 2001). MacKerron et al. (1988) suggested that when water stress is expected it would be wise to select those otherwise equal yielding cultivars that tend to produce lower number of tubers so as to get higher marketable tuber yield.

The reduction of marketable tuber yield because of drought also occurs due to the formation of deformed, knobby, spindly, and cracked tubers (Robins and Domingo, 1956; Shock et al., 1992). Such deformities, however, are not only due to drought but also because of other stress factors, such as heat (Bodlaender et al., 1964). Kleinkopf (1983) indicated that early cultivars, and cultivars with round or less elongated tubers, resisted irregular tuber-growth patterns caused by drought.

### *Root Characteristics*

The most frequently reported reason for higher sensitivity of potato to water stress is its shallower root system (Harris, 1978; Loon, 1981). Gregory and Simmonds (1992) concluded that the depth of rooting in potato crop never exceeded 100 cm even in deep, uniform soils; whereas roots of many temperate cereals go up to a depth of almost 200 cm. However, there are also reports that potato roots could penetrate up to a depth of 120-160 cm (Weaber, 1926; Wolfe et al., 1983; Stalham and Allen, 2001).

Root growth in potato is generally dependent on the maturity class of cultivars, in early cultivars root growth stops earlier (Table 7.1), hence the maximum rooting depth is generally shallower in early than in late cultivars (Iwama, 1998). Stalham and Allen (2001) also suggested that a poor root penetration in potato is a consequence of the short duration of active root growth. In addition, a hard soil pan, which is prevalent in normal tillage, inhibits the root penetration of potato (Loon et al., 1985; Pierce and Burpee, 1995). Bishop and Grimes (1978) also showed that potato root extension was severely reduced at all depths by soil strength exceeding 1 MPa.

The potential of potato to absorb water from soils is lower than in other field crops (Harris, 1978; Ferreira and Carr, 2002). Durrant et al. (1973) showed that even when the rooting depth was nearly 100 cm, potato extracted less amount of water from each soil horizon than sugar beet and barley, especially from below 60 cm. It appears that factor(s) other than rooting depth too affect(s) the ability of potato to absorb water from deep soil layers.

TABLE 7.1. Root dry weight (DW), leaf DW and tuber DW at the maximum stage of shoot growth (almost the end of flowering), and dry-tuber yield at the harvesting stage in 268 unselected clones derived from six crosses grown in rain-fed field.

| Growing period (days) | Range<br>Mean | 68-80<br>76.5 | 81-90<br>84.9 | 91-100<br>94.8 | 101-110<br>105.8 | 111-120<br>114.6 | 121-145<br>133.7 |
|---|---|---|---|---|---|---|---|
| Number of clones | | 24 | 99 | 41 | 45 | 31 | 28 |
| Root DW ($g{\cdot}pl^{-1}$)[2] | Mean | 1.30 | 1.57 | 1.77 | 1.93 | 2.15 | 2.77 |
| | SD | 0.70 | 0.69 | 0.74 | 0.71 | 0.78 | 1.18 |
| Leaf DW ($g{\cdot}pl^{-1}$) | Mean | 17.2 | 21.2 | 20.8 | 21.8 | 22.9 | 22.8 |
| | SD | 8.4 | 7.8 | 7.3 | 7.1 | 9.0 | 7.8 |
| Tuber DW ($g{\cdot}pl^{-1}$) | Mean | 104 | 109 | 103 | 95 | 73 | 67 |
| | SD | 34 | 38 | 32 | 40 | 21 | 26 |
| Dry tuber yield ($g{\cdot}pl^{-1}$) | Mean | 175 | 210 | 226 | 248 | 263 | 281 |
| | SD | 28 | 37 | 44 | 52 | 49 | 60 |

*Source*: Iwama et al. (1981).

Gregroy and Simmonds (1992) attempted to explain the higher sensitivity of potato to water stress from the perspective of quantitative distribution of potato roots through the soil profile, measured as root-length density (root length per unit soil volume, RLD). The importance of RLD to water absorption in potato was first pointed out in a simulation model by Campbell et al. (1976). They showed that the critical soil WP, below which the restriction of transpiration rate started, became higher (less negative) from –80 kPa to –50 kPa when RLD in the depth of 0-30 cm decreased from 0.25 $cm{\cdot}cm^{3}$ to 0.1 $cm{\cdot}cm^{-3}$. Ferreira and Carr (2002) measured soil water absorption from each soil depth by potato and found that the patterns of water extraction from different depths were linked to the relative distribution of roots. These results imply that the higher sensitivity of potato to water stress are primarily due to its small root length than an inherently small conductance per unit length of root.

Unfortunately, there is no report simultaneously measuring RLD and water absorption throughout the soil profile in potato. However, high concentration of root length in plow layer (80-90%) has been generally reported (Lesczynski and Tanner, 1976; Iwama, 1998). In addition, total amount of roots throughout the soil profile in potato is relatively smaller than in other field crops, especially cereals (Yamaguchi and Tanaka, 1989; Gregory and Simmonds, 1992). Because of these two characteristics of potato roots, RLD below plow layer is often lower than 0.5 $cm \cdot cm^{-3}$ during all periods of potato growth (Figure 7.1). Since the distribution of RLD in the soil profile differs between cultivars (Iwama, 1998), soil conditions (Loon et al., 1985), and growing seasons (Iwama, 1988), a relatively higher concentration of root length and RLD below plow layer are also reported (Vos and Groenwold, 1986; Iwama et al., 1993; Stalham and Allen, 2001). Even in RLD values of these experiments, however, the soil depth below the critical RLD of 0.25 $cm \cdot cm^{-3}$ (Campbell et al., 1976) does not exceed 50 cm before initial flowering stage and 75 cm during tuber-bulking period. These critical depths almost agree with the effective rooting depth of about 60 cm reported for active water absorption during the growing period in potato (Durrant et al., 1973; Shalhevet et al., 1983; Feddes et al., 1988). Ferreira and Carr (2002) found that most of the water came from the top 80 cm of soil, ranging from 73-83 percent under rain-fed to 90-95 percent under well-irrigated condition, for the period between 50 days after planting to final harvest. This result supports the present hypothesis for the critical RLD of water absorption in potato.

Wolfe et al. (1983) reported that the maximum depletion of soil-water content occurred in the soil zone from 90-120 cm below the soil surface in a hot-dry region, where root extension was observed down to a depth of 150 cm in potato and 250–300 cm in maize and sorghum. It is supposed to be due to an exceptionally good soil texture until very deep depth. The water conductivity and holding capacity in soils vary depending on soil conditions (Loon, 1981; Opena and Porter, 1999). We need, of course, more research to be sure of the critical RLD for water absorption in potato.

### *Genetic Improvement*

Surprisingly, little attempt has been made to genetically improve the root traits, even though these are considered to be the most important for regulating water stress in potato (Harris, 1978; Gregory and

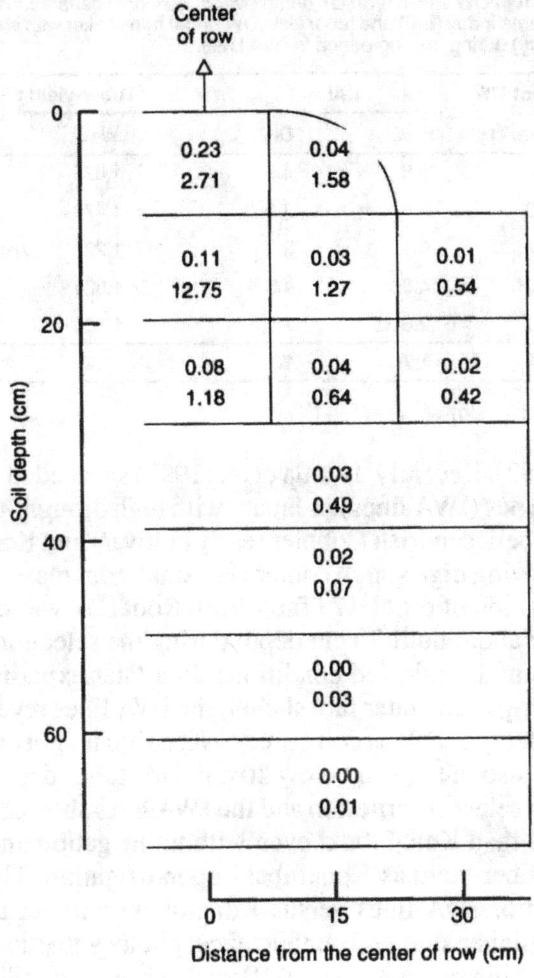


FIGURE 7.1. Root-length density (cm·cm$^{-3}$) throughout the soil profile at the maximum stage of shoot growth (almost the end of flowering) in rain-fed field. The upper and the lower numerals in each soil layer are for nodal roots and the lateral roots, respectively. All values are the mean of four cultivars. *Source:* Iwama (1998).

TABLE 7.2. Mean root DW within 0-30 cm depth over three clonal generations under rain-fed fields, and leaf area index (LAI) and tuber yield DW under rain shelter with irrigation (wet) and no irrigation (dry) during bulking period in IWA lines.

| Genotype | Root DW | LAI | | Tuber-yield DW ($g.m^{-2}$) | |
|---|---|---|---|---|---|
| | ($g{\cdot}m^{-2}$) | Wet | Dry | Wet | Dry |
| IWA-1 | 15.7 | 6.8 | 4.5 | 1,675 | 927 |
| IWA-2 | 9.0 | 3.8 | 3.0 | 1,478 | 1,198 |
| IWA-3 | 8.7 | 5.1 | 3.2 | 1,221 | 1,203 |
| IWA-4 | 8.0 | 4.8 | 4.3 | 1,524 | 1,186 |
| Konafubuki | 4.3 | 2.6 | 2.1 | 1,306 | 828 |
| LSD(0.05) | 1.1 | 1.7 | 1.1 | 230 | 131 |

*Source*: Iwama et al. (1999).

Simmonds, 1992). Recently, Iwama et al. (1999) succeeded in breeding four genotypes (IWA lines) in Japan, with high drought-tolerance from the cross between Irish Cobbler (early cultivar) and Konafubuki (late cultivar having high starch content but small root mass) by using a selection criterion of root DW (Table 7.2). Root DW was measured in a plow layer about until 30 cm depth during the selection process for three years under rain-fed conditions. In a later experiment in a field without irrigation under rain shelter, the IWA lines revealed significantly greater root DW as compared to Konafubuki not only in the plow layer but also in the soil below 30 cm. Dry-tuber-dry yield was used as another selection criterion and the IWA lines showed a higher dry-tuber yield than Konafubuki even without irrigation and almost the same dry-tuber yield as Konafubuki upon irrigation. The higher dry-tuber yield of IWA lines versus Konafubuki without irrigation was related to higher stomatal conductance, photosynthetic rate, and leaf area index (Iwama et al., 1999). Broad-sense heritability in the second clonal generation ($h^2$ = 87%) and simple correlation coefficient between two clonal generations ($r$ = 0.71) for root DW were high, and these values were comparable to those for leaf DW and stem DW (Iwama et al., 2001). The relatively high broad-sense heritability for root DW was also found in other two populations derived from different crosses (Kashiwagi, 2000).

The method used for the root mass measurement in this breeding program was, however, laborious and time-consuming. It cannot be widely used in usual breeding programs, hence we need to find a simple method to estimate the root traits. Ekanayake and Midmore (1992) tested pulling resistance of root (PR) as a selection criterion of root penetration ability to deep-soil layer, at two drought sites in Peru. PR showed a significant positive correlation with maximum length and DW of both pulled roots and residual roots in soils ($r$ = 0.51–0.69). Although PR method can estimate root growth mainly in the plow layer, it may be suitable for rough selection of root mass in the early generations in potato breeding, because it seems to be less laborious compared with the direct measurement of root mass in field.

Iwama et al. (1998) classified the potato seedlings raised from true seeds for root mass by visual inspection of root appearance on the pot-soil surface at the transplanting stage. The visual inspection score related well with the measured root DW and length at the transplanting stage, and also root DW at the maximum shoot growth stage in the field. Kashiwagi (2000) also found similar relationship of root mass between the seedling and the first clonal generations raised by multiplying the plants through stem cutting after measuring the root mass in the seedling generation. In both these experiments, root DW at the transplanting stage in the seedling generation related well with that at the maximum shoot growth stage in the first clonal generation in one cross, but not in another cross, indicating the importance of genetic background for such associations. Relationships between seedling generation and clonal generations are not consistent over crosses even for other tuber and shoot traits, and in most cases the correlations are low (Tai and Young, 1984; Gopal et al., 1992).

A direct measurement of root traits for field-grown plants is destructive and thus impractical in the case of seedlings and first clonal generations when a genotype has only one hill. Since evaluations of root mass in the plants grown in pots are not destructive, it may be possible to use it for an initial rough selection of root mass in the seedling generation. However, reasons for the difference in the inherent performance of root mass among different crosses needs to be worked out.

As mentioned before, extension of potato root to deeper soil layers is frequently restricted by hard-soil pan. A large difference among genotypes in root penetration ability to hard-soil pan has been reported in rice (Yu et al., 1995) and wheat (Kubo et al., 2004), using a pot with paraffin-Vaselin disk. The root penetration ability to hard-soil pan relates to the diameter of root tip (Materechera et al., 1992). Since the diameter of nodal roots varies among potato cultivars and wild relatives (Iwama and Nishibe, 1989), there may be a difference also in the root-penetration ability of potato genotypes. If so, by this simple evaluation method for root penetration a lot of genotypes can be screened for drought-tolerance in potato.

The differences in drought-sensitivity among potato genotypes have also been reported for shoot traits and some authors proposed using them in screening methods. The reported results were, however, not consistent and varied with genotypes and environmental conditions. Bansal and Nagarajan (1987) suggested that the measurement of leaf growth under controlled stress and during recovery could be a useful technique for selecting genotypes tolerant to drought. Jefferies and MacKerron (1993) proposed that the selection for early leaf appearance combined with the ability to sustain leaf growth with increasing soil-moisture deficit would improve productivity under drought. Deblonde and Ledent (2001), however, found that although number of leaves and leaf length were sensitive to drought, many cultivars did not differ significantly in these traits.

Sukumaran et al. (1989) suggested the possibility of using the ratio of photosynthetic rate to stomatal conductance as a selection criterion for water-use efficiency (WUE), because this ratio increased significantly under drought condition in a tolerant cultivar, but not in drought-sensitive genotypes. Vos and Groenwold (1989b) showed significant relationships of WUE with carbon isotope discrimination and stomatal conductance among potato genotypes, indicating that carbon isotope analysis and porometry have potential as indirect screening methods to identify genetic differences in WUE. Ekanayake and Jong (1992) found a significant difference in relative stomatal resistance (under stress versus without stress) within *S. tuberosum* cultivars (1.5–3.9) and among other tuber-bearing species (3.3–5.0). They concluded that stomatal behavior among tuber-bearing *Solanum* species was sufficiently different to warrant investigations of drought-resistance in potato species under dry hot conditions. Dwelle et al. (1981), however, reported high seasonal fluctuations in correlations between stomatal conductance and photosynthetic rate among potato genotypes. Schapendonk et al. (1989) reported that highly significant differences among Dutch cultivars in photosynthetic rate, mesophyll conductance, stomatal conductance, and transpiration rate detected under severe water-stress condition in pots were not related to their drought-tolerance in fields. They concluded that gas-exchange measurements on leaves during a certain developmental stage of the crop could explain, at best, only a part of the variation in drought-tolerance observed under field conditions. According to Jefferies and MacKerron (1997) carbon isotope discrimination is a complex trait influenced by many physiomorphological factors and environmental conditions, and thus does not provide a simple, indirect method of selecting for dry matter production under water stress.

To answer these contrasting results, Weisz et al. (1994) proposed a fraction of transpirable soil water (FTSW) as an indicator for soil-water deficit. In this concept, the water present in the soil is referred to as the extractable (transpirable) water in a targeted crop. The extractable water is defined as the quantity of soil water in the entire profile (measured over the entire soil depth through which water extraction occurs) in excess of the minimum quantity observed in the soil when plant transpiration practically stops because of dry soil (Ritchie, 1973). FTSW is considered as 1 at field capacity and 0 when transpiration of drought-affected plants reached 10 percent or less of potential transpiration rate (Sinclair et al., 1986). Using this concept, Weisz et al. (1994) showed that potato transpiration responded to changes in FTSW similar to other crops. This indicates that the drought-sensitivity of potato in terms of transpiration would appear to be due to an inability to extract as much total transpirable soil water as extracted by other crops due to potato's shallow rooting system. On the other hand, in terms of leaf growth, the critical FTSW value in potato (0.6) was higher than the reported values for other crops (generally 0.4 or 0.5), indicating that some other process besides root extension or water extraction is inhibiting potato leaf growth under drought conditions. These findings imply that in potato, root factor is responsible for drought-sensitivity of some shoot traits, such as transpiration, but not in the others, such as leaf growth.

In many tropical climates, where the infrastructure for irrigation is inadequate, water stress increases with time (Midmore, 1992). The best option for these environments is to grow early cultivars (Malik et al, 1992; Demagante et al., 1996), because early tuber formation and bulking allow an escape from severe drought. In a review for potato production in the tropics, Midmore (1992) concluded that field screening for tuber yield is likely to be more useful to the breeder in identifying tolerant clones than any other single, or a combination of physiological indices. However, maintenance of controlled drought conditions in a field is a difficult task. As an alternative to field evaluation, Demagante et al. (1996) proposed using apical cuttings in a nursery bed as a promising method for screening for drought-tolerance. They reported that the relative order and reduction in percentage of total DW and tuber-dry yield at maturity under water stress was almost the same in the plants from apical cuttings grown in nursery beds and from normal tuber seeds sown in fields. Although the nursery bed method saves space, time, and labor compared with direct measurement of tuber yield in fields, the processes involved for drought-tolerance and their interactions with environments are much more complicated in fields than in the nursery bed. Thus it is necessary to develop practical and reliable tools for screening large populations for drought-tolerance so as to hasten the breeding process.

### *Molecular Aspects*

The use of molecular markers in mapping the traits of agronomic importance holds great promise for increasing our understanding of the physiological or molecular mechanisms underlying the expression of these traits (Price and Courtois, 1999; Ewing et al., 2004). Some quantitative trait loci (QTL) for characters related to drought-tolerance have been reported in rice, wheat, barley, sorghum, and maize (Bennett, 2003). There are many reports especially in rice for QTLs for root (Champoux et al., 1995) and shoot traits (Price and

Courtois, 1999). Shen et al. (2001) bred several elite rice genotypes for root length and DW with marker-assisted selection.

Several molecular linkage maps of potatoes have been constructed in Europe, the United States, and Japan (Gebhardt et al., 1989; Tanksley et al., 1992; Jacobs et al., 1995; Hosaka, 1999). These maps have more than 350 markers and more than 25 single dominant genes, and some QTLs controlling tuber yield and quality traits have been localized on these maps (Barone, 2004). It has also been reported that several QTLs for tuberization, dormancy, and sprouting are interrelated, and these are also associated with QTLs for levels of plant hormones (Ewing et al., 2004). Though no QTL related to drought-tolerance has been reported, some genes that are highly inducible by dehydration have been identified (Broin et al., 2000; Doczi et al., 2002). As mentioned earlier, there are phenotypically close correlations between root growth and earliness, and also between stomatal conductance and photosynthetic rate among potato genotypes. If QTLs for these traits are localized and differences among genotypes for them are clarified, our understanding of drought-tolerance and of the genetic interactions between individual traits will be greatly advanced.

Most of the physiological traits considered to be important to drought-tolerance, such as root traits or photosynthetic rate, are tedious and time-consuming to measure. Field evaluation of these traits for QTL analysis is thus rarely practical. Gopal and Minoha (1998) and Gopal (2001) proposed in vitro selection as an alternative to field evaluation for some traits, such as tuber color, tuber size, earliness, eye depth, etc. The relative orders of potato genotypes for such traits in the clonal generation were the same for the plants grown in vitro and in pots or fields, although the absolute values were greatly different. Kashiwagi (2000) also found that the differences among segregating populations for root length in the seedling generation raised from true seeds in pots or fields were similar to those observed in vitro. These results suggest that evaluations under in vitro condition for phenotypic differences among potato genotypes for traits related to drought-tolerance may be an alternative to evaluations under field or pot conditions. This may facilitate an evaluation of a large number of genotypes needed in QTL analysis for the targeted traits.

A brief mention of the use of transgenics in potatoes to improve drought-tolerance is important here. Yeo et al. (2000) reported high drought-tolerance in potato plants transformed by introducing trehalose-6-phosphate synthase *(TPS1)* gene from *Saccharomyces cervisiae.* Lee et al. (2003) expressed the *Arabidopsis CBF1* gene driven by three copies of an ABA-responsive complex (ABRC1) from the barley *HAV22* gene in transgenic tomato plants. Under stress, these transgenic tomato plants exhibited enhanced tolerance to chilling, salt stress, and water deficit in comparison to untransformed plants. Under normal growing conditions the ABRC1-*CBF1* tomato plants maintained normal growth and yield similar to the untransformed plants. These reports indicate the promise of using transgenic procedures for improving drought-tolerance in potato. High homology is reported between potato and tomato genomes (Tanksley et al., 1992).

However, some transgenic potato plants developed to inhibit starch biosynthesis (Rober et al., 1996) and to change cell metabolites (Keller et al., 1998) exhibited a variety of morphological and physiological changes that could not be explained. In the attempts to transfer $C_4$ photosynthetic mechanism to potatoes, which are expected to introduce high photosynthetic rate and consequently high WUE (Rademacher et al., 2002), overproduction of a single $C_4$ enzyme could alter the carbon metabolism of transgenic potato plants, but did not show any positive effects on photosynthesis (Miyao, 2003). Development and commercialization of transgenic potato plants with improved drought-tolerance or dry matter production, which are controlled by many genes and by their interactions, may need a long process (Sinclair et al., 2004). Cultivated and wild potatoes have a large genetic variation for root traits (Iwama and Nishibe, 1989; Sattelmacher, Kuene et al., 1990) and stomatal conductance (Ekanayake and Jong, 1992). It would be more practical and cost-effective to introduce targeted traits from these gene sources to elite genotypes through conventional breeding assisted by molecular markers.

## *HEAT STRESS*

During the past decade, the area under potato crop increased by 65 percent in Asia and 39 percent in Africa, and these two regions together constitute 44 percent of the total potato area in the world (FAO, 2001). Though in many areas of these regions potato is grown during colder months of the year, night temperatures are close to 20°C and day temperatures are sometimes as high as 40°C. Thus potato growth is adversely affected, resulting in low tuber yield and quality (Van der Zaag and Demagante, 1987; Malik et al., 1992). In the following sections, we review potato growth in relation to heat stress so that adverse effects of high temperatures can be mitigated by means of cultural practices and genetic improvement. Although heat stress frequently coexists with drought stress, we discuss heat stress per se, in order to avoid repetition of the facts described in the previous section.

### *Photosynthesis*

Optimum leaf temperature for photosynthesis in potato is reported to range between 16°C and 25°C (Ku et al., 1977; Hammes and Jager, 1990) and the increase in leaf temperature beyond this drastically decreases the photosynthetic rate. Winkler (1971, cited in Burton, 1989) found that dark respiration of leaves increased rapidly with increase in temperature, roughly doubling for each 10° C increase between 10°C and 35°C. He estimated compensation points at which the loss of carbon in respiration equals its assimilation in photosynthesis and thus net assimilation becomes zero, but these varied with leaf age and radiation level. These findings are frequently used to explain the low adaptability of potato plants to high temperatures. Burton (1981) estimated that if photosynthesis and respiration are affected only by temperature, tuber yields in lowland tropics should be 20-30 $t \cdot ha^{-1}$. On the other hand, Wolf, Olesinski et al. (1990) reported that air temperatures up to 38°C did not cause a reduction in net photosynthetic rate of potato that had been grown at these temperatures for long periods prior to measurement. Thornton et al. (1996) also reported that the net photosynthetic rate of immature and mature leaves tended to be lower at 35°C than at 25°C, however, the differences were significant in only two out of nine comparisons.

Reasons for the large discrepancy in the optimum temperature of photosynthesis may be the adaptation of certain genotypes to high temperature. Wolf, Olesinski, et al. (1990) and Thornton et al. (1996) measured the photosynthetic rate in the plants grown for relatively longer period (several weeks) at the targeted temperatures. Wolf, Olesinski, et al. (1990) pointed out that the photosynthetic system of potato leaves has the capacity to adapt to high temperature. Midmore and Prange (1992) also suggested the biochemical/enzymic adaptation of photosynthesis to high temperature and the greater stomatal density at the higher temperature. Such understanding will help to improve the heat-tolerance of potato cultivars.

### *Leaf and Shoot Growth*

Heat stress also results in a reduction in leaf area in potato (Gawronska et al., 1992). The early work by Bushnell (1925) demonstrated that compared to temperatures below 26°C, leaflet size and leaflet weight decreased drastically at 29°C and the ratio of leaf to stem also decreased. Benoit et al. (1983) reported that plant elongation was maximum at 30°C during early growth and at 25°C during tuber bulking, whereas maximum growth of leaf (as measured by percent change in leaf area) occurred at 25°C at both the stages. High temperature hastens development rate of shoot and senescence rate of individual leaves (Marinus and Bodlaender, 1975). In the hot environment during short-day conditions in Philippines, potato plants showed a fast but not a large canopy development, which after reaching the maximum, decreased rapidly (Van der Zaag and Demagante, 1987). The inability to achieve a large canopy and to maintain it for longer period is the major limitation for obtaining high crop yields and high dry matter in tropical lowlands (Midmore, 1992).

### *Tuber Growth and Yield*

High temperatures affect tuber growth and tuber yield more drastically than shoot growth and total dry matter production. At high temperatures, more dry matter is partitioned to shoots, and to a lesser extent to roots and stolons (Wolf, Marani et al., 1990; Gawronska et al., 1992). High soil temperatures per se decrease starch content of tubers directly by inhibiting the conversion of sugars into starch (Krauss and Marschner, 1984). However, by exposing shoots, roots, and stolons (tubers) separately to different temperatures, Struik et al. (1989a,b,c) demonstrated the complexity of temperature effects on the development of the potato plant. Tuber yields of high quality were maximized when temperatures in all plant parts were low (18°C). When the air temperature was high (28°C), it was possible to obtain acceptable tuber yields of good quality only when the temperatures around the stolons and the roots were much lower than the air temperature. The increase in shoot temperature reduced tuber yield and dry matter concentration, but the increase in temperature around the stolons maximized the induction of secondary growth of tubers. They concluded that stolon and tuber development, dry matter partitioning, and initiation of secondary growth responded differently to variation in temperatures around different plant parts, suggesting that these processes have different regulation mechanisms.

### *Cultural Practices for Management of Heat Stress*

Intercropping is a traditional practice adopted for a wide range of tropical crops to reduce heat stress both at emergence and during seedling stage. In hot tropical sites in Peru, Midmore, Roca, et al. (1988) showed that by planting potato within the established maize canopy, daily mean soil temperature at 7 cm depth decreased by about 2-3.5°C due to about 30 percent reduction in radiation on the soil surface. Since mean daytime soil temperature higher than 28°C delayed emergence, the reduction of soil temperature hastened plant emergence. In addition, Midmore, Berrios et al. (1988) found that tuber yield of potato intercropped with maize was not significantly less than the yield of a sole potato crop until maize population reduced the transmission of solar radiation to potato crop by more than 30 percent. However, they also pointed out that a delicate balance for duration and intensity of shade by maize intercropping has to be achieved to provide sufficient light energy to maximize the photosynthetic capacity of potato crop, and a slight shift in that balance to favor maize would lead to marked reduction in potato yield.

Mulches of rice straw, rice hulls, and sugarcane bagasse are also used to reduce soil temperature. Van der Zaag et al. (1986) reported that these mulches reduced the daily maximum soil temperature by 2-4°C, resulting in faster emergence and canopy development, earlier tuber initiation, and tuber-yield increase by 50 percent during the hottest season in Philippines. They also pointed out that the composted sugarcane bagasse tended to produce superior tuber yields due to its fertility input by way of N, P, K, Mg, and Ca. Tawfik et al. (1996) showed that potato plants given Ca and N during heat stress had higher leaf Ca concentration and transpiration rate, and resulted in greater biomass production and tuber yield than those without additional Ca application. Kleinhenz and Palta (2002) reported that leaf area of potato plants under heat stress (35/25°C day/night temperature) was reduced to 5 percent of control plants (20/15°C) at 5 and 25 $\mu$ mol.$l^{-1}$ Ca but remained at 70 percent of control plants at 125 and 600 $\mu$ mol.$l^{-1}$ Ca. It suggests that heat stress–induced reduction in leaf area may be partially overcome at specific, elevated levels of $Ca^{2+}$ in the root zone. In addition, Sattelmacher, Marschner, (1990) reported that size of the root system in potato clones was reduced by supra-optimum (30°C) root-zone temperature. These results provide an insight into the mechanism by which high soil temperature reduces potato growth and may create an easy practical method to mitigate heat stress.

### *Genetic Improvement*

After the introduction of potato into Europe and North America, potato was bred to adapt to cool environments. However, some cultivars, for example Gineke (Marinus and Bodlaender, 1975), Désirée (Levy, 1986), and Norchip (Khedher and Ewing, 1985) have been found to be tolerant to higher temperatures. These cultivars have the capacity to initiate tubers at high temperatures (Khedher and Ewing, 1985) and exhibit relatively small yield losses in hot seasons (Levy, 1986). Attempts to breed new cultivars adapted to high temperatures have also been made in several tropical and semitropical countries (Pushkarnath, 1969; Ewing et al., 1983; CIP,

1984; Frusciante et al., 1999). Some newly bred genotypes, such as DTO-28, have shown relatively high-tuber yield in tropic and semitropic regions (Van der Zaag and Demagante 1988; Malik et al., 1992). Levy et al. (2001) also reported new potato cultivars adapted to high temperatures and dry conditions in Israel.

A difference in heat-tolerance has also been reported among *Solanum* species. Reynolds and Ewing (1989) screened 319 accessions of 59 tuber-bearing species of *Solanum* for two aspects of heat-tolerance: (1) shoot growth under high temperatures (30-40°C) and long photoperiods (18 h), and (2) ability to produce tubers under high temperatures and short photoperiods (9 h). They identified heat-tolerant accessions of several diploid species including *S. berthaultii, S. chacoense, S. demissum,* and *S. stoloniferum* suitable for utilization in breeding programs. Veilleux et al. (1997) used *S. andigena, S. phureja, S. berthaultii, S. chacoense,* and *S. microdontum* to develop hybrids exhibiting good tuberization potential under high temperatures.

Upon comparison between heat-tolerant and sensitive clones, which were selected by Reynolds and Ewing (1989), Reynolds et al. (1990) found a significant difference in photosynthesis. Hammes and Jager (1990) too reported that two cultivars, which tend to perform well in fields under hot climates in South Africa, showed higher net photosynthetic rate than a heat-sensitive cultivar Up-to-Date at day temperatures in the range of 35-40°C. Wolf, Olesinski et al. (1990) also reported genotypic differences in net photosynthetic rate at temperatures up to 40°C between a clone and commonly grown cultivars, Desiree and Up-to-Date. It was mainly due to the smaller increase in dark respiration in the clone at high temperatures. On the other hand, Prange et al. (1990) reported that among 10 potato cultivars the reduced dry matter production at high temperatures was not related to increased maintenance of dark respiration consuming photosynthates, but related to reduced efficiency in Photosystem II. Thornton et al. (1996) reported that heat-tolerant cultivars (Desiree and DTO-28) and heat-sensitive cultivar (Russet Burbank) had similar response in net photosynthesis and respiration with increasing leaf temperature. Thus factors other than leaf gas exchange rates may be responsible for the genotypic differences in tuber-growth rate under high temperature conditions. In addition, Gawronska et al. (1992), who did not find a correlation between $^{14}C$ assimilate export and partitioning in the genotypes with difference in heat-tolerance, observed that the direct effect of heat-stress on assimilate partitioning is only one aspect of the complex phenomenon of heat-tolerance.

One cycle of phenotypic recurrent selection for tuber yield in diploid potatoes exposed to high temperatures at two locations in eastern North Carolina, USA (Gautney and Haynes, 1983), resulted in increases of 3 percent for survival (shoot growth), 15 percent for tuberization, and 27 percent for tuber yield. Heritability of tuber yield, estimated from selection response, was low, 0.07 over the locations. In addition, the tuberization response of the families differed from location to location and generation to generation.

Van der Zaag and Demagante (1988) observed genetic variability for tuber dry matter among 97 genotypes from temperate and tropical sources, under the high temperature condition in Philippines, but none were in the acceptable range for temperate-climate processors (i.e., dry matter 20 percent or more). Veilleux et al. (1997) found that under severe heat stress in Israel, the complex potato hybrids derived from the crosses among *S. tuberosum* cv. Atlantic, *S. phureja,* and *S. tuberosum* ssp. *andigena,* and *S. chacoense, S. berthaultii,* and *S. microdontum,* had lower frequency of tubers with hollow heart and heat necrosis as compared to the cultivar Atlantic. However, secondary growth of tubers was more common in the complex hybrids suggesting that stress-induced physiological disorders of tubers are under multiple genetic controls (Rickey and Belknap, 1991).

### *Molecular Aspects*

Some genes that control tuber traits and yield have been mapped as QTLs (Barone, 2004), but only some showed stable effects across different genetic backgrounds and environments. Further, molecular and genetic factors underlying the initiation of tuberization and partitioning of photosynthates to the tubers are poorly understood. Thus, at present heat-stress can be best addressed by empirical selection (Frusciante et al., 1999).

Ewing et al. (2004), however, found ten QTLs for the ability to tuberize under long days and eight QTLs for tuber dormancy in a hybrid population derived from a cross between haploid *S. tuberosum* and *S. berthaultii.* Since long photoperiod is a critical component of heat-tolerance (Ewing, 1981), the QTLs for the ability to tuberize under long days may relate to those for the ability to tuberize under high temperatures. Some QTLs may relate to the ability to prevent secondary growth and misshapened tuber shape under high temperatures. Hannapel et al. (2004) found that transgenic plants that overexpressed BEL and KNOX proteins exhibited enhanced tuber formation even under long-day conditions. These transgenic clones exhibited a decrease in GA levels and an increase in cytokinin levels. In addition, it was reported that a slow increase in temperatures induced an increase in heat shock proteins and their mRNAs in tubers (Rickey and Belknap, 1991). Some genes for heat-stress transcription factors have also been identified in tomato gene maps (Tanksley et al., 1992). Further progress in our knowledge may make it possible to perform a marker-assisted selection for heat-tolerance in potato crop.

## *OTHER ABIOTIC STRESSES*

### *Salinity*

Soil salinity occurs due to natural conditions as well as inappropriate water management (both in terms of quantity and quality), land development, and also sometimes by excess use of fertilizers. Due to limitation of fresh water, farmers often use saline water for irrigation (Shannon and Grieve, 1999). Such practices thus easily lead to soil salinity.

Estimated acreage of salinized cropped lands at present varies greatly among studies: that is, less than 4 percent (Oldeman et al., 1991) to over 10 percent (Shannon, 1997) of the total arable land. Although the range depends primarily on how the critical level is defined, the salinized area in which crop productivity is greatly reduced would not be so extensive. Nevertheless, in irrigated agriculture, especially in arid and semi-arid regions, salinization is a big issue, because it is artificially induced and is avoidable with rational management.

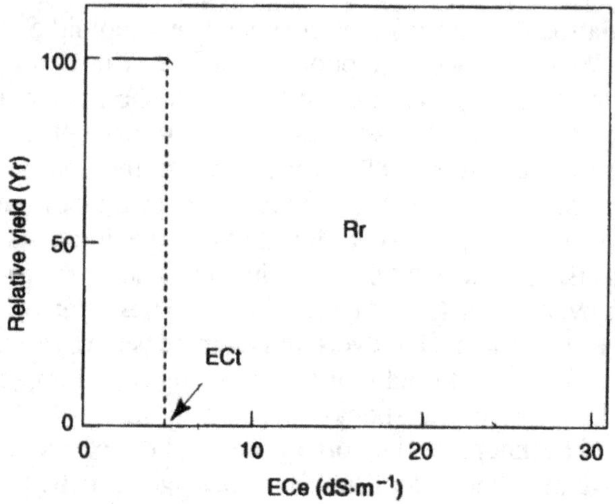


FIGURE 7.2. Schematic illustration of crop yield (or growth) response to salinity. Refer to text for the equation.

The soil salinity of crop fields is commonly measured with the electrical conductivity of a water-saturated soil extract (ECe, $dS{\cdot}m^{-1}$). The following equation is widely accepted to measure crop-tolerance to salinity (Figure 7.2):

$$Y_r = 100 - R_r\,(\mathrm{ECe} - \mathrm{ECt})$$

where $Y_r$ is the relative yield (percentage of the yield under saline conditions to that under nonsaline condition), $R_r$ the reduction rate (the relative-yield decrease for each unit of ECe increase) and ECt the threshold electrical conductivity beyond which the $Y_r$ begins to decrease from the maximum yield ($Y_r = 100$ percent).

Under low or light saline conditions, metabolic processes and plant growth are apparently not affected as the plants are capable of osmotic (or turgor) adjustment. When salt concentration is further enhanced, osmotic changes caused by salinity enhance leaf conductance and reduce photosynthesis and thus retard plant growth. With further severity of salinity, plants eventually reach the stage of complete loss of crop yield or at least economic yield.

TABLE 7.3. Some examples of ECt (threshold ECe ) and the reduction rate ($R_r$) of selected crops, and the ECe at which the relative yield ($Y_r$) is 80 and 20 percent.

| Crop | ECt ($dS{\cdot}m^{-1}$) | $R_r$ | ECe ($dS{\cdot}m^{-1}$) $Y_r$ = 80% | ECe ($dS{\cdot}m^{-1}$) $Y_r$ = 20% | Reference |
|---|---|---|---|---|---|
| Barley | 8.0 | 5 | 12 | 24 | Maas and Hoffman (1977) |
| Sorghum | 6.8 | 16 | 8 | 12 | Maas and Hoffman (1977) |
| Wheat | 6.0 | 7 | 9 | 17 | Maas and Hoffman (1977) |
| Soybean | 5.0 | 20 | 6 | 9 | Maas and Hoffman (1977) |
| Potato | 1.6-2.0 | 13-17 | 3 | 7-8 | Levy (1992) |
| | 1.7 | 12 | 3 | 8 | Maas and Hoffman (1977) |
| | 1.6 | 5 | 5 | 16 | Bustan et al. (2004) |
| | 1.0 | 9 | 3 | 10 | Bernstein et al. (1951) |
| | 0.2 | 22 | 1 | 4 | Paliwal and Yadav (1980) |
| Maize | 1.7 | 12 | 3 | 8 | Maas and Hoffman (1977) |

The two variables $R_r$ and $CE_t$ are used for classifying the tolerance of species and cultivars. Table 7.3 shows some information on selected crops from Mass and Hoffman (1977) and examples of potato by recalculating the original data. Potato is sensitive to salinity, but large differences exist among cultivars in tolerance levels. Salinity effect on potato has been studied not only for total tuber yield but also for marketable tuber yield, yield components, cultivars' growth period, and interaction with drought stress. Patel et al. (2001) showed that the weight of Grade A tubers (tuber length greater than 45 mm) was decreased when crop was irrigated with brackish water (9 $dS{\cdot}m^{-1}$; $EC_e$ = 3.2-3.7 $dS{\cdot}m^{-1}$), whereas there was no significant effect on total tuber yield. Nadler and Heuer (1995) also reported that total tuber yield was not affected by salinity and water deficit, while the percentage of nonmarketable tubers was significantly reduced by high salinity (EC = 6 $dS{\cdot}m^{-1}$) and water stress (288 mm irrigation throughout the growth period, that is 60 percent of the nonstress condition). Tuber size was more adversely affected than tuber number with an increase in salinity (Paliwal and Yadav, 1980). However, Bernstein et al. (1951) showed that salinity reduced both size and number of tubers. The variation in the results of these studies may be due to the cultivar × environment interaction. Starch concentration of tubers was reduced or not affected in the salinized plots (Bernstein et al., 1951).

In the early growth stage, potato is more sensitive to salinity. Seed tubers aged for 68 days at 6°C were more tolerant to salinity compared with those aged for 33 days (Levy et al., 1993). In addition, genetic variation exists for salt-tolerance in potato. Jefferies (1996) showed a positive correlation between the performance of the parent to long-term salinity and the survival of progeny in the seedling generation. Late-maturing cultivars were capable of recovering from adverse effects of salinity. However, Levy (1992) noted no association between the maturity type and the response to salinity among 21 cultivars. Under the initial and mild saline conditions, owing to disturbance of water uptake, salinity induces imbalance of mineral nutrition. Several studies on the nutritional disturbance by salinity, the alleviation from it, and the interaction with cultivars have been reported in potato. Shaterian, Waterer et al. (2005) found that in diploid wild potatoes, early-maturing clones accumulated $Na^+$ in the leaf tissues whereas late-maturing clones generally excluded $Na^+$ from the leaf tissue under stress condition (5 days with 150 $mmol \cdot l^{-1}$ NaCl concentration in culture solution). This fact suggests that salinity-tolerance is affected by maturity period of a genotype. Alhagdow et al. (1999) indicated that potassium (K) deficiency promoted salt ($Na^+$) damage, but K application beyond the usual recommendations was not beneficial.

Levy et al. (1988) reported that salinity tended to lower the water and the osmotic potentials of leaves and tubers while increasing the content of total soluble solids and proline. Under the salt-stress condition (100 $mmol \cdot l^{-1}$ NaCl concentration), the endogenous proline concentration in a potato clone transformed with osmotin-like protein was slightly higher than that of the nontransformed genotype (Evers et al., 1999). This provided evidence for the involvement of osmotin-like proteins in the mechanisms of salt-tolerance in potato. Both salinity and water stresses increased the accumulation of dry matter in the tuber, but only salinity increased proline concentration and water deficit increased the concentration of reducing sugars (Nadler and Heuer, 1995). However, Shaterian, Georges et al. (2005) showed that proline levels were not clearly associated with salinity-tolerance of diploid potatoes. Heuer and Nadler (1998) reported that under salinity, the maintenance of turgor was through the accumulation of chloride and proline, whereas under water deficit it was probably achieved either by altering cell wall elasticity or cell size. Since salinity affects water metabolism, it is uncertain whether increase in proline level in stressed plant is only due to salinity or because of water stress caused by salinity. Shaterian, Waterer et al. (2005) reported that Calreticulin (CR, $Ca^{2+}$-storage protein) is involved in ABA-induced salt-tolerance and that both salt-stress tolerance and CR expression are regulated by the roots.

Interaction between salinity and heat stress is quite strong in potato. As expected, tuber yield declined when saline irrigation and prolonged heat wave occurred together during crop development (Bustan et al., 2004); the effect was highest when both stresses were present at 40-60 days after emergence. Salinity combined with heat stress apparently leads to the collapse of mechanisms controlling salt accumulation in young expanding leaves, resulting in failure of vegetative growth recovery and a consequent reduction in the leaf area index as well as canopy functioning. However, even today many aspects remain obscure in relation to salinity-tolerance, particularly in an interaction with drought and heat stress.

### *Frost*

Frost during the crop season can cause partial or complete loss of leaf area, leading to a reduction in photosynthesis and hence yield loss (Hijmans et al., 2003). The threat of economic losses due to frost injury is of major importance at high latitudes and altitudes. In Andean countries nearly 400,000 ha are threatened every year by frost (Estrada et al., 1993), caused by radiative cooling on clear nights that can occur any time during the growing season (Francois et al., 1999). Hijmans et al. (2003) simulated the effect of frost on potato cultivars in the Altiplano (3,800-4,000 m above sea level distributed in Peru and Bolivia) with a georeferenced database; they estimated that when frost resistance increases from –1°C (current level) to –2°C or –3°C, average potato yield increases by 26 or 40 percent, respectively.

In general, cultural or chemical treatments are likely to be helpful in combating frost to a limited extent only (Vega and Bamberg, 1995). To prevent or reduce frost injury, pre-Hispanic cultures in the Lake Titicaca region constructed more than 82,000 ha of raised fields, which consisted of platforms elevated relative to the original soil surface (up to 1.2 m high and 2-20 m wide) and surrounded by canals (1.6-4.5 m wide) (Smith et al., 1968, cited in Lozada et al., 1998). Minimum temperatures during clear nights with fairly calm conditions were about 1°C higher at 50 cm above the soil surface in the middle of raised field platforms compared with adjacent flat fields (Lozada et al., 1998). The use of irrigation for frost protection is profitable especially where frosts in the growing season are infrequent, such as winter frosts in Mediterranean climates, and spring and autumn frosts in temperate regions (Landers and Witte, 1967 cited in Harris, 1978).

Wild potato species exhibit frost resistance far superior to cultivated potatoes. The frost-resistant species *S. ajanhuiri, S. x juzepczukii* and *S. x curtilobum* are grown from about 3,000-4,000 m above sea level, whilst *S. tuberosum* ssp. *andigena, S. stenotomum,* and *S. goniocalyx* yield best at 2,500-3,500 m or a little higher (Hawkes, 1978). Although *S. tuberosum* ssp. *tuberosum* cannot usually survive below 0 to –1°C, some *Solanum* species, for example *S. acaule,* can survive exposure to temperatures of –5 to –10°C (Richardson and Weiser, 1972). Ross and Rowe (1965) evaluated frost resistance in 586 wild and cultivated introductions representing 82 *Solanum* species, maintained in the IR-1 Potato Collection, in fields. The most resistant introductions were those of *S. acaule, S. chomatophilum, S. commersonii, S. x curtilobum, S. demissum,* and *S. sanctae-rosae.* Most of the hybrids involving *S. acaule* were resistant to temperature of –2.8°C. Chen et al. (1976) also reported that cultivars Red Pontiac and Alaska Frostless were killed at –3°C, *S. multidissectum* and *S. commersonii* at –5°C, *S. chomatophilum* at –5.5°C, and *S. acaule* at –6°C. Vega and

Bamberg (1995) visually screened over 2,600 accessions representing 101 species in the U.S. potato collection at Sturgeon Bay, Wisconsin, after natural frost in a uniform field. In this study also, the hardiest species were *S. acaule, S. albicans, S. commersonii,* and *S. demissum.* In addition, a previously unreported extremely frost-hardy species *S. paucissectum* was also discovered.

Field evaluation is not always dependable since freezing temperatures may be too low to allow genotypic difference readings (Ross and Rowe, 1965). In addition, time of occurrence and severity of frost cannot be forecast, and the damage often varies among plants of the same genotype because of small but important differences in microclimate, physiological conditions of the plant, or both (Mastenbroek, 1956). Therefore, some laboratory techniques have been developed, such as a top-lid freezing chamber equipped with a refrigerator (Mastenbroek, 1956), an excised leaflet test using an electrolyte leaching technique (Sukumaran and Weiser, 1972a,b), and the method based on pulsed nuclear magnetic resonance spectrometer (Chen et al., 1976). Dearborn (1969) reported that refractometer readings of foliage juice showed higher soluble solids content in frost-resistant genotypes than in frost-susceptible genotypes. Palta and Li (1979) found in a survey of 24 *Solanum* species and 7 cultivars of *S. tuberosum,* a good correlation between the presence of two palisade layers (thicker leaves) and frost-hardiness, suggesting that this trait may be used as a selection criterion for identifying frost-hardy genotypes.

However, limited success has been achieved in breeding frost-resistant potato cultivars. Dearborn (1969) bred Alaska Frostless from the progeny of crosses involving *S. acaule* and commercial Alaskan cultivars. Its nearly mature shoots were able to withstand field frosting at –3°C for 2 h and the plant could recover from the effect of several frosts of –2°C. Its tuber yield was comparable to those of standard Alaskan cultivars. Misra et al. (1984) reported that Indian cultivars Kufri Dewa and Kufri Sheetman can tolerate frost under northern plains in India.

Stone et al. (1993) indicated a failure to distinguish two major components for frost resistance, that is, freezing-tolerance in the nonacclimated state (NA) and cold acclimation ability (ACC). They defined the NA as the ability of a plant to survive extracellular ice, and the ACC as the ability to increase its freezing-tolerance (survive lower freezing temperatures) after a brief period of low-temperature exposure. No significant correlation was found between these two traits in segregating populations, suggesting their independent genetic control. Chen et al. (1999) reported that NA, ACC, and acclimated freezing-tolerance (NA plus ACC) evaluated under controlled environments were significantly correlated to the frost-resistance exhibited in fields. Vega et al. (2003) found that using a mapping population derived from *S. commersonii* (frost-resistant) and *S. cardiophyllum* (frost-sensitive), QTLs for NA and ACC were detected at separate genomic regions, suggesting independent genetic control of the two traits.

### *Ozone Injury*

Ozone ($O_3$) is one of the oxidants in photochemical smog, and 90 percent or more of the oxidant value is believed to be of ozone (Heggestad, 1973). NO*x* (mainly NO and $NO_2$) and volatile organic compounds (VOCs) are considered to be precursors of ozone. Ozone can be transported at high concentrations over long distances, and under certain synoptic conditions, elevated ozone concentrations may be formed over large areas (Fuhrer et al., 1997). Levels of tropospheric ozone have more than doubled in the past 100 years and are predicted to continue rising at a faster rate in the future (Houghton et al., 2001). Ozone is therefore considered to be the most important air pollutant in many parts of Europe, North and Central America, and the Far East (Donnelly et al., 2001).

The ozone injury to potato was first recognized in the eastern United States in the middle of last century (Heggestad, 1973). The adverse effect of ozone to plant growth and tuber yield, and the altered sensitivity of plant to biotic and abiotic stresses, has also become a major concern in Europe since the 1980s (Kollner and Krause, 2000). The visual symptom on leaves called "speckle leaf" includes chlorotic yellowing, bronzing, upward rolling, necrotic spotting of intervenial leaf tissue, and frequently extensive necrosis of leaf lamina (Hooker et al., 1973). The injury starts in the palisade mesophyll, the spongy mesophyll cells become necrotic and dead, and this is later extended to the entire leaf. As a result, plants show early maturity and often considerable yield reduction. Later works have clarified that elevated atmospheric ozone has frequently harmful effects on Rubisco activity and photosynthetic performance (Pell et al., 1992), pigment and carbohydrate content in leaves (Kollner and Krause, 2000), and induces a faster reduction of leaf area during crop senescence (Hacour et al., 2002).

Large cultivar differences in ozone injury have been reported in potato. The duration over which plants are exposed to ozone influences the extent of injury. In a comparison of eight cultivars in greenhouses with carbon-filtered and unfiltered air, tuber yields of sensitive cultivars were reduced by as much as 50 percent in unfiltered air, whereas tolerant cultivars produced about the same tuber yields in both the environments (Heggestad, 1973). In a field evaluation of leaf injury among 59 cultivars and selections, a high correlation ($r = 0.96$) was found between maturity type and average ozone injury index; early cultivars were the most sensitive whereas late cultivars were highly resistant (Mosley et al., 1978). Donnelly et al. (2001) reported that tuber yield in Bintje was generally less affected by elevated ozone concentration in open top chambers and Free Air Carbon dioxide Enrichment facilities (FACE) across a range of European agroclimatic conditions. They speculated that this may be because Bintje was bred in the Netherlands where ambient ozone concentrations are relatively high, and that the cultivar is late maturing.

Cultivars sensitive to ozone may be injured in 0.1 h by 0.4 ppm of ozone, and over an 8 h exposure to 0.03 ppm ozone may be equally effective (Heck, 1968, cited in Hooker et al., 1973). In recent experiments with open top chamber and FACE, ozone was artificially added to elevate it until about 0.06 ppm (8 h per day in seasonal means) (Donnelly et al., 2001; Craigon et al., 2002). In order to define critical levels for ozone based on the effects of the pollutant on trees, crops, and natural vegetation, the cumulative exposure index with a cut-off concentration of 0.04 ppm (AOT40) has been standardized in Europe (Fuhrer et al., 1997). However, it has also been reported that ozone peaks induce much stronger effects than constant concentrations using the same total dose (Kollner and Krause, 2000).

Breeding of potato cultivars tolerant to ozone injury has not been reported. However, Foster et al. (1983) developed a technique to estimate the effects of ozone on tuber yield under field conditions. They used N-[-2-(2-oxo-1-imidazolidinyl)ethyl]-N'-phenylurea (EDU or DPX-4891) as an antioxidant compound, which was side-dressed at plant emergence and sprayed on foliage in an ozone-affected area in California, USA. The EDU application increased total and marketable tuber yield of Centennial Russet (ozone-sensitive) by more than twofold and fourfold, respectively, whereas this did not affect tuber yield of White Rose (ozone-resistant). Vos et al. (1982) examined the inheritance of leaf injury due to ozone using two sets of diallel crosses involving resistant, intermediate, and susceptible parents. The general and specific combining ability were significant in both diallels, accounting for over 70 and 20 percent, respectively, of the variation among hybrids, suggesting that the resistance was determined by multiple genes that tended to be dominant. Hence phenotypic mass selection for ozone resistance should be effective in identifying useful genotypes in the initial stages of a potato-breeding program.

### *Effect of Carbon Dioxide*

Due to global climate change, $CO_2$ concentration in atmosphere has increased from 280 ppm (in volume) in preindustrial times to the present level of about 370 ppm, and this is expected to reach about 500-600 ppm by the middle of this century (Houghton et al., 2001). Since the photosynthetic rate in $C_3$ species increases almost linearly with the increase of $CO_2$ concentration at least until the level of 500-600 ppm (Wolfe and Erickson, 1993), the enriched $CO_2$ concentration is supposed to increase dry matter production and yield in $C_3$ crops. In fact, the experiments conducted in open top chamber and FACE during these ten years demonstrated the increase of photosynthetic rate, dry matter production and yield in many $C_3$ crops under these elevated $CO_2$ concentration (Long et al., 2004). In potato, Finnan et al. (2002) reported 32 percent increase in tuber yield of cultivar Bintje in the experiments with open top chamber of elevated

$CO_2$ concentration (680 ppm) in Ireland. Magliulo et al. (2003) also reported about 50 percent increase of below ground dry matter under elevated $CO_2$ (550 ppm) in FACE in Italy. Craigon et al. (2002) summarized the results of effect of $CO_2$ enrichment (550-680 ppm) in open top chamber at six sites and in FACE at two sites for two years in European countries, and concluded that on average, $CO_2$ elevation increased the dry weight of marketable size tubers by 17 percent.

The effect of various climatic changes, however, needs to be considered in totality because various factors are interrelated. The increase in $CO_2$ concentration is coupled with an increase in temperature. Daily mean temperature is expected to increase by about 1.4-5.8°C by year 2100 (Houghton et al., 2001). In addition, climate is expected to become drier during summer months for large portions of North and Central America, western Europe, central Asia, eastern Brazil, and north and western Africa (Parry, 1990, cited in Wolfe and Erickson, 1993). The predicted change in precipitation coupled with $CO_2$ increase will most probably restrict the growth and tuber yield of potato. However, the combined effects of $CO_2$, temperature, and drought on potato have not been evaluated so far. More research is thus needed to make an accurate estimation of the effect of global climate change on potato production. Further, physiological, genetical, and molecular studies related to the effects of abiotic stresses on potato needs to be accelerated for developing improved genotypes and designing effective cultural practices to combat the expected adverse effect of global climate change on potato production.

## REFERENCES

Alhagdow, M.M., Bartharkur, N.N., and Donnelly, D.J. (1999). Salinity stress and sodium-potassium interactions in micropropagated potatoes. *Potato Research 42:*73-78.

Bansal, K.C. and Nagarajan, S. (1987). Reduction of leaf growth by water stress and its recovery in relation to transpiration and stomatal conductance in some potato (*Solanum tuberosum* L.) genotypes. *Potato Research 30:*497-506.

Barone, A. (2004). Molecular marker-assisted selection for potato breeding. *American Journal of Potato Research 81:*111-117.

Belanger, G., Walsh, J.R., Richards, J.E., Milburn, P.H., and Ziadi, N. (2000). Yield response of two potato cultivars to supplemental irrigation and N fertilization in New Brunswick. *American Journal of Potato Research 77:*11-21.

Bennett, J. (2003). Opportunities for increasing water productivity of CGIAR crops through plant breeding and molecular biology. In Kijne, J.W., Barker, R., and Molden, D. (Eds.), *Water Productivity in Agriculture: Limits and Opportunities for Improvement* (pp. 103-126). UK: CAB International.

Benoit, G.R., Stanley, C.D., Grant, W.J., and Torrey, D.B. (1983). Potato top growth as influenced by temperatures. *American Potato Journal 60:*489-501.

Bernstein, L., Ayers, A.D., and Wadleicgh, F.H. (1951). The salt tolerance of White Rose potatoes. *Proceedings of the American Society of Horticultural Science 57:*231-236.

Bishop, J.C. and Grimes, D.W. (1978). Precision tillage effects on potato root and tuber production. *American Potato Journal 55:*65-71.

Blackman, P.G. and Davies, W.J. (1986). Root to shoot communication in maize plants of the effects of soil drying. *Journal of Experimental Botany 36:*39-48.

Bodlaender, K.B.A., Lugt, C., and Marinus, J. (1964). The induction of second-growth in potato tubers. *European Potato Journal 7:*57-71.

Bowen, W.T. (2003). Water productivity and potato cultivation. In Kijne, J.W., Barker, R., and Molden, D. (Eds.), *Water Productivity in Agriculture: Limits and Opportunities for Improvement* (pp. 229-238). UK: CAB International.

Broin, M., Cuine, S., Peltier, G., and Rey, P. (2000). Involvement of CDSP 32, a drought-induced thioredoxin, in the response to oxidative stress in potato plants. *FEBS Letters 467:*245-248.

Burton, W.G. (1981). Challenges for stress physiology in potato. *American Potato Journal 58:*3-14.

———. (1989). *The Potato,* Third edition. England: Longman Scientific & Technical.

Bushnell, J. (1925). The relation of temperature to growth and respiration in the potato plant. *Technical Bulletin Minnesota Agricultural Experiment Station 34:*1-29.

Bustan, A., Sagi, M., de Malach, Y., and Pasternak, D. (2004). Effects of saline irrigation water and heat waves on potato production in an arid environment. *Field Crops Research 90:*275-285.

Campbell, M.D., Campbell, G.S., Kunkel, R., and Papendick, R.I. (1976). A model describing soil-plant-water relations for potatoes. *American Potato Journal 53:*431-441.

Champoux, M.C., Wang, G., Sarkarung, S., Mackill, D.J., O'Toole, J.C., Hang, N., and McCouch, S. (1995). Locating genes associated with root morphology and drought avoidance in rice via linkage to molecular markers. *Theoretical and Applied Genetics 90:*969-981.

Chen, P.M., Burke, M.J., and Li, P.H. (1976). The frost hardiness of several *Solanum* species in relation to the freezing of water, melting point depression, and tissue water content. *Botanical Gazet 137:*313-317.

Chen, Y.K.H., Bamberg, J.B., and Palta, J.P. (1999). Expression of freezing tolerance in the interspecific $F_1$ and somatic hybrids of potatoes. *Theoretical and Applied Genetics 98:*995-1004.

CIP (1984). *Potatoes for the Developing World.* Lima, Peru: International Potato Center.

Craigon, J., Fangmeier, A., Jones, M., Donnelly, A., Bindi, M., Temmerman, L.D., Persson, K., and Ojanpera, K. (2002). Growth and marketable-yield responses of potato to increased $CO_2$ and ozone. *European Journal of Agronomy 17:*273-289.

Cutter, E.G. (1992). Structure and development of the potato plant. In Harris, P.M. (Ed.), *The Potato Crop,* Second edition (pp. 65-161). London: Chapman & Hall.

Dearbone, C.H. (1969). Alaska Frostless, an inherently frost resistant potato variety. *American Potato Journal 46:*1-4.

Deblonde, P.M.K. and Ledent, J.F. (2001). Effects of moderate drought conditions on green leaf number, stem height, leaf length and tuber yield of potato cultivars. *European Journal of Agronomy 14:*31-41.

Demagante, A.L., Harris, P.M., and Van der Zaag, P. (1996). Variation among clonal lines of potato in response to irrigation regimes in an isohyperthemic environment. *Field Crops Research 47:*191-200.

Doczi, R., Csanaki, C., and Banfalvi, Z. (2002). Expression and promoter activity of the desiccation-specific *Solanum tuberosum* gene, *StDS2. Plant, Cell and Environment 25:*1197-2003.

Donnelly, A., Craigon, J., Black, C.R., Colls, J.J., and Landon, G. (2001). Elevated $CO_2$ increases biomass and tuber yield in potato even at high ozone concentrations. *New Phytologist 149:*265-274.

Durrant, M.J., Love, B.J.G., Messem, A.B., and Draycott, A.P. (1973). Growth of crop roots in relation to soil moisture extraction. *Annals of Applied Biology 74:*387-394.

Dwelle, R.B., Kleinkoph, G.E., Steinhrst, R.K., Pavek, J.J., and Hurley, P.J. (1981). The influence of physiological processes on tuber yield of potato clones (*Solanum tuberosum* L.): Stomatal diffusive resistance, stomatal conductance, gross photosynthetic rate, leaf canopy, tissue nutrient levels, and tuber enzymes activities. *Potato Research 24:*33-47.

Earl, H.J. and Davis, R.F. (2003). Effect of drought stress on leaf and whole canopy radiation use efficiency and yield of maize. *Agronomy Journal 95:*688-696.

Ekanayake, I.J. and Jong, J.P.D. (1992). Stomatal response of some cultivated and wild tuber-bearing potatoes in warm tropics as influenced by water deficits. *Annals of Botany 70:*53-60.

Ekanayake, I.J. and Midmore, D.J. (1992). Genotypic variation for root pulling resistance in potato and its relationship with yield under water-deficit stress. *Euphytica 61:*43-53.

Estrada, N., Garcia, W., Gabriel, J., and Carrasco, E. (1993). Breeding potatoes for tolerance to frost and resistance to late blight. *American Potato Journal 70:*809-810.

Evers, D., Overney, S., Simon, P., Greppin, I., and Hausman, J.F. (1999). Salt tolerance of *Solanum tuberosum* L. overexpressing a heterologous osmotin-like protein. *Biologia Plantarum 42:*105-112.

Ewing, E.E. (1981). Heat stress and the tuberization stimulus. *American Potato Journal 58:*31-49.

Ewing, E.E., Kahn, B.A., Lazin, M.B., Ben Kheder, M., Mendoza, H.A., and Plaisted, R.L. (1983). Selecting for heat tolerance within populations derived from the Cornell *andigena* collection. In Hooker, W.J. (Ed.), *Research for the Potato in the Year 2000* (pp. 81-82). Lima, Peru: International Potato Center.

Ewing, E.E., Simko, I., Omer, E.A., and Davies, P.J. (2004). Polygene mapping as a tool to study the physiology of potato tuberization and dormancy. *American Journal of Potato Research 81:*281-289.

Fabeiro, C., de Santa Olalla, F. M., and de Juan, J.A. (2001). Yield and size of deficit irrigated potatoes. *Agricultural Water Management 48:*255-266.

FAO (2001). *FAO Production Yearbook* 55.

Feddes, R.A., Graaf, M.D., Bouma, J., and van Loon, C.D. (1988). Simulation of water use and production of potatoes as affected by soil compaction. *Potato Research 31:*225-239.

Ferreira, T.C. and Carr, M.K.V. (2002). Responses of potatoes (*Solanum tuberosum* L.) to irrigation and nitrogen in a hot, dry climate. *Field Crops Research 78:*51-64.

Finnan, J.M., Donnelly, A., Burke, J.I., and Jones, M.B. (2002). The effects elevated concentrations of carbon dioxide and ozone on potato (*Solanum tuberosum* L.) yield. *Agriculture, Ecosystems and Environment 88:*11-22.

Foster, K.W., Guerard, J.P., Oshima, R.J., Bishop, J.C., and Timm, H. (1983). Differential ozone susceptibility of Centennial Russet and White Rose potato as demonstrated by fumigation and antioxidant treatments. *American Potato Journal 60:*127-139.

Francois, C., Bosseno, R., Vacher, J.J., and Seguin, B. (1999). Frost risk mapping derived from satellite and surface data over the Bolivian Altiplano. *Agricultural and Forest Meteorology 95:*113-137.

Frusciante, L., Barone, A., Carputo, D., and Ranalli, P. (1999). Breeding and physiological aspects of potato cultivation in the Mediterranean region. *Potato Research 42:*265-277.

Fuhrer, J., Skarby, L., and Ashmore, M.R. (1997). Critical levels for ozone effects on vegetation in Europe. *Environmental Pollution 97:*91-106.

Gandar, P.W. and Tanner, C.B. (1976). Leaf growth, tuber growth, and water potential in potatoes. *Crop Science 16:*534-538.

Gauntney, T.L. and Haynes, F.L. (1983). Recurrent selection for heat tolerance in diploid potatoes *(Solanum tuberosum* subsp. *phureja* and *stenotomum). American Potato Journal 60:*537-542.

Gawronska, H., Thornton, M.K., and Dwelle, R.B. (1992). Influence of heat stress on dry matter production and photoassimilate partitioning by four potato clones. *American Potato Journal 69:*653-665.

Gebhardt, C., Titter, E., Debener, T., Schachtschabel, U., Walkemeier, B., Uhrig, H., and Salamini, F. (1989). RFLP analysis and linkage mapping in *Solanum tuberosum. Theoretical and Applied Genetics 78:*65-75.

Gopal, J. (2001). In vitro and in vivo genetic parameters and character association in potato. *Euphytica 118:*145-151.

Gopal, J., Chamail, A., and Sarkar, D. (2004). In vitro production of microtubers for conservation of potato germplasm: Effect of genotype, abscisic acid and sucrose. *In vitro Cellular and Developmental Biology—Plant 40:*285-290.

Gopal, J., Gaur, P.C., and Rana, M.S. (1992). Early generation selection for agronomic characters in a potato breeding programme. *Theoretical and Applied Genetics 84:*709-713.

Gopal, J. and Minocha, J.L. (1998). Effectiveness of in vitro selection for agronomic characters in potato. *Euphytica 103:*67-74.

Gopal, J., Minocha, J.L., and Dhaliwal, H.S. (1998). Microtuberization in potato (*Solanum tuberosum* L.). *Plant Cell Reports 17:*794-798.

Gregory, P.J. and Simmonds, L.P. (1992). Water relations and growth of potatoes. In Harris, P.M. (Ed.), *The Potato Crop,* Second edition (pp. 214-246). London: Chapman & Hall.

Hacour, A., Craigon, J., Van dermeiren, K., Ojanpera, K., Pleijel, H., Danielsson, H., Hogy, P., Finnan, J., and Bindi, M. (2002). $CO_2$ and ozone effects on canopy development of potato crops across Europe. *European Journal of Agronomy 17:*257-272.

Hammes, P.S. and De Jager, J.A. (1990). Net photosynthetic rate of potato at high temperature. *Potato Research 33:*515-520.

Hannapel, D.J., Chen, H., Rosin, F.M., Banerjee, A.K., and Davies, P.J. (2004). Molecular controls of tuberization. *American Journal of Potato Research 81:*263-274.

Harris, P.M. (1978). Water. In Harris, P.M. (Ed.), *The Potato Crop* (pp. 244-277). London: Chapman & Hall.

Hawkes, J.G. (1978). History of the potato. In Harris, P.M. (Ed.), *The Potato Crop* (pp. 1-14). London: Chapman & Hall.

Heggestad, H.E. (1973). Photochemical air pollution injury to potatoes in the Atlantic Coastal Stages. *American Potato Journal 50:*315-328.

Heuer, B. and Nadler, A. (1998). Physiological response of potato plants to soil salinity and water deficit. *Plant Science 137:*43-51.

Hijmans, R.J. (2001). Global distribution of the potato crop. *American Journal of Potato Research 78:*403-412.

Hijmans, R.J., Condori, B., Carrillo, R., and Kropff, M.J. (2003). A quantitative and constraint-specific method to assess the potential impact of new agricultural technology: The case of frost resistant potato for the Altiplano (Peru and Bolivia). *Agricultural Systems 76:*895-911.

Hooker, W.J., Yang, T.C., and Potter, H.S. (1973). Air pollution injury of potato in Michigan. *American Potato Journal 50:*151-161.

Hosaka, K. (1999). A genetic map of *Solanum phureja* Clone 1.22 constructed using RFLP and RAPD markers. *American Journal of Potato Research 75:*97-102.

Houghton, J.T., Ding, Y., Griggs, D.J., Noguer, M., Van der Linden, P.J., Xiaosu, D., Maskell, K., and Johnson, C.A. (2001). *Climate Change 2001. The Scientific Basis*. Cambridge, UK: Cambridge University Press.

Hsiao, T.C. (1973). Plant responses to water stress. *Annual Review of Plant Physiology 24:*519-570.

Isoda, A., Nakaseko, K., and Gotoh, K. (1988). Production potential of two high-yielding hybrid strains between ssp. *andigena* and ssp. *tuberosum* in potato (*Solanum tuberosum* L.). *Japanese Journal of Crop Science 57:*790-791.

Iwama, K. (1988). Difference in root growth of potato plants among years and cropping seasons. *Japanese Journal of Crop Science 57:*346-354.

———. (1998). Development of nodal and lateral roots in potato under field conditions. *Journal of the Faculty of Agriculture, Hokkaido University 68:*33-44.

Iwama, K., Hasegawa, T., and Jitsuyama, Y. (2001). Genetic evaluation of root mass to breed a drought tolerant variety in potato (*Solanum tuberosum* L.). In *Proceedings of the 6th Symposium of the International Society of Root Research* (pp. 144-145). Nagoya, Japan.

Iwama, K., Hasegawa, T., and Nakaseko, K. (1999). New potato lines with high productivity and drought tolerance. In Horie, H., Geng, S., Amano, T., Inamura, T., and Shiraiwa, T. (Eds.), *World Food Security and Crop Production Technologies for Tomorrow* (pp. 189-193). Kyoto, Japan: The Crop Science Society of Japan.

Iwama, K., Hukushima, T., Yoshimura, T., and Nakaseko, K. (1993). Influence of planting density on root growth and yield in potato. *Japanese Journal of Crop Science 62:*628-635.

Iwama, K., Nakaseko, K., Isoda, A., Gotoh, K., and Nishibe, S. (1981). Relations between root system and tuber yield in the hybrid population of the potato plants. *Japanese Journal of Crop Science 50:*233-238.

Iwama, K. and Nishibe, S. (1989). Comparison of root characters among cultivated potatoes *(Solanum tuberosum)* and their wild relatives. *Japanese Journal of Crop Science 58:*126-132.

Iwama, K., Takata, O., Ohnami, M., and Nakaseko, K. (1995). Effects of scion and stock on root growth of grafting plants between two potato varieties with different root mass. *Japanese Journal of Crop Science 64:*86-92.

Iwama, K., Uemura, T., and Umemura, Y. (1998). A simple method for selection of potato lines with a higher root/total ratio at an early stage in the seedling generation. *Plant Production Science 1:*286-287.

Jackson, S.D. and Willmizer, L. (1994). Jasmonic acid spraying does not induce tuberization in short-day-requiring potato species kept in non-inducing condition. *Planta 194:*155-159.

Jacobs, J.M.E., Eck, H.J.V., Arens, P., Verkerk-Bakker, B., Hekkert, B. te L., Bastiaanssen, H.J.M., El-Kharbotly, A., Pereira, A., Jacobsen, E., and Stiekema, W.J. (1995). A genetic map of potato *(Solanum tuberosum)* integrating molecular markers, including transposons and classical markers. *Theoretical and Applied Genetics 91:*289-300.

Jefferies, R.A. (1989). Water-stress and leaf growth in field-grown crops of potato (*Solanum tuberosum* L.). *Journal of Experimental Botany 40:*1375-1381.

———. (1994). Drought and chlorophyll fluorescence in field-grown potato *(Solanum tuberosum). Physiologia Plantarum 90:*93-97.

———. (1995). Physiology of crop response to drought. In Haverkort, A.J. and Mackerron, D.K. (Eds.), *Potato Ecology and Modelling of Crops Under Conditions Limiting Growth* (pp. 61-74). The Netherlands: Kluwer Academic Publications.

———. (1996). Evaluation of seedling selection for salinity tolerance in potato (*Solanum tuberosum* L.). *Euphytica 88:*207-213.

Jefferies, R.A. and MacKerron, D.K.L. (1987). Aspects of the physiological basis of cultivar differences in yield of potato under droughted and irrigated conditions. *Potato Research 30:*201-217.

———. (1989). Radiation Interception and growth of irrigated and droughted potato *(Solanum tuberosum). Field Crops Research 22:*101-112.

———. (1993). Response of potato genotypes to drought. II. Leaf area index, growth and yield. *Annals of Applied Biology 122:*105-112.

———. (1994). Genotypic difference in water use efficiency in potato. *Aspects of Applied Biology 38:*63-70.

———. (1997). Carbon isotope discrimination in irrigated and droughted potato (*Solanum tuberosum* L.). *Plant, Cell and Environment 20:*124-130.

Kashiwagi, J. (2000). *Phenotypic Variation and Selection Method of Root Mass in the Hybrid Populations of Potato* (*Solanum tuberosum* L.). Ph.D. Thesis, Hokkaido University, Sapporo, Japan.

Kawakami, J., Iwama, K., Hasegawa, T., and Jitsuyama, Y. (2003). Growth and yield of potato plants grown from microtubers in fields. *American Journal of Potato Research 80:*371-378.

Keller, R., Brearley, C.A., Tretheway, R.N., and Muller-Rober, B. (1998). Reduced inositol content and altered morphology in transgenic potato plants inhibited for 1D-myo-inositol 3-phosphate synthase. *Plant Journal 16:*403-410.

Khedher, M.B. and Ewing, E.E. (1985). Growth analysis of eleven potato cultivars grown in the greenhouse under long photoperiods with and without heat stress. *American Potato Journal 62:*537-554.

Kleinhenz, M.D. and Palta, J.P. (2002). Root zone calcium modulates the response of potato plants to heat stress. *Physiologia Plantarum 115:*111-118.

Kleinkopf, G.E. (1983). Potato. In Teare, I.D. and Peet, M.M. (Eds.), *Crop-water Relations* (pp. 287-305). New York: John Wiley & Sons.

Koda, Y., Omer, E.A., Yoshihara, T., Shibata, H., Sakamura, S., and Okazawa, Y. (1988). Isolation of a specific potato tuber-inducing substance from potato leaves. *Plant Cell Physiology 29:*1047-1051.

Kollner, B. and Krause, G.H.M. (2000). Changes in carbohydrates, leaf pigments and yield in potatoes induced by different ozone exposure regimes. *Agriculture Ecosystems and Environment 78:*149-158.

Krauss, A. (1978). Tuberization and abscisic acid content in *Solanum tuberosum* as affected by nitrogen nutrition. *Potato Research 21:*183-193.

Krauss, A. and Marschner, H. (1982). Influence of nitrogen nutrition, daylength and temperature on contents of gibberellic acid and on tuberization in potato plants. *Potato Research 25:*13-21.

———. (1984). Growth rate and carbohydrate metabolism of potato tubers exposed to high temperatures. *Potato Research 27:*297-303.

Ku, Sun-Ben, Edwards, G.E., and Tanner, C.B. (1977). Effects of light, carbon dioxide, and temperature on photosynthesis, oxygen inhibition of photosynthesis, and transpiration in *Solanum tuberosum. Plant Physiology 59:*868-872.

Kubo, K., Jitsuyama, Y., Iwama, K., Hasegawa, T., and Watanabe, N. (2004). Genotypic difference in root penetration ability by durum wheat (*Triticum turgidum* L. var. *durum*) evaluated by a pot with paraffin-Vaseline discs. *Plant and Soil 262:*169-177.

Lee, J.T., Prasad, V., Yang, P .T., Wu, J.F., Ho, T.H.D., Charng, Y.Y., and Chan, M.T. (2003). Expression of Arabidopsis CBF1 regulated by an ABA/stress inducible promoter in transgenic tomato confers stress tolerance without affecting yield. *Plant, Cell and Environment 26:*1181-1190.

Lesczynski, D.B. and Tanner, C.B. (1976). Seasonal variation of root distribution of irrigated, field-grown Russet Burbank potato. *American Potato Journal 53:*69-78.

Levy, D. (1983). Varietal differences in the responses of potatoes to repeated short periods of water stress in hot climates. 2. Tuber yield and dry matter accumulation and other tuber properties. *Potato Research 26:*315-321.

———. (1986). Genotypic variation in the response of potatoes (*Solanum tuberosum* L.) to high ambient temperatures and water deficit. *Field Crops Research 15:*85-96.

———. (1992). The response of potatoes (*Solanum tuberosum* L.) to salinity: Plant growth and tuber yields in the arid desert of Israel. *Annals of Applied Biology 120:*547-555.

Levy, D., Fogelman, E., and Itzhak, Y. (1988). The effect of water salinity on potatoes (*Solanum tuberosum* L.): Physiological indices and yielding capacity. *Potato Research 31:*601-610.

———. (1993). Influence of water and soil salinity on emergence and early development of potato (*Solanum tuberosum* L.) cultivars and effect of physiological age of seed tubers. *Potato Research 36:*335-340.

Levy, D., Itzhak, Y., Fogelman, E., Margalit, E., and Veilleux, R.E. (2001). Ori, Idit, Zohar and Zahov: Tablestock and chipstock cultivars bred for adaptation to Israel. *American Journal of Potato Research 78:*167-173.

Long, S.P., Ainsworth, E.A., Rogers, A., and Ort, D.R. (2004). Rising atmospheric carbon dioxide: Plants face the future. *Annual Review of Plant Biology 55:*591-628.

Loon, C.D. van (1981). The effect of water stress on potato growth, development, and yield. *American Potato Journal 58:*51-69.

Loon, C.D. van, Smet, L.A.H.D., and Boone, F.R. (1985). The effect of a ploughpan in marine loam soils on potato growth. 2. Potato plant response. *Potato Research 28:*315-330.

Lozada, D.S. de, Baveye, P., and Riha, S. (1998). Heat and moisture dynamics in raised field systems of the lake Titicaca region (Bolivia). *Agricultural and Forest Meteorology 92:*251-265.

Lynch, D.R. and Tai, G.C.C. (1989). Yield and yield component response of eight potato genotypes to water stress. *Crop Science 29:*1207-1211.

Maas, E.V. and Hoffman, G.J. (1977). Crop salt tolerance- current assessment. *Journal of Irrigation and Drainage Division, ASCE 103:*115-134.

MacKerron, D.K.L. and Jefferies, R.A. (1986). The influence of early soil moisture stress on tuber numbers in potato. *Potato Research 29:*299-312.

MacKerron, D.K.L., Marshall, B., and Jefferies, R.A. (1988). The distributions of tuber sizes in droughted and irrigated crops of potato. II. Relation between size and weight of tubers and the variability of tuber-size distributions. *Potato Research 31:*279-288.

MacKerron, D.K.L, Waister, P.D., and Thompson, R. (1982). Why we fall short of potential potato yields. *Arable Farming (March issue):*28-29.

Magliulo, M., Bindi, M., and Rana, G. (2003). Water use and irrigated potato (*Solanum tuberosum* L.) grown under free air carbon dioxide enrichment in central Italy. *Agriculture, Ecosystems and Environment 97:*65-80.

Malik, N.J., Dwelle, R.B., Thornton, M.K., and Pavek, J.J. (1992). Dry matter accumulation in potato clones under seasonal high temperature conditions in Pakistan. *American Potato Journal 69:*667-676.

Mares D.J., Marschner, H., and Krauss, A. (1981). Effect of gibberellic acid on growth and carbohydrate metabolism of developing tubers of potato (*Solanum tuberosum* L.). *Physiologia Plantarum 52:*267-274.

Marinus, J. and Bodlaender, K.B.A. (1975). Response of some potato varieties to temperature. *Potato Research 18:*189-204.

Mastenbroek, C. (1956). Some experiences in breeding frost-tolerant potatoes. *Euphytica 5:*289-297.

Materechera, S.A., Alston, A.M., Kirby, J.M., and Dexter, A.R. (1992). Influence of root diameter on the penetration of seminal roots into a compacted subsoil. *Plant and Soil 144:*297-303.

Midmore, D.J. (1992). Potato production in the tropics. In Harris, P.M. (Ed.), *The Potato Crop,* Second edition (pp. 728-793). London: Chapman & Hall.

Midmore, D.J., Berrios, D., and Roca, J. (1988). Potato (*Solanum* spp.) in the hot tropics. V. Intercropping with maize and the influence of shade on tuber yields. *Field Crops Research 18:*159-176.

Midmore, D.J. and Prange, R.K. (1992). Growth responses of two *Solanum* species to contrasting temperatures and irradiance levels: Relations to photosynthesis, dark respiration and chlorophyll fluorescence. *Annals of Botany 69:*13-20.

Midmore, D.J., Roca, J., and Berrios, D. (1988). Potato (*Solanum* spp.) in the hot tropics. IV. Intercropping with maize and the influence of shade on potato microenvironment and crop growth. *Field Crops Research 18:*141-157.

Miller, D.E. and Martin, M.W. (1985). Effect of water stress during tuber formation on subsequent growth and internal defects in Russet Burbank potatoes. *American Potato Journal 62:*83-89.

Misra, P.C., Gaur, P.C., and Nayar, N.M. (1984). *Potato cultivars released by the Central Potato Research Institute.* India: Technical Bulletin 12, Shimla, CPRI.

Miyao, M. (2003). Molecular evolution and genetic engineering of $C_4$ photosynthetic enzymes. *Journal of Experimental Botany 54:*179-189.

Mosley, A.R., Rowe, R.C., and Weidensaul, T.C. (1978). Relationship of foliar ozone injury to maturity classification and yield of potatoes. *American Potato Journal 55:*147-153.

Nadler, A. and Heuer, B. (1995). Effect of saline irrigation and water deficit on tuber quality. *Potato Research 38:*119-123.

Oldeman, L.R., Hakbeling, R.T.A., and Sombroek, W.G. (1991). *World Map of the Status of Human-induced Soil Degradation. An Explanatory Note. Global Assessment of Soil Degradation (GLASOD),* Second revised edition. Washington, DC: ISRIC/UNEP.

Opena, G.B. and Porter, G.A. (1999). Soil management and supplemental irrigation effects on potato: II. Root growth. *Agronomy Journal 91:*426-431.

Paliwal, K.V. and Yadav, R.R. (1980). Effect of saline irrigation water on the yield of potato. *Indian Journal of Agricultural Science 50:*31-33.

Palta, J.P. and Li, P.H. (1979). Frost-hardiness in relation to leaf anatomy and natural distribution of several *Solanum* species. *Crop Science 19:*665-671.

Patel, R.M., Prasher, S.O., Donnelly, D., and Bonnell, R.B. (2001). Effect of initial soil salinity and subirrigation water salinity on potato tuber yield and size. *Agricultural Water Management 46:*231-239.

Pell, E.J., Eckardt, N.A., and Enyedi, A.J. (1992). Timing of ozone stress and resulting status of ribulose bisphosphate carboxylase/oxygenase and associated net photosynthesis. *New Phytologist 120:*397-405.

Pierce, F.J. and Burpee, C.G. (1995). Zone tillage effects on soil properties and yield and quality of potatoes (*Solanum tuberosum* L.). *Soil and Tillage Research 35:*135-146.

Prange, R.K., McRae, K.B., Midmore, D.J., and Deng, R. (1990). Reduction in potato growth at high temperature: Role of photosynthesis and dark respiration. *American Potato Journal 67:*357-369.

Price, A. and Courtois, B. (1999). Mapping QTLs associated with drought resistance in rice: Progress, problems and prospects. *Plant Growth Regulation 29:*123-133.

Pushkarnath (1969). *Potato in India-Varieties.* New Delhi, India: Indian Council of Agricultural Research.

Rademacher, T., Hausler, R.E., Hrsch, H.J., Zhang, L., Lipka, V., Weier, D., Kreuzaler, F., and Peterhansel, C. (2002). An engineered phosphoenolpyruvate carboxylase redirects carbon and nitrogen flow in transgenic potato plants. *Plant Journal 32:*25-39.

Reynolds, M.P. and Ewing, E.E. (1989). Heat tolerance in tuber bearing *Solanum* species: A protocol for screening. *American Potato Journal 66:*63-74.

Reynolds, M.P., Ewing, E.E., and Owens, T.G. (1990). Photosynthesis at high temperature in tuber-bearing *Solanum* species. *Plant Physiology 93:*791-797.

Richardson, D.G. and Weiser, C.J. (1972). Foliage frost resistance in tuber-bearing *Solanums. HortSicence 7:*19-22.

Rickey, T. and Belknap, W.R. (1991). Comparison of the expression of several stress-responsive genes in potato tubers. *Plant Molecular Biology 16:*1009-1018.

Ritchie, J.T. (1973). Influence of soil water status and meteorological conditions on evaporation from a corn canopy. *Agronomy Journal 65:*893-897.

Rober, M., Geider, K., Muller-Rober, B., and Willmitzer, L. (1996). Synthesis of fructans in tubers of transgenic starch-deficient potato plants does not result in an increased allocation of carbohydrates. *Planta 199:*528-536.

Robins, J.S. and Domingo, C.E. (1956). Potato yield and tuber shape as affected by severe soil-moisture deficits and plant spacing. *Agronomy Journal 48:*488-495.

Ross, R.W. and Rowe, P.R. (1965). Frost resistance among the *Solanum* species in the IR-1 Potato Collection. *American Potato Journal 42:*177-185.

Sattelmacher, B., Kuene, R., Malagamba, P., and Moreno, U. (1990). Evaluation of tuber bearing *Solanum* species belonging to different ploidy levels for its yielding potential at low soil fertility. *Plant and Soil 129:*227-233.

Sattelmacher, B., Marschner, H., and Kuhne, R. (1990). Effects of the temperature of the rooting zone on the growth and development of roots of potato *(Solanum tuberosum). Annals of Botany 65:*27-36.

Schapendonk, A.H.C.M., Spitters, C.J.T., and Groot, P.J. (1989). Effects of water stress on photosynthesis and chlorophyll fluorescence of five potato cultivars. *Potato Research 32:*17-32.

Shalhevet, J., Shimshi, D., and Meir, T. (1983). Potato irrigation requirements in a hot climate using sprinkler and drip methods. *Agronomy Journal 75:*13-16.

Shannon, M.C. (1997). Adaptation of plants to salinity. *Advances in Agronomy 60:*75-120.

Shannon, M.C. and Grieve, C.M. (1999). Tolerance of vegetable crops to salinity (review). *Scientia Horticulturae 78:*5-38.

Shaterian, J., Georges, F., Hussain, A., Waterer, D., Jong, H.D., and Tanino, K.K. (2005). Root to shoot communication and abscisic acid in calreticulin (*CR*) gene expression and salt-stress tolerance in grafted diploid potato clones. *Environmental and Experimental Botany 53:*323-332.

Shaterian, J., Waterer, D., De Jong, H., and Tanino, K.K. (2005). Differential stress responses to NaCl salt application in early- and late-maturing diploid potato (*Solanum* sp.) clones. *Environmental and Experimental Botany 54:*202-212.

Shen, L., Courtois, B., McNally, K.L., Robin, S., and Li, Z. (2001). Evaluation of near-isogenic lines of rice introgressed with QTLs for root depth through marker-aided selection. *Theoretical and Applied Genetics 103:*75-83.

Shimshi, D. and Susnoschi, M. (1985). Growth and yield studies of potato development in a semi-arid region. 3. Effect of water stress and amounts of nitrogen top dressing on physiological indices and on tuber yield and quality of several cultivars. *Potato Research 28:*177-191.

Shock, C.C., Feibert, E.B.G., and Saunders, L.D. (1998). Potato yield and quality response to deficit irrigation. *HortScience 33:*655-659.

Shock, C.C., Zalewski, J.C., Stieber, T.D., and Burnett, D.S. (1992). Impact of early-season water deficits on Russet Burbank plant development, tuber yield and quality. *American Potato Journal 69:*793-803.

Sinclair, T.R., Muchow, R.C., Bennett, J.M., and Hammond, L.C. (1986). Relative sensitivity of nitrogen and biomass accumulation to drought in field-grown soybean. *Agronomy Journal 79:*986-991.

Sinclair, T.R., Purcell, L.C., and Sneller, C.H. (2004). Crop transformation and the challenge to increase yield potential. *Trends in Plant Science 9:*70-75.

Slater, J.W. (1968). The effect of night temperature on tuber initiation of the potato. *European Potato Journal 11:*14-22.

Stalham, M.A. and Allen, E.J. (2001). Effect of variety, irrigation regime and planting date on depth, rate, duration and density of root growth in the potato *(Solanum tuberosum)* crop. *Journal of Agricultural Science, Cambridge 137:*251-270.

Stone, J. M., Palta, J.P., Bamberg, J.B., Weiss, L.S., and Harbage, J.H. (1993). Inheritance of freezing resistance in tuber-bearing *Solanum* species: Evidence for independent genetic control of nonacclimated freezing tolerance and cold acclimation capacity. *Proceedings of the National Academy of Sciences (USA) 90:*7869-7873.

Struik, P.C. and Ewing, E.E. (1995). Crop physiology of potato *(Solanum tuberosum)*: Responses to photoperiod and temperature relevant to crop modeling. In Haverkort, A.J. and Mackerron, D.K. (Eds.), *Potato Ecology and Modeling of Crops under Conditions Limiting Growth* (pp. 19-40). The Netherlands: Kluwer Academic Publishers.

Struik, P.C., Geertsema, J., and Custers, C.H.M.G. (1989a). Effects of shoot, root and stolon temperature on the development of the potato (*Solanum tuberosum* L.) plants. I. Development of the haulm. *Potato Research 32:*133-141.

———. (1989b). Effects of shoot, root and stolon temperature on the development of the potato (*Solanum tuberosum* L.) plants. II. Development of stolons. *Potato Research 32:*143-149.

———. (1989c). Effects of shoot, root and stolon temperature on the development of the potato (*Solanum tuberosum* L.) plants. III. Development of tubers. *Potato Research 32:*151-158.

Sukumaran, N.P., Ezekiel, R., and Perumal, N.K. (1989). Response of net photosynthetic rate and stomatal conductance to water deficit in different potato cultivars. *Photosynthetica 23:*664-666.

Sukumaran, N.P. and Weiser, C.J. (1972a). An excised leaflet test for evaluating potato frost tolerance. *HortScience 7:*467-468.

———. (1972b). Freezing injury in potato leaves. *Plant Physiology 50:*564-567.

Tai, G.C.C. and Young, D.A. (1984). Early generation selection for important agronomic characteristics in a potato breeding population. *American Potato Journal 61:*419-434.

Takahashi, K., Fujino, K., Kikuta, Y., and Koda, Y. (1994). Expansion of potato cells in response to jasmonic acid. *Plant Science 100:*3-8.

Tanksley, S.D., Ganal, M.W., Prince, J.P., de Vicente, M.C., Bonierbale, M.W., Broun, P., Fulton, T.M., Giovannoni, J.J., Grandillo, S., Martin, G.B. et al. (1992). High density molecular linkage maps of the tomato and potato genomes. *Genetics 132:*1141-1160.

Tawfik, A.A., Kleinhenz, M.D., and Palta, J.P. (1996). Application of calcium and nitrogen for mitigating heat stress effects on potatoes. *American Potato Journal 73:*261-273.

Thornton, M.K., Malik, N.J., and Dwelle, R.B. (1996). Relationship between leaf gas exchange characteristics and productivity of potato clones grown at different temperatures. *American Potato Journal 73:*63-77.

Trebejo, I. and Midmore, D.J. (1990). Effect of water stress on potato growth, yield and water use in a hot and a cool tropical climate. *Journal of Agricultural Science, Cambridge 114:*321-334.

Van der Zaag, P. and Demagante, A.L. (1987). Potato (*Solanum* spp.) in an isohyperthermic environment. I. Agronomic management. *Field Crops Research 17:*199-217.

———. (1988). Potato (*Solanum* spp.) in an isohyperthermic environment. III. Evaluation of clones. *Field Crops Research 19:*167-181.

Van der Zaag, P., Demagante, A., and Acasio, A. (1986). Response of *Solanum* potatoes to mulching during different seasons in an isohyperthermic environment in the Philippines. *Tropical Agriculture (Trinidad) 63:*229-239.

Vega, S.E. and Bamberg, J.M. (1995). Screening the U.S. potato collection for frost hardiness. *American Potato Journal 72:*13-21.

Vega, S.E., del Rio, A.H., Jung, G., Bamberg, J.B., and Palta, J.P. (2003). Marker-assisted genetic analysis of non-acclimated freezing tolerance and cold acclimation capacity in a backcross *Solanum* population. *American Journal of Potato Research 80:*359-369.

Veilleux, R.E., Paz, M.M., and Levy, D. (1997). Potato germplasm development for warm climates: Genetic enhancement of tolerance to heat stress. *Euphytica 98:*83-92.

Vos, J. and Groenwold, J. (1986). Root growth of potato crops on a marine-clay soil. *Plant and Soil 94:*17-33.

———. (1989a). Characteristics of photosynthesis and conductance of potato canopies and the effects of cultivar and transient drought. *Field Crops Research 20:*237-250.

———. (1989b). Genetic differences in water-use efficiency, stomatal conductance and carbon isotope fractionation in potato. *Potato Research 32:*113-121.

Vos, N.E. De, Hill, R.R., Jr., Pell, E.J., and Cole, R.H. (1982). Quantitative inheritance of ozone resistance in potato. *Crop Science 22:*992-995.

Waddell, J.T., Gupta, S.C., Moncrief, J.F., Rosen, C.J., and Steele, D.D. (1999). Irrigation and nitrogen management effects on potato yield, tuber quality and nitrogen uptake. *Agronomy Journal 91:*991-997.

Walker, T.S., Schmiediche, P.E., and Hijmans, R.J. (1999). World trends and patterns in the potato crop: An economic and geographic survey. *Potato Research 42:*241-264.

Weaber, J.E. (1926). Root habits of the potato. In *Root Development of Field Crops* (pp. 240-246). New York: McGraw-Hill Book Co.

Weisz, R., Kaminski, J., and Smilowitz, Z. (1994). Water deficit effects on potato leaf growth and transpiration: Utilizing fraction extractable soil water for comparison with other crops. *American Potato Journal 71*:829-840.

Wolf, S., Marani, A., and Rudich, J. (1990). Effect of temperature and photoperiod on assimilate partioning in potato plants. *Annals of Botany 66*:513-520.

Wolf, S., Olesinski, A.A., Rudich, J., and Marani, A. (1990). Effect of high temperature on photosynthesis in potatoes. *Annals of Botany 65*:179-185.

Wolfe, D.W. and Erickson, J.D. (1993). Carbon dioxide effects on plants: Uncertainties and implications for modeling crop response to climate change. In Kaiser, H.M.K. and Drennen, T.E. (Eds.), *Agricultural Dimensions of Global Climate Change* (pp. 153-178). Delray Beach, FL: St. Lucie Press.

Wolfe, D.W., Fereres, E., and Voss, R.E. (1983). Growth and yield response of two potato cultivars to various levels of applied water. *Irrigation Science 3*:211-222.

Yamaguchi, J. and Tanaka, A. (1989). Quantitative observation on the root system of various crops growing in the field. *Soil Science and Plant Nutrition 36*:483-493.

Yeo, E.T., Kwon, H.B., Han, S.E., Lee, J.T., Rue, J.C., and Byu, M.O. (2000). Genetic engineering of drought resistant potato plants by introduction of the threhalose-6-phosphate synthase (*TPS1*) gene from *Saccharomyces cerevisiae. Molecular Cell 10*:263-268.

Yu, L.X., Ray, J.D., O'Toole, J.C., and Nguyen, H.T. (1995). Use of wax-petrolatum layers for screening rice root penetration. *Crop Science 35*:684-687.

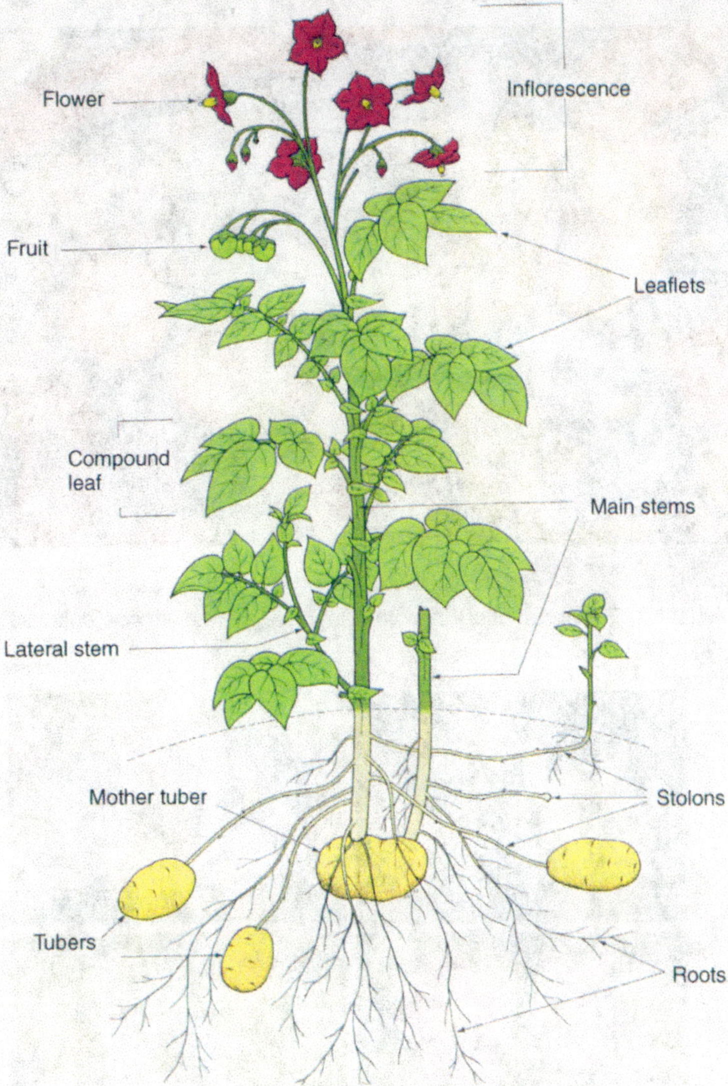


PLATE 1.1. Stylized illustration of a cultivated potato plant. Drawing courtesy of the International Potato Center.

PLATE 1.2. Tuber variation in potato land races (indigenous cultivars) from South America.

PLATE 1.4. Wild and cultivated potato flowers. (A) *Solanum bulbocastanum;* *(B) S. paucijugum;* (C) *S. tuberosum* (cultivated species); (D) *S. colombianum.*

PLATE 1.5. Wild potato fruits. (A) *Solanum verrucosum;* (B) *S. andreanum;* (C) *S. schenckii;* (D) *S. moscopanum.*

PLATE 1.6. Wild potato habitats. (A) Dry deciduous forest in Jalisco, Mexico (2,080 m); (B) paramo in Mérida State, Venezuela (3,050 m); (C) seashore in the Chonos Archipelago, Aisén, Chile; (D) human highlands in La Paz Department, Bolivia (3,900 m).

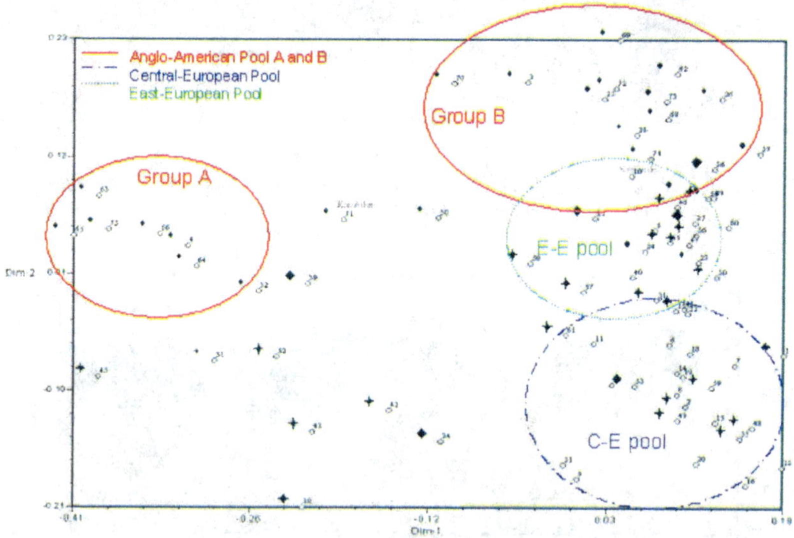


PLATE 4.2. Principal coordinate analysis of 75 cultivars from different locations of the world using AFLPs produced by 6 primer combinations. *Source:* Braun, Schullehner, and Wenzel (2005).

PLATE 8.1. Symptoms of late blight caused by *Phytophthora infestans* on (A) leaves, (B) stems, and (C) tubers of potato. Photos courtesy J. Miller.

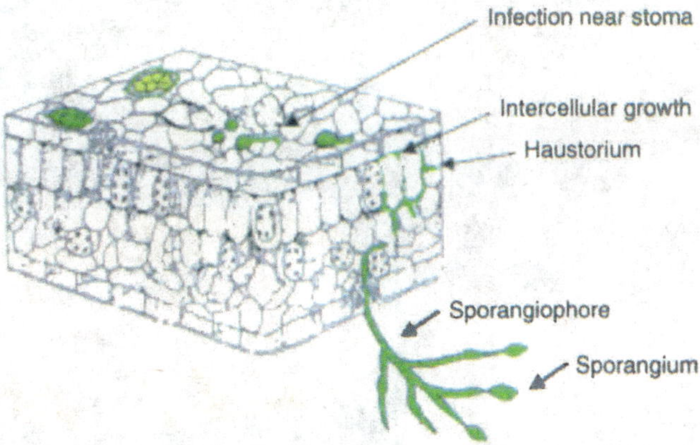


PLATE 8.2. Cross section of leaf showing different stages of infection cycle of *Phythphthora infestans*. Modified from Fernández-Northcote (2001).

PLATE 9.1. Typical symptoms of various fungal and oomycete diseases of potato (tuber symptoms of common scab—a bacterial disease is shown here for comparison with other tuberborne diseases).

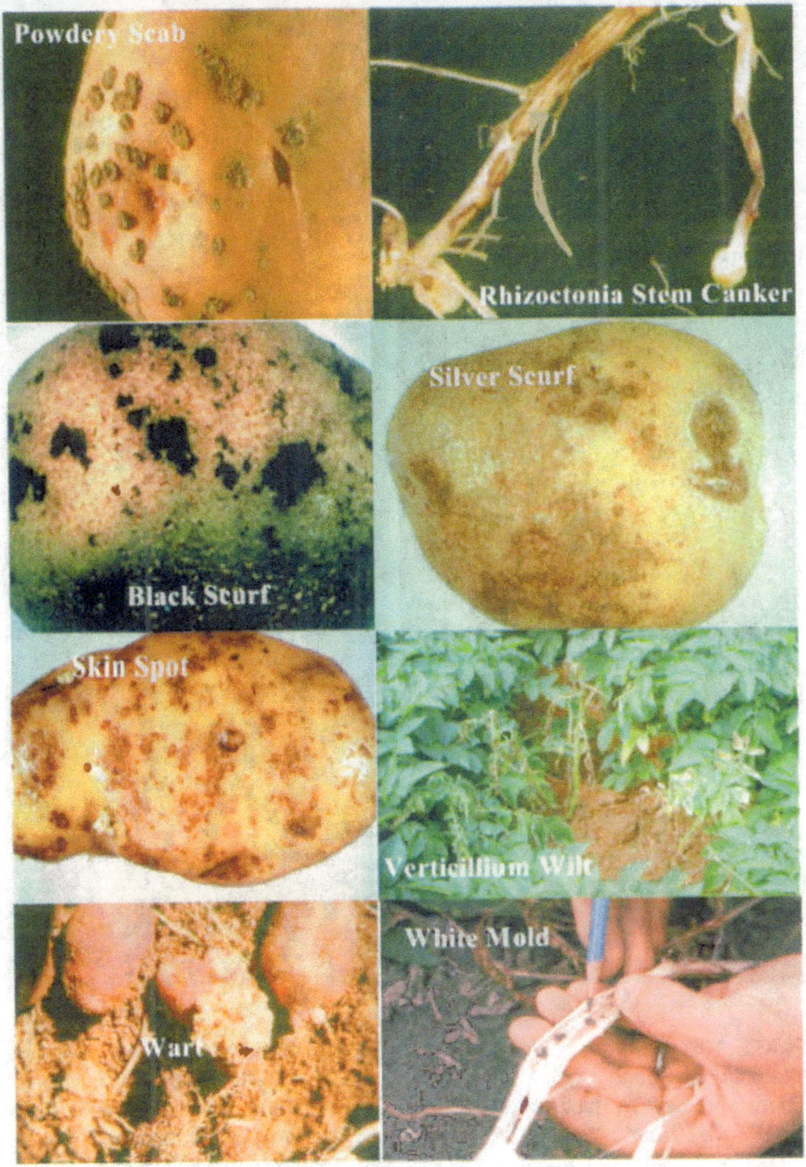


PLATE 9.1 *(continued)*

PLATE 10.1. (A) Potato plant with symptoms of bacterial wilt. Photo courtesy International Potato Center, Peru. (B) Potato plant with symptoms of ring rot. Photo courtesy T. Pastuszewska, Plant Breeding and Acclimatization Institute (IHAR). (C) Potato tuber with symptoms of ring rot. Photo courtesy T. Pastuszewska, IHAR. (D) Potato plant with symptoms of blackleg. Photo courtesy S. Sobkowiak, IHAR. (E) Potato tubers with symptoms of soft rot. Photo courtesy S. Sobkowiak, IHAR (for common scab symptoms please see Plate 9.1).

PLATE 11.1. Symptoms caused by some important viruses. (A) Leafroll caused by potato leafroll virus. Photo courtesy Scottish Agricultural Science Agency. (B) Severe mosaic caused by potato virus Y. Photo courtesy Scottish Agricultural Science Agency. (C) Calico caused by alfalfa mosaic virus. Photo courtesy L. F. Salazar. (D) Ring spots caused by tomato black ring virus. Photo courtesy C. J. Jeffries. (E) Mild mottle caused by potato virus X. Photo courtesy Scottish Agricultural Science Agency. (F) Necrotic rings caused by potato mop-top virus. Photo courtesy Scottish Agricultural Science Agency. (G) Internal "spraing" caused by potato mop-top virus. Photo courtesy H. Barker. (H) Surface lesion caused by tobacco necrosis virus. Photo courtesy Scottish Agricultural Science Agency.

PLATE 12.1. Important insect pests of potato. (A) Green peach aphid *(Myzus persicae)*—apterae. (B) Green peach aphid *(Myzus persicae)*—alatae. (C) Potato aphid *(Macrosiphum euphorbiae)*—alatae. (D) Leaf miner *(Liriomyza huidobrensis)*—adult. (E) Potato leafhopper *(Emposca fabae)*—larva and pupa. (F) Potato leafhopper *(Emposca fabae)*—adult. (G) Spider mite *(Tetranychus urticae)*—adult. (H) Colorado beetle *(Leptinotorsa decemlineat)*—larvae. (I) Colorado beetle *(Leptinotorsa decemlineat)*—adult. (J) Whitefly *(Bemisia tabaci)*—adult. (K) Wireworm *(Agriotes lineatus)* larval damage to potato tuber. (L) Wireworm *(Agriotes lineatus)*—adult. (M) White grub *(Melolontha melolontha)* larval damage to potato tuber. (N) White grub *(Melolontha melolontha)*—adult. (O) Cutworm *(Agrotis segetum)*—moth. (P) Potato tuber moth *(Phthorimaea operculella)* larval damage to potato tuber. (Q) Potato tuber moth *(Phthorimaea operculella)*—pupa. (R) Potato tuber moth (*Phthorimaea operculella)*—adult.

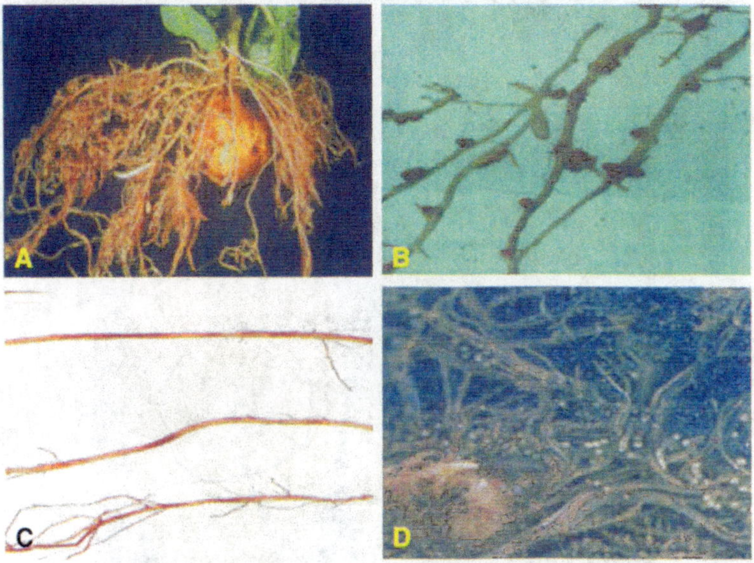


PLATE 12.2. Some nematode pests of potato. (A) Northern root-knot nematode *(Meloidogyne hapla)* infested potato roots. (B) Columbia root-knot nematode *(Meloidogyne chitwoodi)* infested potato roots. (C) Root-lesion nematode *(Pratylenchus penetrans)* infested potato roots. (D) Potato cyst nematode (*Globodera* spp.) infested potato roots.

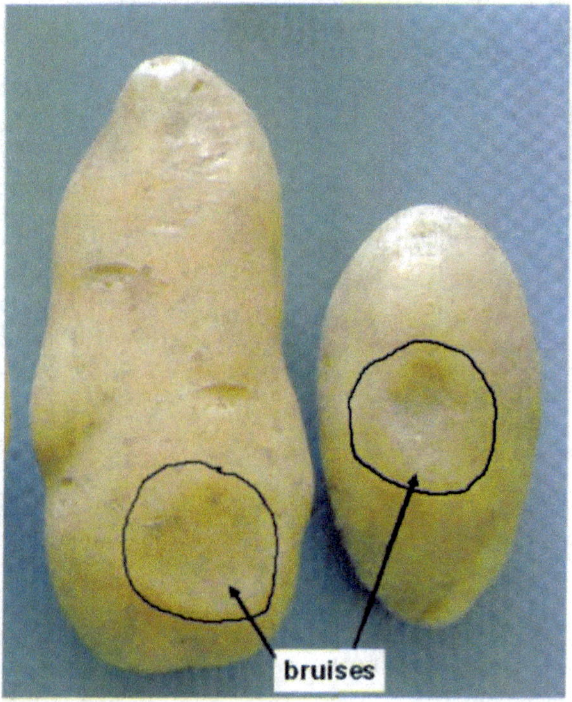


PLATE 13.3. Development of bruises or pressure spots as a result of pressure exerted on tubers in the lower layer by tubers in the upper layer.

PLATE 15.1. Raising potato crop from true seeds. (A) True potato seeds. (B) Seedlings raised and left in bed. (C) Seedlings transplanted in field. (D) Produce (seedling tubers) from a transplant crop of Indian cultivar TPS C-3.

# Chapter 8

# Late Blight

G. A. Forbes
J. A. Landeo

Late blight (*Phytophthora infestans* [Mont.] de Bary) is the most devastating disease of potato worldwide. CAB International (Anonymous, 2003) currently lists 122 countries where late blight has been reported, but it certainly occurs in more countries for which no report has been made. The economic and social impact of this disease was best experienced during the Irish famine in the 1840s when millions of Irish people either died or emigrated (Bourke, 1993) due to starvation caused by killing of the potato crop by *P. infestans.* At present, about $77 million are spent only on fungicides per season throughout the United States for control of this disease (Guenther et al., 2001). Yearly fungicide usage for late blight control in Europe is estimated to be about $150 million (H. Schepers, personal communication). The International Potato Center (CIP) estimated fungicide use in developing countries at $750 million (Anonymous, 1997). Based on these estimates, about $1 billion per year is spent on the fungicides to control late blight in the United States, Europe, and developing countries. These huge social and economic consequences of late blight have attracted considerable attention from research workers. This chapter reviews the current state of knowledge and advances made in understanding and controlling this most dreaded malady of potatoes.

## *CAUSAL AGENT*

### *Morphology*

*Phytophthora infestans* (Mont.) de Bary—the causal agent of potato late blight—is a coenocytic oomycete, which produces ellipsoid to lemon shaped sporangia with a small pedicel. Sporangium length ranges between 29 and 36 μm and width between 19 and 22 μm. Zoospores (ca. 7-12 per sporangium) have two flagella, one tinsel type that is directed forward and a second whiplash type that is directed backward. Zoospores are usually uninucleate, but binucleate zoospores have also been detected.

The pathogen is hetrothallic and requires two mating types, A1 and A2, for sexual reproduction. In culture the mycelium is white and the colony is somewhat slow growing. Growth rates can vary dramatically among isolates, but fast growing isolates can cover a 9 cm plate within about one week. Some isolates grow irregularly, producing a lumpy appearance in agar, which has sometimes been associated with A2 mating type. *Phytophthora infestans* can grow on a variety of culture media, but not all isolates will grow on all media. Commonly used media are rye agar, V-8 juice agar, pea agar, cornmeal agar, corn seed agar, and lima bean agar. Recipes for many media have been presented by Erwin and Ribeiro (1996).

### *Taxonomy*

The genus *Phytophthora*, its closely related genus *Pythium*, and the downy mildews (i.e., *Bremia, Sclerospora, Plasmopara, Peronspora,* etc.) belong to the oomycetes, which are unrelated to true fungi, such as ascomycetes and basidiomycetes. Oomycetes are stramenopiles, a group that includes golden and brown algae and that are characterized by zoospores propelled by heterokont flagella (of unequal length). Nuclei in vegetative cells are generally diploid even though polyploidy has been detected (Tooley and Therrien, 1991). Sexual reproduction occurs via oospores that develop from the joining of oogonia and antheridia. Oomycetes have cellulose in their cell walls. Erwin and Ribeiro (1996) discussed the changes in the understanding of taxonomic position of this group of organisms. Many pathologists now consider the following as the correct taxonomic classification of *P. infestans:*

> *Phytophthora infestans* (Mont.) de Bary—Kingdom Chromista, Phylum Oomycota, Order *Peronosporales,* Family *Peronosporaceae,* Genus *Phytophthora,* of which it is the type species (Birch and Whisson, 2001).

### *Origin and Spread*

*Phytophthora infestans* is thought by most researchers to have originated in the highlands of central Mexico (Grünwald et al., 2001) because both mating types (A1 and A2) were present when the disease was discovered in that location and there is a high genetic diversity for the pathogen in this region (Goodwin and Drenth, 1997; Grünwald et al., 2001). In contrast, only the A1 mating type and low genetic diversity had been found outside Mexico until the mid-1980s (Hohl and Iselin, 1984; Goodwin et al., 1992). An alternative hypothesis proposed the Andes, center of origin of the cultivated potato, as the center of origin for *P. infestans* (Abad and Abad, 1997). This hypothesis is based primarily on historical accounts of potato diseases in the Andes, but appears to have no support from the scientific literature.

Until recently, most isolates of *P. infestans* found outside North America belonged to the US-1 clonal lineage. This led to a hypothesis that US-1 had caused the original epidemics in Europe in the 1800s (Goodwin et al., 1994) and then spread globally, presumably with seed trade (Fry et al., 1993). Originally researchers proposed that US-1 had spread from Mexico to the United States and then from the United States to Europe (Goodwin et al., 1994). Alternatively, some researchers proposed that US-1 was introduced into Europe directly from South America (Andrivon, 1996). However, recent analyses of mitochondrial DNA of *P. infestans* in herbarium material presented evidence that a genotype different from US-1 was involved in the original epidemics in Europe (Ristaino et al., 2001). During the 1980s, the A2 mating type of *P. infestans* was detected in Europe (Hohl and Iselin, 1984) along with several new alleles for known markers. Global migration of new races of *P. infestans* has taken place and their status in selected regions is presented in Table 8.1.

TABLE 8.1. Distribution of A1 and A2 mating types of *Phytophthora infestans.*

| Region | Country | Population |
|---|---|---|
| Africa | Ethiopia | A1, new population |
| | Kenya | A1, clonal, host specific, old population |
| | Uganda | A1, clonal, host specific, old population |
| | Egypt | A1 + A2 |
| | Morocco | A1 + A2 |
| | South Africa | A1, clonal, old population |
| Asia and the Pacific | China | A1 + A2 |
| | India | A1 + A2 |
| | Japan | A1 + A2 |
| | Korea | A1 + A2 |
| | Nepal | A1 + A2 |
| | Pakistan | A1 + A2 |
| | Thailand | A1 + A2 |
| | Bangladesh | A1 |
| | Philippines | A1 |
| | Sri Lanka | A1 |
| | Taiwan | A1 |
| | Vietnam | A1 |
| | Indonesia | A2 |
| | Russia | A1 + A2, clonal (Siberia) recombinant (near Moscow) |
| Latin America | Argentina | A1 + A2, clonal, host specific |
| | Brazil | A1 + A2, clonal, host specific |
| | Uruguay | A2, clonal |
| | Bolivia | A2, clonal, host specific |
| | Ecuador | A1, clonal, host specific |
| | Peru | A1, clonal, host specific |
| | Chile | A1, clonal |
| | Colombia | A1, clonal |
| | Costa Rica | A1, clonal |

TABLE 8.1 *(continued)*

| Region | Country | Population |
|---|---|---|
| | Venezuela | A1, clonal |
| | Mexico | A1 + A2 recombinant |

*Note:* References for many descriptions of pathogen populations in different countries can be found on the Web site of the Global Initiative on Late Blight (GILB): http://gilb.cip. cgiar.org.

The pathogen population in Europe is now highly diverse and there is evidence for sexual reproduction in several European countries (Drenth, 1994; Andersson et al., 1998; Flier, Kessel et al., 2002).

## *Hosts*

Erwin and Ribeiro (1996) list 89 host species of *P. infestans,* but more than 25 percent of these were based only on artificial inoculations. Some of the hosts mentioned have only been reported once and others may have been attacked by morphologically similar species of *Phytophthora,* which have been identified only recently. Species of both *Ipomea* and *Mirabilis* are listed as hosts by Erwin and Ribeiro (1996), but disease on these may have been caused by *P. ipomeae* (Flier, Grünwald et al., 2002) and *P. mirabilis* (Goodwin et al., 1999), respectively.

Of the nontuber-bearing *Solanum* hosts, tomato, *S. lycopersicum,* is the most important economically. Other domesticated *Solanum* species are also hosts of *P. infestans,* including pepino *(S. muricatum)*, tree tomato *(S. betaceum)* (Oliva et al., 2002), and naranjilla *(S. quitoense)* (Unpublished data). Eggplant has also been reported as an occasional host of *P. infestans* (Hooker, 1981), but lack of disease reports in this very cosmopolitan crop puts its host status in doubt. *P. infestans* is undoubtedly also a pathogen of the wild *Solanum* species closely related to tomato (Abad et al., 1995), but there are few reports on these. A recent study in South America listed 6 wild nontuber-bearing *Solanum* species and one species complex attacked in nature by *P. infestans.* However, some isolates of the pathogen associated with certain hosts are so unusual that they may actually be new species of *Phytophthora* (Adler et al., 2004). A study in Canada showed that the nightshades, *S. dulcamara* and *S. sarrachoides* were infected while black nightshade, *S. nigrum,* was not (Platt, 1999). In Europe, natural infections were found on *S. nigrum, S. dulcamara,* and *S. sisymbriifolium* (Flier et al., 2003b). One study in California showed that 11 *Solanum* species were infected by inoculation in a greenhouse, however, only one of these, *S. sarrachodes,* was found naturally infected in the field (Vartanian and Endo, 1985).

There have also been references to infection in solanaceous plants outside the genus *Solanum. Nolana* species are apparently attacked on the coast of Peru during the wet winter season (Turkensteen, 1978; Abad et al., 1995). *Datura* species have been reported as hosts of *P. infestans* (Sunita and Sen, 1997), but it is not clear if infections occur naturally in the field. Extensive sampling in Ecuador, where *Datura* spp. are common weeds, has not produced any indication that this genus is a host (Adler et al., 2004). *Brugmansia sanguinea* is also attacked by *P. infestans* in South America, but lesions have only been found on flower petals (Adler et al., 2004). *P. infestans* also attacks *Petunia* (Platt, 1999).

Several species from Africa have been cited as hosts of *P. infestans.* Kori (1972) reported that *S. indicum, S. incanum,* and *S. aculosturm* were all infected in the field. Natural infection in *S. incanum* and *S. indicum* were confirmed by Nattrass and Ryan (1951) as well as infection in one unidentified species of *Solanum.* A cultivated crop, garden huckleberry *(S. scabrum)* was reported as a host of *P. infestans* in Cameroon (Nkengaka, 2000).

## *DISEASE SYMPTOMS*

The disease appears first as water-soaked irregular pale-green lesions mostly near tips and margins of leaves. These lesions rapidly grow into large brown to purplish black necrotic spots. During morning hours a white mildew, which consists of sporangia and spores of the pathogen, can be seen on the lower surface of infected leaves, especially around the edges of the necrotic lesions (Figure 8.1A). The symptoms thus can vary depending on the age of the lesion and the environmental conditions prior to observation and the tissue infected. Very young lesions are small (2-10 mm) and irregularly shaped, and

FIGURE 8.1. Symptoms of late blight caused by *Phytophthora infestans*. See Plate 8.1 in the color gallery.

may be surrounded by a small halo (collapsed, but green tissue bordering the dark necrotic lesion). As lesions grow they become more circular until they are limited by the leaflet margins. They are usually not delimited by the veins and older lesions are typically surrounded by a chlorotic halo.

When late blight attacks the stem it can cause girdling and the leaves wilt above the point of infection. Light to dark brown lesions on stems or petioles elongate and encircle the stems (Figure 8.1B). Stem lesions become brittle and the stem frequently breaks at that point. Rain-borne sporangia from the diseased foliage can also infect tubers in the soil. The infected tubers show irregular reddish brown to purplish slightly depressed areas that extend deep into internal tissue of the tubers (Figure 8.1C). The infected tubers are initially hard, dry, and firm but may be invaded by other pathogens, mainly bacteria, leading to soft rot. A pungent putrid smell is often associated with heavily infected fields. This is due to rotting of dead tissue and is not a direct consequence of late blight.

Late blight lesions can be mistaken for several other diseases. Under especially wet conditions, *Botrytis cinerea* can cause rot of lower leaves of potato plants in such a manner that they resemble late blight lesions, however sporulation from *B. cinerea* is gray (see Figure 9.1 Chapter 9). When conditions become quite dry and late blight lesions dry out, they can be mistaken for dried lesions caused by *B. cinerea* or sometimes by lesions caused by *Alternaria solani* (early blight). However, early blight lesions are typically zonate with a definite outer margin, whereas active late blight lesions are definitely not zonate and typically without a definite outer margin. Certain types of frost damage as it occurs in the highland tropics can resemble late blight.

## EPIDEMIOLOGY AND DISEASE CYCLE

The occurrence of late blight varies spatially and temporally depending on inoculum sources and weather. *P. infestans* survives on tubers in storage, or within soil or a refuse pile. The two most important sources of initial infection in the temperate zone are refuse piles (Zwankhuizen et al., 1998) and infected seed (van der Zaag, 1956; Marshall and Stevenson, 1996). It is not clear which of these is more important. However, seed tubers play an important role in the long-distance dispersal of *P. infestans* (Fry et al., 1993).

In most temperate zones, *P. infestans* generally reproduces asexually, which means there are no overwintering spores. New epidemics each year must start from the above mentioned inoculum sources. In subtropical and highland tropical areas, potatoes are produced round the year and the pathogen is always present, hence the disease is also always present, even though often limited to only humid microenvironments. For this reason, late blight generally occurs in the tropics as soon as the rains begin.

Most potato growers know the general conditions that favor late blight development. Within months of the appearance of late blight in Ireland in 1845, David Moore, curator of the Royal Dublin Society's Botanic Gardens, (cited in Nelson, 1995, page 3) wrote that "damp is conducive to the progress of the disease ... dryness retards it." The disease has been associated historically with "cool wet weather," but optimum temperatures are probably near 20°C (Harrison, 1992), although the disease can occur over a range from about 5°C to 30°C.

A number of authors have reviewed the effects of the external environment on one or more aspects of the biology of *P. infestans* (Harrison, 1992; Mizubuti et al., 2000). The principal factor determining sporangial germination appears to be temperature, although the exact effect of temperature probably depends on the pathogen's genotype being evaluated. Regardless, 15°C appears to be a point of differentiation, below which germination is indirect, and above which it is direct. Interestingly, germination and zoospore activity can occur at very low temperatures, near 0°C, although at a very slow rate. Above 30°C, sporangia do not germinate (Crosier, 1933). Most phases of *P. infestans* cannot survive temperatures above 30°C (Harrison, 1992), but the pathogen is more robust when it is inside plant tissue and has survived temperatures of 40°C in stem tissue (Kable and Mackenzie, 1980).

Similarly, the effect of humidity on pathogen development, and subsequent disease, is complex. Generally, saturated air or leaf wetness is required for sporangia to germinate and for zoospore motility (Harrison, 1992). After infection has occurred, the mycelium is relatively protected from low humidity, but high ambient humidity, near saturation, is needed for sporangia formation (Harrison, 1992).

There is inconsistency in the literature relating to sporangial sensitivity to drying. Most studies found that once dried, sporangia were no longer viable (Warren and Colhoun, 1975). However, Minogue and Fry (1981) showed that sporangia could regain viability after drying if rehydrated gradually. Because of the putative sensitivity of sporangia to drying, early workers considered rain splashing as one of the primary dispersal modes for this pathogen (Hirst, 1958). More recent studies indicate that sporangia are dispersed in air and are generally released during the day when relative humidity is lower (Aylor et al., 2001). Because of this wind-borne mode of dispersal, the sensitivity of sporangia to ultraviolet light is an important issue. Under high light conditions, most sporangia are killed within 1 h, but sporangia can last for very long periods on cloudy days (Mizubuti et al., 2000). Long-distance wind dispersal can virtually be inferred by the extremely rapid rate of overland and oversea movement detected when *P. infestans* was introduced to Europe in 1845 (Bourke, 1991). Factors other than temperature and humidity are also known to affect sporangia germination, for example, pH and osmotic potential of water.

*P. infestans* has a complex pre-infection process involving sporangial germination, zoospore movement, encystment, cyst germination, germ tube development, appresorium formation and penetration. These steps may be further divided into substeps. For example, penetration may involve both physical pressure from the infection peg, as well as enzymatic activity. Sporangia of *P. infestans* germinate either directly with a germ tube or indirectly by liberating zoospores. Germ tubes can also form secondary sporangia, which may serve to increase the longevity of the spore (Harrison, 1992). Germ tubes can penetrate most leaf epidermal cells, but various authors have detected that infection on the leaf surface is not random and that successful infections more readily occur near the stomatal complexes (Coffey and Gees, 1991). Germ tubes do not penetrate host tissue directly, but rather via formation of an appresorium and subsequently an infection peg. After penetration, the pathogen forms a specialized hyphal structure, sometimes referred to as an infection vessel. Hyphae extend from this and begin colonization of plant tissue intercellularly. Intercellular hyphae form haustoria that penetrate cells to absorb nutrients (Hohl and Suter, 1976). After a certain amount of time, sporangiophores grow out of stomatal openings (Figure 8.2).

Lesions become visible within a few days, the time taken depends on temperature, and the genetics of the host and pathogen. Under optimal conditions (18-22°C), and with a susceptible potato cultivar, infections can be visible in less than three days. Within a day or two after the lesion first becomes visible, the pathogen is capable of sporulation. Moderate temperatures (10-25°C) and wet conditions (100 percent RH) are required for sporulation. Within 8-12 h of favorable conditions sporangia are borne on indeterminate sporangiophores. Sporangia dislodge during changing relative humidity and can be captured in air currents or splash dispersed. As sporangia can survive for hours in unsaturated atmospheres when protected from solar radiation (Mizubuti et al., 2000), dispersal to

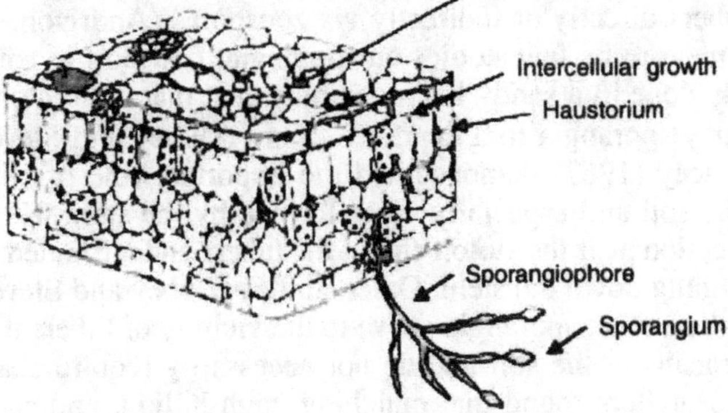


FIGURE 8.2. Cross section of leaf showing different stages of infection cycle of *Phythphthora infestans*. See Plate 8.2 in the color gallery.

hundreds of kilometers is possible. Under favorable conditions, large numbers of sporangia can be produced from a single lesion (more than 100,000 sporangia/lesion).

When individuals of opposite mating type (A1 and A2) colonize the same substrate (plant tissue or agar), sexual structures (antheridia and oogonia) are produced by each thallus and meiosis occurs in the gametes. After fertilization, the oogonium develops into an oospore, which can survive adverse conditions better than other forms of the pathogen. After a period of dormancy (weeks or months) oospores become capable of germination. Germination in the laboratory occurs in water or 1.5 percent agar at 18°C in the presence of blue light. It is clear that oospores can survive winter in temperate zones (Turkensteen et al., 2000), but the conditions stimulating germination are not yet precisely known. Oospores germinate via a germ sporangium. This sporangium may then either liberate zoospores or may germinate directly by giving out a germ tube.

The problem of tuber blight predates the relatively recent international spread of oospores and is thought to develop from sporangia and zoospores coming from foliage infection. There appears to be little evidence that oospores present in the soil are significant sources of tuber infection (Flier, Kessel et al., 2002). Infection in tubers generally occurs at eyes, lenticels, or through wounds (Adams, 1975).

Most researchers agree that sporangia wash down through soil and infect tubers directly or indirectly via zoospores (Andrivon, 1995). There appear to be few studies on sporangial transport in soils, but one study done in a sandy loam demonstrated that 0.6 cm of water could carry sporangia to a depth of 40 cm (Dubey and Stevenson, 1996). Lacey (1967) demonstrated the important role of channels within the soil and especially those formed by the stem. He found more infection near the stolon end of the tubers and attributed this to water running down the stem. Other studies (Dubey and Stevenson, 1996) indicate that sporangia arrive to the vicinity of tubers through various means in the soil and do not necessarily require channels. These researchers found that mulching, high hilling, and chemical treatment of soil could reduce but did not eliminate tuber infection when foliage infection was present.

## *DISEASE MANAGEMENT*

### *Cultural*

Eliminating the sources of initial inoculum is the first step in the control of any disease. So, use of healthy seed and destruction of infected foliage prior to harvest may help in reducing the incidence of late blight in the next season. Tubers left in the ground sprout leading to an early emergence of the volunteer plants that may become infected very early. Lesions on these plants sporulate and can cause disease in potato crop. Removal or destruction of volunteer plants is a common component of most late blight management strategies. Alternative hosts are other potential sources of initial infection (Flier et al., 2003b). However, alternative hosts frequently have specialized pathogen populations and therefore may not always be involved in potato disease development (Oyarzún et al., 1998).

Areas that remain wet because of excess soil moisture or heavy shading are potential hot spots for late blight. Planting only in sunny and well-drained fields can thus reduce the incidence of late blight. High hilling has been associated with reduced foliage severity (Boyd, 1980). This is probably due to improved drainage and better aeration, which dries foliage quickly. Hilling also reduces tuber blight by increasing the barrier between tubers and spores washing out of foliage. Some traditional techniques in the Andes such as raised beds or a similar technique in Colombia and Ecuador called "Huacho Rozado" (unpublished), have been associated with reduced late blight. Within a given season, the time of greatest late blight can often be avoided by opportune planting (Devaux and Haverkort, 1987). In the highland tropics, a large number of farmers escape the major late blight season by either planting the crop before or after rains. Planting date can reduce the time the plants are exposed to the disease potential. Green sprouting also known as diffused light sprouting, presprouting, or chitting, can create a more uniform crop that can be harvested earlier thereby reducing the period of exposure to late blight.

Data generated on the effects of plant density and plant nutrition on late blight incidence are not consistent. According to Garrett and Dendy (2001), the effect of plant density on potato late blight appears to be minimal. Several authors found that increased doses of phosphorous and potassium reduce late blight (Awan and Struchtemeyer, 1957) whereas nitrogen increases late blight incidence (Carnegie and Colhoun, 1983). Phosphorous and nitrogen apparently have contrasting effects on tuber blight. Nitrogen delays tuber maturation, resulting in more blight whereas phosphorous reduces incidence by accelerating maturation (Herlihy, 1970). A study in the highland tropics (Juarez et al., 2001) demonstrated that fertilization effects on late blight were much smaller than yield effects and therefore farmers would make decisions based on yield.

Acid soils and high aluminum availability in soils, commonly found in the tropics, have been shown to inhibit *P. infestans* in soil (Andrivon, 1995). Tuber blight incidence is also checked by early removal or destruction of foliage prior to harvest, and harvesting and storing of tubers under dry conditions.

### *Chemical*

The use of chemicals to control late blight began about a century ago. Initially materials such as sodium chloride, lime, and sulfur were used, but these were not very effective (Schwinn and Margot, 1991). The first effective compound was Bordeaux mixture, which was invented in the 1880s and is made of copper sulfate and lime. Bordeaux mixture was widely used on potatoes until other copper compounds were shown to be more effective in the 1930s. One of these, copper oxychloride is still used for control of blight.

In the 1940s the ethylenebisdithiocarbamates (often referred to as EBDCs), were first introduced into the market. Several of these—zineb, maneb, metiran, mancozeb, and propineb—make up the group of fungicides most frequently used against late blight in many parts of the world.

Another important group of fungicides are the tin compounds, such as fentin acetate and fentin hydroxide. These fungicides are effective and last for a long time on foliage. The main disadvantage is that they can be phytotoxic so they should only be used on mature plants. Chlorothalonil is another effective protectant fungicide and it has low plant toxicity.

In the 1970s systemic fungicides greatly increased the efficacy of chemical control. The most effective of these are the phenylamides, such as metalaxyl, ofurace, oxadyxil, and benalaxyl. Of these, the most effective is metalaxyl (Schwinn and Margot, 1991). These fungicides have a strong curative effect, that is, they can kill the pathogen even after it is in the plant. The main disadvantage of thephenylamides is that resistance readily develops in the pathogen population. To avoid this problem guidelines have been developed for the use of phenylamide fungicides.

Some more protective and systemic fungicides have been released in the past 20 years that offer growers a much larger array of options (Table 8.2). Many of the protective fungicides have similar levels of effectiveness and durability on the leaf. The best period to use them, therefore, may depend on their levels of phytotoxicity.

The most commonly used means of protecting tubers against infection is by applying fungicides to foliage. In theory, this approach can reduce tuber blight in three ways: (1) reducing sporulation; (2) reducing viability of spores on leaves; and (3) when repeated applications leave residues in the soil, formation or motility of zoospores is inhibited (Schepers and van Soesbergen, 1995). As may be expected, not all fungicides are equally effective in controlling

TABLE 8.2. The effectiveness of the most important fungicide active ingredients used for the control of *P. infestans* in Europe (Bradshaw, 2003).

| Active ingredient[a] | Effectiveness | | | | Action mode | | | Rain fast-ness | Mobility in the plant |
|---|---|---|---|---|---|---|---|---|---|
| | Leaf blight | New growing point | Stem blight | Tuber blight | Protectant | Curative | Eradicant | | |
| Chlorothalonil | + + | 0 | (+) | 0 | + + | 0 | 0 | + + (+) | contact |
| Copper | + | 0 | + | + | + (+) | 0 | 0 | + | contact |
| Cyazofamid | + + + | 0 | + | + + +[c] | + + + | 0 | 0 | + + + | contact |
| Dithiocarbamate[b] | + + | 0 | + | 0 | + + | 0 | 0 | + (+) | contact |
| Famoxadone | + + | 0 | + (+)[e] | N/A | + + | + +[d] | +[d] | + + | contact |
| Fentin acetate | + + | 0 | + | + + | + + | 0 | 0 | + + | contact |
| Fentin hydroxide | + + | 0 | + | + + | + + | 0 | 0 | + + | contact |
| Fluazinam | + + + | 0 | + | + + (+) | + + + | 0 | 0 | + + (+) | contact |
| Zoxamide | + + + | 0 | +[e] | + + | + + + | 0 | 0 | + + (+) | contact |
| Cymoxanil | + + (+) | 0 | +(+) | 0 | + + | + + | + | + + | Trans-laminar |

TABLE 8.2 *(continued)*

| Active ingredient[a] | Effectiveness | | | | Action mode | | | Rain fastness | Mobility in the plant |
|---|---|---|---|---|---|---|---|---|---|
| | Leaf blight | New growing point | Stem blight | Tuber blight | Protectant | Curative | Eradicant | | |
| Dimethomorph | + + (+) | 0 | + (+) | + + | + + (+) | + | + + | + + (+) | Trans-laminar |
| Fenamidone | + + (+) | 0 | + (+)[e] | + + | + + (+) | 0 | 0 | + + | Trans-laminar |
| Benalaxyl[c] | + + | + + | + + | N/A | + + (+) | + + (+) | + + (+) | + + + | systemic |
| Metalaxyl-M (mefenoxam)[c] | + + + | + + | + + | N/A | + + (+) | + + (+) | + + (+) | + + + | systemic |
| Oxadixyl[c] | + + (+) | + + | + + | N/A | + + (+) | + + (+) | + + (+) | + + + | systemic |
| Propamocarb-HCl | + + (+) | + (+) | + + | + + | + + (+) | + + | + + | + + + | systemic |

*Notes:* [a]The scores of individual active ingredients are based on the label recommendation and are NOT additive for mixtures of active ingredients. Inclusion of an active substance in the list is NOT indicative of its registration status either in the EU or elsewhere in Europe; [b]Includes maneb, mancozeb, propineb, and metiram; [c]See text for comments on phenylamide resistance; [d]Provided by the active ingredient cymoxanil in Tanos (famoxadone + cymoxanil); [e]Based on limited data. Key to ratings: 0 = no effect; + = reasonable effect; ++ = good effect; +++ = very good effect; N/A = not recommended for control of tuber blight.

tuber blight via foliar application (Schepers and van Soesbergen, 1995).

### *Forecasting*

One rather old but still active area of research is that of forecasting of disease or decision support systems (DSS) for timing prophylactic fungicidal sprays. Forecasting models were developed as early as the 1940s and numerous models now exist. A description of this research in general chronological order has been made by Fry and Doster (1991).

Forecasting models have also been tested in developing countries (Katsurayama and Boneti, 1996; Gómez et al., 1999; Grünwald et al., 2000; Singh et al., 2000) but there appears to be little evidence that they are being used effectively. This is probably because most of these systems require weather-measuring devices and/or computers, which are not within the economic reach of poor farmers.

### *Host Resistance*

Reports of potato cultivars with resistance to late blight date back to the 1800s, and apparently breeding was an important activity in Ireland during the decades following the famine (Dowley, 1995). A number of cultivars with resistance to late blight were available in Europe by the late 1800s (Umaerus et al., 1983). In 1882, the French Professor Pierre Millardet discovered that a mixture of copper sulfate and lime (later named Bordeaux mixture), used to deter people from stealing grapes, also protected plants from fungal and oomycete pathogens. According to Robinson (1996), this discovery had a profound negative effect on potato breeding for resistance to late blight. Robinson's hypothesis is supported by the fact that during the Bordeaux mix era, from the 1880s until the First World War, several highly susceptible varieties (e.g., Bintje, Russet Burbank) were developed, which would have been difficult to produce without chemical protection.

Potato genotypes also differ in their resistance to tuber blight (Wastie et al., 1987), but the level of resistance currently available is not sufficient to protect plants from tuber blight. Resistance mechanisms apparently differ for foliage and tuber blight and in parts of Europe the pathogen population varies in its aggressiveness on tubers (Flier et al., 2003a).

Traditional Andean farmers have used high host diversity for centuries, either in the form of cultivar mixtures or by intercropping (Thurston, 1990) to reduce late blight severity. Research to test the effect of host diversity on late blight severity has given surprising results. Cultivar mixtures have consistently reduced blight severity in North America and Europe (Andrivon and Lucas, 1998; Garrett and Mundt, 2000) but the effect is much smaller in the highland tropics (Garrett et al., 2001). This has been attributed to a much higher amount of outside (coming from other fields) inoculum in the highland tropics that tends to cause uniform infection foci throughout the field.

### *Nature of Host Resistance*

Until recently, there was a fairly general consensus that most resistance to *P. infestans* could be classified into two types. The first governed by single dominant genes with major effects, and a clear discontinuous segregation in progeny; and the second governed by several or many genes, called minor genes, with small cumulative effects and continuous distribution of resistant genotypes. Major gene resistance has also been described as vertical resistance, *R*-gene resistance, qualitative resistance, specific resistance, race-specific re sistance, unstable resistance, and complete resistance. Minor gene resistance has been described with contrasting names, as horizontal resistance, polygenic resistance, quantitative resistance, general resistance, nonrace-specific resistance, stable resistance, partial resistance, field resistance, and rate-reducing resistance. Some reviews discuss terminology used to describe these types of resistance (Thurston, 1971; Umaerus et al., 1983). For consistency, we will use the terms "specific resistance" and "general resistance" in this chapter.

By definition, general resistance was considered effective against all races of the pathogen (Vanderplank, 1984). Some authors, however, proposed an alternative hypothesis for resistance, in which some degree of race specificity exists for all resistance types (Parlevliet and Zadoks, 1977), and pathogen adaptation to plant genotypes with general resistance has been identified for some diseases (Parlevliet, 1989).

Basic research aimed at elucidating the genetic and physiological basis of resistance has not demonstrated clear mechanistic differences between general and specific resistance. Until recently, specific resistance was closely associated with the hypersensitive response (HR), which is a particular type of programmed cell death that enables plants to limit invading pathogens. Several cells near the infection point die rapidly, severely restricting or stopping the infection process. The most commonly accepted model suggests that *R* genes in the plant trigger HR when they recognize elicitors (signal molecules) coming from the pathogen, which are encoded by avirulence *(Avr)* genes (Kamoun et al., 1999). It is the specificity between the *Avr* gene in the pathogen and the *R* gene in the host that causes the race-specificity of specific resistance. The exact way in which *R* genes and *Avr* genes interact is not yet known but different models have been proposed (Dangl and Jones, 2001).

*R* genes were not traditionally associated with general resistance, and therefore neither was HR. However, histological studies, some done decades ago, have consistently found rapid necrosis occurring during the infection process in plants with general resistance (Gees and Hohl, 1988; Coffey and Gees, 1991; Vleeshouwers et al., 2000b). HR has also been associated with nonhost resistance in potato (Vleeshouwers et al., 2000b). Kamoun et al. (1999) proposed that resistance in all plants against oomycetes might always involve HR. In general resistance, HR occurs somewhat later and allows the pathogen to escape, with HR often trailing behind colonization by the pathogen (Vleeshouwers et al., 2000b). Based on histological observations, Vleeshouwers et al. (2000b) proposed that general resistance is conferred by weak interactions between *R* genes and *Avr* genes. Weak *R* genes producing a phenotype that mimics general resistance had been identified in the past (Landeo and Turkensteen, 1989). At the cellular level, resistance of weak *R* genes resembles general resistance (Vleeshouwers et al., 2000b).

General resistance in potato is also associated with basal levels of pathogenesis-related (PR) proteins (Vleeshouwers et al., 2000a), as well as with other defense-related genes (Trognitz et al., 2002).

Callose formation on cell walls, which may hinder colonization of the plant tissue by the pathogen, was observed in histological studies of resistant cultivars (Wilson and Coffey, 1980; Gees and Hohl, 1988).

Another area of research that has tended to blur the apparently clear field effects of general and specific resistance is that carried out on residual effects of *R* genes. Some plant pathologists have hypothesized that "defeated" (no longer effective) *R* genes have residual effects, that is they contribute additively to general resistance. Putative residual effects were demonstrated for some diseases (Pedersen and Leath, 1988), but not for others (Young et al., 1994). Recently, potato progeny of several crosses were compared in the field in Scotland. On the average, those genotypes with a defeated *R* gene were slightly more resistant than those without the *R* gene (Stewart and Bradshaw, 2001). In an evaluation of 882 breeding clones, Darsow et al. (1987) found that the clones that had *R* genes also had more general resistance. These authors interpreted their data as an indication of linkage between *R* genes and general resistance. It is difficult to differentiate between a direct effect of the defeated *R* gene and an effect of genes that might be tightly linked to the *R* gene (Anderson, 1982). In a large population in which an *R* gene was segregating, there was no relationship between lesion expansion rate, and the presence of the *R* gene (Ordoñez et al., 1998) when inoculated with a compatible isolate. Similar to the effects of pathogen specificity to general resistance discussed earlier, residual effects, if they exist, are probably small and difficult to measure.

### *Pathogen's Adaptation to Host Resistance*

Studies on pathogen adaptation to host resistance have produced inconclusive results. Pathogen adaptation to general resistance to *P. infestans* was observed after successive culturing on potato tubers by some authors (Caten, 1974), but in one study host specialization did not occur after 90 cycles of subculturing on tubers (Paxman, 1963). Host specialization was demonstrated for incubation period on detached leaves (Jeffrey et al., 1962) and field studies demonstrated a small hostpathogen interaction for epidemic rate (Latin et al., 1981), but the latter did not prove consistent over years. In field studies carried out by James and Fry (1983), host specialization was not measured. Recently, a series of studies on tuber and foliage infection concluded that plant genotypes carrying different levels of general resistance react differentially against new aggressive genotypes of the pathogen (Flier et al., 2003a). These authors hypothesized that they were able to measure host specificity because the pathogen population they studied involves both asexual and sexual reproduction (Turkensteen et al., 2000). General resistance might be more stable in areas where clonal pathogen populations exist and less stable with sexual populations. However, a recent assessment of historical data of Mexican cultivars grown in Toluca, an area of high pathogen variability resulting from sexual reproduction, demonstrated that the resistance was generally stable (Grünwald et al., 2002).

Inconsistency in studies on host specialization to general resistance may be due to the fact that specialization is partial and difficult to measure (Turkensteen, 1993). The constancy of resistance ratings of potato cultivars over time (Colon et al., 1995; Inglis et al., 1996; Grünwald et al., 2002; Landeo, 2002) probably indicates that the host specificity, to the extent that it occurs, is smaller than the general effects of host resistance.

### *Breeding*

Specific resistance due to single genes (*R* genes) was first discovered in the wild *S. demissum,* or the closely related species *S. edinese* and *S. stoloniferum* (Black, 1954) from Mexico. The first *R* genes were transferred from *S. demissum* to cultivated potato at the beginning of the last century (Wastie, 1991), but they were not really used in breeding programs until the 1920s (Bradshaw et al., 1995). Originally, four *R* genes were transferred into cultivated potatoes, but subsequently more *R* genes were discovered. A differential set of potato varieties containing 11 *R* genes (alone and in combinations) is commonly used to classify races of the pathogen (Perez, Gamboa et al., 2001). However, most breeding for *R* genes involved only the first four to be discovered, because it soon became evident that the resistance they conferred was not at all durable. Virulent races of the pathogen rapidly appeared to which the cultivars were highly susceptible. The ephemeral nature of *R*-gene mediated resistance became evident by the 1950s and 1960s (Bradshaw et al., 1995) and researchers recognized the need to emphasize general resistance (Vanderplank, 1956).

Breeding potatoes with general resistance is technically more difficult than breeding for resistance based on a few *R* genes with large effects. As mentioned earlier, general resistance is quantitatively inherited, which makes the achievement of high levels of resistance a slow process, usually requiring several cycles of recombination. This problem is compounded by the fact that *R* genes can easily mask, or otherwise interfere with general resistance (Turkensteen, 1993). Thus, screening for general resistance in plant populations in which one or more, large-effect *R* genes are segregating is particularly tricky.

One way potato breeders overcome this problem is by screening with a highly complex race of the pathogen. This is generally done by culturing large quantities of inoculum and inoculating plants in the field. Robinson (1996) used this technique in Africa and the International Potato Center (CIP) used inoculation with a complex race to screen for resistance in wild potatoes (Perez, Salas et al., 2001). There is a theoretical assumption underlying this practical application of plant pathology that should be considered when interpreting results. Breeders inoculate the populations with complex races to overcome any *R* genes that actually may have large effects that mask other resistance factors. The efficacy of this approach is based on the compatibility of the race with the parental material used to produce the potato genotypes being screened. Compatibility of a complex pathogen race on the plants used as parents should, in theory, ensure compatibility with all the progeny. This, however, is not always the case because some parental plants can harbor major genes that have been rendered ineffective by modifier genes (El Kharbotly et al., 1996; Ordonez et al., 1997). When these parental plants are crossed, the *R* gene might become effective by being separated from the modifier gene in the progeny.

Another way that potato breeders have attempted to overcome the problem of major *R* genes interfering with general resistance is by screening progeny in areas where the naturally occurring pathogen population is variable, at least for specific virulence (the ability to infect plants with different strong *R* genes). The principal location for this type of screening has historically been near Toluca, Mexico, which is the putative center of origin of the pathogen (Goodwin, 1997). The sexually derived population of *P. infestans* in Toluca is so variable that several different genotypes are often found within a single field (Flier et al., 2003). The utility of central Mexico as a "hot spot" for resistance screening has been borne out by a large number of resistant cultivars that have been developed fully or in part there (Bradshaw et al., 1992; Robinson, 1996). Because of the very high pathogen diversity in this location it is likely that any potato cultivar grown over a relatively large area will eventually become infected, regardless of the *R* gene(s) it may carry. How large this area must be for disease to occur would depend on the frequency of compatible isolates in the pathogen population. In small experimental plots, however, potato genotypes sometimes remain immune in Mexico throughout the duration of the season (unpublished data). For this reason, high levels of resistance in Mexico must be interpreted with caution if the plot size is small.

A third technique used to breed for general resistance in potato is the elimination of certain *R* genes from the host population. Turkensteen (1993) described the theoretical feasibility of this approach for potato and it has been implemented by the CIP (Landeo, 2002) since the early 1990s. This approach was actually implemented by the CIP in two different ways. The first was by using only parents from the putatively *R* gene-free tetraploid native cultivars from the Andes that are generally classified as *Solanum tuberosum* ssp. *andigena* (Landeo et al., 1995). Even though this species is considered free of *R* genes, selected parents are checked for the presence of strong *R* genes by greenhouse inoculation. Two inoculations are done separately, one with a highly compatible race of *P. infestans* and one with a race that does not infect any of the known 11 *R*-gene differentials. If a potato genotype reacts in a similar fashion to both races, then it is considered free of *demissum R* genes that react differentially to the avirulence genes in the avirulent isolate.

The dual inoculation method is also used for potato populations in which known *R* genes with qualitative effects are segregating. By the dual inoculation technique, individual plant genotypes are identified that do not react differentially to the two pathogen isolates. These plant genotypes are referred to as "*R* gene-free" (Landeo et al., 1995) but in fact they should be considered free of *R* genes that react to avirulence genes in the avirulent isolate.

Stability of general resistance across different locations is also difficult to study and has only been reported for late blight of potato on a few occasions and over a limited geographical area. Recently, sets of potato genotypes were evaluated at several locations in the United States in two separate studies ( Haynes et al., 1998, 2002). These authors found that the most resistant materials were generally stable but some of the intermediate clones were less stable. In an earlier study, the genotype Alpha was more resistant in New York State than in Toluca, Mexico, compared to three other varieties that behaved similarly at both locations (Parker et al., 1992).

There is anecdotal evidence for and against the geographic stability of general resistance in potato to *P. infestans*. Several genotypes with appreciable levels of resistance are grown in different continents and generally maintain their resistance (Forbes and Jarvis, 1994). However, other genotypes have gained a reputation for being resistant in one location whereas being susceptible in another. One such genotype is Atzimba, which is thought to have an intermediate level of resistance in Mexico but is considered highly susceptible in Costa Rica. The study on Alpha, mentioned earlier, was inspired by the generalized opinion that this genotype is highly susceptible in Mexico but moderately resistant in upstate New York (W. E. Fry, personal communication). Anecdotal information, however, is extremely difficult to interpret because it seldom involves experimental comparison of two or more plant genotypes in the same region, which would permit an appreciation of a genotype by environment (G × E) interaction.

### *Molecular Biology*

The advent of molecular biology has enabled researchers to elucidate many aspects of the host-pathogen relationship, but much uncertainty remains about underlying mechanisms of putatively different resistance types. To date, at least five *R* genes that confer specific resistance to *P. infestans* have been located on a potato genetic map

(Birch and Whisson, 2001). Several of these *R* genes are clustered and others are located near defense genes for other pathogens and nematodes (Meksem et al., 1995; De Jong et al., 1997). Quantitative trait loci (QTL) segregating in several different breeding populations have also been located on the potato map (Ewing et al., 2000; Ghislain et al., 2001). Originally some QTL were associated with *R* genes and this led researchers to speculate that *R* genes may be highly effective allelic forms of QTL (Gebhardt, 1994). However, QTL that are not known to be associated with *R* genes have been identified in subsequent studies using plant material evaluated for resistance to late blight in the field (Collins et al., 1999; Oberhagemann et al., 1999; Ewing et al., 2000; Ghislain et al., 2001). QTL expression may be related to day length (Ghislain et al., 2001) and appears to be influenced by overall plant physiology, as QTL for resistance can also vary with vigor and rate of physiological development of the plant (Collins et al., 1999).

*R* genes have also been identified in species other than *S. demissum.* Two genes (apparently different) from the diploid species *S. bulbocastanum* were cloned and expressed after introduction into a susceptible potato genotype (Van der Vossen et al., 2002; Song et al., 2003). The sources of these genes were resistant against all known races of the pathogen, which has led to great speculation on their durability. Should these genes prove to be durable, they could provide a novel and extremely effective disease control mechanism.

Gene expression probably constitutes the most active area of research on mechanisms of resistance. The complex recognition process between pathogen elicitors and host receptors is thought to up- and down-regulate many genes. Different techniques are being used to evaluate gene expression during the infection process. Using high throughput DNA sequencing, scientists are rapidly developing libraries of sequenced cDNAs (Expressed Sequence Tag or ESTs). Thousands of ESTs from *P. infestans* and potato are now housed at the *Phytophthora* Genome Initiative database (Waugh et al., 2000) and the Institute for Genomic Research , respectively. Analyses of ESTs recently led to the discovery of novel necrosis inducing cDNAs from *P. infestans* (Torto et al., 2003). This research relied on high throughput bioassays to link a particular sequence with a phenotype.

Suppression subtractive hybridization (SSH), a PCR-based method for studying expressed genes, has the advantage of requiring only small amounts of biological material (Birch and Whisson, 2001). SSH has been used to study genes that are up-regulated in both compatible (Dellagi et al., 2000) and incompatible (Birch et al., 1999) reactions. PCR-based techniques were also used to identify genes differentially expressed in resistant and susceptible genotypes grown in the field in Peru (Evers et al., 2003). Some up-regulated genes associated with general resistance in this study had previously been identified in association with specific resistance (Birch and Whisson, 2001).

Microarray technology will enhance the ability of researchers to evaluate gene expression at different stages of the infection process. This technology has already been used to evaluate factors affecting adaptation of *P. infestans* to tomato (Smart et al., 2003).

The above account shows that there has been tremendous progress in understanding the late blight disease of potato. Although the disease has been controlled through the development and deployment of cultural and chemical means as well as host resistance, it continues to be the most devastating disease of potato crop worldwide.

## REFERENCES

Abad, Z.G. and Abad, J.A. (1997). Another look at the origin of late blight of potatoes, tomatoes, and pear melon in the Andes of South America. *Plant Disease 81:*682-688.

Abad, Z.G., Abad, J.A., Fernandez-Northcote, E.N., and Ochoa, C. (1995). Host range of *Phytophthora infestans* in central Peru and list in the world since 1840's. Resistance on wild tomatoes. *Phytopathology 85:*1173.

Adams, M.J. (1975). Potato tuber lenticels: Susceptibility to infection by *Erwinia carotovora* var. *atroseptica* and *Phytophthora infestans. Annals of Applied Biology 79:*275-282.

Adler, N.E., Erselius, L.J., Chacón, M.G., Flier, W.G., Ordoñez, M.E., Kroon, L.P.N.M., and Forbes, G.A. (2004). Genetic diversity of *Phytophthora infestans sensu lato* in Ecuador provides new insight into the origin of this important plant pathogen. *Phytopathology 94:*154-162.

Anderson, M.G. (1982). Interpreting residual effects of "defeated" resistance genes. *Phytopathology 72:*1383-1384.

Andersson, B., Sandstrom, M., and Stromberg, A. (1998). Indications of soil borne inoculum of *Phytophthora infestans. Potato Research 41:*305-310.

Andrivon, D. (1995). Biology, ecology, and epidemiology of the potato late blight pathogen *Phytophthora infestans* in soil. *Phytopathology 85:*1053-1056.

———. (1996). The origin of *Phytophthora infestans* populations present in Europe in the 1840s: A critical review of historical and scientific evidence. *Plant Pathology 45:*1027-1035.

Andrivon, D. and Lucas, J.M. (1998). Mixtures of varieties for control of potato late blight: Is it possible? *First Transnational Workshop on Biological, Integrated and Rational Control: Status and Perspectives with Regard to Regional and European Experiences* (pp. 55-56). Lille France.

Anonymous. (1997). CIP in 1996. *Annual Report.* Lima, Peru: The International Potato Center.

———. (2003). *Crop Protection Compendium,* Wallingford, UK: CABI Publishing.

Awan, A.B. and Struchtemeyer, R.A. (1957). The effect of fertilization on the susceptibility of potatoes to late blight. *American Potato Journal 34:*315-319.

Aylor, D.E., Fry, W.E., Mayton, H., and Andrade-Piedra, J.L. (2001). Quantifying the rate of release and escape of *Phytophthora infestans* sporangia from a potato canopy. *Phytopathology 91:*1189-1196.

Birch, P.R.J., Avrova, A.O., Duncan, J.M., Lyon, G.D., and Toth, R.L. (1999). Isolation of potato genes that are induced during an early stage of the hypersensitive response to *Phytophthora infestans. Molecular Plant-Microbe Interactions 12:*356-361.

Birch, P.R.J. and Whisson, S.C. (2001). *Phytophthora infestans* enters the genomics era. *Molecular Plant Pathology 2:*257-263.

Black, W. (1954). Late blight resistance work in Scotland. *American Potato Journal 31:*93-100.

Bourke, A. (1991). Potato blight in Europe in 1845: The scientific controversy. In Lucas, J.A., Shattock, R.C., Shaw, D.S., and Cooke, L.R. (Eds.), *Phytophthora* (pp. 12-24). Cambridge, UK: Cambridge University Press.

———. (1993). *'The Visitation of God'? The Potato and the Great Irish Famine.* Dublin, Ireland: Lilliput Press, Ltd.

Boyd, A.E.W. (1980). Development of potato blight *(Phytophthora infestans)* after planting infected seed tubers. *Annals of Applied Biology 95:*301-309.

Bradshaw, J.E. (2003). Subgroup fungicides: Review ratings given to various fungicides at Edinburgh workshop and rate efficacy of fungicides to *Alternaria.* In Schepers, H.T.A.M. and Westerdijk, C.E. (Eds.), *Proceedings of the Seventh Workshop of an European Network on Integrated Control of Potato Late Blight,* Poznan, Poland: PPO Special Report no. 9.

Bradshaw, J.E., Wastie, R.L., Stewart, H.E., and Mackay, G.R. (1995). Breeding for resistance to late blight in Scotland. In Dowley, L.J., Bannon, E., Cooke, L.R., Keane, T., and O'Sullivan, E. (Eds.), *Phytophthora infestans* (pp. 246-253). Dublin, Ireland: Boole Press.

Bradshaw, J.E., Wastie, R.L., Stewart, H.E., Solomon-Blackburn, R.M., and Wilkinson, M.J. (1992). Breeding for disease resistance in potatoes. *Aspects of Applied Biology 33:*135-142.

Carnegie, S.F. and Colhoun, J. (1983). Effects of plant nutrition on susceptibility of potato leaves to *Phytophthora infestans. Phytopathologische Zeitschrift 108:*242-250.

Caten, C.E. (1974). Intra-racial variation in *Phytophthora infestans* and adaptation to field resistance for potato blight. *Annals of Applied Biology 77:*259-270.

Coffey, M.D. and Gees, R. (1991). The cytology of development. In Ingram, D.S. and Williams, P.H. (Eds.), *Advances in Plant Pathology* (pp. 31-51). San Diego: Academic Press Limited.

Collins, A., Milbourne, D., Ramsay, L., Meyer, R., Chatot-Balandras, C., Oberhagemann, P., De Jong, W., Gebhardt, C., Bonnel, E., and Waugh, R. (1999). QTL for field resistance to late blight in potato are strongly correlated with maturity and vigour. *Molecular Breeding 5:*387-398.

Colon, L.T., Turkensteen, L.J., Prummel, W., Budding, D.J., and Hoogendoorn, J. (1995). Durable resistance to late blight *(Phytophthora infestans)* in old potato cultivars. *European Journal of Plant Pathology 101:*387-397.

Crosier, W. (1933). Studies in the biology of *Phytophthora infestans* (Mont.) de Bary. Memoir—Cornell University Agricultural Experiment Station (USA) 155:1-40.

Dangl, J.L. and Jones, J.D.G. (2001). Plant pathogens and integrated defense responses to infection. *Nature 114:*826-833.

Darsow, U., Göbel, S., Götz, E., Oertel, H., and Schüler, K. (1987). *R* genes and relative resistance of potato tubers to *Phytophthora infestans* (Mont.) de Bary. *Archiv Züchtungsforschung 6:*387-397.

De Jong, W., Forsyth, A., Leister, D., Gebhardt, C., and Baulcombe, D.C. (1997). A potato hypersensitive resistance gene against potato virus X maps to a resistance gene cluster on chromosome 5. *Theoretical and Applied Genetics 95:*153-162.

Dellagi, A., Heilbronn, J., Avrova, A.O., Montesano, M., Palva, E.T., Stewart, H.E., Toth, I.K., Cooke, D.E.L., Lyon, G.D., and Birch, P.R.J. (2000). A potato gene encoding a WRKY-like transcription factor is induced in interactions with *Erwinia carotovora* subsp. *atroseptica* and *Phytophthora infestans* and is coregulated with class I endochitinase expression. *Molecular Plant-Microbe Interactions 13:*1092-1101.

Devaux, A. and Haverkort, A.J. (1987). The effects of shifting planting dates and mulching on late blight *(Phytophthora infestans)* and drought stress on potato crops grown under tropical highland conditions. *Experimental Agriculture 23:*325-333.

Dowley, L.J. (1995). Research on *Phytophthora infestans* in Ireland: A short historical view. In Dowley, L.J., Bannon, E., Cooke, L.R., Keane, T., and O'Sullivan, E. (Eds.), *Phytophthora infestans* (pp. 12-29). Dublin, Ireland: Boole Press.

Drenth, A. (1994). *Molecular Genetic Evidence for a New Sexually Reproducing Population of* Phytophthora infestans *in Europe.* Ph.D. Thesis, Wageningen, The Netherlands: Wageningen Agricultural University.

Dubey, T. and Stevenson, W.R. (1996). Factors affecting the movement and viability of sporangia of *Phytophthora infestans* in soil. *Phytopathology 86:*S61-S62.

El Kharbotly, A., Pereira, A., Stiekema, W.J., and Jacobsen, E. (1996). Race specific resistance against *Phytophthora infestans* in potato is controlled by more genetic factors than only *R*-genes. *Euphytica 90:*331-336.

Erwin, D.C. and Ribeiro, O.K. (1996). *Phytophthora infestans* (Mont.) de Bary (1876), In Erwin. D.C. and Ribeiro, O.K. (Eds.), *Phytophthora Diseases Worldwide* (pp. 346-353). St. Paul, MN: APS Press.

Evers, D., Ghislain, M., Hausman, J.F., and Dommes, J. (2003). Differential gene expression in two potato lines differing in their resistance to *Phytophthora infestans. Journal of Plant Physiology 160:*709-712.

Ewing, E.E., Simko, I., Smart, C.D., Bonierbale, M.W., Mizubuti, E.S.G., May, G.D., and Fry, W.E. (2000). Genetic mapping from field tests of qualitative and quantitative resistance to *Phytophthora infestans* in a population derived from *Solanum tuberosum* and *Solanum berthaultii. Molecular Breeding 6:*25-36.

Flier, W.G., Grünwald, N.J., Kroon, L.P.N.M., Sturbaum, A.K., van den Bosch, T.B.M., Garay-Serrano, E., Lozoya-Saldaña, H., Fry, W.E., and Turkensteen, L.J. (2003). The population structure of *Phytophthora infestans* from the Toluca Valley of Central Mexico suggests genetic differentiation between populations from cultivated potato and wild *Solanum* spp. *Phytopathology 93:*382-390.

Flier, W.G., Grünwald, N.J., Kroon, L.P.N.M., van den Bosch, T.B.M., Garay-Serrano, E., Lozoya-Saldaña, H., Bonants, P.J.M., and Turkensteen, L.J. (2002). *Phytophthora ipomoeae* sp. nov., a new homothallic species causing leaf blight on *Ipomoea longipendunculata* in the Toluca Valley of central Mexico. - *Mycological Research 106:*848-856.

Flier, W.G., Kessel, G.J.T., Schepers, H.T.A.M., and Turkensteen, L.J. (2002). The impact of oospores of *Phytophthora infestans* on late blight epidemics. In Lizarraga, C. (Ed.), *Late Blight: Managing the Global Threat* (pp. 18-22). Hamburg, Germany: European Association for Potato Research.

Flier, W.G., van den Bosch, G.B.M., and Turkensteen, L.J. (2003a). Stability of partial resistance in potato cultivars exposed to aggressive strains of *Phytophthora infestans. Plant Pathology 52:*326-337.

———. (2003b). Epidemiological importance of *Solanum sisymbriifolium, S. nigrum* and *S. dulcamara* as alternative hosts for *Phytophthora infestans. Plant Pathology 52:*595-603.

Forbes, G.A. and Jarvis, M.C. (1994). Host resistance for management of potato late blight, In Zehnder, G., Jansson, R., and Raman, K.V. (Eds.), *Advances in Potato Pest Biology and Management* (pp. 439-457). St. Paul, MN: American Phytopathological Society.

Fry, W.E. and Doster, M.A. (1991). Potato late blight: Forecasts and disease suppression, In Lucas, J.A., Shattock, R.C., Shaw, D.S., and Cooke, L.R. (Eds.), *Phytophthora* (pp. 326-336). Cambridge, UK: Cambridge University Press.

Fry, W.E., Goodwin, S.B., Dyer, A.T., Matuszak, J.M., Drenth, A., Tooley, P.W., Sujkowski, L.S., Koh, Y.J. Cohen, B.A., Spielman, L.J. et al. (1993). Historical and recent migrations of *Phytophthora infestans:* Chronology, pathways, and implications. *Plant Disease 77:*653-661.

Garrett, K.A. and Dendy, S.P. (2001). Cultural Practices in Potato Late Blight Management. In *Complementing Resistance to Late Blight in the Andes,* Cochabamba, Bolivia.

Garrett, K.A. and Mundt, C.C. (2000). Host diversity can reduce potato late blight severity for focal and general patterns of primary inoculum. *Phytopathology 90:*1307-1312.

Garrett, K.A., Nelson, R.J., Mundt, C.C., Chacon, G., Jaramillo, R.E., and Forbes, G.A. (2001). The effects of host diversity and other management components on epidemics of potato late blight in the humid highland tropics. *Phytopathology 91:*993-1000.

Gebhardt, C. (1994). RFLP mapping in potato of qualitative and quantitative genetic loci conferring resistance to potato pathogens. *American Potato Journal 71:*339-345.

Gees, R. and Hohl, H.R. (1988). Cytological comparison of specific (*R3*) and general resistance to late blight in potato leaf tissue. *Phytopathology 78:*350-357.

Ghislain, M., Trognitz, B., Herrera, M.d. R., Solis, J., Casallo, G., Vasquez, C., Hurtado, O., Castillo, R., Portal, L., and Orrillo, M. (2001). Genetic loci associated with field resistance to late blight in offspring of *Solanum phureja* and *S. tuberosum* grown under short-day conditions. *Theoretical and Applied Genetics 103:*433-442.

Gómez, G., Rodríguez, J., Castellanos, L., Gonzáles, M.L., Figueroa, I., Sarmientos, A., Alujas, D., and Martín, E.L. (1999). Naumova modificado: Ajuste de un método de pronóstico para el tizón de la papa y el tomate en Cuba. *Fitosanidad 3:*95-100.

Goodwin, S.B. (1997). The population genetics of *Phytophthora. Phytopathology 87:*462-473.

Goodwin, S.B., Cohen, B.A., and Fry, W.E. (1994). Panglobal distribution of a single clonal lineage of the Irish potato famine fungus. *Proceedings of the National Academy of Science (USA) 91:*11591-11595.

Goodwin, S.B. and Drenth, A. (1997). Origin of the A2 mating type of *Phytophthora infestans* outside Mexico. *Phytopathology 87:*992-999.

Goodwin, S.B., Legard, D.E., Smart, C.D., Levy, M., and Fry, W.E. (1999). Gene flow analysis of molecular markers confirms that *Phytophthora mirabilis* and *P. infestans* are separate species. *Mycologia 91:*796-810.

Goodwin, S.B., Spielman, L.J., Matuszak, J.M., Bergeron, S.N., and Fry, W.E. (1992). Clonal diversity and genetic differentiation of *Phytophthora infestans* populations in northern and central Mexico. *Phytopathology 82:*955-961.

Grünwald, N.J., Cadena Hinojosa, M.A., Rubio Covarrubias, O.A., Rivera Pena, A., Niederhauser, J.A., and Fry, W.E. (2002). Potato cultivars from the Mexican national program: Sources and durability of resistance against late blight. *Phytopathology 92:*688-693.

Grünwald, N.J., Flier, W.G., Sturbaum, A.K., Garay-Serrano, E., van den Bosch, T.B.M., Smart, C.D., Matuszak, J.M., Lozoya-Saldaña, H., Turkensteen, L.J., and Fry, W.E. (2001). Population structure of *Phytophthora infestans* in the Toluca valley region of Central Mexico. *Phytopathology 91:*882-890.

Grünwald, N.J., Rubio-Covarrubias, O.A., and Fry, W.E. (2000). Potato late-blight management in the Toluca Valley: Forecasts and resistant cultivars. *Plant Disease 84:*410-416.

Guenther, J.F., Michael, K.C., and Nolte, P. (2001). The economic impact of potato late blight on US growers. *Potato Research 44:*121-125.

Harrison, J.G. (1992). Effects of the aerial environment on late blight of potato foliage—a review. *Plant Pathology 41:*384-416.

Haynes, K.G., Christ, B.J., Weingartner, D.P., Douches, D.S., Thill, C.A., Secor, G., Fry, W.E., and Lambert, D.H. (2002). Foliar resistance to late blight in potato clones evaluated in national trials in 1997. *American Journal of Potato Research 79:*451-457.

Haynes, K.G., Lambert, D.H., Christ, B.J., Weingartner, D.P., Douches, D.S., Backlund, J.E., Secor, G., Fry, W., and Stevenson, W. (1998). Phenotypic stability of resistance to late blight in potato clones evaluated at eight sites in the United States. *American Journal of Potato Research 75:*211-217.

Herlihy, M. (1970). Contrasting effects of nitrogen and phosphorous on potato tuber blight. *Plant Pathology 19:*65-68.

Hirst, J.M. (1958). New methods of studying plant disease epidemics. *Outlook on Agriculture 2:*16-26.

Hohl, H.R. and Iselin, K. (1984). Strains of *Phytophthora infestans* from Switzerland with A2 mating type behaviour. *Transactions of the British Mycological Society 83:*529-530.

Hohl, H.R. and Suter, E. (1976). Host–parasite interfaces in a resistant and a susceptible cultivar of *Solanum tuberosum* inoculated with *Phytophthora infestans:* Leaf tissue. *Canadian Journal of Botany 54:*1956-1970.

Hooker, W.J. (1981). *Compendium of Potato Diseases.* St. Paul, MN: APS Press.

Inglis, D.A., Johnson, D.A., Legard, D.E., Fry, W.E., and Hamm, P.B. (1996). Relative resistances of potato clones in response to new and old populations of *Phytophthora infestans. Plant Disease 80:*575-578.

James, R.V. and Fry, W.E. (1983). The potential for *Phytophthora infestans* populations to adapt to potato cultivars with rate-reducing resistance. *Phytopathology 73:*984-988.

Jeffrey, S.I.B., Jinks, J.L., and Grindle, M. (1962). Intraracial variation in *Phytophthora infestans* and field resistance to potato blight. *Genetica 32:*323-338.

Juarez, H.S., Amaro, J.R., Rivera, M.D., Parraga, A., and Hijmans, R.J. (2001). The effect of nitrogen fertilization on potato late blight in the field. Scientist and farmer: Partners in research for the 21st Century. *Program Report 1999-2000* (pp. 69-76). Lima, Peru: International Potato Center.

Kable, P.F. and Mackenzie, D.R. (1980). Survival of *Phytophthora infestans* in potato stem lesions at high temperatures and implications for disease forecasting. *Plant Disease 64:*165-167.

Kamoun, S., Huitema, E., and Vleeshouwers, V.G.A.A. (1999). Resistance to oomycetes: A general role for the hypersensitive response? *Elsevier Trends Journals 4:*196-200.

Katsurayama, Y. and Boneti, J.I.d.S. (1996). Viability of the BLITECAST system for the potato late blight control at Sao Joaquim, SC, Brazil. *Fitopatologia Brasileira 21:*105-109.

Kori, J. (1972). A survey of races of *Phytophthora infestans* in East Africa and the development of late blight resistant varieties. Thesis MS, Makerere University, Kampala, Uganda.

Lacey, J. (1967). The role of water in the spread of *Phytophthora infestans* in the potato crop. *Annals of Applied Biology 59:*245-255.

Landeo, J.A. (2002). Durable resistance: Quantitative/qualitative resistance. Global Initiative on Late Blight (GILB) Conference (p. 6). Hamburg, Germany: GILB.

Landeo, J.A., Gastelo, M., Pinedo, H., and Flores, F. (1995). Breeding for horizontal resistance to late blight in potato free of *R* genes, In Dowley, L.J. Bannon, E., Cooke, L.R., Keane, T., and O'Sullivan, E. (Eds.), *Phytophthora infestans* (pp. 268-274). Dublin, Ireland: Boole Press.

Landeo, J. and Turkensteen, L. (1989). Assessment of partial resistance to late blight *(Phytophthora infestans)* of major genes in potato. *American Potato Journal 66:*530.

Latin, R.X., MacKenzie, D.R., and Cole, J.H. (1981). The influence of host and pathogen genotypes on the apparent infection rates of potato late blight epidemics. *Phytopathology 71:*82-85.

Marshall, K.D. and Stevenson, W.R. (1996). Transmission of *Phytophthora infestans* from infected seed potato tubers to developing sprouts. *American Potato Journal 73:*370-371.

Meksem, K., Leister, D., Peleman, J., Zabeau, M., Salamini, F., and Gebhardt, C. (1995). A high resolution map of the vicinity of the *R1* locus on chromosome V of potato based on RFLP and AFLP markers. *Molecular and General Genetics 249:*74-81.

Minogue, K.P. and Fry, W.E. (1981). Effect of temperature, relative humidity and rehydration rate on germination of dried sporangia of *Phytophthora infestans. Phytopathology 71:*1181-1184.

Mizubuti, E.S.G., Aylor, D.E., and Fry, W.E. (2000). Survival of *Phytophthora infestans* sporangia exposed to solar radiation. *Phytopathology 90:*78-84.

Nattrass, R.M. and Ryan, M. (1951). New hosts of *Phytophthora infestans* in Kenya. *Nature 168:*85-86.

Nelson, E.C. (1995). The cause of the calamity: The discovery of the potato blight in Ireland, 1845-1847, and the role of the National Botanic Gardens, Glasnevin, Dublin, In Dowley, L.J., Bannon, E., Cooke, L.R., Keane, T., and O'Sullivan, E. (Eds.), *Phytophthora infestans* (pp. 1-11). Dublin, Ireland: Boole Press.

Nkengaka, P. (2000). *Inventory and Assessment of Major Diseases of Indigenous Vegetables in Fouimbot.* Maitrise Memoir, University of Dschang, Dschang.

Oberhagemann, P., Chatot-Balandras, C., Schäfer-Pregl, R., Wegener, D., Palomino, C., Salamini, F., Bonnel, E., and Gebhardt, C. (1999). A genetic analysis of quantitative resistance to late blight in potato: Towards marker-assisted selection. *Molecular Breeding 5:*399-415.

Oliva, R.F., Erselius, L.J., Adler, N.E., and Forbes, G.A. (2002). Potential of sexual reproduction among host-adapted populations of *Phytophthora infestans sensu lato* in Ecuador. *Plant Pathology 51:*710-719.

Ordoñez, M.E., Forbes, G.A., and Trognitz, B.R. (1997). Resistance to late blight in potato. A putative gene that suppresses *R* genes and is elicited by specific isolates. *Euphytica 95:*167-172.

———. (1998). Relationship between ineffective *R*-genes and expansion rate of lesions on potato leaves, caused by *Phytophthora infestans. Plant Pathology 47:*130-136.

Oyarzún, P.J., Pozo, A., Ordoñez, M.E., Doucett, K., and Forbes G.A. (1998). Host specificity of *Phytophthora infestans* on tomato and potato in Ecuador. *Phytopathology 88:*265-271.

Parker, J.M., Thurston, H.D., Villareal-Gonzalez, M.J., and Fry W.E. (1992). Stability of disease expression in the potato late blight pathosystem: A preliminary field study. *American Potato Journal 69:*635-644.

Parlevliet, J.E. (1989). Identification and evaluation of quantitative resistance, In Leonard K.J. and Fry, W.E. (Eds.), *Plant Disease Epidemiology. Vol. 2: Genetics, Resistance and Management* (pp. 215-247). New York: MacMillan.

Parlevliet, J.E. and Zadoks, J.C. (1977). The integrated concept of disease resistance; a new view including horizontal and vertical resistance in plants. *Euphytica 26:*5-21.

Paxman, G.J. (1963). Variation in *Phytophthora infestans. European Potato Journal 6:*14-23.

Pedersen, W.L. and Leath, S. (1988). Pyramiding major genes for resistance to maintain residual effects. *Annual Review of Phytopathology 26:*369-378.

Perez, W.G., Gamboa, J.S., Falcon, Y.V., Coca, M., Raymundo, R.M., and Nelson, R.J. (2001). Genetic structure of Peruvian populations of *Phytophthora infestans. Phytopathology 91:*956-965.

Perez, W., Salas, A., Raymundo, R., Huamán, Z., Nelson, R., and Bonierbale, M. (2001). Evaluation of wild potato species for resistance to late blight, *Program Report 1999-2000* (pp. 49-62). Lima, Peru: International Potato Center.

Platt, H.W. (1999). Response of solanaceous cultivated plants and weed species to inoculation with A1 or A2 mating type strains of *Phytophthora infestans. Canadian Journal of Plant Pathology 21:*301-307.

Ristaino, J.B., Groves, C.T., and Parra, G.R. (2001). PCR amplification of the Irish potato famine pathogen from historic specimens. *Nature 411:*695-697.

Robinson, R.A. (1996). *Return to Resistance. Breeding Crops to Reduce Pesticide Dependence.* Davis, CA: agAccess.

Schepers, H.T.A.M. and van Soesbergen, M.A.T. (1995). Factors affecting the occurrence and control of tuber blight. In Dowley, L.J., Bannon, E., Cooke, L.R., Keane, T., and O'Sullivan, E. (Eds.), *Phytophthora infestans* (pp. 171-176). Dublin, Ireland: Boole Press.

Schwinn, F.J. and Margot, P. (1991). Control with chemicals. In Ingram, D.S. and Williams, P.H. (Eds.), *Advances in Plant Pathology* (pp. 225-265). San Diego, CA: Academic Press.

Singh, B.P., Ahmad, I., Sharma, V.C., and Shekhawat, G.S. (2000). JHULSACAST: A computerized forecast of potato late blight in western Uttar Pradesh. *Journal of the Indian Potato Association 27:*25-34.

Smart, C.D., Myers, K.L., Restrepo, S., Martin, G.B., and Fry, W.E. (2003). Partial resistance of tomato to *Phytophthora infestans* is not dependent upon ethylene, jasmonic acid, or salicylic acid signaling pathways. *Molecular Plant Microbe Interactions 16:*141-148.

Song, J., Bradeen, J.M., Naess, S.K., Raasch, J.A., Wielgus, S.M., Haberlach, G.T., Liu, J., Kuang, H., Austin-Phillips, S., Buell, C.R. et al. (2003). Gene *RB* cloned from *Solanum bulbocastanum* confers broad spectrum resistance to potato late blight. *Proceedings of the National Academy of Science (USA) 100:*9128-9133.

Stewart, H.E. and Bradshaw, J.E. (2001). Assessment of the field resistance of potato genotypes with major gene resistance to late blight (*Phytophthora infestans* (Mont.) de Bary) using inoculum comprised of two complementary races of the fungus. *Potato Research 44:*41-52.

Sunita, S. and Sen, S. (1997). Host range of *Phytophthora infestans*, the incitant of leaf blight of Kangaroo apple. *Journal of Mycology and Plant Pathology 27:*224.

Thurston, H.D. (1971). Relationship of general resistance: Late blight of potato. *Phytopathology 61:*620-626.

———. (1990). Plant disease management practices of traditional farmers. *Plant Disease 74:* 96-102.

Tooley, P.W. and Therrien, C.D. (1991). Variation in ploidy in *Phytophthora infestans*, In Lucas, J.A., Shattock, R.C., Shaw, D.S., and Cooke, L.R. (Eds.), *Phytophthora* (p. 447). Cambridge, UK: Cambridge University Press.

Torto, T.A., Shuang, L., Styer, A., Huitema, E., Testa, A., Gow, N.A.R., West, P.V., and Kamoun, S. (2003). EST mining and functional expression assays identify extracellular effect of proteins from the plant pathogen *Phytophthora. Genome Research 13:*1675-1685.

Trognitz, F., Manosalva, P., Gysin, R., Niño-Liu, D., Simon, R., Herrera, M.R., Trognitz, B., Ghislain, M., and Nelson, R. (2002). Plant defense genes associated with quantitative resistance to potato late blight in *Solanum phureja* × dihaploid *S. tuberosum* hybrids. *Molecular Plant–Microbe Interactions 15:*587-597.

Turkensteen, L.J. (1978). *Phytophthora infestans:* Three new hosts and a specialized form causing a foliar blight of *Solanum muricatum* in Peru. *Plant Disease Reporter 62:*829.

———. (1993). Durable resistance of potatoes against *Phytophthora infestans*. In Jacobs, T. and Parlevliet, J.E. (Eds.), *Durability of Disease Resistance* (pp. 115-124). Dordrecht, The Netherlands: Kluwer Academic Publishers.

Turkensteen, L.J., Flier, W.G., Wanningen, R., and Mulder, A. (2000). Production, survival and infectivity of oospores of *Phytophthora infestans*. *Plant Pathology 49:*688-696.

Umaerus, V., Umaerus, M., Erjefalt, L., and Nilsson B.A. (1983). *Phytophthora, its Biology, Taxonomy, Ecology and Pathology*. St. Paul, USA: The American Phytopathological Society.

Vanderplank, J.E. (1956). Horizontial (polygenic) and vertical (oligogenic) resistance against blight. *American Potato Journal 42:*306.

———. (1984). *Disease Resistance in Plants*. San Diego, CA: Academic Press.

Van der Vossen, E., Sikkema, A., Hekkert, B.T.L., Gros, J., Musken, M., Stiekema, W.J., and Allefs, S. (2002). Cloning of an *R* gene from *Solanum bulbocastanum* conferring complete resistance to *Phythophthora infestans*. Late Blight: Managing the Global Threat. *Proceedings Global Initiative on Late Blight Conference* (p. 155). Hamburg, Germany. GILB.

van der Zaag, D.E. (1956). Overwintering en epidemiologie van *Phytophthora infestans*, tevens enige nieuwe bestrijdingsmogelijkheden. *Tijdschrift over Planteziekten 62:*89-156.

Vartanian, V.G. and Endo, R.M. (1985). Overwintering hosts, compatibility types, and races of *Phytophthora infestans* on tomato in southern California. *Plant Disease 69:*516-519.

Vleeshouwers, V.G.A.A., Dooijeweert, W.V., Govers, F., Kamoun, S., and Colon, L.T. (2000a). Does basal *PR* gene expression in *Solanum* species contribute to non-specific resistance to *Phytophthora infestans? Physiological and Molecular Plant Pathology 57:*35-42.

———. (2000b). The hypersensitive response is associated with host and nonhost resistance to *Phytophthora infestans*. *Planta 210:*853-864.

Warren, R.C. and Colhoun, J. (1975). Viability of sporangia of *Phytophthora infestans* in relation to drying. *Transactions British Mycological Society 64:*73-78.

Wastie, R.L. (1991). Breeding for resistance. In Ingram, D.S. and Williams, P.H. (Eds.), *Phytophthora infestans: The Cause of Late Blight of Potato* (pp. 193-224). San Diego, CA: Academic Press.

Wastie, R.L., Caligari, P.D.S., Stewart, H.E., and Mackay, G.R. (1987). A glasshouse progeny test for resistance to tuber blight *(Phytophthora infestans)*. *Potato Research 30:*533-538.

Waugh, M., Hraber, P., Weller, J., Wu, Y., Chen, G., Inman, J., Kiphart, D., Sobral, B., Wu, Y.H., and Chen, G.H. (2000). The *Phytophthora* Genome Initiative database: Informatics and analysis for distributed pathogenomic research. *Nucleic Acids Research 28:*87-90.

Wilson, U.E. and Coffey, M.D. (1980). Cytological evaluation of general resistance to *Phytophthora infestans* in potato foliage. *Annals of Botany 45:*81-90.

Young, B.A., St. Martin, S.K., Schmitthenner, A.F., Buzzell, R.I., and McBlain, B.A. (1994). Absence of residual effects of defeated resistance genes on *Phytophthora* rot of soybean. *Crop Science* 34:409-414.

Zwankhuizen, M.J., Govers, F., and Zadoks, J.C. (1998). Development of potato late blight epidemics: Disease foci, disease gradients and infection sources. *Phytopathology 88:*754-763.

# Chapter 9

# Fungal and Oomycete Diseases

H. W. (Bud) Platt
R. D. Peters

Potato diseases caused by fungal and *Oomycete* pathogens pose a great threat to potato production, and demand significant inputs of time and money by the potato industry. These occur in all parts of the world and lead to annual losses in billions of dollars in terms of crop production, quality, and marketing as well as soil and water health. In agricultural settings, populations of pathogenic microorganisms often become unnaturally high in numbers altering the "ecological balance" and affecting environmental stability and health. This chapter provides recent research findings and relevant general information for the most important potato diseases caused by fungi and *Oomycetes,* except for late blight and its causal agent *Phytophthora infestans.* Aspects of the epidemiology, characterization, and control of late blight have been well described in Chapter 8 "Late Blight."

The diseases (and causal agents) included in this chapter are: Early Blight (*Alternaria* species); Gray Mold *(Botrytis cinerea)*; Black Dot *(Colletotrichum coccodes)*; Fusarium Dry Rot and Wilt (*Fusarium* species); Gangrene and Pocket Rot (*Phoma* species); Leak (*Pythium* species); Pink Rot *(Phytophthora erythroseptica)*; Powdery Scab *(Spongospora subterranea)*; Rhizoctonia Canker and Black Scurf *(Rhizoctonia solani)*; Silver Scurf *(Helminthosporium solani)*; Skin

FIGURE 9.1. Typical symptoms of various fungal and oomycete diseases of potato (tuber symptoms of common scab—a bacterial disease is shown here for comparison with other tuberborne diseases). See Plate 9.1 in the color gallery.

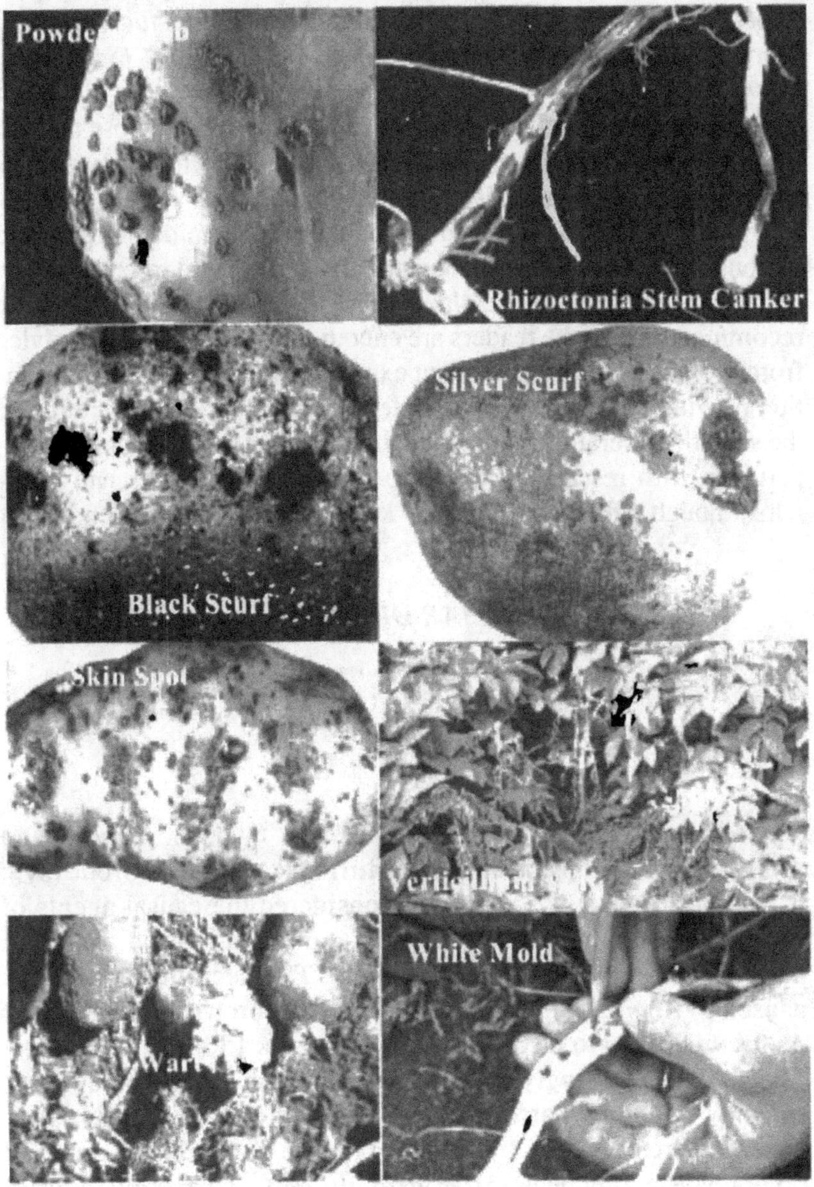


FIGURE 9.1 *(continued)*

Spot *(Polyscytalum pustulans)*; Verticillium Wilt (*Verticillium* species); Wart *(Synchytrium endobioticum)*; and White Mold *(Sclerotinia sclerotiorum)*. To aid the reader, these diseases have been classified as those that are mainly considered foliar diseases and those that are most generally classified as soilborne or tuberborne diseases. For each disease, information is presented on the causal agents, distribution and impact, disease symptoms (Figure 9.1), epidemiology, and various aspects of disease management. Only general information on various disease-control treatments has been included. For specific recommendations, the readers are encouraged to seek specific advice from crop and land management experts in their area. The information provided in this chapter is based on an extensive exploration of the scientific literature and practical knowledge gathered through experience with research and potato-production activities and problems, though only a few selected "key" references have been cited.

## *FOLIAR DISEASES*

### *Early Blight* (Alternaria *species)*

#### *Causal Agents*

*Alternaria alternata* (Fries.) Keissler (syn. *A. tenuis*) and *Alternaria solani* (Ellis & G. Martin) L.R. Jones & Grout (syn. *A. solani* Sorauer) are generally considered the causal agents of potato early blight. These two fungi are found in crop and other host debris, soils, and infected tubers. The necrotic leaf lesions caused by *A. alternata* are also referred to as brown spot. Spores of *A. solani* are tan to olive-brown and 15-19 × 150-300 μm in size. They are formed singly and are 8-10 celled, ellipsoid to oblong in shape with a long, narrowing terminal cell or beak. Spores of *A. alternata* are only 20-63 × 9-18 μm in size, formed in chains, and do not have a long beak. *Alternaria alternata* is generally considered a weaker pathogen than *A. solani.* These pathogens affect a wide range of native and "cultivated" plant species.

### *Distribution and Impact*

All potato-production areas have early blight present to some degree but major impacts on tuber yield and quality are most often found under warm, wet conditions causing rapid disease development early in the season. Disease epidemics cause significant defoliation and tuber yield losses of 20-30 percent. Late season or slow development of the disease generally does not result in tuber yield and quality losses. Late season disease enhances crop maturation and defoliation. The impact of early blight is perhaps most significant in areas of dryland agriculture where frequent irrigation creates the wet/dry cycles ideal for sporulation and dissemination of the fungus.

### *Symptoms*

Plant foliar symptoms generally involve mature leaf and on occasion, stem lesions but are also associated with tissues weakened by other diseases or environmental and nutritional stresses. Leaf lesions are 1-2 mm in size and black in color, but develop into large, brown to black, circular to ovoid lesions that expand until limited by leaf veins. Lesion expansion involves concentric rings of raised and depressed necrotic tissue that appear like "target boards" but with a chlorotic "halo" around the lesion. Necrotic tissues of the lesions will have dark mycelium and spores formed. Expansion of initial lesions and the development of new lesions result in chlorosis and then necrosis of the whole leaf which can remain attached to the stem as a desiccated, brown to black parchment. Roots and stolons do not typically exhibit disease symptoms, but tubers can have irregularly shaped, darkened, and sunken lesions with raised dark-colored borders. Tuber tissues underneath the skin are leathery to corky in texture with a "wet," yellowish appearance. Infected tubers develop dry rot slowly during storage.

### *Epidemiology*

Early blight can develop throughout the growing season when plants are under stress, wounded, or diseased. The primary inoculum sources are the spores formed in spring that are carried by wind, water, and equipment from field to field. Spore dissemination often peaks during the morning dry periods. Spores germinate and penetrate leaf tissues directly through the epidermis, stomata, or wounds. Mature or weakened (diseased or stressed) foliage is more susceptible than newly formed and actively expanding foliage. Infected plant tissues provide inoculum for further disease spread during the growing season, giving rise to characteristics typical of a polycyclic disease. Foliar lesions expand until the whole leaf becomes necrotic leading to plant defoliation and death in extreme situations. Belowground, potato tubers with immature skin or damaged tuber surfaces are more susceptible to infection than mature tubers. When free moisture is present, spores germinate and penetrate tuber lenticels and wound sites. Spores on tuber surfaces can germinate and cause a slowly developing dry rot during storage.

### *Disease Management*

Cultivar resistance, foliar fungicides, agronomic amendments, and harvesting when tuber skin is mature are the major methods for disease management. Late-maturing cultivars are generally less susceptible to disease than early-maturing ones. Genetics of host resistance indicates the preponderance of nonadditive gene action (Gopal, 1998) and a few disease-resistant commercial cultivars are currently available. When resistant cultivars are grown, fungicide usage can be reduced. However, the application of fungicides is currently the major method for disease control. These include both protectant (such as chlorothalonil, dithiocarbamates, and fentin hydroxide) and systemic (such as azoxystrobin, difenoconazole, and tebuconazole) fungicides for control of the initial stages of the disease (Platt, 1983, 1994). Application of fungicides should be timed according to forecasted disease threats as related to spore-production periods (Stevenson, 1993; Shtienberg et al., 1995). Disease occurrence and impact can also be reduced by proper crop rotation, crop debris disposal, and effective weed management. In addition, crop management practices that maintain adequate nutrition, moisture, and control other diseases will lessen the risk of plant stress and early blight occurrence. Similarly, harvest timing and tuber care can limit or prevent tuber disease by minimizing tuber wounding and enhancing skin development. A postharvest tuber treatment fungicide (e.g., iprodione) can also reduce black pit disease incidence during storage.

## *Gray Mold* (Botrytis cinerea)

### *Causal Agent*

*Botrytis cinerea* Pers.:Fr. (Teleomorph: *Botryotinia fuckeliana* (de Bary) Whetz.; syn: *Sclerotinia fuckeliana* (de Bary) Fuckel) is the causal agent of gray mold. It produces conidia in grape-like clusters on conidophores that are more or less straight, branched, and brown in color. Single-celled conidia are obovate or ellipsoidal, hyaline to brown, or in-mass, gray-brown and measure 9-15 × 6.5-10 μm in size. The fungus overwinters as small, 1-15 mm black sclerotia. The teleomorph stage is rarely seen. This pathogen is found on many native and cultivated plant species (Coley-Smith et al., 1980).

### *Distribution and Impact*

The fungus is ubiquitous and present in all potato-production areas of the world. It is a weak pathogen, but a primary saprophyte (Jarvis, 1977). It attacks stressed, wounded or diseased plant tissues. It does not normally cause yield losses, but can contribute to poor plant growth and tuber-quality (appearance) losses.

### *Symptoms*

Plant stems, roots, and stolons do not normally exhibit disease symptoms, but leaf margins and edges can have chlorotic and necrotic lesions with concentric zonation, often bordered by major veins. Leaf lesions can have a "late blight" appearance, but do not have the white mycelial growth and sporulation on the undersurface of the leaf that is so characteristic of late blight. Mature flowers are often infected and when they land on leaves, can cause circular lesions on the leaf. Stem infections result in water-soaked lesions that develop into wet rots, particularly when secondary organisms invade. Tuber infections are rare, but wounds can become infected causing a wrinkled skin and softening of the tissues that further develops into a dark, sunken dry rot unless secondary organisms intervene to cause a wet rot.

### *Epidemiology*

Mature or stressed plant tissues usually exhibit symptoms first, particularly where plant shading or excessive moisture occurs. Spores produced in darkened mycelial mats on necrotic tissues are spread by wind and water. Infection and disease development is limited by warm, dry conditions but favored by cool, moist conditions. The potential for tuber infection is increased by storage of unhealed tubers in cool, humid conditions.

### *Disease Management*

Due to the ubiquitous nature of the causal agent and the wide variety of plant stress-related conditions that favor infection and disease development, disease management is difficult. Control of gray mold requires actions that minimize plant stress, wounding and other diseases. But disease management measures such as crop rotation, resistant cultivars, healthy seed, etc. are not very effective. In addition, growing conditions that favor excessive crop growth can create microclimatic conditions conducive to disease development; so overfertilization and excessive irrigation should be avoided. Most prophylactic fungicides (e.g., chlorothalonil, maneb, mancozeb, etc.) are effective if plant foliage is properly protected before the spores arrive. Disease forecasting systems provide benefits to disease management efforts (Shtienberg and Elad, 1997).

## *SOILBORNE DISEASES*

### *Black Dot* **(Colletotrichum coccodes)**

### *Causal Agent*

*Colletotrichum coccodes* (Wallr.) Hughes (syn.: *C. atramentarium* (Berk. & Broome) Taub.) is a soil- and tuberborne fungus with white mycelium. Sclerotia survive in plant debris, soil, and tubers. Conidia are formed in acervuli on the surface of sclerotia (100 μm to 0.5 mm in size). Acervuli contain septate, pointed setae that vary in length from 80 to 350 μm. Conidia are hyaline and 18-22 μm × 3-8 μm in size. Depending on growth medium, conidial masses appear yellow to pink and have an optimum temperature for germination between 22°C and 28°C. The fungus also attacks other Solanaceous plants and common weeds.

### *Distribution and Impact*

The fungus is common in most potato-production areas of the world but the relative economic impact of black dot is often not understood. Yield effects are rare, but tuber quality (appearance) losses can occur. Symptoms of black dot are commonly seen in potato stems that are senescing due to physiological plant stress or other diseases including Verticillium wilt (early-dying). However, its contribution to the late season disease complex and early plant death requires further investigation.

### *Symptoms*

The black sclerotia of the fungus give the name, black dot, to this disease. Plant stems, roots, stolons, and tubers can harbor sclerotia on or just below the tissue surface (Andrivon et al., 1998). Foliar symptoms include chlorosis, wilting, and, in extreme cases, necrosis and premature plant death. Plant wilt often begins at the top of the plant and can be confused with or be found in conjunction with Verticillium wilt. Root and stolon lesions have an appearance similar to disease caused by *Rhizoctonia solani* with cortical tissue invasion, browning, and skin sloughing. As infected tissues, particularly stems, mature and dry, numerous sclerotia become evident. Tubers may also develop dry and gray patches somewhat similar to silver scurf, except that sclerotia are usually visible in the lesion. So, the two diseases are commonly mistaken for each other.

### *Epidemiology*

Sclerotia overwinter in plant debris, soil, and potato tuber surfaces and can germinate to form mycelium that attacks weakened host tissue or germinate to produce spores that are carried by wind to host plants (Denner et al., 1998). *C. coccodes* is a weak pathogen and requires plant stress, wounding, and other diseases to allow it to overcome plant defenses and successfully establish the disease. Within the plant, the fungus grows and produces more sclerotia that allow continuation of the disease cycle. The disease is favored by light soils, low nitrogen, high temperatures, and poor soil drainage.

### *Disease Management*

Commercially grown cultivars are not resistant to black dot, but variation in the response of cultivars to infection does occur and early cultivars are generally more susceptible than late-maturing types. Thin-skinned cultivars also appear to be more susceptible than thick-skinned cultivars. In terms of cultural control strategies, use of disease-free seed, crop rotation, and proper crop management practices are important measures to prevent or limit disease occurrence and impact. Irrigation can contribute to disease severity whereas crop rotations of at least five years between potato reduce the disease problems. Good soil structure, drainage, and fertility and weed control are also important. Fungicides, such as prochloraz, inhibit mycelial growth of *C. coccodes* and may be useful as seed treatments, but common seed treatments are not very effective, particularly when soil inoculum levels are high (Dillard, 1988; Marais, 1990; Uribe and Loria, 1994; Read and Hide, 1995a,b). Soil fumigation can reduce levels of soilborne inoculum and the incidence of black dot, but this is an expensive and environmentally unacceptable practice. Biological control measures are being investigated but are not commonly used (Okhovat et al., 1994). Early harvest reduces black dot in tubers and postharvest studies on tuber washing and drying for two weeks at 5-10°C prior to storage at 5°C have documented decrease in disease. Molecular tools (real-time polymerase chain reaction [PCR] techniques) have been developed to quantify inoculum levels in soil and will be useful for identification of infested fields and facilitate research on the impact of cropping practices on soil inoculum (Cullen et al., 2002).

### *Fusarium Dry Rot and Wilt* (**Fusarium** *species)*

#### *Causal Agents*

*Fusarium* species causing dry rot include: *F. coeruleum* (Lib.) ex Sacc. ([syn. *F. solani* {Mart}. Sacc. var. *coeruleum* {Lib. ex Sacc.} C. Booth]); *F. sambucinum* Fuckel ([syn. *F. sulphureum* Schlechtend.; teleomorph—*Giberella pulicaris* {Fr.:Fr.} Sacc]); *F. avenaceum* (Fr.:Fr.) Sacc.; *F. culmorum* (Wm. G. Sm.) Sacc.; *F. oxysporum* Schlechtend.:Fr.; *F. acuminatum* Ellis & Everh. and *F. crookwellense* L.W. Burgess, P.E. Nelson, & T.A. Toussoun. Those causing wilt include: *F. oxysporum* f.sp. *tuberosi* Snyder & Hansen (syn. *F. oxysporum*); *F. eumartii* (C.W. Carp.) (syn. *F. solani* f.sp. *eumartii* (C. Carpenter) W.C. Snyder & H.N. Hans.; teleomorph—*Nectria haematococca* var. *haematococca*); *F. javanicum* Koord. (teleomorph—*Nectria ipomoeae* Halsted); *F. solani* (Martius) Sacc. (teleomorph—*Nectria haematococca* var. *brevicona* (Wollenw.) Gerlach.) and *F. avenaceum* (Corda) Sacc. (teleomorph—*Gibberella avenacea* Cooke). Other *Fusarium species* associated with disease of potato include: *F. arthrosporioides* Sherbakoff (similar to *F. avenaceum*); *F. sporotrichioides* Sherbakoff; *F. trichothecioides* Wollenw, and *F. equiseti* (Corda) Sacc. (teleomorph—*Gibberella intricans* Wollenw). These fungi have hyaline hyphae and condiophores that produce multiple, single to multicelled oblong to sickle-shaped conidia (macro and/or microconidia). In culture, colonies of the different species vary from white to red or orange-brown to blue or purple. Teleomorph stages of these fungi do not appear to be involved with disease situations. These fungi generally attack only potatoes, but morphologically similar *Fusarium* species cause diseases on a number of other plant species.

#### *Distribution and Impact*

Fusarium dry rot and wilt can occur in almost all potato-production areas around the world, but wilt is more widespread and severe where hot, dry weather occurs (Boyd, 1952). These fungi can cause seed-tuber rot resulting in weak plant development or seed-germination failures, susceptibility to other diseases, wilt and premature plant death, and in extreme cases, some yield and tuber-quality losses, including postharvest losses. *F. coeruleum* and *F. sambucinum* are the most important dry rot pathogens. *F. coeruleum* is more common in cooler climates causing most of the dry rot in northern Europe. In North America, seed piece decay and storage rot is frequently caused by *F. sambucinum.* The increase in mechanization in potato production has provided opportunities for these wound pathogens to incite more disease. Fusarium wilt occurrence and impact vary from region to region. While *F. oxysporum* f. sp. *tuberosi* is a problem in hot, humid regions, *F. eumartii* and *F. javanicum* are prevalent in cooler and drier regions. *F. solani* and *F. avenaceum* are weak pathogens requiring plant stresses, wounds, or other diseases for opportunities to infect.

### *Symptoms*

Fusarium dry rot in seed tubers can result in germination gaps or severely stunted, chlorotic, and necrotic stems as well as abnormal growth of roots and stolons. Following harvest and in storage, infected tubers have sunken lesions with concentric rings of shrivelled skin and white mycelium on the lesion surface that can merge into irregular mats. Other diseases (e.g., Phoma rot) can produce similar symptoms. *F. sambucinum* causes light to dark brown lesions that are dry in texture with a distinct dark brown, but irregular margin. Mycelium within the lesion can vary in color (white to yellow to red). *F. coeruleum* causes lesions that are dark brown at the initial site to light brown at the edges and the lesions have a more regular edge than those of *F. sambucinum.* Cavities may develop within older lesions having mycelium and spores that continue to invade the tissue so that tubers shrivel to a dried "mummy" state. Secondary organisms often occur in lesions and sometimes result in the development of wet rot problems.

Fusarium wilt symptoms generally involve decay of roots and lower parts of stems and vascular discoloration is frequently visible. Symptoms usually start on the lower leaves with veinal chlorosis and/or yellowing followed by wilting. Leaf browning or purpling and curling can occur and are often seen on one side of a plant due to partial stem infection. Further wilt development results in stem streaking, cortical tissue invasion, leaf and stem chlorosis, and necrosis. *F. oxysporum f. sp. tuberosi* can cause a rapid premature stem and/or plant death and vascular discoloration is often seen in stems at or below the soil line and occasionally in tuber stem-end vascular tissues. *F. eumartii* and *F. javanicum* cause stem pith discoloration at the nodes. Belowground stem tissues may rot completely and discoloration of tuber stem-end vascular tissues can also occur. *F. avenaceum* wilt symptoms occur during the second half of the growing season and, as with the other wilt pathogens, symptoms are more severe in hot and dry conditions. In general, wilt symptoms are often more severe on one side of a plant, but the remainder of the plant is also often weak and stunted and vascular discoloration in the stem can be seen at or below the soil surface. Root rot and belowground stem rot occur with *F. solani,* yet vascular discoloration is rare, although aerial tubers may develop. Tuber stem-end rot and vascular discoloration can also occur.

*Epidemiology*

*Fusarium* species invade tuber through wounds created by seed-cutting, sprout damage at planting, other disease or pest damage, and during harvest or tuber-handling operations. Soilborne or seedborne inoculum provides a continuous source of propagules for infection of plant wounds and stressed tissues throughout the production, storage, and handling phases. Infected seed tubers and plant tissues release spores into the soil which contaminate progeny tubers at harvest. *Fusarium coeruleum* and *F. sambucinum* can infect and develop over a wide temperature range (3-25°C) with symptoms and rapid disease development at higher temperatures.

Chlamydospores of *Fusarium oxysporum* f. sp. *tuberosi, F. eumartii* and *F. javanicum* can persist in plant debris and soil for several years. Other inoculum sources include movement of soil containing resting spores via farm machinery, wind, etc. *Fusarium. oxysporum* f. sp. *tuberosi* and *F. eumartii* attack roots at the elongation zone and also roots that are damaged by nematodes. These pathogens move upward in the plant through the vascular tissue as mycelium and microconidia causing blockages that result in plant wilt. Crop stress(es) also enhance wilt occurrence, development, and impact.

### *Disease Management*

Use of disease-free, undamaged seed tubers, adherence to proper land and crop management practices, and prevention of tuber wounding during harvest and tuber-handling operations are essential for disease control. Postharvest tuber storage management requires wound healing with ventilation to remove moisture at temperatures of 10°C. Warming seed tubers before grading will lessen tuber damage and disease risk. Equipment and facility hygiene will also reduce inoculum sources. Fungicide treatments (imazalil and thiabendazole, mancozeb, maneb, etc.) are available for cut seed or as postharvest treatments to prevent infection of wounds and restrict disease development. However, as some pathogens (e.g., *F. sambucinum*) have developed resistance to some fungicides (e.g., thiabendazole), it is important to manage fungicide usage to prevent resistance development within pathogen populations and avoid any unnecessary application of ineffective fungicides (Hide et al., 1992; Kawchuk et al., 1994; Platt, 1997). Biological controls using beneficial bacteria applied to the potatoes prior to tuber storage have been developed but are not yet in common usage (Schisler et al., 2000). Postharvest treatments are also being investigated (Tweddell et al., 2003). The application of organic or inorganic salts to tubers entering storage can also suppress disease development. Crop rotation (e.g., potatoes once every 3-5 years) is a vital method for long-term reduction of soil-borne inoculum and disease control. In addition, effective weed and volunteer potato control are critical factors as are harvest methods (Carnegie et al., 2001). While the mechanisms of disease control are not fully understood, soils suppressive to *Fusarium* wilt pathogens due to the activities of antagonistic microorganisms have been reported. Although some cultivars are resistant to races of some pathogen species, no commercially acceptable cultivars are highly resistant to all species of *Fusarium.*

### *Gangrene and Pocket Rots* (Phoma *species)*

#### *Causal Agents*

*Phoma exigua* var. *foveata* (Foister) Boerema (syn: *P. solanicola f. foveata, P. foveata, P. exigua* f. sp. *foveata*) causes gangrene while *Phoma exigua* Desm. var. *exigua* causes "pocket" or "button" rot of potatoes. *Phoma exigua* var. *foveata* is more aggressive and a much more persistent soilborne fungus than *P. exigua* var. *exigua,* but the two fungi are morphologically similar producing globoid, dark brown pycnidia, 90-200 μm in size. Pycnidia are initially formed beneath the surface of infected tissue, but erupt and exude hyaline, nonseptate, cylindrical pycnidiospores, 4-5 × 2-3 μm in size. In culture, *P. exigua* var. *foveata* forms nonzonate colonies while *P. exigua* var. *exigua* has zonated colonies that are red when exposed to ammonia vapor. These pathogens attack a wide variety of plants including weed species.

#### *Distribution and Impact*

*Phoma exigua* var. *foveata* predominates throughout northern Europe whereas *P. exigua* var. *exigua* is found in all potato-production areas of the world. Whereas pocket rot causes only minor blemishes on tuber surfaces, gangrene is a more severe disease that can decimate whole crops when conditions favor the disease epidemics.

#### *Symptoms*

For both gangrene and pocket rot, potato stems, leaves, roots, and stolons do not normally exhibit disease symptoms, but when stems are dead, pycnidia can be visible on the decaying surfaces. Tuber surfaces infected with *P. exigua* var. *foveata* have small, thumb-mark depressions that are dark brown or black in color that expand to reveal small black pycnidia. Cross sections through these lesions reveal a dark-brown, purple, or black rot that extends into the flesh and may be quite shallow or extremely deep. Lesion margins are clearly demarcated from healthy tissue and pycnidia form on the lining of lesion cavities. For pocket rot, infected tubers have small dark and usually circular depressions, most frequently at wounds, eyes, or lenticels. The hard lining covering the depressions can easily be removed to leave a smooth cavity delimited by the wound cork that prevents further spread of the fungus.

### *Epidemiology*

Infected or contaminated tubers are the most important source of inoculum as pycnidia can persist in dust in store, especially on a grading line, and contaminate otherwise healthy tubers. After planting, the fungus on the tuber surface can infect stem bases and other underground parts via pycnidiospores that can spread by wind and water. Aboveground plant infection can progress unseen during the growing season, but as plants senesce or are killed, lesions fully develop and produce clusters of pycnidia that release pycnidiospores into the soil. Latent infection of tubers can occur before harvest when the fungus penetrates the periderm and forms a small pad beneath it. Infections develop further as the fungus enters wounds caused during harvest and tuber grading. Delayed harvest seems to increase spore levels around the daughter tubers and consequently disease incidence, particularly when soil conditions are wet and tuber temperatures low. Susceptibility to *P. exigua* var. *foveata* has been found to increase during the storage period. The optimal temperature range for the germination of conidia is 25-28°C while optimal growth occurs at 22-26°C, but the range of fungal activity is 4-30°C.

### *Disease Management*

As these diseases are both tuber- and soilborne, successful management depends on utilizing a number of control strategies. Use of disease-free seed and planting in fields without a history of disease are vital practices, but since infested soil and crop debris can contaminate equipment and facilities for potato storage, seed preparation, etc., farm hygiene is also important. Early plant kill and harvest with effective antiwound measures will reduce tuber exposure to pathogen spores and limit disease initiation. Drying of tubers for wound healing and soil removal after harvest with warm (15°C), but dry

(<80 percent RH) conditions reduces disease. Fungicides (e.g., imazalil and thiabendazole) can also reduce disease problems when applied as spray applications at or soon after harvest. Fumigants (e.g., 2-aminobutane) are also available for use after wound healing is complete (Carnegie et al., 1988; Hide and Cayley, 1989).

### *Leak (*Pythium ultimum *and Other* Pythium species*)*

#### *Causal Agents*

*Pythium ultimum* Trow. and other *Pythium* species cause leak or watery rot of potato tubers. *Pythium aphanidermatum* (Edson) Fitzp. causes leak in warmer regions (Triki et al., 2001). These common soilborne inhabitants are *Oomycetes* with some characteristics of true fungi. They survive in soil as smooth, thick-walled oospores, which are spherical, 14.2-19.5 μm in size, and formed on coenocytic hyphae. Sporangia produced along hyphae are oblong and 17-27 × 14-24 μm in size whereas those formed at the end of the hyphae are spherical and 12-29 μm in size. Sporangia do not always produce zoospores. Germ tubes from fungal propagules grow to invade host tissue and/or produce hyphal mats. *Pythium* species have a wide host range, including many vegetable crops, causing damping-off, root rot, and wet rot diseases.

#### *Distribution and Impact*

Leak or watery wound rot is found in almost all potato-production areas globally, but it occurs infrequently. Disease occurrence is usually observed shortly after harvest. Crop losses can be dramatic when disease levels and development rates are high.

#### *Symptoms*

Root and stolon symptoms are not distinct. Tuber skins become discolored and develop lesions with a water-soaked appearance. If the skin is ruptured or diseased tissues are squeezed, a watery fluid is exuded. Internally, regions of healthy and diseased tissues are clearly visible. Infected tissue is moist, spongy, and has a cream color, but after exposure to the air, progressively changes from a gray to brown to black, sometimes with pinkish tones. Diseased tissue can have cavities and completely disintegrates within a few days. A strong odor due to the presence of bacteria and other microorganisms is often apparent. The final stages of the rot can result in thin-skinned tuber shells.

### *Epidemiology*

Soilborne hyphae and germinating sporangia or oospores invade freshly cut seed tubers or wounds in tubers at harvest. In some cases, the pathogens can invade sunburned or frosted tubers. Warm (>15°C), wet conditions at harvest dramatically increase the rate of infection and disease development, whereas cool conditions slow or halt the disease.

### *Disease Management*

Resistant cultivars are not commonly available (Priou et al., 1997; Salas et al., 2003). Seed-piece rot can be avoided or reduced by planting well-healed cut tubers and/or the use of proper seed-treatment fungicides. In addition, fields or areas in a field with a history of disease should be avoided at planting. Fields should be well drained and crop rotations should include a minimum of three years between potato crops and the use of rotational crops (e.g., cereals) that reduce soilborne populations of the pathogens. At harvest, disease can be reduced by digging potatoes in cool, dry conditions, and when tuber skins are as mature as possible. Furthermore, harvest systems that minimize or eliminate tuber wounding are essential for disease management. If disease is detected after harvest in stored potatoes, the temperature should be lowered (<12-15°C), air circulation increased and where possible, dehumidifiers employed immediately. The systemic fungicide mefenoxam, applied as a foliar spray during the growing season or as an in-furrow application at planting, has been shown to suppress tuber disease. To date, populations of *P. ultimum* in North America are still largely sensitive to mefenoxam, even though instances of resistant strains have been reported (Taylor et al., 2002). Various biological and chemical disease management treatments have been tested for control of *P. aphanidermatum* (Triki and Priou, 1997).

## *Pink Rot* (Phytophthora erythroseptica)

### *Causal Agents*

*Phytophthora erythroseptica* Pethybr. is a soilborne *oomycete* (Ho and Jong, 1989) that asexually reproduces via sporangia (39-47 × 25-29 μm) that are hyaline, variable in shape, mainly ellipsoid, thin-walled, and nonpapillate. These can germinate directly or cleave to form zoospores that move through soil via water films. The pathogen also reproduces sexually via oogonia that form readily in single isolate culture on V8 agar and are globose and 34-36 μm in diameter. Antheridia are strictly amphigynous, spherical to ovoid, and 14-16 × 11-13 μm in size. Oospores are aplerotic, 28-31 μm in diameter with walls about 3 μm thick. A few other *Phytophthora* spp. have been shown to induce tuber symptoms similar to pink rot, including *P. cryptogea* Pethybr. & Laff. and *P. parasitica* Dastur, but with apparently lesser significance.

### *Distribution and Impact*

*Phytophthora erythroseptica* occurs in soils in all potato-production areas of the world (Erwin and Ribeiro, 1996). It is one of the most damaging potato-tuber diseases, since harvested tubers that appear healthy can rot completely within a few weeks after being placed in store (Evans and Howard, 1994). Strains of this pathogen can cause root rots, leaf spots, crown rots, fruit rots, wilt, and leaf and petiole blights on several plant species including lilies, peas, raspberry, sugarcane, tomatoes, tulips, vetch, wild rice, etc. and have been isolated from the roots of many nonsolanaceous plants, including wheat and rye. Potato-infecting strains of *P. erythroseptica* have been reported as pathogens of tulip, spinach, and tomato. Favored by warm, moist environmental conditions, pink rot is enhanced in monoculture or short-crop rotation systems. Recent increases in disease occurrence and severity have been related to the development of strains of the pathogen resistant to the fungicide mefenoxam (metalaxyl).

### Symptoms

Potato leaves and stems do not normally exhibit symptoms except when extensive infection of underground plant parts results in plant wilt, chlorosis and necrosis. Stem base lesions cause infected stems to turn brown with water-soaked and light brown vascular discoloration, whereas severely infected plants may wilt and die prematurely. Potato root and stolon symptoms often involve a light brown discoloration or streaking along the affected tissue. Tubers become infected via stolons, eyes, lenticels, or wounds. Infection of tubers via diseased stolons has been shown to be the primary route of infection, except when extremely wet conditions facilitate infection via natural openings and wounds (Lonsdale et al., 1980). Outer surfaces of infected tuber tissues appear dull brown in color with irregularly shaped margins and dark skin tones around the eyes and lenticels. A dark line on the tuber surface delineates the extent of infection and the infected tuber tissues have a soft, wet, spongy texture. Rotting of the internal tuber tissue often begins at the stem end and is creamy or light brown; cut tubers ooze a clear, odorless liquid when squeezed. Surfaces of cut tubers change to pink, brown, and then black following exposure to air for several hours and continued disease development results in a wet-rot and complete tuber tissue breakdown.

### Epidemiology

The disease is most damaging at soil temperatures above 20°C in water-logged soils, as the pathogen is active at temperatures of 5-33°C (optimum 20-27°C). Germ tubes from sporangia, zoospores, or oospores infect stolons or enter the host via open lenticels, wounds, and tuber eyes. Sporangia form on sporangiophores and mycelium growing on the surface of the host tissue at the edge of the diseased area. *Phytophthora erythroseptica* can persist in soil for many years as oospores, which are disseminated from diseased potato stems, roots, and stolons. The main means of pathogen movement and attack in soil is via zoospores that swim through soil water to the host where they encyst and germinate to form a germ tube that penetrates the host. Infection of stolons creates a pathway for mycelium to infect daughter tubers. When infected plant parts decay, previously formed oospores are released into the soil. Tubers with eye infections caused by zoospore attack may rot prior to harvest, whereas tubers with superficial infections may harbor viable mycelium and oospores and appear to be healthy when placed in storage, only to rot a few weeks later. Alternatively, infected seed tubers may appear healthy, but carry oospores or mycelium in the skin or eyes from which disease can develop after planting, resulting in the pathogen's spread to new sites. *P. erythroseptica* can also spread to healthy tubers in storage or during tuber handling.

*Disease Management*

Successful strategies for disease management often involve the use of many disease-control tools. Pathogen detection is very important for effective reduction of inoculum sources and new DNA-based methods for pathogen detection have been developed. These tests can also distinguish *P. erythroseptica* from *P. infestans* (Mont.) de Bary (causal agent of late blight) that are often found coinfecting potato tubers (Tooley et al., 1998). Where oospores survive in soils, plant debris, or weeds or where buried tubers survive between growing seasons ("volunteer" plants or "groundkeepers"), crop rotations of four or more years and/or the use of soil treatments (e.g., solarization, chemicals) may be needed. Fields or sites with a history of pink rot should be avoided and cropped with nonhosts to reduce soilborne pathogen population levels. Use of potato cultivars with resistance to pink rot is recommended, but many of the commercially available cultivars vary only slightly in their susceptibility to the disease (Peters and Sturz, 2001; Salas et al., 2003). In general, cultivars with late-season field maturity are less susceptible to the disease than those with early or mid-season maturity. Seed tubers should be disease-free and handled to reduce or eliminate the spread of the pathogen by preventing moisture conditions that activate mycelial growth, spore production and germination, and tissue invasion. Seed treatment with fungicides (e.g., mancozeb, etc.) may limit spore formation and germination but these are not 100 percent effective. Tubers should be planted in warm, well-drained soils to hasten healthy plant growth. As for leak, the systemic fungicide mefenoxam, applied as a foliar spray during the growing season or as an in-furrow application at planting, has been shown to suppress tuber pink rot. To date, populations of *P. erythroseptica* in many parts of the world are still largely sensitive to mefenoxam (Peters et al., 2003), however, high incidence of resistant strains have been reported in some areas of the United States, including Maine and Idaho (Taylor et al., 2002). Harvesting when the foliage is completely dead, either naturally or following application of desiccants, will encourage tuber skin development and reduce the disease. Moist soil often enhances late-season tuber infections. Tuber damage should be prevented and the amount of soil adhering to tuber surfaces reduced during harvest and tuber-handling operations, and wet areas in the field should be harvested separately to reduce pathogen spread. Grading out diseased tubers prior to storage will reduce storage rot. Tubers from "questionable" fields should be stored last. Postharvest tuber wound healing and dry storage conditions are required to limit disease spread and development.

### *Powdery Scab (***Spongospora subterranea** *f.sp.* **subterranea***)*

#### *Causal Agent*

*Spongospora subterranea* (Wallr.) Lagerh. f.sp. *subterranea* Toml. causes powdery scab. The thick-walled resting spores of the pathogen form cystosori that are 19-85 μm in diameter and may survive for more than ten years in soil. Cystosori germinate releasing relatively short-lived uninucleate, flagellated zoospores that swim to the surfaces of root hairs and surface cells of roots, encyst, and penetrate. Inside the root, multinucleate sporangial plasmodia are formed by mitotic division and produce zoosporangia that release secondary zoospores. Zoospores can also infect tubers, especially during early tuber formation. The pathogen also infects other solanaceous plants, but does not form resting spores when infecting various dicotyledonous and graminaceous hosts. The pathogen serves as vector of the potato mop-top virus (Jones and Harrison, 1969).

### *Distribution and Impact*

Powdery scab occurs wherever potatoes are grown, but is of importance mainly in cool temperate climates (Harrison et al., 1997). It is an increasing problem in northern Europe, Australia, New Zealand, and Pakistan. The disease does not normally affect the yield, but significantly affects tuber quality and market value. The demand for high tuber quality with little or no tuber surface blemishes and the use of susceptible cultivars has led to significant impacts by this disease on the potato industry. When planted, infected tubers produce fewer stems that may reduce the yield. In addition, where the pathogen vectors potato mop-top virus, it can have a dramatic effect on potato production and marketing.

### *Symptoms*

Potato stems and leaves do not typically express disease symptoms. Zoospores of the pathogen can infect root hairs, and root and stolon surface cells causing small lesions and the subsequent formation of small galls. Symptoms on tubers include both a scab and a canker. Scabs form due to infection of lenticels in unsuberized tissue of the developing tuber, where the pathogen grows slowly and is not visible for several weeks. Small, purple-brown, pimple spots form on the surface of the tuber that expand, turn brown, and burst releasing a brown, powdery mass of sporeballs (cystosori). Edges of the lesion have small flaps of loose skin, but in more severe cases, lesions may have a smooth border with scabs merged to form large areas of infected tissue. Lesions penetrate into the tuber only up to a few millimeters. Cankers form as a result of the infection of eyes, a process that stimulates new tuber tissue growth that is initially unsuberized and frequently becomes severely infected with the scab form.

### *Epidemiology*

Resting spores in soil, crop debris, or cystosori formed on infected plant surfaces germinate in the presence of a host crop releasing zoospores that swim to the host, encyst, and penetrate. Inside the host, the pathogen multiplies and causes the host tissue to grow forming galls, cankers, and scabs in which secondary zoospores form and are released leading to further disease spread. Infected plant cells are stimulated to multiply forming cankers and scabs that affect tuber appearance due to tissue swelling from hyperplasia and hypertrophy and ultimately causing rupture of the epidermis. Infection occurs most frequently at cool temperatures (9-17°C) and moist conditions, but disease severity increases when alternating high and low soil water levels occur.

### *Disease Management*

Soil moisture and temperature significantly influence disease occurrence and development. Regions with high soil temperatures at and after tuber initiation tend to have less disease compared to those with cool, moist soil. Frequent use of irrigation can affect disease levels as soil temperatures tend to drop and water content increases (Adams et al., 1987). In addition, planting in well-drained soils, preventing soil compaction, and the use of disease-free seed is beneficial, but of greatest importance is the avoidance of fields with a history of the disease. Fields where manure from animals fed with infected potatoes has been spread should also be avoided or at least tested for the presence of the pathogen. Few highly effective chemical soil and tuber treatments are available, but zinc has been shown to reduce the disease severity when applied to the tuber or soil, whereas fluazinam and flusulphamide have shown some promise as soil treatments. Where possible, use of resistant cultivars is recommended, but many of the commercially acceptable cultivars are susceptible. Molecular, DNA-based techniques have been developed to detect and quantify pathogen propagules and these provide information that aids in the development of disease management strategies (Bell et al., 1999).

## ***Rhizoctonia Canker and Black Scurf* (Rhizoctonia solani)**

### *Causal Agent*

*Rhizoctonia solani* Kühn (teleomorph—*Thanatephorus cucumeris* (A.B. Frank) Donk. is a soilborne fungus with melanized mycelium and sclerotia that survive in crop debris, tuber surfaces, and soil. The mycelium is tan to dark brown with multinucleate, large hyphae (8-10 μm in diameter) that branch at right angles with hyphal narrowing at the base of branches and a prominent septal pore apparatus. Hyphal fusion is common and isolates are grouped according to colony hyphal anastamosis groupings (AG). The teleomorph stage involves formation of basidiospores on basidia that form on host tissues (usually stems) near the soil surface. The fungus attacks a very wide variety of graminaceous and dicotyledonous plant species. However, *R. solani* AG-3 is the predominant pathogen of potatoes and is relatively specific to this crop.

*Distribution and Impact*

Canker and black scurf occur in all potato-production areas around the world. The disease affects sprout, stem, root, stolon, and tuber development resulting in crop losses, mainly due to poor tuber appearance because of the presence of the disease on tuber surfaces. This is particularly important in the seed and tablestock industries. In severe cases, stem and stolon canker can lead to yield losses as well.

*Symptoms*

Along the surface or at the end of shoots, reddish-brown lesions form that can prune young stems as the lesion extends around and through the tissue. Weakened, stunted plants may be killed by stem girdling (Carling et al., 1989). Severe sprout infection causes delayed plant emergence and uneven stands of weakened plants. Continued stem attack leads to sunken cankers girdling stem bases, premature death of smaller stems, chlorosis of main stems, upward curling and purpling of leaves, and the formation of aerial tubers. A white, powdery mold may sometimes appear at stem bases with cankers at or below the soil surface. These are signs of the sexual phase of the fungus. Brown to reddish-brown lesions can occur along the roots or at root ends resulting in pruning of roots as the lesion extends around and through the tissue. This reduces the root mass and function preventing normal plant growth and development. Similar disease symptoms are found along the surface or at the end of stolons, resulting in pruning of stolons as the lesion extends around and through the tissue. This can cause tuber abortion or malformed tubers. Tuber black scurf is seen on the surface of washed or soil-free potatoes. Black or dark brown sclerotia form on the surface of the tuber shortly before harvest and enlarge at and shortly after harvest. Dark reddish-brown mycelial strands or networks can also occur on the surface of the tuber. This does not cause extensive damage to the tuber tissues below the skin and can often be scraped from the tuber leaving no disease symptoms. Diseased roots, stolons, and haulms can result in misshapen tubers, tuber cracking, tuber skin surface pitting or netting, and stolon-end necrosis that affect crop growth and yield (Platt, 1989; Simons and Gilligan, 1997).

### *Epidemiology*

Soilborne populations of *R. solani* are high in fields with little or no crop rotation and planting of infested tubers can further increase disease risk. Cool, moist soils enhance fungal growth and limit crop development, which leads to increased disease. Sclerotia and melanized mycelium can germinate and attack young sprouts, roots, and stolons. Similarly, wounded, stressed, or diseased tissues can also become infected during the growing season. During tuber development, particularly as the harvest period nears, sclerotia and melanized mycelium adhere to tuber surfaces, which are sources of inoculum for disease the following year.

### *Disease Management*

Effective and successful management of disease caused by *R. solani* must include a number of control strategies that reduce levels of the pertinent strains of the fungus in the soil and eliminate or limit the occurrence and disease potential of sclerotia on tuber surfaces. Currently no commercially acceptable potato cultivars are resistant to the disease. Use of disease-free seed reduces the disease risk in the crop. In general, seed with less than 5 percent surface coverage of sclerotia should be planted. Fungicides are available for disease prevention but none have curative activity (Hide and Cayley, 1982). Seed treatments (e.g., thiabendazole, fludioxinil, tolclofos-methyl, pencycuron) will limit the damage caused by sclerotia and mycelium of the fungus on the seed tubers but none are 100 percent effective. Soilborne inoculum is also important (Scholte, 1989) and various treatments (e.g., pentachloro-nitrobenzene, tolclofos-methyl, pencycuron) are used in some countries to reduce inoculum levels in the soil (Wicks et al., 1996), but community acceptance and the cost-effectiveness of these measures is not always favorable. Some in-furrow chemical treatments with reduced environmental risk have recently become available to help in the management of soilborne inoculum. In addition, various organic soil amendments and biocontrol treatments have shown promise for disease control in some regions (Prasad and Gupta, 2002). Long-term crop rotations reduce soil-borne levels of the fungus (Leach et al., 1993). This involves production of potatoes once every 3-4 years and inclusion of cereals, grasses, and other less susceptible crops. Cropping systems that employ minimum tillage practices and three-year rotations have been shown to be more effective in reducing disease severity than shorter rotations managed with conventional tillage. Molecular, DNA-based techniques have also been developed to detect and quantify pathogen propagules in soil (Lees et al., 2002), and this information aids in the development of disease management strategies. *Rhizoctonia solani* is an opportunistic fungus and favored by conditions that slow plant growth or result in weakened plants. Disease management strategies that improve plant growth and health will lessen disease risk and impact. Seed should be warmed, green-sprouted, and planted in warm and well-drained soils. Limiting the depth of the hill (ridge) to encourage rapid emergence of healthy plants reduces early-season disease occurrence and severity. Separation of haulm, roots, and stolons from tubers prior to harvest (e.g., by haulm pulling) and limiting the time tubers remain in the soil after plants begin to mature reduces black scurf development. Early harvest reduces black scurf levels whereas delaying harvest of mature tubers or storage of potatoes coated with soil enhances disease occurrence. Biological control of *R. solani* by application of *Verticillium biguttatum* or *Thrichoderma harzianum* is another option.

## *Silver Scurf* (Helminthosporium solani)

### *Causal Agent*

*Helminthosporium solani* Dur. & Mont. (syn. *Spondylocladium atrovirens* Harz.) has typical hyaline, septate mycelium that turns brown when mature. Conidia are brown, cylindrical, 7-8 × 18-64 μm in size, multiseptate (4-8) and borne in whorls at the distal end of conidiophorous cells. These spores are often formed at the edge of necrotic tissue, which when released, survive in soil and crop debris for an unknown period. The pathogen appears to infect only potatoes.

### *Distribution and Impact*

The pathogen is found in most potato-production areas and disease incidence has dramatically increased in the past number of years (Errampalli et al., 2001). Crop growth and tuber yields are not normally affected by the disease (Jellis and Taylor, 1977). However, tuber-quality reductions have led to significant market losses for seed, table, or ware, and processing potato producers.

### *Symptoms*

No symptoms are normally observed on aboveground parts and roots, but stolons and tubers typically exhibit surface lesions at and after tuber initiation. Lesion development is slow prior to harvest. After harvest, lesions continue to develop and spores that are often formed at the edge of the necrotic tissue can spread via air circulation systems during the storage period. Symptoms are most evident on red-skinned cultivars, but can occur on all commercial cultivars. Lesions are tan to gray and appear silvery when moist. Lesions can coalesce to cover much of the tuber surface. Severe disease results in tuber-water loss and subsequent shrinkage when in storage.

### *Epidemiology*

Disease initiation and development is favored by moist conditions. Seed tuberborne or soilborne conidia germinate and the pathogen enters lenticels and wounded periderm of stolons and tubers at tuber initiation and thereafter. Lesions on which conidia are produced are additional sources of inoculum throughout the growing season. After harvest, high humidity during the storage enhances lesion development, spore production, and spread and initiation of new infections.

*Disease Management*

Disease-free seed and selection of fields without a history of disease are essential for reducing disease risk. Crop rotation may reduce levels of soilborne inoculum. Resistant cultivars are generally not available. Fungicides (benomyl, captan, imazalil, and thiabendazole, etc.) for seed treatment are available but efforts to properly manage the use of fungicide treatments is vital as resistance to thiabendazole is now widespread (Carnegie et al., 1994; Kawchuk et al., 1994, 2002; Platt, 1997). DNA-based techniques have been developed to detect and quantify pathogen propagules and to rapidly distinguish thiabendazole-resistant strains (Cullen et al., 2001; Saunders and Errampalli, 2001). No other practical land or crop management methods or treatments are currently available to reduce or prevent silver scurf during the growing season. At the end of the growing season, disease levels can be reduced by harvesting as soon as possible to avoid tuber exposure to the pathogen in the soil. Similarly, maintaining dry conditions from the beginning in the storage reduces pathogen spread and disease development. Use of chlorine-based air-circulation treatments for stored potatoes has also provided control in some cases (Tweddell et al., 2003), but results have been variable, often due to the difficulty of distributing the active ingredient in the proper concentration through the pile. Postharvest application of organic or inorganic salts has shown potential for disease control in some studies (Hervieux et al., 2002).

***Skin Spot* (Polyscytalum pustulans)**

*Causal Agent*

*Polyscytalum pustulans* (Owen & Wakef.) M.B. Ellis (syn. *Oospora pustulans* Owen & Wakef.) microsclerotia survive in soil and crop debris in fields and stores for 5-8 years. Conidia are cylindrical, 6-18 × 2-3 μm in size, unicellular or with a single septum, and borne in chains, on light brown conidiophores that are up to 140 μm in length. Mycelial colonies are gray in appearance and microsclerotia are up to 1 mm in diameter. The fungus causes tuber skin spots on potato tubers and brown lesions on other solanaceous plants.

### *Distribution and Impact*

The fungus can be found in most cool temperate potato-production areas, but is most prevalent in northern Europe (Read et al., 1995). Tuber skin spots are primary inoculum reservoirs, reduce tuber quality as seed, increase tuber peeling requirements for potato processors, and decrease the appearance and marketability of table or ware potatoes. The disease can also cause tuber eye injury resulting in nonemergence or potato stem disease and poor crop growth and reduced tuber size and appearance.

### *Symptoms*

The fungus infects roots, stems, and stolons causing small, superficial, light brown lesions that coalesce over time. Postharvest tuber lesions are black to dark purple, slightly raised spots about 2 mm in diameter.

### *Epidemiology*

Cool (<15°C), wet growing conditions enhance fungal activity and disease occurrence, and heavy clay loam soils are often associated with more disease than light, sandy soils. Disease is initiated the following growing season predominantly via infected seed tubers; however, soils and crop debris also harbor the pathogen. Roots, stems, and stolons are affected after planting and during the growing season. Conidia germinate and hyphae enter through lenticels, eyes, and skin abrasions, causing cortical tissues to become detached. Tubers become infected prior to, during, and after harvest. Stored, infected tubers are sources of conidia for further disease initiation and development during the storage period.

### *Disease Management*

Control of skin spot begins with the use of disease-free seed and fields without a history of the disease or with at least a five-year rotation. Planting when conditions favor rapid plant growth enhances the development of healthy plants. Harvesting when tuber skin is mature enough, reducing tuber wounding, and fostering conditions that are warm and dry minimize tuber infections. Storage cleanliness reduces sources of inoculum. Initial storage period conditions that enhance tuber wound healing and skin maturation will reduce disease problems. Various chemical and nonchemical treatments have been utilized for disease control (Carnegie et al., 1994; Dashwood et al., 1991; Hide and Cayley, 1987). Postharvest tuber treatments (2-aminobutane, benomyl, imazalil, thiabendazole, etc.) can also be used to limit disease during storage. Fungicide spray applications are preferable to dips to minimize moist conditions that increase disease. In addition, warm (>15°C) and dry (<75 percent RH) conditions in storage reduce tuber infections. Most commercially produced cultivars are susceptible to disease and thin-skinned cultivars tend to be more susceptible than thick-skinned ones (Carnegie and Cameron, 1983; Wastie et al., 1997).

## Verticillium *Wilts (*Verticillium *species)*

### *Causal Agents*

*Verticillium albo-atrum* Reinke & Berth., *V. dahliae* Kleb., and *V. tricorpus* (Issac) are vascular pathogens that can cause wilt in potatoes (Heale, 1988; Pegg, 1974). *Verticillium albo-atrum* and *V. dahliae* are more aggressive and destructive than *V. tricorpus,* but all three are found in soil, crop debris, and infected or infested potato tubers. A fourth, putative species has been found that is morphologically similar to *V. albo-atrum* or *V. tricorpus,* but genetically distinct from either of them (Mahuku and Platt, 2002). Conidiophores form in whorls on hyaline mycelium and produce numerous cylindrical conidia, 3-6 × 2-3 μm in size. Conidia are short-lived compared to the dark mycelium also produced by *V. albo-atrum* and *V. tricorpus* or the microsclerotia (30-90 μm in size) produced by *V. dahliae* and *V. tricorpus*, which can survive for months to years in soil. Soil type, temperature, microbial activity, and other factors greatly affect survival of these structures, which germinate to produce mycelium and then conidia that can be carried by equipment, wind, and water from field to field.

### *Distribution and Impact*

All three species are found in field soils. Some or all of these species may exist in a single infected plant or area in a field, but often only one will predominate due to the effects of climate and soil conditions. *Verticillium albo-atrum* predominates in cool (16-27°C), moist, and heavy soils, while *V. dahliae* predominates in warm (22-27°C), dry, and light soils. *Verticillium tricorpus* occurs over a range of conditions similar to those for the other two species. These pathogens affect most dicotyledonous plant species, although many infected hosts can remain symptomless. Grasses are nonhosts. The disease results in premature plant death, quality reductions, and production losses of 50-60 percent.

### *Symptoms*

Plant symptoms include wilting, chlorosis, necrosis, and premature death within 40-60 days of planting. Stem symptoms often begin at the base and progress upward as the water-supply system of the plant becomes plugged. Leaflet curling, interveinal chlorosis, and edge necrosis is often observed as is vascular discoloration of the stems, leaflets, roots, stolons, and tubers. In severe cases, tuber pitting or blotching of the pith area occurs.

### *Epidemiology*

Planting diseased seed or soilborne inoculum are the primary sources of the disease. Infection is generally through root hairs, root wounds, and sprouts or leaf surfaces where hyphae invade both intra- and intercellularly to reach the xylem in the vascular tissues. Pathogen growth results in the production of conidia and substances that plug the vascular system, resulting in restricted water movement, wilting, and death. Dark mycelium and microsclerotia formation in these diseased potato plant parts and tubers provide a means of survival until the next host crop. In addition, coinfections with root lesion nematodes *(Pratylenchus penetrans)* have been implicated, especially for *V. dahliae.*

*Disease Management*

Most commercially acceptable cultivars are susceptible to Verticillium wilt, but some resistant cultivars (e.g., Desiree, Ranger Russet, Reddale) are available (Powelson and Rowe, 1993). However, many resistant cultivars often succumb to high inoculum levels and may be resistant to only one of the pathogens (Platt, 1986). No single disease control method is 100 percent effective and an integrated disease management approach is required (Rowe et al., 1987). Fungicides for seed treatment (e.g., benomyl, captan, mancozeb, thiophanate-methyl) and soil fumigants (e.g., 1,3-dichloropropene, metham sodium) are available (Easton et al., 1992). Various organic soil amendments, green manures, and other treatments have shown promise for disease control in some regions (Davis et al., 1996; Mol et al., 1996; Soltani et al., 2002; Tenuta et al., 2002). Integrated disease management strategies involve methods that reduce initial inoculum levels over the long term. The use of healthy seed is vital for disease prevention and for reducing primary inoculum sources. This involves selection or testing of seed to eliminate contaminated (pathogen in soil on tuber) and infected (pathogen inside) seed tubers. Molecular techniques have been developed to detect and quantify pathogen propagules of all three *Verticillium* spp. in plant tissues and soils, which aid in the management of wilt (Mahuku et al., 1999; Heinz and Platt, 2000). Fields with a history of disease or soil with large numbers of pathogen propagules should be avoided. Crop rotation restricting cultivation of potatoes and other host crops once every three or four years is very important. Similarly, destruction of "volunteer" plants or "groundkeepers" will eliminate inoculum sources. Cultivation of nonhost crops, such as grasses, cereals, or legumes and control of susceptible weed species such as *Capsella bursa-pastoris, Chenopodium album, Equisetum arvense,* and *Taraxacum* spp. will reduce or eliminate inoculum sources. Disposal of crop debris by removal from the field and their composting or burning will reduce soilborne inoculum levels.

### *Wart* (Synchytrium endobioticum)

#### *Causal Agent*

*Synchytrium endobioticum* (Schilb.) Perc. is a nonhyphae producing pathogen that forms sori, sporangia, and zoospores. In wet conditions, zoospores infect plant tissues and swell into pro-sori that eventually develop into haploid sori inside the host cells. Each sorus contains 1-10 sporangia that are spheroid, 35-80 μm in diameter, golden in color, and thick-walled. Germinating sporangia release haploid zoospores that are pear-shaped, 1.5-2.2 μm in diameter, and motile with a single posterior flagellum. Zoospores can conjugate and form a zygote that develops into a resting sporangium. Resting sporangia are golden-brown with prominent exterior ridges and are capable of surviving for long periods of time (i.e., more than 40 years) under harsh conditions in fields and warehouses or on surfaces of tubers, equipment, etc. A number of physiologic races of this pathogen are known. The pathogen may have originated in the ancestral, Andean home of the potato.

#### *Distribution and Impact*

Wart has been reported in Africa, Asia, Europe and North and South America following its discovery during the late 1890s in parts of Europe (Hampson, 1993). At present, the disease is found in limited and restricted areas in all continents except Australia. The pathogen can attack a number of solanaceous hosts, but is most frequently associated with potato. Though the pathogen does not spread rapidly or cannot cause extensive crop losses during a single growing season, it is a recognized quarantined pest that causes significant negative impacts to potato farms and/or farming areas in which it is found. Tuber quality is seriously affected due to the presence of galls and the occurrence of poor tuber shape and appearance.

*Symptoms*

Even though aboveground plant symptoms are often absent, plants may have less vigorous growth and slightly darker foliage. Infrequently, green to brown galls on upper stems, leaves, and flowers have been reported. The disease is most noted for the white to tan galls that are formed at the stem base at the soil-line or below ground. These galls can also be found on stolons and tubers, but not normally on roots. Galls can range from "pea-size" to that of the "fist" of a human.

*Epidemiology*

Moist conditions are required for the release of zoospores from resting or newly formed sporangia. Soil type, pH, and texture are not very influential, but temperatures ranging from 12°C to 24°C favor infection. Zoospores infect epidermal cells of susceptible host tissues such as sprouts, stolons, and tubers. The host cells surrounding the infected ones display hyperplasia and hypertrophy that leads to the development of the sporangia-laden galls or warts. Sporangia are released into the soil as plant tissues decay. The pathogen is spread via soil and plant debris carried by wind, humans, farm equipments, tubers, etc. (Hampson and Wood, 1997).

*Disease Management*

Control of wart is primarily through quarantine of the infested area and infected hosts to prevent pathogen spread. Eradication of the pathogen is very difficult and expensive. Crushed crab shell soil amendments may reduce inoculum levels in infested soils (Hampson and Coombes, 1995). Efforts are under way in most potato-production areas that deal with wart to develop resistant cultivars based on major gene resistance (Gopal, 1993), and cultivars immune to local races are already available (Phadtare et al., 2000). Molecular pathogen detection and quantification techniques have recently been developed to augment more traditional techniques for quantification of propagules in soil.

## *White Mold (*Sclerotinia sclerotiorum *and Other* Sclerotinia *species)*

### *Causal Agents*

*Sclerotinia sclerotiorum* (Lib.) de Bary (syn. *Whetzelinia sclerotiorum* (Lib.) Korf. & Dumont), *S. minor* Jagger and *S. intermedia* Ramsey are recognized causal agents of white mold, although *S. sclerotiorum* is the predominant pathogen. The fungus produces white mycelium and hard, black resting sclerotia of variable shape and size. Sclerotia can remain viable in soils for 7-10 years. Near soil surfaces, flat or funnel-shaped apothecia that are white, pink, or tan colored and 0.5 cm in diameter are formed on sclerotia. Eight-spore, hyaline, cylindroclavate asci, 80-250 × 4-23 μm in size, are formed with single-celled, ovoid, hyaline ascospores, 6-28 x 2-15.2 μm in size. Ascospores are produced over several weeks and dispersed by wind and soil.

### *Distribution and Impact*

White mold is predominantly found in cool temperate regions. The fungus has a long survival period in soils and crop debris and under cool, moist conditions spreads quickly via airborne or soilborne routes throughout the growing season. The disease can lead to poor crop growth, reduced yield, and tuber quality, and production area declines as fungal sclerotia can infest soils for 7-10 years. The primary pathogen, *S. sclerotiorum*, can attack a very wide variety of dicotyledonous hosts including many weed and solanaceous species.

### *Symptoms*

Infected potato plants express symptoms on stems and only rarely on roots, stolons, or tubers. Potato stems exhibit water-soaked lesions most often at stem bases or where stems are in contact with the soil. White mycelial mats are visible that darken as they mature and dry.

Disease is also found at petiole junctures or wounded areas of the stem. As the disease develops and stems mature, mycelium and sclerotia can be found in the hollow stems.

*Epidemiology*

The pathogen can overwinter in dead or slow-growing plants but primary inoculum is most often from sclerotia. Sclerotia can germinate directly, but generally disease is initiated by the production of apothecia and ascospores that are transported by wind to new host sites and germinate to infect plant tissues. The fungus usually colonizes senescing tissues prior to infection of healthy tissue. These processes are favored by high temperature (28-30°C) and high relative humidity (>90 percent).

*Disease Management*

Rotation with graminaceous crops for 4-7 years can reduce soilborne inoculum levels. Flooding fields to destroy sclerotia is another control measure. Properly timed foliar fungicide (benomyl, fluazinam, iprodione, etc.) applications can reduce disease levels, but repeated sprays coinciding with ascospore releases are needed. Fungicide usage requires care to avoid fungicide-resistance development as has been reported for benomyl. Resistant cultivars are not currently available. Management of crop nutrition and irrigation to avoid excessive canopies and prolonged leaf wetness can aid in disease management. New biocontrol and cultural measures are also being developed (Gerlagh et al., 1999; Lazarovits et al., 2000; Li et al., 2003; Novikova et al., 2003; Van Loenen et al., 2003).

In conclusion, fungal diseases of potato, as a group, comprise the greatest threats to crop production, and management of these diseases is one of the most costly components of potato farming. In addition, due to the inability to fully "see" and understand the complexity of the soil ecosystem and the disease-causing agents that reside there, attempts to deal with these diseases involve great effort, expense, and frustration. A glance at the management strategies outlined for the various diseases presented in this chapter quickly reveals some common themes. Molecular biology is playing an increasing role in the characterization, detection, and quantification of plant pathogens and, from a disease management perspective, will allow the identification of contaminated fields and seed-lots. Cultural practices that include crop rotation, tillage management practices, and organic soil amendments including green manures have been engineered to stimulate the development of disease-suppressive soils and as such are becoming increasingly important as part of an integrated pest management program. The registration of specific biological agents that promote crop growth and suppress disease is finally occurring, particularly for those that can be applied to the seed-piece or to the harvested tubers prior to storage. Increased knowledge of pathogen biology and the epidemiology of disease has allowed basic farm practices such as time of planting and harvest, application of nutrients or irrigation water, etc. to be more finely tuned to exploit vulnerable phases of pathogen life cycles. Improvements in storage technology have allowed storage conditions to be quickly adjusted to manage postharvest problems. Chemical control is still extremely important, but is shifting to the use of "reduced risk" products that target specific pathogens and are less toxic in the environment. Unfortunately, the availability of commercially viable cultivars with multiple disease-resistance, a potentially important tool in the management of many diseases discussed in this chapter, is still quite limited.

With the ever-increasing influence of "community" and "environment" on potato farming, constant cooperative efforts are required to ensure that agriculturally sustainable, environmentally benign, and community-friendly methods are adopted to manage these diseases. The era of "chemical" control as an exclusive option has passed and the availability of high-tech, multi-tool, integrated disease management strategies is critical for long-term, sustainable farming systems.

## REFERENCES

Adams, M.J., Read, P.J., Lapwood, D.H., Cayley, G.R., and Hide, G.A. (1987). The effect of irrigation on powdery scab and other tuber diseases of potatoes. *Annals of Applied Biology 110:*287-294.

Andrivon, D., Lucas, J.M., Guérin, C., and Jouan, B. (1998). Colonization of roots, stolons, tubers and stems of various potato *(Solanum tuberosum)* cultivars by the black-dot fungus *Colletotrichum coccodes. Plant Pathology 47:*440-445.

Bell, K.S., Roberts, J., Verrall, S., Cullen, D.W., Williams, N.A., Harrison, J.G., Toth, I.K., Cooke, D.E.L., Duncan, J.M., and Claxton, J.R. (1999). Detection and quantification of *Spongospora subterranea* f.sp. *subterranea* in soils and on tubers using specific PCR primers. *European Journal of Plant Pathology 105:*905-915.

Boyd, A.E.W. (1952). Dry rot disease of the potato. *Annals of Applied Biology 39:*322-357.

Carling, D.E., Leiner, R.H., and Westphale, P.C. (1989). Symptoms, signs and yield reduction associated with Rhizoctonia disease of potato induced by tuberborne inoculum of *Rhizoctonia solani* AG-3. *American Potato Journal 66:*693-701.

Carnegie, S.F., Cameron, A.H., and Haddon, P. (2001). The effect of date of haulm destruction and harvest on the development of dry rot caused by *Fusarium solani* var. *coeruleum* on potato tubers. *Annals of Applied Biology 139:*209-216.

Carnegie, S.F. and Cameron, A.M. (1983). Testing potato cultivars for susceptibility to skin spot *(Polyscytalum pustulans). Potato Research 26:*69-72.

Carnegie, S.F., Cameron, A.M., Hide, G.A., and Hall, S.M. (1994). The occurrence of thiabendazole-resistant isolates of *Polyscytalum pustulans* and *Helminthosporium solani* on seed potato tubers in relation to fungicide treatment and storage. *Plant Pathology 43:*961-971.

Carnegie, S.F., Hide, G.A., Ruthven, A.D., Lindsay, D.A., and Adams, M.J. (1988). Control of potato gangrene by thiabendazole in relation to time of harvest after haulm destruction and site of production. *Annals of Applied Biology 113:*471-481.

Coley-Smith, J.R., Verhoeff, K., and Jarvis, W.R. (1980). *The Biology of Botrytis.* London, UK: Academic Press.

Cullen, D.W., Lees, A.K., Toth, I.K., and Duncan, J.M. (2001). Conventional PCR and real-time quantitative PCR detection of *Helminthosporium solani* in soil and on potato tubers. *European Journal of Plant Pathology 107:*387-398.

Cullen, D.W., Lees, A.K., Toth, I.K., and Duncan, J.M. (2002). Detection of *Colletotrichum coccodes* from soil and potato tubers by conventional and quantitative real-time PCR. *Plant Pathology 51:*281-292.

Dashwood, E.P., Burnett, E.M., and Perombelon, M.C.M. (1991). Effect of a continuous hot water treatment of potato tubers on seed-borne fungal pathogens. *Potato Research 34:*71-78.

Davis, J.R., Huisman, O.C., Westermann, D.T., Hafez, S.L., Everson, D.O., Sorensen, L.H., and Schneider, A.T. (1996). Effects of green manures on Verticillium wilt of potato. *Phytopathology 86:*444-453.

Denner, F.D.N., Millard, C.P., and Wehner, F.C. (1998). The effect of seed- and soil-borne inoculum of *Colletotrichum coccodes* on the incidence of black dot on potatoes. *Potato Research 41:*51-56.

Dillard, H.R. (1988). Influence of temperature, pH, osmotic potential and fungicide sensitivity on germination of conidia of *Colletotrichum coccodes* in vitro. *Phytopathology 78:*1357-1361.

Easton, G.D., Nagle, M.E., and Seymour, M.D. (1992). Potato production and incidence of *Verticillium dahliae* following rotation to nonhost crops and soil fumigation in the state of Washington. *American Potato Journal 69:*489-502.

Errampalli, D., Saunders, J.M., and Holley, J.D. (2001). Emergence of silver scurf *(Helminthosporium solani)* as an economically important disease of potato. *Plant Pathology 50:*141-153.

Erwins, D.C., and Ribeiro, O.K. (1996). *Phytophthora Diseases Worldwide.* St. Paul, MN: APS Press.

Evans, I.R. and Howard, R.J. (1994). Pink rot. In Howard, R.J., Garland, J.A., and Seaman, W.L. (Eds.), *Diseases and Pests of Vegetable Crops in Canada: An Illustrated Compendium.* Canada: The Canadian Phytopathological Society and Entomological Society of Canada.

Gerlagh, M., Goossen-Van De Geijn, H.M., Fokkema, N.J., and Vereijken, P.F.G. (1999). Long-term biosanitation by application of *Coniothyrium minitans* on *Sclerotinia sclerotiorum* infected crops. *Phytopathology 89:*141-147.

Gopal, J. (1993). Potato breeding for North-Bengal hills. *Journal of Hill Research 6:*77-80.

———. (1998). Heterosis and combining ability analysis for resistance to early blight *(Alternaria solani)* in potato. *Potato Research 41:*311-317.

Hampson, M.C. (1993). History, biology and control of potato wart disease in Canada. *Canadian Journal of Plant Pathology 15:*223-244.

Hampson, M.C. and Coombes, J.W. (1995). Reduction of potato wart with crushed crabshell: Suppression or eradication? *Canadian Journal of Plant Pathology 17:*69-74.

Hampson, M.C. and Wood, S.L. (1997). Detection of infective resting spores of *Synchytrium endobioticum* in vehicles. *Canadian Journal of Plant Pathology 19:*57-59.

Harrison, J.G., Searle, R.J., and Williams, N.A. (1997). Powdery scab disease of potato—a review. *Plant Pathology 46:*1-45.

Heale, J.B. (1988). *Verticillium* species, the cause of vascular wilts in many species. *Advances in Plant Pathology 6:*291-312.

Heinz, R.A. and Platt, H.W. (2000). Improved DNA extraction method for *Verticillium* detection and quantification in large-scale studies using PCR-based techniques. *Canadian Journal of Plant Pathology 22:*117-121.

Hervieux, V., Yaganza, E.S., Arul, J., and Tweddell, R.J. (2002). Effect of organic and inorganic salts on the development of *Helminthosporium solani,* the causal agent of potato silver scurf. *Plant Disease 86:*1014-1018.

Hide, G.A. and Cayley, G.R. (1982). Chemical techniques for the control of stem canker and black scurf *(Rhizoctonia solani)* disease of potatoes. *Annals of Applied Biology 100:*105-116.

———. (1987). Effects of delaying fungicide treatment and of curing and chlorpropham on the incidence of skin spot on stored potatoes. *Annals of Applied Biology 110:*617-627.

———. (1989). Factors influencing the control of potato gangrene by fungicide treatment. *Potato Research 32:*91-99.

Hide, G.A., Read, P.J., and Hall, S.M. (1992). Resistance to thiabendazole in *Fusarium* species isolated from potato tubers affected with dry rot. *Plant Pathology 41:*745-748.

Ho, H.H. and Jong, S.C. (1989). *Phytophthora erythroseptica. Mycotaxon 36:*73-90.

Jarvis, W.R. (1977). *Botryotinia* and *Botrytis* species: Taxonomy, physiology, and pathogenicity: A guide to the literature. *Canada Department of Agriculture, Monograph* No. 15.

Jellis, G.J. and Taylor, G.S. (1977). The development of silver scurf *(Helminthosporium solani)* disease of potato. *Annals of Applied Biology 86:*19-28.

Jones, R.A.C. and Harrison, B.D. (1969). The behaviour of potato mop-top virus in soil and evidence for its transmission by *Spongospora subterranea* (Wallr.) Lagerh. *Annals of Applied Biology 63:*1-17.

Kawchuk, L.M., Holley, J.D., Lynch, D.R., and Clear, R.M. (1994). Resistance to thiabendazole and thiophanate-methyl in Canadian isolates of *Fusarium sambucinum* and *Helminthosporium solani. American Potato Journal 71:*185-192.

Kawchuk, L.M., Hutchison, L.J., Verhaeghe, C.A., Lynch, D.R., Bains, P.S., and Holley, J.D. (2002). Isolation of the β-tubulin gene and characterization of thiabendazole resistance in *Gibberella pulicaris. Canadian Journal of Plant Pathology 24:*233-238.

Lazarovits, G., Starratt, A.N., and Huang, H.C. (2000). The effect of tricyclazole and culture medium on production of the melanin precursor 1,8-dihydroxynaphthalene by *Sclerotinia sclerotiorum* isolate Ss7. *Pesticide Biochemistry and Physiology 67:*54-62.

Leach, S.S., Porter, G.A., Rourke, R.V., and Clapham, W.M. (1993). Effects of moldboard plowing, chisel plowing and rotation crops on the Rhizoctonia disease of white potato. *American Potato Journal 70:*329-337.

Lees, A.K., Cullen, D.W., Sullivan, L., and Nicolson, M.J. (2002). Development of conventional and quantitative real-time PCR assays for the detection and identification of *Rhizoctonia solani* AG-3 in potato and soil. *Plant Pathology 51:*293-302.

Li, G.Q., Huang, H.C., and Acharya, S.N. (2003). Antagonism and biocontrol potential of *Ulocladium atrum* on *Sclerotinia sclerotiorum. Biological Control 28:*11-18.

Lonsdale, D., Cunliffe, C., and Epton, H.A.S. (1980). Possible routes of entry of *Phytophthora erythroseptica* Pethyb. and its growth within potato plants. *Phytopathologische Zeitschrift 97:*109-117.

Mahuku, G.S. and Platt, H.W. (2002). Molecular evidence that *Verticillium albo-atrum* Grp 2 isolates are distinct from *V. albo-atrum* Grp 1 and *V. tricorpus. Molecular Plant Pathology 3:*71-79.

Mahuku, G.S., Platt, H.W., and Maxwell, P. (1999). Comparison of polymerase chain reaction based methods with plating on media to detect and identify Verticillium wilt pathogens of potato. *Canadian Journal of Plant Pathology 21:*125-131.

Marais, L. (1990). Efficacy of fungicides against *Colletotrichum coccodes* on potato tubers. *Potato Research 33:*275-281.

Mol, L., van Halteren, J.M., Scholte, K., and Struik, P.C. (1996). Effects of crop species, crop cultivars and isolates of *Verticillium dahliae* on the populations of microsclerotia in the soil and consequences for crop yield. *Plant Pathology 45:*205-214.

Novikova, I.I., Litvinenko, A.I., Boikova, I.V., Yaroshenko, V.A., and Kalko, G.V. (2003). Biological activity of new microbiological preparations alirins B and S designed for plant protection against diseases. I. Biological activity of alirins against diseases of vegetable crops and potato. *Mikologiya I Fitopatologiya 37:*92-98.

Okhovat, M., Zafari, D.M., Karimi-Roozbahani, A.R., and Rohani. H. (1994). Study on biological control of black dot disease of potato by *Trichoderma* Pers. ex Fr. *Iranian Journal of Agricultural Sciences 25:*35-42.

Pegg, G.F. (1974). Verticillium diseases. *Annual Review of Plant Pathology 53:*157-182.

Peters, R.D. and Sturz, A.V. (2001). A rapid technique for the evaluation of potato germplasm for susceptibility to pink rot. *Plant Disease 85:*833-837.

Peters, R.D., Sturz, A.V., and Arsenault, W.J. (2003). Use of mefenoxam to control pink rot (*Phytophthora erythroseptica* Pethyb.) of potato in Prince Edward Island. *Canadian Journal of Plant Pathology 25:*33-40.

Phadare, S.G., Gopal, J., and Singh, P.H. (2000). Kufri Kanchan-a wart immune and late blight resistant red tuber potato variety. *Journal of Indian Potato Association 27:*113-118.

Platt, H.W. (1983). Effects of metalaxyl, mancozeb, and chlorothalonil on blight, yield, and tuber rot of potato. *Canadian Journal of Plant Pathology 5:*38-42.

———. (1986). Varietal response and crop loss to Verticillium wilt of potato caused by *V. albo-atrum. Phytoprotection 67:*123-127.

———. (1989). Potato growth and tuber production as affected by inoculation of cut and whole seed with *Rhizoctonia solani* (AG-3) and the use of seed treatment fungicides. *American Potato Journal 66:*365-378.

———. (1994). Foliar application of fungicides affects occurrence of potato tuber rots caused by four foliar pathogens. *Canadian Journal of Plant Pathology 16:*341-346.

———. (1997). Resistance to thiabendazole in *Fusarium* species and *Helminthosporium solani* in potato tubers treated commercially in eastern Canada. *Phytoprotection 78:*1-10.

Powelson, M.L. and Rowe, R.C. (1993). Biology and management of early dying of potatoes. *Annual Review of Phytopathology 31:*111-126.

Prasad, C.S. and Gupta, V. (2002). Studies on bio-efficacy of *Trichoderma harzianum* in management of stem rot of potato *(Rhizoctonia solani). National Academy Science Letters 25:*357-359.

Priou, S., Triki, M.A., El Mahjoub, M., Fahem, M., and Souibgui, M. (1997). Assessing potato cultivars in Tunisia for susceptibility to leak caused by *Pythium aphanidermatum. Potato Research 40:*399-406.

Read, P.J. and Hide, G.A. (1995a). Development of black dot disease (*Colletotrichum coccodes* (Wallr.) Hughes) and its effects on the growth and yield of potato plants. *Annals of Applied Biology 127:*57-72.

———. (1995b). Effects of fungicides on the growth and conidial germination of *Colletotrichum coccodes* and on the development of black dot disease of potatoes. *Annals of Applied Biology 126:*437-447.

Read, P.J., Storey, R.M.J., and Hudson, D.R. (1995). A survey of black dot and other fungal tuber blemishing diseases in British potato crops at harvest. *Annals of Applied Biology 126:*249-258.

Rowe, R.C., Davis, J.R., Powelson, M.L., and Rouse, D.I. (1987). Potato early dying: Causal agents and management strategies. *Plant Disease 71:*482-489.

Salas, B., Secor, G.A., Taylor, R.J., and Gudmestad, N.C. (2003). Assessment of resistance of tubers of potato cultivars to *Phytophthora erythroseptica* and *Pythium ultimum. Plant Disease 87:*91-97.

Saunders, J.M. and Errampalli, D. (2001). Comparison of methods to detect resistance of *Helminthosporium solani* to a thiabendazole fungicide. *Phytoprotection 82:*103-112.

Schisler, D.A., Slininger, P.J., Kleinkopf, G., Bothast, R.J., and Ostrowski, R.C. (2000). Biological control of Fusarium dry rot of potato tubers under commercial storage conditions. *American Journal of Potato Research 77:*29-40.

Scholte, K. (1989). Effects of soil-borne *Rhizoctonia solani* Kuhn on yield and quality of ten potato cultivars. *Potato Research 32:*367-376.

Shtienberg, D., Blachinsky, D., Kremer, Y., Beh-Hardor, G., and Dinoor, A. (1995). Integration of genotype and age-related resistances to reduce fungicide use in management of *Alternaria* diseases of cotton and potato. *Phytopathology 85:*995-1002.

Shtienberg, D. and Elad, Y. (1997). Incorporation of weather forecasting in integrated, biological-chemical management of *Botrytis cinerea. Phytopathology 87:*332-340.

Simons, S.A. and Gilligan, C.A. (1997). Relationships between stem canker, stolon canker, black scurf *(Rhizoctonia solani)* and yield of potato *(Solanum tuberosum)* under different agronomic conditions. *Plant Pathology 46:*651-658.

Soltani, N., Conn, K.L., Abbasi, P.A., and Lazarovits, G. (2002). Reduction of potato scab and verticillium wilt with ammonium lignosulfonate soil amendment in four Ontario potato fields. *Canadian Journal of Plant Pathology 24:*332-339.

Stevenson, W.R. (1993). Management of early blight and late blight. In Rowe, R. (Ed.), *Potato Health Management* (pp. 141-148). Minnesota, MN: APS Press.

Taylor, R.J., Salas, B., Secor, G.A., Rivera, V., and Gudmestad, N.C. (2002). Sensitivity of North American isolates of *Phytophthora erythroseptica* and *Pythium ultimum* to mefenoxam (metalaxyl). *Plant Disease 86:*797-802.

Tenuta, M., Conn, K.L., and Lazarovits, G. (2002). Volatile fatty acids in liquid swine manure can kill microsclerotia of *Verticillium dahliae. Phytopathology 92:*548-552.

Tooley, P.W., Carras, M.M., and Lambert, D.H. (1998). Application of a PCR-based test for detection of potato late blight and pink rot in tubers. *American Journal of Potato Research 75:*187-194.

Triki, M.A. and Priou, S. (1997). Using chemical and biological treatments to reduce the potential of potato leak caused by *Pythium aphanidermatum* in Tunisia. *Potato Research 40:*391-398.

Triki, M.A., Priou, S., El Mahjoub, M., and Baudry, A. (2001). Leak syndrome of potato in Tunisia caused by *Pythium aphanidermatum* and *Pythium ultimum. Potato Research 44:*221-231.

Tweddell, R.J., Boulanger, R., and Arul, J. (2003). Effect of chlorine atmospheres on sprouting and development of dry rot, soft rot and silver scurf on potato tubers. *Postharvest Biology and Technology 28:*445-454.

Uribe, E. and Loria, R. (1994). Response of *Colletotrichum coccodes* to fungicides in vitro. *American Potato Journal 71:*455-465.

Van Loenen, M.C.A., Turbett, Y., Mullins, C.E., Feilden, N.E.H., Wilson, M.J., Leifert, C., and Seel, W.E. (2003). Low temperature-short duration steaming of soil kills soil-borne pathogens, nematode pests and weeds. *European Journal of Plant Pathology 109:*993-1002.

Wastie, R.L., Stewart, H.E., Bradshaw, J.E., and Lees, A.K. (1997). Assessing progenies of potato for resistance to skin spot *(Polyscytalum pustulans)* in the glasshouse. *Potato Research 40:* 383-389.

Wicks, T.B., Morgan, B., and Hall, B. (1996). Influence of soil fumigation and seed tuber treatment on the control of *Rhizoctonia solani* on potatoes. *Australian Journal of Experimental Agriculture 36:*339-345.

# Chapter 10

# Bacterial Diseases

R. Lebecka
E. Zimnoch-Guzowska
E. Łojkowska

Potato is afflicted by six important bacterial diseases: bacterial wilt *(Ralstonia solanacearum)*, ring rot *(Clavibacter michiganensis* ssp. *sepedonicus)*, soft rot or blackleg (*Erwinia* spp.), common scab (*Streptomyces* spp.), pink eye (*Pseudomonas* spp.), and leaf spot *(Xanthomonas vesicatoria)*. Pink eye and leaf spot, however, are of minor importance, whereas the other four bacterial diseases can cause substantial losses in potato crops depending on the disease severity, susceptibility of the potato cultivar, and region of cultivation. Bacterial wilt and ring rot are on the quarantine list, and the requirement for zero tolerance for these in seed and ware potatoes, especially exported seed potatoes, leads to severe economic losses (Mills et al., 1997).

This chapter provides general information and a review of research findings on bacterial diseases of potato. In addition to covering their biology, distribution, impact, and epidemiology, it describes various aspects of disease management, including host resistance.

## *BACTERIAL WILT (BROWN ROT)*

### *Causal Agent*

Bacterial wilt or brown rot of potato is caused by *Ralstonia solanacearum* (Smith), formerly called *Pseudomonas solanacearum* (Smith) (Yabuuchi et al., 1995). The pathogen is a gram-negative, rod-shaped, nonspore-forming, noncapsulated, and nitrate-reducing aerobic bacterium. Its 5.8 Mb genome, with two replicons (a 3.7 Mb chromosome and a 2.1 Mb megaplasmid), has been sequenced (Salanoubat et al., 2002). Both replicons have a mosaic structure, indicating the acquisition of genes through horizontal gene transfer.

Based on pathogenicity for specific host plants (Timms-Wilson et al., 2001) and ability to produce acids (Buddenhagen, 1986), five races and five biovars of *R. solanacearum* have been identified. Growth on semiselective media (Englebrecht, 1994), immunofluorescence (IF) staining, enzyme-linked immunosorbent assay (ELISA) (Robinson-Smith et al., 1995), and molecular analysis using polymerase chain reaction (PCR) (Seal et al., 1997; Schaad et al., 2001) have been used to further characterize various isolates. In general, Race 3 is reported to be genetically more homogeneous than Race 1 (French et al., 1997). Fatty acid methyl esters (FAMEs) analysis (Janse, 1996), restriction fragment length polymorphism (RFLP) analysis of DNA regions coding factors involved in virulence and hypersensitive *(hpr)* reaction (Cook and Sequeira, 1991), amplified fragment length polymorphism (AFLP) analysis for *hrp* gene region, sequencing of 16S rDNA (Poussier et al., 2000), and fingerprinting of genomic DNA (Smith et al., 1998) have revealed a high diversity in the population of this pathogen.

### *Distribution and Impact*

*Ralstonia solanacearum* is widespread throughout the world (Janse, 1988; Hayward, 1991). Race 1 (Biovars 3 and 4) is prevalent in the hot tropics, whereas Race 3 (Biovar 2) is restricted to cooler areas. *Ralstonia solanacearum* has an extremely wide host range; some races infect more than 200 species belonging to at least 50 plant families, including several crops such as potato, tomato, chili, eggplant, pepper, and ginger (Hayward, 1991). Within the European Plant Protection Organization (EPPO) region, Race 3 infects potato, tomato, and the weeds *Solanum dulcamara* and *S. nigrum.*

Outbreaks of bacterial wilt have occurred in several European countries: Sweden, Belgium, the United Kingdom, the Netherlands, Italy, France, Germany, Portugal, and Spain (Pastrik and Maiss, 2000). Yield losses as a result of this pathogen have been reported to be as high as 100 percent when potato crops are grown from infected seed tubers in the warm climate of tropical regions such as Africa (Olanya et al., 2002).

### *Symptoms and Detection*

The first symptom is wilting of the leaves at the ends of the branches during the day, with recovery at night. Afterward, streaky brown discoloration of the stems is observed, and leaves may turn bronze (Figure 10.1A). When cut stems are kept in water, bacterial slime exudes from vascular bundles (Smith et al., 1997). Often this is used for the first diagnosis of bacterial wilt in field-grown plants. In cooler, temperate regions the more common symptom is brown staining of the vascular tissue (Stevenson et al., 2001). Tubers mostly show no external symptom except that under certain conditions bacterial slime from the heel end and the eyes, with the development of a brown rot (hence the name "brown rot" for the disease), is observed. Sometimes the symptoms on tubers can be confused with those of ring rot.

A number of methods have been developed to detect latent infection in tubers: dilution plating on a selective medium (Englebrecht, 1994), indirect immunofluorescence antibody staining (IFAS) (Janse, 1988), ELISA (Robinson-Smith et al., 1995), fluorescent in situ hybridization, and bioassay on tomato plant (Anonymous, 1990). Irrespective of the method, the reliability of diagnosis is affected by the inoculum concentration. In several laboratories PCR results were less reliable than those of other methods (Elphinstone et al., 2000). However, Race 3 (Biovar 2) can be detected with real-time PCR (Schaad et al., 2001) within 1-3 hours.

FIGURE 10.1. Typical symptoms of important bacterial diseases (for common scab symptoms please see Figure 9.1). See Plate 10.1 in the color gallery.

## *Epidemiology*

*Ralstonia solanacearum* can survive in soil and water for varying periods. Van Elsas et al. (2000) reported that the density of bacterial populations in the soil after outbreaks of potato brown rot declined progressively over 4-5 months, but bacteria had persisted in two out of three tested fields for 10-12 months. Their study also showed that the bacterium survived longer in deep soil layers than in topsoil. Many weeds, mainly *S. dulcamara* and volunteer potatoes, are important carriers of inoculum into the next growing season (van Elsas et al., 2002). The infection can occur when bacteria in the soil penetrate roots or when the seedborne pathogen moves systematically through the vascular system to stems and daughter tubers. Favorable conditions for disease development are high soil moisture, rainy weather, and temperature between 24 and 35°C. However, different races have different optimum temperature requirements (Swanepol, 1990). In addition to latent infection in seed tubers, irrigation with contaminated water or even carrying infected soil on shoes or tools can lead to the spread of the disease (Kinyua et al., 2001).

### *Disease Management*

#### *Cultural*

Management of bacterial wilt depends on the pathogen's biological properties (temperature requirements, host range) and environmental factors. Bacterial wilt can be most efficiently controlled through a combination of the use of healthy (tested) plant material and good cultivation practices, including crop rotation (of 5-7 years, with nonsusceptible crops such as corn); intercropping with bean, corn, cowpea, rouging volunteers, and wilted plants; disinfection of tools; and maintenance of storage hygiene (Hayward, 1991; French et al., 1997; Smith et al., 1997). Integrated control of bacterial wilt was achieved in Japan by fumigating the soil with chloropicrin and late planting to avoid higher temperatures in the planting season (Hayward, 1991). The application of bleaching powder to the soil also decreased the disease incidence (Sangar, 1993). Biological control through the application of the bacterium *Pseudomonas fluorescens* showed positive results in laboratory as well as field conditions (McLaughlin and Sequeira, 1988), but the large-scale application of this method is not feasible.

### Host Resistance

Cultivars reported to possess resistance to bacterial wilt are Prisca, Cruza 148, Caxamarca, Molinera, and Amapola (French et al., 1997). However, their practical utility in controlling bacterial wilt is yet to be demonstrated. Their resistance to bacterial wilt is perhaps not durable and limited to specific races or biovars. The resistance response of a genotype is known to vary with the race (Fock et al., 2002). Further, foliage symptoms and latent tuber infection are not correlated (Ciampi and Sequeira, 1980; Priou et al., 2001). Therefore, it would be unsafe to grow cultivars harboring latent infection.

Considerable efforts are, however, under way to develop cultivars with the required resistance. Resistance to bacterial wilt is inherited in a polygenic way (Rowe and Sequeira, 1970) and controlled mainly by nonadditive gene action (Chakrabarti et al., 1994). Resistance originating from *S. phureja* was shown to be controlled by four major genes (Fock et al., 2002). Hartman and Elphinstone (1994) listed a number of *Solanum* spp. as sources of resistance to bacterial wilt: *S. acaule, S. acroscopicum, S. berthaultii, S. blanco-galdosii, S. boliviense, S. brachycarpum, S. chacoense, S. chomatophilum, S. commersonii, S. demissum, S. microdontum, S. phureja, S. pinnatisectum, S. polytrichon, S. raphanifolium, S. sparsipilum, S. stenotomum, S. stoloniferum*, and *S. sucrense*. Schmiediche (1986) and Fock et al. (2002) listed species that have been used in breeding for resistance to bacterial wilt: *S. chacoense, S. microdontum, S. multidissectum, S. phureja, S. raphanifolium, S. sparsipilum*, and *S. stenotomum*.

Screening of potato genotypes for resistance to bacterial wilt is carried out primarily by dipping plant roots in bacterial suspension (Lima et al., 1996; Fock et al., 2002) or planting tubers in pathogen-infested fields (Lopes et al., 1993). McLaughlin and Sequeira (1988) used a stem inoculation method. Inoculation of potato seedlings with *R. solanacearum* has been shown to be effective in the selection of genotypes resistant to bacterial wilt in the early stages of a breeding program (Lima et al., 1996). The pathogenicity of two strains of Race 1 (Biovars 3 and 4) prevalent in hot tropics increases at higher temperatures (28°C). Race 3 (Biovar 2) requires cooler temperatures to cause wilting of potato (16-24°C) (Ciampi and Sequeira, 1980).

Different mating schemes ($4x \times 2x$, $2x \times 2x$, and $4x \times 4x$) are being used particularly to transfer the resistance from diploid species to cultivated tetraploid potatoes (Watanabe et al., 1992). Laferriere et al. (1999) introduced resistance to Race 3 (Biovar 2) from *S. commersonii* to cultivated potatoes via somatic hybridization. Transgenic lines of cv. Desiree carrying an antibacterial gene coding for a cecropin peptide analogue have been developed (Montanelli et al., 1995). Some of these had resistance to bacterial wilt comparable to that of cv. Cruza 148, but the majority of them were more susceptible than cv. Desiree.

## *RING ROT*

### *Causal Agent*

Ring rot is caused by *Clavibacter michiganensis* ssp. *sepedonicus* (Spieckermann and Kotthoff), formerly called *Corynebacterium sepedonicum* (Skaptason and Burkholder, 1942). This bacterium is gram-positive, rod-shaped, short, nonmotile, and aerobic (Skaptason and Burkholder, 1942). *Clavibacter michiganensis* has been divided into five subspecies: *sepedonicus, michiganensis, insidiosum, nebraskensis,* and *tesselarius.* This division is based on phenotypic traits such as pigmentation, growth on different media, total protein profile, serological reactions, differences in host range (Louws et al., 1998), and DNA fingerprints created by repetitive extragenic palindromic-PCR with primers homologous to conserved repetitive motifs: REP, BOX, and ERIC (Brown et al., 2002).

### *Distribution and Impact*

*Clavibacter michiganensis* occurs in all major potato-growing areas except Australia (Hooker, 1981). In natural conditions this bacterium is pathogenic only to potato. However, symptomless natural infection of sugar beet and its seeds has also been reported. Tomato *(Lycopersicum esculentum)* and eggplant *(S. melongena)* can also be artificially inoculated with this bacterium and used as indicator plants in bioassay. Losses in potato crop as a result of ring rot have been reported at 50 percent in North America (Easton, 1979) and 47 percent in Russia (Muller and Ficke, 1974).

### *Symptoms and Detection*

Disease usually appears late in the growing season and can be mistaken for late blight *(Phytophthora infestans)*, drought, or natural senescence. Lower leaves wilt and roll inward and upward. The color of leaves changes from light green through gray-green to yellow and the ends become brown and necrotic (Figure 10.1B). A stem or tuber cross section may show discoloration of vascular tissue (Figure 10.1C). The tuber tissue outside the vascular ring can be easily separated from the inner tissue after squeezing. Cheese-like bacterial ooze is exuded. In the advanced infection stage external symptoms can be observed as irregular, often star-shaped cracks. Soft rot causing microorganisms can lead to secondary infection, obscuring the symptoms of ring rot.

Occasionally potato tubers or plants infected with bacteria can be without any visible symptoms (Pastrik, 2000). Symptoms depend on environmental conditions: at temperatures around 15°C plants show atypical symptoms such as stunting and rosetting; at low light intensity plants express severe ring rot symptoms (Laurila et al., 2003).

Serological tests such as ELISA and IFAS (Anonymous, 1993), bioassay on the eggplant, and molecular tests based on PCR (Mills et al., 1997; Pastrik, 2000) are used to confirm that disease symptoms are due to this bacterium. Owing to the limited sensitivity of immunological tests (Pastrik, 2000) and cross-reactions with other bacteria, tests are used in combination.

### *Epidemiology*

Infected seed tubers and potato-handling equipment are the main source of infection for ring rot. Bacteria colonize vascular rings in potato tubers, stolons, and stems. The bacteria can survive for 15-35 days in sterile water and up to 63 months on dried potato stems (Nelson, 1985). The bacterium was able to persist and remain infectious for 18 months on burlap subjected to temperatures between –40°C and –5°C, alternating temperatures of –5°C to 5°C, or constant temperature of 5°C (Nelson, 1985). It could survive for 1 year in soil at 4°C with 50 percent water-holding capacity and for more than 5 months at a temperature of –5°C (van der Wolf et al., 2002).

## Disease Management

### Cultural

In the same way as bacterial wilt, ring rot is best controlled through a combination of the use of healthy (tested) seed tubers and good cultivation practice, including crop rotation, disinfection of tools, and storage hygiene. The last two measures should not be underestimated, as the bacteria are persistent on equipment and in storage for a long time. Disinfection of planting machinery contaminated with bacteria effectively reduces the disease incidence (van der Wolf et al., 2002). Use of antagonistic bacteria, *Arthrobacter* spp., a soil coryneform bacterium (isolated from the surface of potato tubers), was also effective in suppressing ring rot in experimental trials (Gamard and De Boer, 1995), but the biocontrol method is difficult to use on a large scale.

### Host Resistance

Although genetic variability for resistance to ring rot has been found among cultivars and wild species, no cultivar with the desired level of resistance is available. The extreme resistance reported in *S. acaule* (Ishimaru et al., 1994) was shown to be temperature dependent (Laurila et al., 2003). The quarantine requirement of zero tolerance for this pathogen in seed potatoes warrants more than the development of simply tolerant genotypes (Langerfeld and Batz, 1992). Attempts have been made to evaluate resistance of potato to ring rot disease by dipping potato roots in bacterial suspension or planting inoculated seed tubers and by assessing the progeny tubers for the presence of the bacterium (Ishimaru et al., 1994).

## SOFT ROT AND BLACKLEG

### Causal Agents

Soft rot and blackleg of potato are caused by *Erwinia* spp., namely *E. carotovora* ssp. *atroseptica* (van Hall) Dye, synonym: *Pectobacterium carotovorum* ssp. *atrosepticum* (Hauben et al., 1998); *E. carotovora* ssp. *carotovora* (Jones) Bergey et al., synonym: *Pectobacterium carotovorum* ssp. *carotovorum* (Hauben et al., 1998); and *E. chrysanthemi* (Burkholder et al., 1953), synonym: *Pectobacterium chrysanthemi* (Hauben et al., 1998). These bacteria are gram-negative, rod-shaped, nonspore-forming, and facultative anaerobes (Kotoujansky, 1987).

*Erwinia carotovora* ssp. *atroseptica* has nine serogroups, and 60-90 percent of the strains represent serogroup I (De Boer and McNaughton, 1987; Śledź et al., 2000). *Erwinia carotovora* ssp. *carotovora* has 41 serogroups (De Boer and McNaughton, 1987; Pérombelon, 2002). *Erwinia chrysanthemi* is a highly differentiated species containing several groups with different serological and molecular characteristics (Samson et al., 1989). Low diversity in *E. carotovora* ssp. *atroseptica* was observed upon application of molecular markers such as PCR-RFLP (Waleron et al., 2002). However, high diversity in *E. carotovora* ssp. *carotovora* and *E. chrysanthemi* was reported after RFLP analysis of the genes encoding isoenzymes from the pectate lyase family (Nassar et al., 1996) or other genes (Helias et al., 1998; Waleron et al., 2002). The highest diversity after molecular typing was reported for *E. carotovora* ssp. *carotovora* when ERIC-PCR and random amplified polymorphic DNA were applied (Mäki-Valkama and Karjalainen, 1994; Seo et al., 2002).

### Distribution and Impact

Soft rot occurs worldwide, but the distribution of *Erwinia* spp. depends on temperature: *E. carotovora* ssp. *atroseptica* is common in temperate climates; *E. chrysanthemi* predominates in tropical, subtropical, and warm temperate areas; and *E. carotovora* ssp. *carotovora* is present in all these climates (Pérombelon and Kelman, 1980). Strains of *E. chrysanthemi* adapted to lower temperatures have also been isolated in the Netherlands (Janse and Ruissen, 1988).

*Erwinia carotovora* ssp. *atroseptica* infections are restricted mostly to potato (Pérombelon and Kelman, 1980). *Erwinia chrysanthemi* mainly affects potato, African violet *(Saintpaulia ionantha)*, carnation, maize, pineapple, and dahlia. *Erwinia carotovora* ssp. *carotovora* has probably the widest host range, including potato, brussels sprout, carrot, celery, cucumber, capsicum, turnip, and chicory (Pérombelon and Kelman, 1980).

Global annual losses of potato crop as a result of soft rot (blackleg) caused by pectinolytic *Erwinia* spp. were estimated to be worth USD 50-100 million (Pérombelon and Kelman, 1980). In Poland, storage losses in potato as a result of soft rot were estimated to be 6.7 percent in 1976-1979 (Pietkiewicz, 1980). The average incidence of potato blackleg in the field in temperate regions is approximately 2 percent (Bain et al., 1990), though Hide et al. (1996) estimated yield reduction of up to 28 percent.

### *Symptoms and Detection*

Blackleg symptoms in wet conditions are blackening of rotting potato stems at the base; under dry conditions yellowing, wilting, and desiccation of potato plants predominate (Pérombelon and Kelman, 1980) (Figure 10.1D). Tuber soft rot symptoms develop in storage and can vary from small lesions to complete rotting. Rotting tissue is odorless and it has the same color as the tuber tissue soaked with water. Sometimes decayed tissue can become brown or black at the margin (Figure 10.1E). When some other pathogens combine with soft rot infection, unpleasant odor may develop. The type of symptoms also depends on environmental conditions. Bacteria present in tubers can cause early decay of seed tubers after planting, leading to missing plants in the field.

### *Epidemiology*

The disease is primarily seedborne (Pérombelon and Kelman, 1987). Tubers start rotting in storage if the conditions are anaerobic and humid. If both *E. carotovora* ssp. *atroseptica* and *E. carotovora* ssp. *carotovora* are present in rotting tubers at temperature <15°C, *E. carotovora* ssp. *atroseptica* tends to predominate, whereas at higher temperatures the reverse is true. At temperature of approximately 22°C both bacteria develop at equal rates (Pérombelon and Kelman, 1980). The cell wall–degrading enzymes pectin lyase, pectin methyl esterase, and polygalacturonase secreted by bacterial cells degrade the potato tissue (Collmer and Keen, 1986). In low-temperature storage, bacteria can survive in tuber as latent (symptomless) infections. Such latently infected tubers planted in the field may rot or have damaged sprouts or develop blackleg symptoms. Rotten mother tubers are the source of infection for daughter tubers, and disease development is correlated with the level of tuber contamination (Bain et al., 1990). Bacteria penetrate potato tubers through lenticels and wounds. Survival of *Erwinia* spp. in soil and water is limited to a few months (Pérombelon and Kelman, 1987).

To isolate *Erwinia* spp. from the soil, water, or infected tissue, a pectate-based selective medium (CVP) is commonly used. Identification of the species and subspecies requires biochemical, physiological, and immunological tests (ELISA) (De Boer and McNaughton, 1987). PCR-based methods allow the identification of species and subspecies of pectinolytic *Erwinia* (Waleron et al., 2002).

### *Disease Management*

#### *Cultural*

As with other bacterial diseases, control of this disease is based mainly on prevention. Planting healthy seeds, avoiding the risks of damage and wetting of tubers during harvest, and avoiding anaerobic conditions in store and transport are effective control measures. Although other practices, such as dipping seed potatoes in hot water (55°C) for 5 minutes (Robinson and Foster, 1987) or using antagonist bacteria *Pseudomonas fluorescens* (Rhodes and Logan, 1986), resulted in a significant reduction of disease development under laboratory conditions, they have not been used on a commercial scale.

### *Host Resistance*

The level of resistance to *Erwinia* spp. in potato cultivars is relatively low. Evaluation of 617 potato cultivars from the world collection against this pathogen showed only one, cv. Sickingen, to be highly resistant. Krause et al. (1982) did not confirm cv. Sickingen to be resistant but reported another cultivar, Revenu, to be resistant out of 747 cultivars screened by them in 1974-1977.

Resistance to *Erwinia* spp. is inherited in a polygenic fashion (Zimnoch-Guzowska et al., 2000). Though many *Solanum* spp. are known sources of resistance to *Erwinia* spp., the list of species utilized in potato breeding programs is much shorter. These are *S. brevidens* (Austin et al., 1988), *S. chacoense* (Zimnoch-Guzowska and Łojkowska, 1993), *S. commersonii* (Carputo et al., 1997), *S. phureja*, *S. stenotomum* (Iwanaga, 1985), *S. tuberosum* ssp. *andigena* (Huamán et al., 1988), and *S. yungasense* (Zimnoch-Guzowska and Łojkowska, 1993). Tetraploid potato clones resistant to soft rot were obtained after $4x \times 2x$ crosses, where interspecific diploid hybrids were used as donors of resistance (Lebecka et al., 2004).

Somaclonal variations have been explored for resistance to soft rot by Taylor et al. (1993). Resistant somatic hybrids of *S. tuberosum* and *S. brevidens* and their sexual progeny were used in breeding programs by Helgeson et al. (1993).

Transformation of potato plants for resistance to bacteria has been performed with genes coding for lysozyme from fag T4 (Düring et al., 1993), lysozyme from chicken (Serrano et al., 2000), and attacin and analogue peptide cecropin SB-37 (Arce et al., 1999). The expression of these genes was weak in potato tissue and did not impart higher resistance to bacterial soft rot than in the controls. Transformation of potato with genes coding an enzyme generating $H_2O_2$ (Wu et al., 1995) and peroxydase from cucumber (Ray et al., 1998) resulted in higher resistance in transformed lines in laboratory evaluation but were not always effective in field conditions. Four transgenic lines of cv. Desiree, with genes coding pectate lyases, showed higher resistance than cv. Desiree in 4 years of field trials (Wegener, 2002).

Blackleg resistance has been evaluated by inoculating various plant parts, including the stem base (Munzert, 1975), seedlings (Łojkowska and Kelman, 1989), leaves (Bisht et al., 1993), and tubers (Lapwood et al., 1984), with *Erwinia* spp. in a greenhouse, climatic chambers, and under field conditions. Soft rot resistance is assessed by inoculating tubers or slices (Bain and Pérombelon, 1988; Taylor et al., 1993; Łojkowska and Kelman, 1994). Incubation has been carried out mostly under atmospheric conditions with high humidity (oxygen deficiency) but also in anaerobic chambers filled with nitrogen (Łojkowska and Kelman, 1994). The incubation temperature ranges from 22 to 27°C for 3-6 days in various screening methods (Bain and Pérombelon, 1988; Zimnoch-Guzowska and Łojkowska, 1993). Infection is assessed by various measures of rotten tissue: the diameter (millimeters), volume (milliliters), weight (grams), or deepness (millimeters).

Resistance evaluation most frequently uses a scale from 1 to 9 (where 9 is the most resistant). The ranking of tested clones or cultivars, however, varies with the methods of evaluation (Bain and Pérombelon, 1988; Łojkowska and Kelman, 1994). The biggest differences in level of resistance evaluated using various methods were noted for clones or cultivars expressing a medium level of resistance (Lapwood et al., 1984). The ranking of tested clones does not seem to be dependent on the species or subspecies of *Erwinia* used for inoculation (Lapwood et al., 1984; Austin et al., 1988; Zimnoch-Guzowska et al., 1999).

## *COMMON SCAB*

### *Causal Agents*

Common scab is caused by *Streptomyces* spp., namely, *S. scabies* (Thaxter) (Waksman and Henrici), *S. turgidiscabies* (Miyajima et al., 1998), *S. europaeiscabiei* (Bouchek-Mechiche et al., 2000), and *S. stelliscabiei* (Bouchek-Mechiche et al., 2000). Some other species of *Streptomyces* have also been reported to be pathogenic e.g. *S. albidoflavus, S. rochei, S. diastaticus, S. cyaneus,* and *S. alroolivaceus* (Bramwell et al., 1998). However, *S. scabies* is now considered the principal causal agent of this disease on potato (Goyer and Beaulieu, 1997).

Common scab–causing *Streptomyces* spp. produce phytotoxins (thaxtomins) that when applied alone to tubers can produce plant symptoms similar to those caused by the infection. Strains that lack the ability to produce thaxtomins tend to be avirulent. *Streptomyces* spp. are soilborne, aerobic bacteria. They are cylindrical, 0.0-1.7 × 0.5-0.8 μm in size, and have growth characteristics similar to fungi. They form filamentous, branched structures similar to fungi and are classified as actinomycetes. Colonies have gray-colored aerial growth and an "earthy" smell. They have specific biochemical properties, sensitivity to streptomycin, and no growth at pH 4.5 (Goyer and Beaulieu, 1997). Lambert and Loria (1989) differentiated *S. scabies, S. turgidiscabies,* and *S. acidiscabies* by their pigmentation, shape of sporulating mycelium, and ability to utilize sugars. Based on the polymorphism of the gene coding 16S rRNA, Bouchek-Mechiche et al. (2002) found five different RFLP patterns among *Streptomyces* spp. However, *S. europaeiscabiei* and *S. stelliscabiei,* two common scab–causing bacteria in France, were found to be closely related to saprophytic strains *S. distatocromogenes* and *S. botropensis,* respectively (Bouchek-Mechiche et al., 2000). There are actually difficulties of taxonomy within the *Streptomycetes.* Bramwell et al. (1998) examined the diversity of *Streptomycetes* causing potato scab using phenotypic identification, 16S rRNA sequence, and pathogenicity. In their studies, pathogenicity did not correlate well with taxonomic identification or phylogenetic position, and phenotypic characterizations were not fully consistent with phylogenetic relationships.

## *Distribution and Impact*

*Streptomyces scabies* is present worldwide. Virulent strains of the pathogen can even be recovered from soils with no history of crop production. Other pathogen species are more common in certain regions. *Streptomyces acidiscabies* and *S. turgidiscabies* were found in the northeastern United States and Japan; *S. turgidiscabies* in northern Scandinavia (Valkonen et al., 2002); *S. europaeiscabiei,*

*S. stelliscabiei*, and *S. reticuliscabiei* in France (Bouchek-Mechiche et al., 2000); and *S. caviscabies* in Canada (Goyer et al., 1996).

*Streptomyces* spp. have a wide host range and cause diseases of diverse root and tuber crops such as potato, radish, turnip, beet, carrot, sugar beet, weeds with fleshy roots, cabbage, eggplant, and sweet potato (Goyer and Beaulieu, 1997; Halseth and Loria, 1998).

Although tuber yields are not normally affected by this pathogen, tuber quality is reduced because the scab on tuber surfaces worsens tuber appearance and is a source of inoculum for pathogen dissemination via seed tubers. Marketability and seed grade-out losses can cause a major economic impact.

### *Symptoms and Detection*

Whereas potato stems and leaves do not normally exhibit disease symptoms, tubers and less frequently roots and stolons are affected. Infected tubers have lesions on the surface (see Figure 9.1). The extent of surface lesions ranges from a superficial cork-like layer (russet scab) to a raised corky area (deep common scab). Pitted scab lesions are dark brown to black pits of varying depths and diameter surrounded by straw-colored translucent tissue. A tuber can have lesions of both types. Lesions can be either discrete or coalescent, completely covering the tuber surface (Goth et al., 1995). *Streptomyces* spp. can be isolated from infected tissue by dilution plating on water agar (1.2%) (Goyer and Beaulieu, 1997). However, for positive identification it is necessary to grow bacteria up to 10 days, when bacterial filaments and pigment production allow differentiation of the pathogens (Loria et al., 2001). Pathogenicity testing may be necessary for the purposes of confirmation. Detection and identification of pathogenic *Streptomyces* strains can be achieved with PCR-based tests using primers designed for genes coding the putative pathogenicity factors (Cullen et al., 2000).

### *Epidemiology*

The pathogen is dispersed passively by wind, water, farm equipment, and tubers. *Streptomyces scabies* can live on decomposing material in the soil and does not require potato or any root crop to remain alive (Halseth and Loria, 1998). The pathogen enters the lenticels of immature tubers. Tubers are most susceptible for 6-8 weeks following initiation. The optimum temperature for the growth of *S. scabies* is 30°C, but soil conditions affect disease occurrence and symptom development. Common scab can be severe at a soil pH from 5.2 to 8.0 or above but is greatly reduced below pH 5.2, whereas 'acid' scab caused by *S. acidiscabies* occurs at a pH of 4.5. *Streptomyces scabies* is favored by nitrate nitrogen (which is associated with high soil pH) and reduced by phosphorus and calcium fertilizers. High soil moisture inhibits growth of these organisms and disease occurrence. Common and 'acid' scab do not spread after harvest, when the tubers are in storage.

### *Disease Management*

The best control measure is using pathogen-free seed tubers. Crop rotation with small grains, corn, and alfalfa is recommended, but avoiding red clover and alternative hosts (Loria et al., 2001). Methods such as sufficient irrigation during the period of tuber susceptibility and application of chemicals do not always provide satisfactory control (Mishra and Srivastava, 2001). Reducing soil pH with acid-producing fertilizers or sulfur applications can decrease scab severity, but negative effects on plant growth and yield can occur. The recommendation to raise soil pH to 9 to control common scab (Waterer, 2002) requires further studies on the influence on subsequent crops. Applications of animal manures can have variable effects on scab severity, but soil amendments based on swine manures and chicken byproducts have been shown to reduce disease levels effectively (Soltani et al., 2002). Soil fumigation and seed treatments are available but are not fully effective and may be cost prohibitive or unacceptable within the community.

### *Host Resistance*

No potato cultivar is sufficiently resistant to infection by *S. scabies* (Haynes et al., 1997), but cultivar responses vary considerably (Goth et al., 1995). The inheritance of this relative resistance is not clear yet. Cipar and Lawrence (1972) and Howard (1978) suggested that more than one gene is involved in the expression of resistance to common scab. Very high ($h^2$ = 0.9) heritability for scab resistance (Haynes et al., 1997) suggests its relatively simple mode of inheritance. Resistance to common scab has been identified in *S. chacoense* (Dionne and Lawrence, 1961), *S. commersonii, S. yungasense* (Hawkes, 1990), *S. phureja* (De Maine et al., 1993), *S. multidissectum, S. canasense, S. bukasovii,* and *S. kurtzianum* (Hosaka et al., 2000). Diploid interspecific hybrids originating from *S. berthaultii, S. bukasovii, S. chacoense, S. gourlayi, S. phureja, S. sparsipilum,* and *S. tarijense* showed higher resistance to common scab than cvs. Atlantic, Russet Burbank, Russet Norkotah, which are considered to have scab resistance (Jansky, 2003). Resistance to scab has been successfully transferred from the diploid to tetraploid level via $4x \times 2x$ crosses (Murphy et al., 1995).

Screening of potato genotypes for resistance to *S. scabies* is carried out mainly in the field, but greenhouse tests have also been used (Miller et al., 1965). The response depends on environmental conditions and also on the interaction of genotype and environment (Goth et al., 1995; Haynes et al., 1997; Jansky, 2003). Thus, the assessment of resistance should be made in relation to standards. Surface area covered and lesion type are evaluated on different scales (surface area covered: 0-5 or 1-9 grades; lesion type: 0-5 or 1-3 grades) (Bjor and Roer, 1980; Goth et al., 1995).

## *PINK EYE*

The *Pseudomonas* sp. associated with pink eye disease of potato is generally classified as *P. fluorescens* (Folsom and Friedman, 1959) and has been characterized as a weak parasite of potatoes and of little economic meaning. The bacteria are common in soil and composts and usually can cause problems only in susceptible cultivars.

The disease expresses as pink to brown discoloration near the eyes. In more advanced stages of the disease, lesions become sunken, cracked, discolored, and rotted. These bacteria isolated from potato tuber lesions were also responsible for the rotting of potato tubers at the high temperature used to recondition tubers for processing (Huether and McIntyre, 1969). Bacteria enter potato tubers through lenticels or wounds in wet soil conditions. The disease on potato tubers in storage is favored by high humidity and temperature >7°C. It can appear late in the growing season under conditions of high soil moisture (Folsom and Friedman, 1959).

## *LEAF SPOT*

The potato is an incidental host for bacterium *Xanthomonas vesicatoria* (Doidge), which is responsible for leaf spot (Vauterin et al., 1995). The bacterium is aerobic, mobile, gram-negative, rod-shaped, and has a single polar flagellum. It can enter the plant through lenticels and wounds. It can survive the winter in infected debris or on seeds. The principal hosts are *Lycopersicum* and *Capsicum.* The disease occurs widely in the warmer parts of the world.

We conclude that although bacterial diseases are of great significance in potato cultivation, knowledge is lacking for use in conventional breeding programs of the sources of extreme resistance to bacterial pathogens. A high level of resistance to bacterial diseases in potato cultivars is required, particularly for those that are not on the quarantine list. Transgenics might be a solution for developing extreme bacterial resistance. Therefore, at present, controlling bacterial diseases of potato relies mainly on a combination of the use of healthy planting material and good cultivation practices: proper crop rotation, avoiding mechanical bruising and wounding of potatoes, avoiding harvesting tubers in wet conditions, maintenance of proper temperature, and ventilation in storage.

## REFERENCES

Anonymous (1990). Phytosanitary procedures. No, 26. *Pseudomonas solanacearum*, inspection and test methods. *EPPO/OEPP Bulletin 20:*255-262.

Anonymous. (1993). Councile Directive 98/85/EEC of October 4, 1993 on the control of potato ring rot. *Official Journal of the European Communities,* No. L259/1, Annex I-IV.

Arce, P., Moreno, M., Gutierrez, M., Gebauer, M., Dell'Orto, P., Torres, H., Acuna, I., Oliger, P., Venegas, A., Jordana, X. et al. (1999). Enhanced resistance to bacterial infection by *Erwinia carotovora* subsp. *atroseptica* in transgenic potato plants expressing the attacin or the cecropin *SB-37* genes. *American Journal of Potato Research 76:*169-177.

Austin, S., Łojkowska, E., Ehlenfeldt, M.K., Kelman, A., and Helgeson, J.P. (1988). Fertile interspecific somatic hybrids of *Solanum:* A novel source of resistance to *Erwinia* soft rot. *Phytopathology 78:*1216-1220.

Bain, R.A. and Pérombelon, M.C.M. (1988). Methods of testing potato cultivars for resistance to soft rot of tubers caused by *Erwinia carotovora* subsp. *atroseptica. Plant Pathology 37:*431-437.

Bain, R.A., Pérombelon, M.C.M., Tsror, L., and Nachmias, A. (1990). Blackleg development and tuber yield in reaction to numbers of *Erwinia carotovora* subsp. *atroseptica* on seed tubers. *Plant Pathology 39:*125-133.

Bisht, V.S., Bain, P.S., and Letal, J.R. (1993). A simple and efficient method to assess susceptibility of potato to stem rot by *Erwinia carotovora* subspecies. *American Potato Journal 70:*611-616.

Bjor, T. and Roer, L. (1980). Testing the resistance of potato varieties to common scab. *Potato Research 23:*33-47.

Bouchek-Mechiche, K., Gardan, L., Normand, P., and Jouan, B. (2000). DNA relatedness among strains of *Streptomyces* pathogenic to potato in France: Description of three species *Streptomyces europaeiscabiei* sp. nov. and *Streptomyces stelliscabiei* sp. nov. associated with common scab, and *Streptomyces reticuliscabei* sp. nov. associated with netted scab. *International Journal of Systematic and Evolutionary Microbiology 50:*91-99.

Bouchek-Mechiche, K., Normand, P., Pasco, C., and Andrivon, D. (2002). Development of molecular tools for identifying and detecting *Streptomyces* spp. causal agents of potato scab. In Wenzel, G. and Wulfert, I. (Eds.), *Abstracts of Papers and Posters, 15th Triennial Conference of the EAPR: Potatoes Today and Tomorrow* (p. 94). Hamburg, Germany: EAPR.

Bramwell, P.A., Wiener, P., Akkermans, A.D.L., and Wellington, E.M.H. (1998). Phenotypic, genotypic and pathogenic variation among streptomycetes implicated in common scab disease. *Letters in Applied Microbiology 27:*255-260.

Brown, S.E., Reilley, A.A., Knudson, D.L., and Ishimaru, C.A. (2002). Genomic fingerprinting of virulent and avirulent strains of *Clavibacter michiganensis* subspecies *sepedonicus. Current Microbiology 44:*112-119.

Buddenhagen, I.W. (1986). Bacterial wilt revisited. In Persley, G.J. (Ed.), *Proceedings of an International Workshop held at PCARRD: Bacterial Wilt Disease in Asia and the South Pacific* (pp. 126-143). Los Banos, Philippines.

Burkholder, W.H., McFadden, L.H., and Dimock, A.W. (1953). A bacterial blight of chrysanthemums. *Phytopathology 43:*522-525.

Carputo, D., Cardi, T., Speggiorin, M., Zoina, A., and Frusciante, L. (1997). Resistance to blackleg and tuber soft rot in sexual and somatic interspecific hybrids with different background. *American Potato Journal 74:*161-172.

Chakrabarti, S.K., Gadewar, A.V., Gopal, J., and Shekhawat, G.S. (1994). Performance of tetraploid x diploid (TD) crosses of potato for bacterial wilt resistance in India. *Bacterial Wilt Newsletter 10:*7.

Ciampi, L. and Sequeira, L. (1980). Multiplication of *Pseudomonas solanacearum* in resistant potato plants and the establishment of latent infections. *American Potato Journal 57:*319-329.

Cipar, M.S. and Lawrence, C.H. (1972). Scab resistance of haploids from two *Solanum tuberosum* cultivars. *American Potato Journal 49:*117-120.

Collmer, A. and Keen, N.T. (1986). The role of pectic enzymes in plant pathogenesis. *Annual Review of Phytopathology 24:*383-409.

Cook, D. and Sequeira, L. (1991). Genetics and biochemical characterization of a *Pseudomonas solanacearum* gene cluster required for extracellular polysaccharide production and for virulence. *Journal of Bacteriology 173:*1654-1662.

Cullen, D.W., Lees, A.K., Toth, I.K., Bell, K.S., and Duncan, J.M. (2000). Detection and quantification of fungal and bacterial potato pathogens in plants and soil. *EPPO/OEPP Bulletin 30:*485-488.

De Boer, S.H. and McNaughton, M.E. (1987). Monoclonal antibodies to the lipopolysaccharide of *Erwinia carotovora* subsp. *atroseptica* serogroup I. *Phytopathology 77:*828-832.

De Maine, M.J., Caroll, C.P., Stewart, H.E., Solomon, R.M., and Wastie, R.L. (1993). Disease resistance in *Solanum phureja* and diploid and tetraploid *S. tuberosum* x *S. phureja* hybrids. *Potato Research 36:*21-28.

Dionne, L.A. and Lawrence, C.H. (1961). Early scab resistant derivatives of *Solanum chacoense* x *Solanum phureja. American Potato Journal 38:*6-8.

Düring, K., Porsch, P., Fladung, M., and Lörz, H. (1993). Transgenic potato plants resistant to the phytopathogenic bacterium *Erwinia carotovora. The Plant Journal 3:*587-598.

Easton, G.D. (1979). The biology and epidemiology of potato ring rot. *American Potato Journal 56:*459-460.

Elphinstone, J.G, Stead, D.E., Caffier, D., Janse, J.D., López, M.M., Mazzucchi, U., Müller, P., Persson, P., Rauscher, E., Schiessendoppler, E. et al. (2000). Standarization of methods for detection of *Ralstonia solanacearum* in potato. *EPPO/OEPP Bulletin 30:*391-395.

Englebrecht, M.C. (1994). Modification of a semi-selective medium for the isolation and quantification of *Pseudomonas solanacearum. Bacterial Wilt Newsletter 10:*3-5.

Fock, I., Luisetti, J., Collonier, C., Vedel, F., Ducreux, G., Kodja, H., and Sihachakr, D. (2002). *Solanum phureja* and *S. stenotomum* as sources of resistance to *Ralstonia solanacearum* in somatic hybrids of potato. In *Book of Abstracts of 3rd International Bacterial Wilt Symposium* (p. 47). South Africa.

Folsom, D.O. and Friedman. B.A. (1959). *Pseudomonas fluorescens* in relation to certain diseases of potato tubers in Maine. *American Potato Journal 36:*90-97.

French, E.R., Anguiz, R., and Aley, P. (1997). The usefulness of potato resistance to *Ralstonia (Pseudomonas) solanacearum* for the integrated control of bacterial wilt. In *Proceedings of the 2nd Conference IBWS: Disease Management,* Guadelupe.

Gamard, P. and De Boer, S.H. (1995). Evaluation of antagonistic bacteria for suppression of bacterial ring rot. *European Journal of Plant Pathology 101:*519-525.

Goth, R.W., Haynes, K.G., Young, R.J., Wilson, D.R., and Lauer, F.I. (1995). Relative resistance of the potato cultivar Krantz to common scab caused by *Streptomyces scabies* as determined by cluster analysis. *American Potato Journal 72:*505-511.

Goyer, C. and Beaulieu, C. (1997). Host range of streptomycete strains causing common scab. *Plant Disease 81:*901-904.

Goyer, C., Faucher, E., and Beaulieu, C. (1996). *Streptomyces caviscabies* sp. nov., from deep pitted lesions in potatoes in Québec, Canada. *International Journal of Systematic Bacteriology 46:*635-639.

Halseth, D. and Loria, R. (1998). Fundamental facts on common scab of potato in U.S. and world. *The Great Lakes Vegetable Growers News (USA)* (July issue).

Hartman, G.L. and Elphinstone, J.G. (1994). Advances in the control of *Pseudomonas solanacearum* race 1 in major food crops. In Hayward, A.C. and Hartman, G.L. (Eds.), *Bacterial Wilt. The Disease and its Causative Agent, Pseudomonas solanacearum* (pp. 157-177). Wallingford, UK: CAB International.

Hauben, L., Moore, E.R.B., Vauterin, L., Steenackers, M., Verdonck, L., and Swings, J. (1998). Phylogenetic position of phytopathogens within the *Enterobacteriacae. Systematic Applied Microbiology 21:*384-397.

Hawkes. J.G. (1990). *The Potato, Evolution, Biodiversity and Genetic Resources.* London, UK: Belhaven Press.

Haynes, K.G., Goth, R.W., and Young, R.J. (1997). Genotype x environment interactions for resistance to common scab in tetraploid potato. *Crop Science 37:*1163-1167.

Hayward, A.C. (1991). Biology and epidemiology of bacterial wilt caused by *Pseudomonas solanacearum. Annual Review of Phytopathology 29:*65-87.

Helgeson, J.P., Haberlach, G.T., Ehlenfeldt, M.K., Hunt, G., Pohlman, J.D., and Austin, S. (1993). Sexual progeny of somatic hybrids between potato and *Solanum brevidens:* Potential for use in breeding programs. *American Potato Journal 70:*437-452.

Helias, V., Roux, A.C., Bertheau, Y., Andrivon, D., Gauthier, J.P., and Jouan, B. (1998). Characterization of *Erwinia carotovora* subspecies and detection of *Erwinia carotovora* subsp. *atroseptica* in potato plants, soil and water extracts with PCR-based methods. *European Journal of Plant Pathology 104:*685-699.

Hide, G.A., Welham, S.J., Read, P.J., and Ainsley, A.E. (1996). The yield of potato plants as affected by stem cancer *(Rhizoctonia solani)*, blackleg *(Erwinia carotovora* subsp. *atroseptica)* and by neighboring plants. *Journal of Agricultural Science 126:*429-440.

Hooker, W.J. (1981). *Compendium of Potato Diseases.* St. Paul, MN, USA: American Phytopathological Society.

Hosaka, K., Matsunaga, H., and Senda, K. (2000). Evaluation of several wild tuber-bearing *Solanum* species for scab resistance. *American Journal of Potato Research 77:*41-45.

Howard, H.W. (1978). The production of new varieties, In Harris, P.M. (Ed.), *The Potato Crop* (pp. 607-646). London, UK: Chapman & Hall.

Huamán, Z., de Lindo, L., and Elphinstone, J.G. (1988). Resistance to blackleg and soft rot and its potential use in breeding. In French, E.R. (Ed.), *Report of the Planning Conference on bacterial diseases of the potato 1987* (pp. 215-228). Lima, Peru: International Potato Center.

Huether, J.P. and McIntyre, G.A. (1969). Pectic enzyme production by two strains of *Pseudomonas fluorescens* associated with the Pinkeye disease of potato tubers. *American Potato Journal 46:*414-423.

Ishimaru, C.A., Lapitan, N.L., VanBuren, A., Fenwick, A., and Pedas, K. (1994). Identification of parents suitable for molecular mapping of immunity and resistance genes in *Solanum* species. *American Potato Journal 71:*517-533.

Iwanaga, M. (1985). Ploidy level manipulation approach: Development of diploid populations with specific resistance and FDR 2*n* pollen production. In *Report of the Planning Conference: Present and Future Strategies for Potato Breeding and Improvement,* 1983 (pp. 57-70). Lima, Peru: International Potato Center.

Janse, J.D. (1988). A detection method for *Pseudomonas solanacearum* in symptomless potato tubers and some data on its sensitivity and specificity. *EPPO/OEPP Bulletin 18:*343-351.

———. (1996). Potato brown rot in Western Europe—history, present occurrence and some remarks on possible origin, epidemiology and control strategies. *EPPO/OEPP Bulletin 26:*679-695.

Janse, J.D. and Ruissen, M.A. (1988). Characterization and classification of *Erwinia chrysanthemi* strains from several hosts in the Netherlands. *Phytopathology 78:*800-808.

Jansky, S.H. (2003). Multiple disease resistance in interspecific hybrids of potato. *Plant Disease 87:*266-272.

Kinyua, Z.M., Smith, J.J., Lung'aho, C., Olanya, M., and Priou, S. (2001). On-farm successes and challenges of producing bacterial wilt-free tubers in seed plots in Kenya. *African Crop Science Journal* 9:279-285.

Kotoujansky, A. (1987). Molecular genetics of pathogenesis by soft-rot *Erwinia. Annual Review of Phytopathology 25:*405-430.

Krause, B., Koczy, T., Komorowska-Jedrys, J., and Ratuszniak, E. (1982). Laboratory assessment of tuber resistance of world potato cultivar collection to main causes of storage rots. *Biuletyn Instytutu Ziemniaka 27:*111-134.

Laferriere, L.T., Helgeson, J.P., and Allen, C. (1999). Fertile *Solanum tuberosum* plus somatic hybrids as sources of resistance to bacterial wilt caused by *Ralstonia solanacearum. Theoretical and Applied Genetics 98:*1272-1278.

Lambert, D.H and Loria, R. (1989). *Streptomyces acidiscabies* sp. nov. *International Journal of Systematic Bacteriology 39:*393-396.

Langerfeld, E. and Batz, W. (1992). Reaction of potato cultivars to the causal agent of bacterial ring rot of potatoes (*Clavibacter michiganensis* subsp. *sepedonicus* (Spieck. et Kotth.) Davis et al.). *Nachrichtenblatt des Deutschen Pflanzenschutzdienstes 44:*157-161.

Lapwood, D.H., Read, P.J., and Spokes, J. (1984). Methods for assessing the susceptibility of potato tubers of different cultivars to rotting by *Erwinia carotovora* subspecies *atroseptica* and *carotovora. Plant Pathology 33:*13-20.

Laurila, J., Metzler, M.C., Ashimaru, C A., and Rokka, V.M. (2003). Infection of plant material derived from *Solanum acaule* with *Clavibacter michiganensis* subsp. *sepedonicus:* Temperature as a determining factor in immunity of *S. acaule* to bacterial ring rot. *Plant Pathology 52:*496-504.

Lebecka, R., Zimnoch-Guzowska, E., and Kaczmarek, Z. (2004). Resistance to soft rot *(Erwinia carotovora* subsp. *atroseptica)* in tetraploid potato families obtained from 4*x*-2*x* crosses. *American Journal for Potato Research 82:*107-114.

Lima, M.F., Lopes, C.A., and de Melo, P.E. (1996). Selecao de genotipos de batata no estadio de plantulas para resistencia a murcha-bacteriana. (Screening of potato genotypes at seedling stage as for the resistance to bacterial wilt). *Pesquisa Agropecuaria Brasileira 31:*249-257.

Łojkowska, E. and Kelman, A. (1989). Screening of seedlings of wild *Solanum* species for resistance to bacterial stem rot caused by soft rot *Erwinias. American Potato Journal 66:*379-390.

———. (1994). Comparison of the effectiveness of different methods of screening for soft rot resistance of potato tubers. *American Potato Journal 71:*99-113.

Lopes, C.A., Buso, J.A., and Accatino, P. (1993). Screening CIP potato germplasm for resistance to bacterial wilt in Brazil: Methods and preliminary results. *Bacterial Wilt Newsletter 9:*3-5.

Loria, R., Clark, C.A., Bukhalid, R.A., and Fry, B.A. (2001). *Streptomyces.* In Schaad, N.W., Jones, V, and Chun, V. (Eds.), *Laboratory Guide for Identification of Plant Pathogenic Bacteria* (pp. 236-249). St. Paul, MN, USA: American Phytopathological Society.

Louws, F.J., Bell, J., Medina-Mora, C.M., Smart, C.D., Opgenorth, D., Ishimaru, C.A., Hausbeck, M.K., de Bruijn, F.J., and Fulbright, D.W. (1998). Rep-PCR-mediated genomic fingerprinting: A rapid and effective method to identify *Clavibacter michiganensis. Phytopathology 88:* 863-868.

Mäki-Valkama, T. and Karjalainen, R. (1994). Differentiation of *Erwinia carotovora* subsp. *atroseptica* and *carotovora* by RAPD-PCR. *Annals of Applied Biology 125:*301-309.

McLaughlin, R.J. and Sequeira, L. (1988). Evaluation of an avirulent strain of *Pseudomonas solanacearum* for biological control of bacterial wilt of potato. *American Potato Journal 65:*255-268.

Miller, J.C., Fontenot, J.F., and Dainello, F.J. (1965). Techniques used in screening potato clones for resistance to *Streptomyces scabies. American Potato Journal 42:*294 (abstract).

Mills, D., Russell, B.W., and Hanus, J.W. (1997). Specific detection of *Clavibacter michiganensis* subsp. *sepedonicus* by amplification of three unique DNA sequences isolated by subtraction hybridization. *Phytopathology 87:*853-861.

Mishra, K.K. and Srivastava, J.S. (2001). Screening potato cultivars for common scab of potato in naturally infested field. *Potato Research 44:*19-24.

Miyajima, K., Tanaka, F., Takeuchi, T., and Kuninaga, S. (1998). *Streptomyces turgidiscabies* sp. nov. *International Journal of Systematic Bacteriology 48:* 495-502.

Montanelli, C., Stefanini, F.M., Chiari, A., Chiari, T., and Nascari, G. (1995). Variability in the response to *Pseudomonas solanacearum* of transgenic lines of potato carrying a cecropin gene analogue. *Potato Research 38:*371-378.

Muller, H.J. and Ficke, W. (1974). Bacterial ring rot *(Corynebacterium sepedonicum)* a dangerous quarantine disease for potato cultivation. *Nachrichtenblatt für den Pflanzenschutz in der DDR 28:*159-160.

Munzert, M. (1975). Eine Methode zur Prüfung der Resistenz der Kartoffelpflanze gegenüber dem Erreger der Schwarzbeinigkeit (*Erwinia carotovora* var. *atroseptica* /van Hall/ Dye). *Potato Research 18:*308-313.

Murphy, A.M., de Jong, H., and Tai, G.C.C. (1995). Transmission of resistance to common scab from the diploid to the tetraploid level via 4x-2x crosses in potatoes. *Euphytica 82:*227-233.

Nassar, A., Darrasse, A., Lemattre, M., Kotoujansky, A., Dervin, C., Vedel, R., and Bertheau, Y. (1996). Characterisation of *Erwinia chrysanthemi* by pectinolytic isozyme polymorphism and restriction fragment length polymorphism analysis of PCR-amplified fragments of *pel* genes. *Applied and Environmental Microbiology 62:*2228-2235.

Nelson, G.A. (1985). Survival of *Corynebacterium sepedonicum* in potato stems and on surfaces held at freezing and above-freezing temperatures. *American Potato Journal 62:*23-28.

Olanya, O.M., Kinyua, Z.M., El-Bedewy, R., Smith, J.J., and Kihara, S.N. (2002). Incidence and spatial variability of bacterial wilt infection on advanced potato clones in central Kenya. In *Book of Abstracts of 3rd International Bacterial Wilt Symposium* (p. 47). South Africa.

Pastrik, K.H. (2000). Detection of *Clavibacter michiganensis* subsp. *sepedonicus* in potato tubers by multiplex PCR with coamplification of host DNA. *European Journal of Plant Pathology 106:*155-165.

Pastrik, K.H. and Maiss, E. (2000). Detection of *Ralstonia solanacearum* in potato tubers by polymerase chain reaction. *Journal of Phytopathology 148:*619-626.

Pérombelon, M.C.M. (2002). Potato diseases caused by soft rot erwinias: An overview of pathogenesis. *Plant Pathology 51:*1-12.

Pérombelon, M.C.M. and Kelman, A. (1980). Ecology of the soft rot erwinias. *Annual Review of Phytopathology 18:*361-387.

Pérombelon, M.C.M. and Kelman, A. (1987). Blackleg and other potato diseases caused by soft rot erwinias: Proposal for revision of terminology. *Plant Disease 71:*283-285.

Pietkiewicz, J.B. (1980). Variation in the reaction of potato tubers to diseases. *Potato Research 23:*473.

Poussier, S., Trigalet-Damery, D., Vandewalle, P., Goffinet, B., Luisetti, J., and Trigalet, A. (2000). Genetic diversity of *Ralstonia solanacearum* as assessed by

PVR-RFLP of the *hrp* gene region, AFLP and 16S rRNA sequence analysis, and identification of an African subdivision. *Microbiology 146:*1679-1692.

Priou, S., Salas, C., de Mendiburu, F., Aley, P., and Gutarra, L. (2001). Assessment of latent infection frequency in progeny tubers of advanced potato clones resistant to bacterial wilt: A new selection criterion. In *International Potato Center Program Report 1999-2000* (pp. 105-116). Lima, Peru: International Potato Center.

Ray, H., Douches, D.S., and Hammerschmidt, R. (1998). Transformation of potato with cucumber peroxidase: Expression and disease response. *Physiological and Molecular Plant Pathology 53:*93-103.

Rhodes, D.J. and Logan, C. (1986). Effects of fluorescent pseudomonads on the potato blackleg syndrome. *Annals of Applied Biolosgy 108:*511-518.

Robinson, K. and Foster, G. (1987). Quantitative assessment of tuber contamination by pectolytic erwinia and its possible use in the prediction and control of blackleg. *Potato Research 30:*669-674.

Robinson-Smith, A., Jones, P., Elphinstone, J.G., and Forde, S.M.D. (1995). Production of antibodies to *Pseudomonas solanacearum,* the causative agent of bacterial wilt. *Food and Agriculture Immunology 7:*67-79.

Rowe, P.R. and Sequeira, L. (1970) Inheritance of resistance to *Pseudomonas solanacearum* in *Solanum phureja. Phytopathology 60:*1499-1501.

Salanoubat, M., Genin, S., Artiguenave, F., Gouzy, J., Mangenot, S., Arlat, M., Biliaut, A., Brottier, P., Camus, J.C., Cattolico, L. et al. (2002). Genome sequence of the plant pathogen *Ralstonia solanacearum. Nature 415:*497-502.

Samson, R., Ngwira, N., and Rivera, N. (1989). Biochemical and serological diversity of *Erwinia chrysanthemi.* In Klement, Z. (Ed.), *Proceedings of the 7th International Conference on Plant Pathogenic Bacteria* (pp. 895-901). Budapest, Hungary: Akademie Kiadó és Nyomoda Vállalat.

Sangar, R.B.S. (1993). Efficacy of agro-practices to control bacterial wilt of potatoes in central India. *Journal of Indian Potato Association 20:*172-173.

Schaad, N.W., Gaush, P., Postnikova, E., and Frederick, R. (2001). On-site one hour PCR diagnosis of bacterial diseases. APS/MSA/SON Joint Meeting APS Abstracts of Presentations. *Phytopathology 91:*S79.

Schmiediche, P. (1986). Breeding potatoes for resistance to bacterial wilt caused by *Pseudomonas solanacearum.* In Persley, G.J. (Ed.), *Bacterial Wilt Disease in Asia and the South Pacific. Proceedings of an International Workshop held at PCARRD* (pp. 105-111). Los Banos, Philippines.

Seal, S.E., Jackson, L.A., Toung, J.P.W., and Daniels, M.J. (1997). Differentiaion of *Pseudomonas solanacearum, Pseudomonas syzygii, Pseudomonas pickettii* and the blood disease bacterium by partial 16S rRNA sequencing: Construction of oligonucleotide primers for sensitive detection by polymerase chain reaction. *Journal of General Microbiology 139:*1587-1594.

Seo, S.T., Furuya, N., Lim, C.K., Takanami, Y., and Tsuchiy, K. (2002). Phenotypic and genetic diversity of *Erwinia carotovora* subsp. *carotovora* strains from Asia. *Journal of Phytopathology 150:*120-127.

Serrano, C., Arce-Johnson, P., Torres, H., Gebauer, M., Gutierrez, M., Moreno, M., Jordana, X., Venegas, A., Kalazich, J., and Holuique, L. (2000). Expression of the chicken lysozyme gene in potato enhances resistance to infection by *Erwinia carotovora* subsp. *atroseptica. American Journal of Potato Research 77:*191-199.

Skaptason, J.B. and Burkholder, W.H. (1942). Classification and nomenclature of the pathogene causing "ring rot" of potatoes. *Phytopathology 32:*439-441.

Śledź, W., Jafra, S., Waleron, M., and Łojkowska, E. (2000). Genetic diversity of *Erwinia carotovora* strains isolated from infected plants grown in Poland. *EPPO/OEPP Bulletin 30:*403-407.

Smith, I.M., McNamara, D.G., Scott, P.R., and Holderness, M. (1997). *Ralstonia solanacearum.* In *Quarantine Pest for Europe,* Second edition. Wallingford, U.K.: EPPO/CABI.

Smith, J.J., van der Wolf, J.M., Feuillade, R., Trigalet, A., Offord, L.C., and Saddler, G.S. (1998). Genetic diversity amongst *Ralstonia solanacearum* isolates of potato in Europe. *EPPO/OEPP Bulletin 28:*83-84.

Soltani, N., Conn, K.L., Abbasi, P.A., and Lazarovits, G. (2002). Reduction of potato scab and verticillium wilt with ammonium lignosulfonate soil amendment in four Ontario potato fields. *Canadian Journal of Plant Pathology 24:*332-339.

Stevenson, W.R., Loria, R., Franc, G.D., and Weingartner, D.P. (2001). *Compendium of Potato Diseases,* Second edition. St. Paul, MN: American Phytopathological Society.

Swanepol, A.E. (1990). The effect of temperature on the development of wilting and on progeny tuber infection of potatoes inoculated with South African strains of biovar 2 and 3 of *Pseudomonas solanacearum. Potato Research 33:*287-290.

Taylor, R.J., Secor, G.A., Ruby, C.L., and Orr, P.H. (1993). Tuber yield, soft rot resistance, bruising resistance and processing quality in a population of potato (cv. Crystal) somaclones. *American Potato Journal 70:*117-130.

Timms-Wilson, T.M., Bryant, K., and Bailey, M.J. (2001). Strain characterization and 16S-23S probe development for differentiating geographically dispersed isolates of the phytopathogen *Ralstonia solanacearum. Environmental Microbiology 3:*785-797.

Valkonen, J.P.T., Bång, H., Jaakola, T., Kortemaa, H., Kreuze, J.F., Lehtonen, M., Palohuhta, J.P., Rantala, H., Vilham, K., Virtanen, E. et al. (2002). Common scab pathogens in northern Scandinavia. In Wenzel, G. and Wulfert, I. (Eds.), *Abstracts of Papers and Posters, 15th Triennial Conference of the EAPR: Potatoes Today and Tomorrow* (p. 140). Hamburg, Germany: EAPR.

van der Wolf, J.M., Mansfeld-Giese, K., Müller, P., Karjalainen, R., and Stead, V. (2002). Epidemiological studies for control of *Clavibacter michiganensis* ssp. *sepedonicus,* the causal agent of bacterial ring rot in potato. In Wenzel, G. and Wulfert, I. (Eds.), *Abstracts of Papers and Posters, 15th Triennial Conference of the EAPR: Potatoes Today and Tomorrow* (p. 142). Hamburg, Germany: EAPR.

van Elsas, J.D., Kastelein, P., van Bekkum, P., van der Wolf, J.M., de Vries, P.M., and van Overbeek, L.S. (2000). Survival of *Ralstonia solanacearum* biovar 2, the causative agent of potato brown rot, in field and microcosm soils in temperate climates. *Phytopathology 90:*1358-1366.

van Elsas, J.D., van Overbeek, L.S., Smalla, K., Bailey, M.J., Trigalet, A. Marco, Y., and Lopes, M.M. (2002). The fate, activity and spread of *R. solanacearum* biovar 2 in European habitats. In *Book* of *Abstracts of 3rd International Bacterial Wilt Symposium* (p. 5). South Africa.

Vauterin, L., Hoste, B., Kersters, K., Swings, J. (1995). Reclassification of *Xanthomonas. International Journal of Systematic Bacteriology 45:*472-489.

Waleron, M., Waleron, K., Podhajska, A.J., and Łojkowska, E. (2002). Genotyping of bacteria belonging to the former *Erwinia* genus by PCR-RFLP analysis of *rec*A gene fragment. *Microbiology 148:*583-595.

Watanabe, K.N., El-Nashaar, H.M., and Iwanaga, M. (1992). Transmission of bacterial wilt resistance by First Division Restitution (FDR) 2n pollen via 4x-2x crosses in potatoes. *Euphytica 60:*21-26.

Waterer, D. (2002). Impact of high soil pH on potato yields and grade losses to common scab. *Canadian Journal of Plant Science 82:*583-586.

Wegener, C.D. (2002). Induction of defence responses against *Erwinia* soft rot by an endogenous pectate lyase in potatoes. *Physiological and Molecular Plant Pathology 60:*91-100.

Wu, G., Shortt, B.J., Lawrence, E.B., Levine, E.B., Fitzsimmons, K.C., and Shah, D.M. (1995). Disease resistance conferred by expression of a gene encoding $H_2O_2$-generating glucose oxidase in transgenic potato plants. *Plant Cell 7:*1357-1368.

Yabuuchi, E., Y. Kosako, I. Yano, H. Hotta, and Nishiuchi, Y. (1995). Transfer of two *Burkholderia* and an *Alcaligenes* species to *Ralstonia* gen. nov.: Proposal of *Ralstonia pickettii* (Ralston, Palleroni and Doudoroff, 1973) comb. nov., *Ralstonia solanacearum* (Smith 1896) comb. nov. and *Ralstonia eutropha* (Davis 1969) comb. nov. *Microbiology and Immunology 39:*897-904.

Zimnoch-Guzowska, E., Lebecka, R., and Pietrak, J. (1999). Soft rot and blackleg reactions in diploid potato hybrids inoculated with *Erwinia* ssp. *American Journal of Potato Research 76:*199-207.

Zimnoch-Guzowska, E. and Łojkowska, E. (1993). Resistance to *Erwinia* in diploid potatoes with high starch content. *Potato Research 36:*177-182.

Zimnoch-Guzowska, E., Marczewski, W., Lebecka, R., Flis, B., Schäfer-Pregl, R., Salamini, F., and Gebhardt, C. (2000). QTL analysis of new sources of resistance to *Erwinia carotovora* ssp. *atroseptica* in potato done by AFLP, RFLP and resistance-gene-like markers. *Crop Science 40:*1156-1167.

# Chapter 11

# Viruses and Viroids

Colin Jeffries
H. Barker
S. M. Paul Khurana

Potato is infected by at least 40 different viruses and two viroids naturally, and experimentally it may be infected by at least a further 22 viruses and four viroids (Khurana, 2004; Verhoeven et al., 2004). Since viruses are systemic, they are transmitted through tubers or plants and thus spread from one generation to the next. As infection levels increase, crop productivity and quality decline. Historically, this phenomenon was attributed to physiological rather than pathological causes and referred to as "decline," "degeneration," "wearing out," or "senility." Records describing this process date back to the mid-eighteenth century in England (Salaman, 1926). Symptoms were described as leafroll but probably included mosaic and crinkle diseases also, and severe reductions in yield were observed. In this chapter we describe aspects of viruses and viroids important to potato production.

## *VIRUSES (AND VIROIDS) AND THEIR TRANSMISSION*

The principal features of most of the characterized viruses and viroids and uncharacterized virus-like diseases that have been found infecting potato naturally are shown in Table 11.1 and their distribution and particle morphology, including nucleic acid type, are summarized in Table 11.2.

In the majority of countries, viruses transmitted by aphid vectors are the most important; however, viruses transmitted by beetles, leafhoppers, nematodes, thrips, whitefly, and fungi and protista can also cause problems. Some viruses do not require a vector and can be spread by contact or mechanical means. Until fairly recently it was considered that true (botanical) potato seed (TPS) presented little risk in the transmission of potato pathogens, but the discovery of *Potato spindle tuber viroid* (PSTVd) (Singh, 1970) and transmission of some viruses by TPS (Table 11.3) changed this view. Viruses can be grouped according to whether they are dependent or independent of the potato for survival and spread (Salazar, 1996a; Table 11.1). Dependent viruses usually show a restricted host range and the potato is the main source of inoculum, whereas independent viruses have a wide host range and other plant species may be the primary inoculum source.

### *Aphid-Transmitted Viruses*

Aphids transmit more potato viruses than any other vector (Table 11.1), including those in the genera *Alfamovirus, Carlavirus, Cucumovirus, Polerovirus, Potexvirus,* and *Potyvirus.* The most important aphid-transmitted viruses are *Potato leafroll virus* (*PLRV*) (genus *Polerovirus*) and *Potato virus Y* (PVY) (genus *Potyvirus*). *Potato virus S* (PVS) and *Potato virus M* (PVM) (genus *Carlavirus*) are also important in some areas.

Within the PVY species a number of strain groups (PVY$^{C}$, PVY$^{O}$, PVY$^{N}$) have been determined based on local and systemic symptoms in potato and indicator plant species and also by serological and molecular methods. Within the PVY$^{N}$ strain there are subgroups (pathotypes according to Blanco-Urgoiti et al. [1998]): PVY$^{NTN}$ and PVY$^{N}$-W. Another strain, PVY$^{Z}$, has been proposed by Jones (1990), but Blanco-Urgoiti et al. (1998) consider this to be a pathotype of PVY$^{O}$. The relative incidence of various strains appears to be changing. In western Europe, PVY$^{O}$ has long been considered to be the

TABLE 11.1. Potato viruses (and viroids) reported as naturally infecting potato, their mode of transmission, self-elimination, and whether dependent on potato for survival and spread.

| Pathogen[a] | Acronym | Genus | Transmission | | | Dependent on potato for spread and survival? | Reference for natural infection in potato |
|---|---|---|---|---|---|---|---|
| | | | Vector | Persistence in vector | Other means of transmission and self-elimination | | |
| Virus | | | | | | | |
| *Alfalfa mosaic virus* | AMV | *Alfamovirus* | Aphid | Nonpersistent | TPS. Self-eliminating | No | Jeffries (1998) |
| *Andean potato latent virus* | APLV | *Tymovirus* | Beetle | Nonpersistent | Mechanical, Contact, Pollen, TPS | Yes | Jeffries (1998) |
| *Andean potato mottle virus* | APMoV | *Comovirus* | Beetle | Nonpersistent? | Mechanical? Contact | Yes | Jeffries (1998) |
| *Arracacha virus B (oca strain)* | AVB-O | *Nepovirus* | Unknown | | Pollen, TPS | No? | Jeffries (1998) |
| *Beet curly top virus* | BCTV | *Curtovirus* | Leafhopper | Persistent | | No | Jeffries (1998) |
| *Belladona mottle virus* | BeMV | *Tymovirus* | Beetle | Nonpersistent | Mechanical | No | Salazar (1996a) |
| *Cucumber mosaic virus* | CMV | *Cucumovirus* | Aphid | Nonpersistent | | No | Jeffries (1998) |
| *Eggplant mottled dwarf virus* | EMDV | *Nucleorhabdovirus* | Aphid? | | Mechanical | No | Jeffries (1998) |
| *Groundnut bud necrosis virus* (synonym: Peanut bud necrosis virus ) | GBNV | *Tospovirus* | Thrip | Persistent | | No | Jain et al. (2004) |

TABLE 11.1 *(continued)*

| Pathogen[a] | Acronym | Genus | Transmission | | | Dependent on potato for spread and survival? | Reference for natural infection in potato |
|---|---|---|---|---|---|---|---|
| | | | Vector | Persistence in vector | Other means of transmission and self-elimination | | |
| *Groundnut ringspot virus* | GRSV | *Tospovirus* | Thrip | Persistent | | No | Jeffries (1998) |
| *Impatiens necrotic spot virus* | INSV | *Tospovirus* | Thrip | Persistent | | No | Jeffries (1998) |
| *Potato aucuba mosaic virus* | PAMV | *Potexvirus* | Aphid[b] | Nonpersistent | Mechanical, Contact | Yes | Jeffries (1998) |
| *Potato black ringspot virus* | PBRSV | *Nepovirus* | Nematode? | Persistent | Pollen, TPS | Yes | Jeffries (1998) |
| Potato deforming mosaic virus | PDMV | *Begomovirus* | Whitefly | Persistent | | Not known | Jeffries (1998) |
| *Potato latent virus* | PotLV | *Carlavirus* | Aphid | Nonpersistent? | Mechanical? | Yes? | Jeffries (1998) |
| *Potato leafroll virus* | PLRV | *Polerovirus* | Aphid | Persistent | | Yes | Jeffries (1998) |
| *Potato mop-top virus* | PMTV | *Pomovirus* | Protista | Persistent | Self-eliminating | Yes | Jeffries (1998) |
| Potato rough dwarf virus | PRDV | *Carlavirus* | Aphid | Nonpersistent? | Mechanical? Contact? | Yes? | Jeffries (1998) |
| *Potato virus A* | PVA | *Potyvirus* | Aphid | Nonpersistent | | Yes | Jeffries (1998) |
| *Potato virus M* | PVM | *Carlavirus* | Aphid | Nonpersistent | Mechanical, Contact (some isolates) | Yes | Jeffries (1998) |
| Potato virus P | PVP | *Carlavirus* | Aphid | Nonpersistent? | Mechanical? Contact? | Yes? | Jeffries (1998) |
| *Potato virus S* | PVS | *Carlavirus* | Aphid (some isolates) | Nonpersistent | Mechanical, Contact | Yes | Jeffries (1998) |
| *Potato virus T* | PVT | *Trichovirus* | None | | Mechanical, Contact, Pollen, TPS | Yes | Jeffries (1998) |

| | | | | | | | |
|---|---|---|---|---|---|---|---|
| *Potato virus U* | PVU | *Nepovirus* | Nematode | Persistent? | | Not known | Jeffries (1998) |
| *Potato virus V* | PVV | *Potyvirus* | Aphid | Nonpersistent | | Yes? | Jeffries (1998) |
| *Potato virus X* | PVX | *Potexvirus* | None | | Mechanical, Contact | Yes | Jeffries (1998) |
| *Potato virus Y* | PVY | *Potyvirus* | Aphid | Nonpersistent | Mechanical? Contact? | Yes | Jeffries (1998) |
| *Potato yellow dwarf virus* | PYDV | *Nucleorhabdo virus* | Leafhopper | Persistent | Self-eliminating | No | Jeffries (1998) |
| *Potato yellow mosaic virus*[c] | PYMV | *Begomovirus* | Whitefly | Persistent? | | No | Jeffries (1998) |
| Potato yellow vein virus | PYVV | *Crinivirus* | Whitefly | Semipersistent | | No? | Jeffries (1998) |
| Potato yellowing virus | PYV | *Alfamovirus* | Aphid | Semipersistent | Pollen, TPS | Yes | Jeffries (1998) |
| Solanum apical leaf curling virus | SALCV | *Begomovirus* | Unknown | | | No | Jeffries (1998) |
| *Sowbane mosaic virus* | SoMV | *Sobemovirus* | Leafhopper, Leafminer | Nonpersistent | Mechanical | No | Jeffries (1998) |
| *Tobacco mosaic virus* | TMV | *Tobamovirus* | None | | Mechanical, Contact | No | Jeffries (1998) |
| *Tobacco necrosis virus* | TNV | *Necrovirus* | Fungus | Nonpersistent | | No | Jeffries (1998) |
| *Tobacco rattle virus* | TRV | *Tobravirus* | Nematode | Persistent | TPS? | No | Jeffries (1998) |
| *Tobacco streak virus* | TSV | *Ilarvirus* | Thrip? | Persistent? | Self-eliminating | No | Jeffries (1998) |
| *Tomato black ring virus* | TBRV | *Nepovirus* | Nematode | Persistent | Pollen? TPS | No | Jeffries (1998) |
| *Tomato chlorotic spot virus* | TCSV | *Tospovirus* | Thrip | Persistent | | No | Jeffries (1998) |
| *Tomato leaf curl New Dehli virus* | ToLCNDV | *Begomovirus* | Whitefly | Persistent | Mechanical | No | Garg et al. (2001); Usharani et al. (2004) |
| *Tomato mosaic virus* | ToMV | *Tobamovirus* | None | | Mechanical, Contact | No | Jeffries (1998) |
| *Tomato mottle Taino virus* | ToMoTV | *Begomovirus* | Whitefly | Persistent? | | No | Cordero et al. (2004) |
| *Tomato spotted wilt virus* | TSWV | *Tospovirus* | Thrip | Persistent | | No | Jeffries (1998) |
| *Tomato yellow leaf curl virus* | TYLCV | *Begomovirus* | Whitefly | Persistent | | No | Abdel-Salam (1991) |

TABLE 11.1 *(continued)*

| Pathogen[a] | Acronym | Genus | Transmission | | | Dependent on potato for spread and survival? | Reference for natural infection in potato |
|---|---|---|---|---|---|---|---|
| | | | Vector | Persistence in vector | Other means of transmission and **self-elimination** | | |
| *Tomato yellow mosaic virus*[c] | ToYMV | *Begomovirus* | Whitefly | Pesistent | | No | Debrot and Centeno (1985) |
| *Wild potato mosaic virus* | WPMV | *Potyvirus* | Aphid | Nonpersistent | | Yes | Jeffries (1998) |
| Viroid | | | | | | | |
| *Mexican papita viroid* | MPVd | *Pospiviroid* | | | Contact, Pollen, TPS | Yes? (*Solanum cardiophyllum*) | Martinez-Soriano et al. (1996) |
| Potato spindle tuber viroid | PSTVd | *Pospiviroid* | Aphid[d] | | Contact, Pollen, TPS | Yes | Jeffries (1998) |
| Uncharacterized virus and virus-like diseases | | | | | | | |
| Potato deforming mosaic disease | PDMD | *Unknown* | Unknown | | | | Jeffries (1998 |

*Notes:* Potato stunt disease (Cockerham and McGhee, 1953) is not included since it has not been characterized and has not been reported again since its original discovery. An uncharacterized virus isolated from *Solanum hannemannii* (Horvath et al., 1993) that is thought to be related to *Belladonna mottle virus* (Salazar, 1996a), SB-23, which is serologically related to *Sowbane mosaic virus* (Salazar, 1996a), and Solanum yellows virus or Southern potato latent *virus* (Brunt et al., 1996) are not included, since little information is available. Potato deforming mosaic virus = *Tomato yellow vein streak virus* (ToYVSV) (Ribero et al. 2005). Potato rough dwarf virus is a strain of Potato virus P. (Nisbet et al., 2005).

[a]For virus names and acronyms the nomenclature approved by the International Committee on Taxonomy of Viruses (ICTV) is used (see http://www.ncbi.nlm.nih.gov/ICTVdb). Names in italic are approved by the ICTV.

[b]Helper component required for transmission (i.e., PVA or PVY).

[c]Morales et al. (2001) concluded that PYMV is a synonym of ToYMV. However, currently both are listed as distinct viruses in the ICTV database (http://www._Hlt70391853._Hlt70391853ncbi.nlm.nih.gov/ICTVdb/).

[d]Aphid-borne when plants co-infected with PLRV.

? = not conclusive.

TABLE 11.2. Summary of viroid and virus properties and their distribution.

| Pathogen | Genus | 1 | 2 | 3 | 4 | 5 | Geographical distribution (Jeffries, 1998) |
|---|---|---|---|---|---|---|---|
| Virus | | | | | | | |
| *Alfalfa mosaic virus* | *Alfamovirus* | Bac | RNA | 4-5 | 18 | 19-58 | Worldwide where alfalfa is grown, but uncommon in potatoes |
| *Andean potato latent virus* | *Tymovirus* | Iso | RNA | 1 | 28-30 | N/A | Andean region of S. America (Bolivia, Colombia, Ecuador, Peru) |
| *Andean potato mottle virus* | *Comovirus* | Iso | RNA | 1 | 28 | N/A | Andean region of S. America (Chile, Ecuador, Peru, Brazil). A potato and tabasco pepper infecting strain has been found in Honduras, Nicaragua |
| *Arracacha virus B* (oca strain) | *Nepovirus* | Iso | RNA | 2 | 26 | N/A | Bolivia and Peru in oca (*Oxalis* tuberosa) and Peru in potatoes. Rare in potatoes |
| *Beet curly top virus* | *Curtovirus* | Seg | DNA | 2 | 18-20 | 32-35 | Arid and semiarid regions of Africa, Australia, Middle East, the eastern Mediterranean basin, N. (Mexico to Canada), Central, and S. America (Argentina to Costa Rica). Usually causes only localized problems in potato in southeastern S. America and western N. America. It occurs only rarely in potatoes in the United States because most potato production occurs outside the range of the vector (Radcliffe et al., 1993) |

TABLE 11.2 *(continued)*

| Pathogen | Genus | 1 | 2 | 3 | 4 | 5 | Geographical distribution (Jeffries, 1998) |
|---|---|---|---|---|---|---|---|
| *Belladona mottle virus* | *Tymovirus* | Iso | RNA | 1 | 30 | N/A | Europe, S. America. Very rare in potato |
| *Cucumber mosaic virus* | *Cucumovirus* | Iso | RNA | 1 | 30 | N/A | Worldwide, but only occasionally reported in potato from Europe, Egypt, India, Saudi Arabia |
| *Eggplant mottled dwarf virus* | *Nucleorhabdovirus* | Bac | RNA | 1 | 76 | 190 | Mediterranean basin and the Middle East in solanaceous hosts. In potato only from Iran |
| *Groundnut bud necrosis virus* | *Tospovirus* | Env | RNA | 3 | 80-110 | N/A | India, Southeast Asia, but has only been reported infecting potato in India |
| *Groundnut ringspot virus* | *Tospovirus* | Env | RNA | 3 | 80-110 | N/A | Argentina, Brazil, South Africa, but has only been reported infecting potato in Argentina |
| *Impatiens necrotic ringspot virus* | *Tospovirus* | Env | RNA | 3 | 80-110 | N/A | Europe, United States, but has only been reported infecting various glasshouse grown potato species in the Netherlands |
| *Potato aucuba mosaic virus* | *Potexvirus* | Fil | RNA | 1 | 11 | 580 | Worldwide, but rare in potato |
| *Potato black ringspot virus* | *Nepovirus* | Iso | RNA | 2 | 26 | N/A | Peru |
| Potato deforming mosaic virus | *Begomovirus* | Seg | RNA | 2 | 18 | 30 | Brazil |
| *Potato latent virus* | *Carlavirus* | Fil | RNA | 2 | 12 | 525 and 625 | N. and S. America (L. Salazar, personal communication) |

| | | | | | | | |
|---|---|---|---|---|---|---|---|
| *Potato leafroll virus* | *Luteovirus* | Iso | RNA | 1 | 24 | N/A | Worldwide |
| *Potato mop-top virus* | *Pomovirus* | Rod | RNA | 3 | 18-20 | 100-150 250-300 | Mainly cooler climates of the Andean region of S. America (e.g., Peru), China, Japan, northern Europe (e.g., United Kingdom, Denmark), N. America (Lambert et al., 2003) |
| Potato rough dwarf virus | *Carlavirus* | Fil | RNA | 1 | 12 | 638 | Argentina |
| *Potato virus A* | *Potyvirus* | Fil | RNA | 1 | 11 | 730 | Worldwide, but not the Andean region of S. America and uncommon in some other countries such as India |
| *Potato virus M* | *Carlavirus* | Fil | RNA | 1 | 12 | 650 | Worldwide, but not the Andes |
| Potato virus P | *Carlavirus* | Fil | RNA | 1 | 12 | 640 | Brazil |
| *Potato virus S* | *Carlavirus* | Fil | RNA | 1 | 12 | 660 | PVS$^{O}$ worldwide. PVS$^{A}$ Andean region of S. America, Germany, the Netherlands, New Zealand, United States |
| *Potato virus T* | *Trichovirus* | Fil | RNA | 1 | 12 | 640 | Bolivia, Peru |
| *Potato virus U* | *Nepovirus* | Iso | RNA | 2 | 28 | N/A | Peru. Only found once in potato. |
| *Potato virus V* | *Potyvirus* | Fil | RNA | 1 | 12-13 | 700-720 | Bolivian highlands, Peru, northern Europe (e.g. France, Germany, the Netherlands), United States |

TABLE 11.2 *(continued)*

| Pathogen | Genus | 1 | 2 | 3 | 4 | 5 | Geographical distribution (Jeffries, 1998) |
|---|---|---|---|---|---|---|---|
| *Potato virus X* | *Potexvirus* | Fil | RNA | 1 | 13 | 515 | Serotype $PVX^O$ worldwide and common in potato. Serotype $PVX^A$ restricted: CP strain to the south central Andes of Peru and the HB strain mainly to the Bolivian plateau around Lake Titicaca. |
| *Potato virus Y* | *Potyvirus* | Fil | RNA | 1 | 11 | 740 | $PVY^O$ worldwide. $PVY^C$ Australia, Ecuador, Europe, India, N. America, South Africa, New Zealand. $PVY^N$ in Africa, Europe, S. America (Argentina, Chile, Colombia, Peru), New Zealand, with localized outbreaks in N. America. $PVY^{NTN}$ Europe, Israel, Lebanon. |
| *Potato yellow dwarf virus* | *Nucleorhabdovirus* | Bac | RNA | 1 | 75 | 380 | Canada and northern United States (although there have been no reports in potatoes for 50 years), Saudi Arabia. A rhabdovirus similar to PYDV has been reported from South Russia and the Ukraine |
| *Potato yellow mosaic virus* | *Begomovirus* | Seg | DNA | 2 | 18-20 | 30 | Venezuela in potato. Guadeloupe, Martinique, Puerto Rico, Trinidad, Tobago, Venezuela in tomato |
| Potato yellow vein virus | *Crinivirus* | Fil | RNA | 2 | 10-13 | 950-2000 | Colombia, Ecuador, Peru, Venezuela |

| | | | | | | | |
|---|---|---|---|---|---|---|---|
| Potato yellowing virus | *Alfamovirus* | Bac | RNA | 4-5 | 18 | 21-60 | Peru, Chile |
| Solanum apical leaf curling virus | *Begomovirus* | Seg | DNA | 3 | 18 | .50 | Peru. Rare in potato, best adapted to tropical conditions. |
| *Sowbane mosaic virus* | *Sobemovirus* | Iso | RNA | 1 | 26-28 | N/A | Worldwide in Chenopodiaceae. Extremely rare in potato |
| *Tobacco mosaic virus* | *Tobamovirus* | Rod | RNA | 1 | 18 | 300 | Worldwide in crops other than potato. In potato occasionally found in Argentina (*Solanum commersonii*), China, India, Peruvian Andes, Saudi Arabia |
| *Tobacco necrosis virus* | *Necrovirus* | Iso | RNA | 1 | 26 | N/A | Worldwide distribution, but in potato only occasionally reported from Europe and N. America. |
| *Tobacco rattle virus* | *Tobravirus* | Rod | RNA | 2 | 22 | 50-115, 190 | Europe, China, Japan, New Zealand, the former USSR, N., Central, and S. America |
| *Tobacco streak virus* | *Ilarvirus* | Quasi-iso | RNA | 3 | 22-35 | N/A | Probably worldwide, but only reported in potato from Brazil, Peru |
| *Tomato black ring virus* | *Nepovirus* | Iso | RNA | 2 | 28 | N/A | Various crop plants infected, mainly Europe but also Asia, Africa, N. and S. America. Occasionally reported infecting potatoes in Europe (e.g., Germany, Poland, the United Kingdom) |
| *Tomato chlorotic spot virus* | *Tospovirus* | | | | | | Brazil, Argentina |

TABLE 11.2 *(continued)*

| Pathogen | Genus | 1 | 2 | 3 | 4 | 5 | Geographical distribution (Jeffries, 1998) |
|---|---|---|---|---|---|---|---|
| *Tomato leaf curl New Delhi virus* | *Begomovirus* | Seg | DNA | 1 | 18 | 30 | India, Pakistan, but has only been reported infecting potato in India (Usharani et al., 2004) |
| *Tomato mosaic virus* | *Tobamovirus* | Rod | RNA | 1 | 18 | 300 | Worldwide, but only occasionally reported from potato and *Solanum* spp. |
| *Tomato mottle Taino virus* | *Begomovirus* | | DNA | 2 | Unknown | Unknown | Cuba (Cordero, 2003) |
| *Tomato spotted wilt virus* | *Tospovirus* | Env | RNA | 3 | 70-110 | N/A | Widespread. On potatoes reported mainly from Argentina, Australia, Brazil, India, South Africa. It has also been observed in Portugal, the Netherlands and the United States |
| *Tomato yellow leaf curl virus* | *Begomovirus* | Seg | DNA | 1 | 18 | 30 | Widespread, but has only been reported infecting potato in Egypt (Abdel-Salem, 1991) |
| *Wild potato mosaic virus* | *Potyvirus* | Fil | RNA | 1 | 11 | 685-800 | Peru |
| Viroid | | | | | | | |
| *Mexican papita viroid* | *Pospiviroid* | None | RNA | Circular ssRNA only | | | Mexico (Martinez-Soriano et al., 1996) |

| | | | | | |
|---|---|---|---|---|---|
| *Potato spindle tuber viroid* | *Pospiviroid* | None | RNA | Circular ssRNA only | China, parts of Eastern Europe including the former USSR, India, N. America. Systematic indexing over the past 10-20 years has apparently eliminated or significantly reduced PSTVd from commercial seed and ware production schemes in N. America and parts of Eastern Europe. |

*Columns:*

1. Particle morphology: Bac = bacilliform; Env = enveloped particles; Fil = filamentous; Iso = isometric; Quasi-iso = quasi-isometric; Rod = rod or tubular shaped; Seg = segmented (i.e., geminiviruses that occur as bi- or trigeminate particles in addition to the single particles; bigeminate forms are usually predominant and the most stable).
2. Type of nucleic acid.
3. Number of distinct particle sizes.
4. Particle diameter in nanometers.
5. Particle length for rods, filaments, segmented (geminiviruses), and bacilliform particles in nanometers. N/A = not applicable.

most abundant (Robert et al., 2000), but in some areas it has been almost completely replaced by $PVY^{N}$ (Weidemann, 1988), $PVY^{NTN}$ (Wolf and Horvath, 2001), or $PVY^{N}$-W (Chrzanowska, 2002).

The Andean strain of PVS ($PVS^{A}$), which has a restricted distribution (Table 11.2), may cause more severe symptoms than the ordinary strain ($PVS^{O}$) (Dolby and Jones, 1987) and it has been reported to be more efficiently transmitted by aphids than $PVS^{O}$ (Slack, 1983).

Depending on the length of time that aphids are able to transmit a virus after acquisition, they transmit viruses in a persistent, semipersistent, or nonpersistent manner (Hull, 2002). The only potato virus transmitted in a persistent manner is PLRV, which is circulative but does not propagate within the vector. There are three stages in the transmission cycle of persistently transmitted viruses: (1) the acquisition phase, in which the aphid feeds on an infected plant (the phloem) to acquire sufficient virus; (2) the latent period, in which the aphid is not able to transmit the virus; (3) the retention period, during which the aphid can transmit the virus to a healthy host. The acquisition phase for PLRV may last minutes, but days are required for maximum transmission efficiency. The latent period is approximately 24-72 hours, during which the virus passes through the haemocoel and is taken up by the accessory salivary gland. The virus may then be transmitted by the aphid throughout its life, even through molts (Robert and Bourdin, 2001).

For nonpersistently transmitted viruses, such as PVY, only brief and superficial probes of the epidermal cells are required for virus transmission. The aphid acquires virus within seconds to minutes of probing infected plants and may immediately infect other plants. The aphid looses its infectiousness usually within an hour.

Semipersistently transmitted viruses have properties intermediate between persistent and nonpersistent viruses; vectors require probes of around 30 min, and infectivity of the aphid increases with increased feeding times up to approximately 12 hours. Persistence of the virus in the vector usually lasts a few days. Potato yellowing virus (PYV) is believed to be transmitted in this manner (Fuentes, 1991).

For nonpersistent virus transmission, two phases in vector-virus interaction are retention of the virus at a specific site on the aphid (i.e., the stylet) and the release of the virus. For retention, two types of

TABLE 11.3. Rate of PSTVd and virus transmission through TPS.

| Pest[a] | Experimental | Natural |
|---|---|---|
| PSTVd | | 100% (Singh et al., 1992) |
| AMV | <1% in *S. etuberosum* (Valkonen et al., 1992) | |
| APLV | 0<1% in *S. andigena* (Jones, 1982) | < 1% in *S. acaule* (J. Roenhorst and J. Verhoeven, personal communication) |
| AVB-O | 4-12% (Jones, 1982) | <1% (L. Salazar, personal communication) |
| PBRSV (TRSV-Ca) | 2-9% (Jones, 1982) | |
| PVT | 0-59% (Jones, 1982) | |
| PYV | 20% in *S. brevidens* (Valkonen et al. , 1992) | |
| TBRV | | 0-59% (Jeffries, 1993) |
| TRV | | 1-7% (Horvath et al., 1996); 0% (Dale et al., 2000) |

*Notes:* Transmission through TPS of *Solanum tuberosum* unless stated otherwise. [a]Full names as given in Table 11.1. TRSV-Ca (Tobacco ring spot virus-calico strain) is a strain of PBRSV.

interaction have been identified: a direct interaction with the virus capsid and the involvement of a nonstructural protein (the helper component). For the coat protein (CP), the amino acid triplet DAG (Asp-Ala-Gly) located near the N-terminus and its relationship with surrounding amino acids affect transmissibility (Hull, 2002). Helper components are also required for aphid transmission of potyviruses. Helper components may form a bridge between the capsid and the aphid stylet to permit virus retention (Robert and Bourdin, 2001; Hull, 2002). Some viruses or virus strains (*Potato aucuba mosaic virus* [PAMV] and $PVY^{C}$) appear to have lost their ability to be transmitted by aphids unless the helper component is present, that is, unless aphids first feed on a plant infected with an aphid-transmissible strain or a plant coinfected with both viruses (Salazar, 1996a). In potato, helper-dependent aphid transmission of PSTVd may occur when plants are coinfected with PLRV (Salazar et al., 1995). This is due to encapsidation of the viroid in PLRV particles (Querci et al., 1997).

Aphid species important in the spread of PLRV, because of the acquisition phase and latent period required for transmission, are those that can settle, feed and reproduce on potato (resident aphids) (Ragsdale et al., 2001). There are at least 12 resident aphid species on potato and these are also believed to be the most important vectors of nonpersistently transmitted viruses (e.g., PVY) (Robert and Bourdin, 2001). Among aphids, the most important (and efficient) vector is *Myzus persicae,* although *Aphis gossypii* is also important in subtropical conditions (Rongai et al., 1998). *Macrosiphum euphorbiae* and *Aulacorthum solani* are frequently abundant on potato crops but are less efficient than *M. persicae* as vectors of PLRV and nonpersistent viruses. In addition, 30 more aphid species that are temporary migrants from other crops (such as cereals) and do not colonize and reproduce on potato may transmit nonpersistent viruses because of stylet probings of epidermal cells (Ragsdale et al., 2001).

### *Nematode-Transmitted Viruses*

*Tobacco rattle virus* (TRV), transmitted by free-living nematodes in the genera *Paratrichodorus* and *Trichodorus,* has the widest host range of any plant virus, particularly among weed species. In a few plant species this virus can be transmitted through seed, with up to 10 percent transmission rates (Cooper and Harrison, 1973). Although Horvath et al. (1996) found that TRV could be transmitted by TPS, Dale et al. (2000) found no evidence for this. TRV-infected weeds perpetuate the virus within sites and within nematode populations, and thus TRV can survive for several decades in soil in the absence of the potato crop. No evidence is available to indicate that the virus multiplies in the vector, is retained through the molt, or is passed to the eggs (Robinson and Harrison, 2003).

There are two types of TRV: TRV-M, or normal multiplying type, and TRV-NM, or nonmultiplying type. In the United Kingdom, "stem mottle" plants are usually infected with the TRV-NM type, but in other parts of the world the M type predominates. Depending on the virus-plant interaction reactions to TRV, potatoes are described as resistant (not infected), susceptible (asymptomatic infection), or sensitive (developing spraing symptoms). It was believed that TRV in potato was self-eliminating because of the small number of stem mottle plants growing from spraing-infected seed (Dale and Solomon, 1988). This, however, needs to be reassessed because of the discovery of the M-type TRV infecting asymptomatic tubers and daughter plants (Xenophontos et al., 1998). Also, it was thought that TRV caused losses in only sensitive cultivars, where the appearance of as few as 2 percent spraing-affected tubers could lead to rejection of the consignment by the processing industry and supermarkets. However, recently it was shown that in susceptible potato cultivars, TRV infection can also adversely affect yield and quality traits such as dry matter and after-cooking blackening (Dale et al., 2000).

### *Protistal- and Fungal-Transmitted Viruses*

*Potato mop-top virus* (PMTV) is transmitted in a persistent manner in soil by *Spongospora subterranea* (kingdom *Protista,* which causes powdery scab, tuber cankers, and root galls (Adams, 1991). Similar to TRV, PMTV causes spraing symptoms in tubers.

PMTV-infected *S. subterranea* spores can survive in the soil for at least 10-18 years (Jones and Harrison, 1972). Spores germinate to release zoospores without the presence of a host, but root exudates from some hosts may promote germination (Harrison et al., 1997). PMTV has been detected in primary zoospores (Arif et al., 1995). Infection by zoospores occurs within the range 9-20°C, but the optimum temperature for development of powdery scab is approximately 12°C. Root gall formation is favored by temperatures of 17°C or higher (van de Graaf et al., 2005). Environmental factors, including soil type, soil moisture, and temperature, are thought to be more important than the level of tuber- or soilborne inoculum in influencing levels of powdery scab as, once infection has occurred, the number of secondary zoospores reduces the importance of other sources of inoculum (van de Graaf et al., 2005).

The importance of alternative hosts as sources of PMTV is thought to be low in temperate climates because of the limited number of plant species (e.g., *Solanum nigrum*) that are hosts of both *S. subterranea* and PMTV (Jones, 1988). Alternative hosts may be more important in countries that have a greater range of *Solanum* spp. and host plants such as tobacco and pepper (Jones, 1988). Only a proportion of tubers from PMTV-infected plants become infected, and on replanting only a few develop into infected plants; that is, in the absence of the vector the virus is self-eliminating (Calvert, 1968).

*Tobacco necrosis virus* (TNV), which infects potato only occasionally, is transmitted by zoospores of *Olpidium brassicae* in a nonpersistent manner (Table 11.1). Various strains of this virus have been reported (Adam et al., 1990), representing two serotypes, A and D (Babos and Kassanis, 1963). Both virus and vector have a very wide host range (Harrison, 1977). The virus seems to be carried externally by the zoospores and is not retained by the resting sporangia. Therefore, infection is thought to occur every year from infected hosts. Both vector and virus are favored by low soil temperatures (10-20°C) and high soil moisture (Harrison, 1977), and therefore stocks that are irrigated may be more prone to infection.

### *Thrip-Transmitted Viruses*

A number of thrip-transmitted tospoviruses are known to infect potato (Table 11.1), of which the most important is *Tomato spotted wilt virus* (TSWV). It has a wide host range and has spread worldwide along with its vectors, but on potatoes it still has a restricted distribution (Table 11.2). TSWV is important in areas where vector and virus occur primarily in other crops (e.g., *Lycopersicon esculentum* [tomato] and *Capsicum annuum* [pepper]) or in natural vegetation but then move to potato after harvest or deterioration of the primary host.

*Frankliniella occidentalis, F. schultzei, Thrips palmi,* and *T. tabaci* are the important vectors on potato but other thrip species (i.e., *F. bispinosa, F. fusca, F. intonsa,* and *T. setosus*) can also act as vectors. Tospoviruses are transmitted in a persistent (latent phase of 3-10 days) and propagative manner by thrips but are not passed through the egg. Only the larvae can acquire virus, and acquisition ability decreases with age (Hull, 2002).

### *Leafhopper-Transmitted Viruses*

Two viruses that infect potato are transmitted by leafhoppers: *Beet curly top virus* (BCTV) and *Potato yellow dwarf virus* (PYDV). They are usually of little economic importance (Radcliffe et al., 1993) and PYDV self-eliminates. Neither virus depends on potato for survival, but unfavorable conditions (e.g., drought) force the leafhoppers off the primary host onto potato.

BCTV is transmitted by *Agallia albidula, Agalliana ensigera, Circulifer opacipennis,* C. *tenellus,* and *Orosius argentatus* (Klein, 2001) in a persistent manner. It circulates but does not multiply within the leafhopper and is not transmitted through the egg. The minimum latent period is rarely less than 4 hours, and frequency of transmission is greatest with acquisition times of several days (Thomas and Mink, 1979). The virus has a wide host range, which includes *Phaseolus* spp. (bean) , *Capsicum* spp. (pepper), potato, and tomato. Weed species can be virus reservoirs (Creamer et al., 1996). Various strains exist, which differ in virulence, symptoms, and host range (Thomas and Mink, 1979), and not all strains may be able to infect potato. The virus is phloem limited.

Two serotypes of PYDV have been described: Sanguinolenta yellow dwarf virus (SYDV) transmitted by *Aceratagallia sanguinolenta* and Constricta yellow dwarf virus (CYDV) transmitted by *Agallia constricta. Agallia quadripunctata* is able to transmit both strains (Klein, 2001). The leafhoppers transmit PYDV in a persistent and propagative manner, and the virus is able to overwinter in adult leafhoppers. PYDV has a narrow host range. The common *Chrysanthemum leucanthemum* var. *pinnatifidum* (weed ox-eye daisy) is thought to be the principal source of virus for infecting potato crops (Younkin, 1942). The increasing acreage of alfalfa (a leafhopper host but immune to infection by PYDV) has been implicated in the decline of PYDV in the United States (Radcliffe et al., 1993).

### *Whitefly-Transmitted Viruses*

The begomoviruses Potato deforming mosaic virus (PDMV), *Potato yellow mosaic virus* (PYMV), *Tomato leaf curl New Delhi virus* (ToLCNDV), and *Tomato mottle Taino virus* (ToMoTV) are transmitted by *Bemisia tabaci.* Transmission of begomoviruses by whiteflies is considered to be persistent and circulative, with an acquisition feed of 2-24 hours followed by a latent period of 4-10 hours before transmission occurs. The virus may be retained for 5-20 days, with a gradual loss in transmission efficiency over time (Brown and Bird, 1992). The crinivirus, potato yellow vein virus (PYVV) is transmitted by *Trialeurodes vaporariorum* (Saldarriaga et al., 1988) in a semipersistent manner. Transmission may occur after a minimum feeding period of 30 min with no incubation period (Saldarriaga et al., 1988), and whiteflies lose infectivity after 6 hours if they are not given access to infected plants (Salazar, 1996b).

Other than potato, the natural host range for PDMV (Brazil) is *Solanum chacoense* and *S. sisymbrifolium;* for PYMV *Solanum* spp., tomato, *Manihot esculenta* (cassava), and *P. vulgaris* (bean); and for ToLCNDV and ToMoTV, tomato. In Colombia *Polygonum mepalense* has been identified as an important reservoir for PYVV, with lower virus concentrations found in *Rumex obtusifolium* (broad leaf dock), *Tagetes spp.* (marigold), and *Catharanthus roseus* (Madagascar periwinkle) (Salazar et al., 2000).

In recent years, increases in the number of whitefly-transmitted viruses on potato and in the geographic range of geminiviruses have been reported (Polston and Anderson, 1997). A number of reasons lie behind this. All the viruses have alternative hosts, and potatoes are now more likely to be grown in association with these alternative hosts. In addition, for viruses transmitted by *B. tabaci,* biotype B has emerged, which can exploit solanaceous hosts for both feeding and reproduction more successfully than local species. Until the mid-1980s, many of the *B. tabaci* biotypes fed and reproduced on tomato to only a limited extent, minimizing transmission of geminiviruses from tomato plants. Since then the *B. tabaci* biotype B has spread from the Mediterranean, probably on traded plants, to infest countries in North, Central, and South America and the Caribbean

(Polston and Anderson, 1997). In the north Indian plains, ToLCNDV is thought to have spread to potato from tomato because of an extension of the potato growing season into the tomato growing season, and this has resulted in a buildup of whitefly populations on potato (Garg et al., 2001; Usharani et al., 2004).

The spread of PYVV in Colombia has followed the rise in vector population in potato fields, which has been exacerbated by the intercropping of *P. vulgaris* (bean) with potato. Although not a host of PYVV, bean is a favorite host for *T. vaporariorum.* In Colombia, the disease and vector occur under a wide range of environmental conditions from the cool Bogota savana to the warmer Rionegro (Salazar, 1996b).

### *Beetle-Transmitted Viruses*

*Andean potato latent virus* (APLV) is transmitted by *Epitrix* spp. (flea beetle) and *Andean potato mottle virus* (APMoV) has been shown to be transmitted by *Diabrotica* spp. in glasshouse experiments (Salazar, 1996a). Leaf-feeding beetles do not have salivary glands and feed by biting leaf tissue and regurgitating, which bathes the mouthparts with infective plant sap. Transmission is not a simple mechanical process since there is a substantial degree of specificity between the virus and the vector, and the virus may be retained for long periods. Fulton et al. (1987) have reviewed beetle transmission of viruses.

### *Mechanical Transmission*

The most important potato viruses transmitted through contact are *Potato virus X* (PVX), PVS, and the viroid PSTVd. APLV and APMoV are also transmitted in this manner (Salazar, 1996a). This type of transmission occurs when infective sap contaminates wounds in healthy plants. These viruses can be spread by clothing, rabbits and dogs, contaminated tractor wheels, cultivating and ridging equipment (Manzer and Merriam, 1961), and tuber cutting knives (Franc and Banttari, 1984). Contaminated surfaces can retain infectivity from hours to weeks depending on the type of surface, virus, and environmental conditions (Franc and Banttari, 1984).

### Seed Transmission

The viruses transmitted through TPS are listed in Table 11.3. The most important pathogen transmitted in this way is PSTVd. Seed-borne transmission is an important issue for gene banks (Jeffries et al., 1993) and producers of TPS.

## SYMPTOMS AND DETECTION

Some typical symptoms of various viruses (viroids) are illustrated in Figure 11.1 and summarized in Table 11.4. The type of symptom is, however, dependent on a number of factors including the virus strain, the cultivar, whether the infection is primary or secondary, and interactions with other viruses and growing conditions. This means that the pathogen may give distinct symptoms in some circumstances but not in others. In addition, different potato species may react differently to virus infection. For example PLRV in *S. tuberosum* ssp. *tuberosum* induces leafrolling. In ssp. *andigena*, however, chlorotic dwarfing is produced, known as "enanismo amarillo" in the Andes (Salazar, 1996a). A number of methods are used in the detection and diagnosis of viruses and viroids. The choice of method depends upon the resources available and the level of certainty required in the result. Ideally, detection methods should be sensitive, specific, reliable, reproducible, and cheap. In practice, however, the methods used are a compromise of these characteristics.

### Visual Inspection

Visual inspection is the simplest detection method. Large quantities of material can be screened rapidly. Where symptoms are produced reliably, visual inspection may be more sensitive than laboratory tests. Seed potato certification still relies heavily on visual inspection by experienced, trained inspectors.

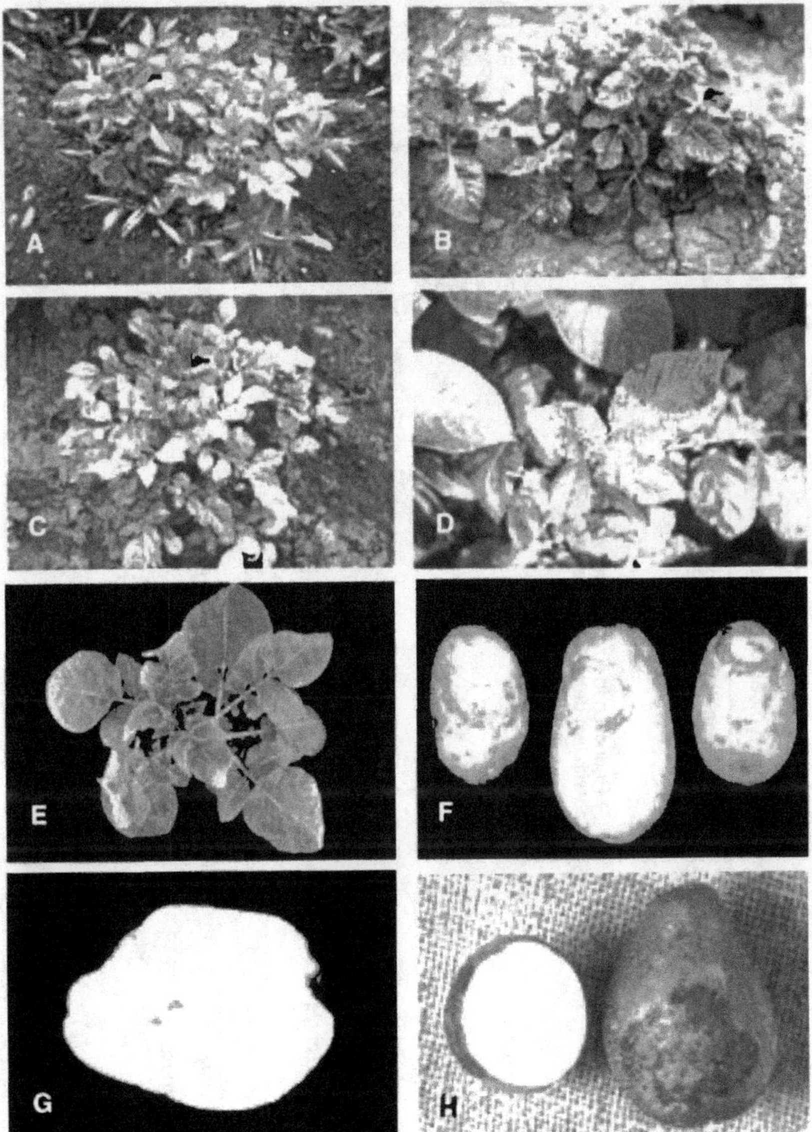


FIGURE 11.1. Symptoms caused by some important viruses. See Plate 11.1 in the color gallery.

TABLE 11.4. Key of some typical disease symptoms caused by virus (viroid) infection.

| Disease | Disease description | Causal virus and comments |
|---|---|---|
| A. Foliage | | |
| 1. Stunting | Reduction in size of plant. When this is combined with leaf deformation it is referred to as dwarfing | |
| 1.1 Whole plant | | BCTV, EMDV, PAMV, PLRV, PRDV, PSTVd (severe strains), PVM (with leaf deformation), PYDV, PYMV, TBRV (some isolates), TMV, ToLCNDV, ToYLCV |
| 1.2 Some shoots | | PMTV, PVM, TBRV, TRV |
| 2. Changes in angle of leaf to main stem | | |
| 2.1 Epinasty | Leaves form obtuse angles with the main stem | EMDV |
| 2.2 Erectness | Leaves form acute angles with stem | PSTVd (for severe strains), BCTV, PLRV |

| | | |
|---|---|---|
| 3. Color changes | | |
| 3.1 Mild mosaic (or mottle) | Mild pale green or pale chlorotic areas in the leaf that may be best observed under cloudy rather than sunny conditions | PSTVd, AMV, APLV (also sometimes chlorotic netting of minor veins), CMV, PVA, PVM (also crinkling of leaves), PVS (also leaf bronzing), PVV, PVX, PVY (depends on strain and potato cv.), PYV (whole plant may become yellow), SoMV, TMV, TRV (few stems), ToMV, ToLCNDV |
| 3.2 Severe mosaic (or mottle) | Very obvious pale green or chlorotic areas in the leaf. Plants may sometimes be stunted with leaf rugosity | PSTVd, APMV, EMDV, PVM (some isolates in some cvs.), PVV (some cvs.), PVX (some isolates in some cvs. and particularly if in combination with PVA or PVY), PVY (depends on strain and potato cv.), PYDV, PYMV, WPMV, PDMD |
| 3.3 Pigmentation and yellowing | | |
| Pigmentation | Anthocyanin accumulation. Leaves may show red, purple, and blue colors and may change shape | PLRV, Saq'O |
| Aucuba | Small, usually round spots or flecks of bright yellow color occur irregularly in leaves. They may coalesce but may still be recognized as individual spots | PAMV, PMTV, TBRV (pseudo aucuba strain) |
| Calico | Large, bright yellow areas with irregular borders usually dispersed throughout foliage | AMV, PBRSV (TRSV-Ca strain) |
| Chlorosis | The normal green color lacks uniformity in part or all of the leaf | BCTV, EMDV, PYDV, Saq'O |
| Yellow patterns | These include yellow lines forming defined shapes such as rings, incomplete rings, or arcs or chevrons | PMTV, TMV, TRV |

TABLE 11.4 *(continued)*

| Disease | Disease description | Causal virus and comments |
|---|---|---|
| Yellow mosaic or mottle | A bright yellow coloration of the veins which contrasts sharply with the green leaf lamina | PDMV, PYMV (dwarfing rugosity), PYV, ToMoT, ToYLCV |
| Vein yellowing | | PYVV (whole plant may become yellow) |
| 4. Changes in leaf shape | | |
| 4.1 Bunchiness | Leaves are small and so severely crinkled that they grow tightly together along the stem | PSTVd, BCTV (rosette appearance, leaves cupped and deformed), |
| 4.2 Leaf deformation | Leaves lose shape by elongation or widening of the leaf lamina, often with thickening of the main veins | BCTV, PDMV, PYDV, TSV, WPMV, PDMD |
| 4.3 Leafroll | Severe upward rolling of leaflets, which may be leathery and break easily when crushed. May be associated with stunting in secondary infected plants | BCTV, PVM (paracrinkle), PLRV, SALCV, TBRV (not all stems affected), Saq'O, PLRV (leathery leaves), ToYLCV, ToLCNDV (apical leaves) |
| 4.4 Rugosity | The leaf surface is uneven, giving the appearance of blisters | PSTVd, APMoV, CMV, PRDV, PVA, PVS, PYMV |
| 5. Necrosis | | |
| 5.1 Leaflet necrosis | Necrosis that affects the leaf | AMV (may extend into stems) |
| 5.2 Leaf drop | Necrosis that affects the whole leaf, causing it to drop | PVV, PVY |
| 5.3 Stem necrosis/apical necrosis | Necrosis that starts at the top of the plant, possibly extending to the whole plant | PAMV, PYDV, PVX (some isolates), PVV, TSV, TSWV |
| 5.4 Systemic necrosis | Necrotic streaks, spots, or rings on leaves | EMDV, PAMV, PBRSV (ringspots), PVV, TSV (concentric rings), TBRV (ringspots or spots), TSV, TSWV (also chlorotic ringspots or spots), PVY |

| | | |
|---|---|---|
| 6. Symptomless | | AVB-O, PotLV, PVP, PVS, PVT, PVU, PVV |
| B. Tubers | | |
| 1. Misshapen | | PSTVd (spindle tubers), AMV, PMTV, PYDV, TBRV, TRV |
| 2. Cracked | | PSTVd, AMV, PMTV, TBRV, TRV |
| 3. External necrosis | Necrotic lines, arcs, rings, and lesions (usually circular and dry) | AMV, PAMV (some cvs. at high storage temperature 20-24°C), PMTV (rings and lines on tuber surface), PVY N , PVY NTN (raised rings becoming sunken), TNV (surface lesion with cracks, may become sunken in storage), TRV, TSWV |
| 4. Internal necrosis | Necrotic lines, arcs, rings, and lesions Net necrosis (necrotic lines form a net) | AMV (some strains starting at stolon end, leaves corky dry areas), PAMV (some cvs. at high storage temperature 20-24°C), PMTV (spraing), PLRV (net necrosis), PVX (some isolates in some cvs), TRV (spraing), TSWV (lesions) |

*Source:* Based on definitions given by Salazar (1996a) and virus symptoms given by Jeffries (1998) and Stevenson et al. (2001).
[a]Full names as given in Table 11.1.

### *Bioassay*

Bioassay is the mechanical inoculation of extracted sap to indicator plants, which develop symptoms that are often distinctive for the virus, either alone or when used with other species of indicator plant. Bioassay is an important test for postentry quarantine and in the production of pathogen-tested nuclear stocks because it is relatively nonspecific and may detect unknown viruses. However, bioassay has the disadvantage that it cannot be used for the detection of viruses that are not mechanically transmitted (e.g., PLRV). In addition, it is relatively slow, may give variable results if growing conditions are not precisely controlled, and requires either considerable experience for an accurate diagnosis or use of an alternative diagnostic method for confirmation of the virus.

For postentry quarantine and testing of potato nuclear stock, indicator plant lists have been recommended (EPPO, 1999; EPPO, 2004b). Important indicator plants for potato viruses and their reactions are described by Jeffries (1998). In an attempt to reduce the number of indicator plants that need to be used, and hence the cost of the test, recently the three-indicator set of *Nicotiana occidentalis*-P1, *N. hesperis*-67A, and *Chenopodium quinoa* has been proposed as sufficient to detect most viruses of concern (Verhoeven and Roenhorst, 2003). Since differences in susceptibility to virus infection between accessions of the same species have been reported (Van Dijk and Cuperus, 1989), care should be exercised in the selection of indicators.

Inoculum can usually be prepared by grinding leaf tissue in tap water, but to ensure reproducibility of results, distilled or deionized water should be used. Phosphate buffer can also be used (e.g., 0.02M phosphate buffer, pH 7.4 + 2% w/v polyvinylpyrrolidone, MW 10,000), and this can improve the reliability of the test for some viruses (e.g., TSWV) (Jeffries, 1998). Indicator plants are inoculated by rubbing plant extracts on leaves lightly dusted with carborundum (400-600 mesh). Celite can also be used. Indicator plants are most susceptible when young and actively growing; shading them (e.g., by covering with paper) for a short period prior to and after inoculation may enhance susceptibility and symptom development.

Where viruses cannot be mechanically transmitted, grafting to susceptible indicators may be used for testing (Hill, 1984).

### *Electron Microscopy (EM)*

Electron microscopy procedures for virus detection are described by Roberts (1986). The major disadvantages of EM are the small number of samples that can be screened; EM's relative insensitivity, which can cause difficulties with the detection of low concentrations of viruses; and spherical virus particles, which can be more difficult to detect than rod-shaped particles. Nevertheless, it is a useful test for confirmation of positive results obtained by other means, and although it may not always be possible to diagnose the virus, it should be possible to deduce the genus to which it belongs. EM is a particularly useful test in postentry quarantine because it is perceived as a multitarget assay, and at least one country (New Zealand) requires the testing of all samples by EM (Anonymous, 2003). Immunosorbent EM using virus-specific antiserum-coated grids can improve sensitivity and further aid diagnosis (Roberts, 1986; Garg and Khurana, 1992; Khurana et al., 1993).

### *Antibody-Based Methods*

Two major kinds of antibody preparations are used for virus detection and diagnosis: polyclonal antisera from immunized animals and monoclonal antibodies produced from cells cultured in vitro. More recently, phage display technology has been developed to produce novel monoclonal antibody-like proteins.

#### *Polyclonal Antibodies*

Polyclonal antibodies are usually raised in rabbits, and large quantities of sera can be produced. Sometimes the antibodies may not be very specific, however, reacting with plant proteins or to a related virus that carries the same antigenic determinants (epitopes) as the original antigen. Antibody specificity can be increased in various ways, such as cross-adsorption (Beltz and Burd, 1989). However, the supply is finite and even if the same virus preparation is used, new sera may have different titer and binding characteristics, requiring time-consuming validation. Despite these drawbacks, polyclonal antibodies are easier and quicker to produce than monoclonal antibodies. In addition they have tended to be the preferred choice for quarantine testing laboratories because they are less likely to miss serotypes than a monoclonal antibody as detection does not depend on a single epitope.

### *Monoclonal Antibodies*

A monoclonal antibody (MAb) is secreted in vitro from a clonal population of hybridoma cells (the product of a somatic fusion between a myeloma cell and a single spleen cell) from an immunized animal. It consists of antibody molecules (Kohler and Milstein, 1975). By careful screening it is possible to select MAbs that can distinguish epitopes unique to individual virus strains or those that are common within a virus genus (e.g., the Agdia Potyvirus Group Test). Hybridoma cell lines stored at −195°C are "immortal" and can be revived and cultured to produce more MAb indefinitely and in unlimited quantities. These factors make MAbs suitable for use in accredited testing programs where standardization and uniformity are important (Karande et al., 1998).

### *Production of Novel Antibody-Like Proteins by Phage Display Technology*

This method uses recombinant DNA technology and involves the expression of fragments of antibody genes in bacterial systems (Orlandi et al., 1989) without the need to immunize animals. These synthetic antibody fragments are called single-chain variable fragments (scFVs). Diverse synthetic antibody expression libraries have been constructed and used to obtain antibody fragments specific for a wide range of antigens. Recombinant antibodies have been produced to a number of viruses, including *Cucumber mosaic virus* (CMV) (Ziegler et al., 1995) and PLRV (Harper et al., 1999). The antibody can be genetically fused to reporter molecules (such as alkaline phosphatase or biotin), which is advantageous because conventional chemical coupling of reporter molecules to monoclonal antibodies can be detrimental to activity. Using these fusion proteins, Toth et al. (1999) showed that a fully recombinant assay was as effective as triple antibody sandwich (TAS) enzyme-linked immunosorbant assay (ELISA) for detection of PLRV. Despite the prospect of producing recombinant antibody reagents much more cheaply than by conventional hybridoma culture, this technology has not yet been taken up commercially, and few if any plant diagnostic laboratories use recombinant antibodies on a routine basis.

*Antibody-Based Tests*

Enzyme immunoassays fulfill many of the requirements of an ideal test. Numerous formats, enzymes, and substrates have been employed for antibody-based tests (Kemeny, 1991), but the most commonly used is probably the double antibody sandwich (DAS) ELISA using the enzyme and substrate, alkaline phosphatase and p-nitrophenol (Clark and Adams, 1977; Jordan, 2005). It is important that the enzyme immunoassay is optimized for each pathogen, by determining the dilutions of antibodies, enzyme conjugates, extraction buffers, blocking solutions, and incubation times that maximize absorbance values from virus-containing samples and minimize absorbance values from virus-free samples. For many potato viruses, antibodies are available commercially and ELISA protocols (Copeland, 1998) are often provided together with quality control data.

Dot-ELISA on nitrocellulose membranes (NCMs) and other (filter) papers is reported to be less specific and more sensitive for some viruses with lower backgrounds than DAS-ELISA (Banttari and Goodwin, 1985; Kumar and Singh, 1999).

Despite the development of other detection methods, particularly hybridization and polymerase chain reaction (PCR) based tests, ELISA remains the method of choice for detection in most laboratories because it is robust (failures tend to be infrequent) and tolerant of minor changes (inadvertent or deliberate) to the protocol. In addition, sample preparation is simple, it is a relatively quick and cheap technique, and importantly ELISA lends itself to high throughput. For most viruses, the tissue extract is prepared in a standard sample buffer (PBS-Tween, +2% polyvinylpyrrolidone). Furthermore, little specialized equipment is required and relatively unskilled personnel can be trained to do the tests. Despite its relative lack of sensitivity in comparison with PCR, ELISA is likely to continue as an important tool for virus detection and diagnosis for many years to come.

### *Lateral Flow Devices*

Lateral flow devices were originally developed for pregnancy testing in the home, but the principle has been developed for testing plant pathogenic viruses, bacteria, and fungi in the field. They are based on the flow of plant extract, buffer, and antibody colored latex beads across a membrane impregnated with two or more lines of trapping antibodies. Results are displayed as visible colored lines in a window within 2-10 min. Two visible lines indicate a positive result, and a single line indicates that the device is working correctly but no infection is present. Devices are currently available for *Potato virus A* (PVA), PVS, *Potato virus V* (PVV), PVX, PVY, and TSWV, and an up-to-date list can be obtained at http://www.pocketdiagnostic.com. The method may be more cost effective than laboratory ELISA tests when only a small number of samples are to be tested, but for a large number of samples laboratory-based ELISA is still the method of choice. The main use of the technique is rapid detection or confirmation of a pathogen in the field by the grower, extension specialist, or potato inspector without the need for a laboratory test.

### ***Nucleic Acid–Based Detection***

A number of nucleic acid detection techniques relying on very different principles are available.

### *Return-Polyacrylamide Gel Electrophoresis (r-PAGE)*

For potato, gel electrophoresis has been used primarily for the detection and diagnosis of PSTVd. Serological methods cannot be used since viroids do not have a protein coat. r-PAGE is based on the difference in conformation of the viroid molecule under native and denaturing conditions (Reisner, 1987), which results in a clear separation of viroid and plant nucleic acids (Schumacher et al., 1986; Roenhorst et al., 2000).

The method is not specific for PSTVd and can detect all viroids. This makes it particularly suitable for postentry quarantine testing and the production of pathogen-tested nuclear stocks. PSTVd is the only viroid known to infect cultivated species of potato naturally. Wild potato species have been found to be infected with PSTVd (Owens et al., 1992) and *Mexican papita viroid* (Martinez-Soriano et al., 1996).

Although r-PAGE is less sensitive than other molecular techniques such as digoxigenin- or $^{32}$P-labeled complementary RNA (cRNA) probes and reverse transcription (RT) PCR for detection of PSTVd, its sensitivity is sufficient for routine use provided that plants for test are grown under optimal conditions for viroid concentration and limited pooling of plants for testing is done (EPPO, 2004a; Jeffries and James, 2005).

### *Nucleic Acid Hybridization*

The technique involves the use of labeled complementary DNA (cDNA) or cRNA probes prepared from purified or cloned viroid/viral nucleic acid. When the probe is incubated with viroid or viral nucleic acid extracts, double-stranded hybrids form by specific base pairing, which can be detected by the presence of labeled nucleotides incorporated into the nucleic acid probe. Labeling can be by incorporation of $^{32}$P or nonradioactive markers such as digoxigenin or fluorescein, which have been shown to have comparable levels of sensitivity to radioactive probes and a much longer shelf life (Querci and Salazar, 1998; Verma et al., 2003).

Although cDNA and cRNA probes are available for many potato viruses, their use in diagnostic laboratories is in general limited to those viruses and viroids that lack a protein coat (e.g., TRV-NM particles and PSTVd). To avoid sample deterioration in transit, dot/squash blots or NASH (nucleic acid spot hybridization) methods have been developed where the virus/nucleic acid is trapped on membranes, which can then be mailed to a diagnostic laboratory for probing.

### *Microarrays*

Microarray technology allows the detection and diagnosis of all viruses (and other pests) for which sequence information is available. The potential of this technology for the detection of potato viruses has been demonstrated by Boonham et al. (2003). Microarrays (also known as gene chips) involve printing thousands of probes onto a substrate (e.g., glass) using a robot. The probes are DNA fragments (500-5,000 bases long) typically from PCR products for the target organism or oligonucleotides (20-80 bases long). Up to 30,000 probes can be arrayed onto a single slide. Virus RNA is extracted and reverse transcribed to cDNA, at which stage fluorescent dyes are added. The labeled cDNAs (targets) are then hybridized to the array. Hybridization signals are detected using a laser scanner, resulting in the production of digital images. These images are analyzed by special software that takes into account, for example, variations in spot intensity and background signals (Hess et al., 2001) to produce a figure for the level of hybridization.

### *Polymerase Chain Reaction*

Virus and viroid RNA can be detected using RT-PCR. cDNA synthesized from viral RNA is used as template for the PCR, which amplifies a specific sequence of the DNA in a cyclic process using oligonucleotide primers. PCR allows the selective amplification of a particular DNA sequence by a factor of $10^6$ to $10^7$, which enables detection of a single copy or a few copies of DNA. Primers may be designed in a number of ways, but in general the process involves obtaining the sequence of interest (either from a public database or sequence obtained by the researcher); design of primer pairs to the target sequence; and sometimes examining the binding sites of the primers selected on the sequence using a computer program to ensure suitability for the purpose (e.g., to minimize mispriming). PCR assays have been developed for many viruses and viroids that infect potato, and virus genus specific methods are also available (Jeffries, 1998). Broader-spectrum RT-PCR detection of related but not homologous sequences can be achieved using degenerate primers, designed by looking at an alignment of several sequences. Assays can also be designed to detect a number of viruses at once (multiplex assays). In this procedure multiple PCR target sequences are amplified in a single tube using multiple primer sets (Nie and Singh, 2000).

Although PCR is already very sensitive, increased sensitivity has been achieved by blotting the PCR products with a labeled complementary nucleic acid probe (Vunsh et al., 1990), transcriptional amplification, and immunocapture PCR (Wetzel et al., 1992). Increased sensitivity may be useful for testing pooled tissue samples from, for example, postharvest tuber tests and surveys.

PCR suffers from a number of disadvantages. The nucleic acid extractions are time consuming and thus costly. In addition, poor-quality nucleic acid extract or simply not putting any extract into the tube may cause the PCR reaction to fail, giving a false negative result. Although the value of an internal amplification control (IAC) is recognized, this has not yet been used routinely in most laboratories. An IAC designed against plant cytochrome oxidase (COX) or ribosomal 5S gene sequences has been used in detection of PSTVd (EPPO, 2004a). These genes are expressed in all plant tissue types and failure to detect them means that the test should be repeated.

### *Fluorogenic 5′-Nuclease Assay (TaqMan)*

In the mid-1990s techniques were developed to allow immediate detection of PCR products. The most successful of these is the fluorogenic 5′-nuclease assay, which has been patented using probes marketed under the trade name TaqMan. The technique combines a probe labeled at the 5′ end with a reporter (fluorescent) dye and at the 3′ end with a quencher dye, which is designed to anneal to a sequence internal to the PCR primers. While the probe is intact, fluorescence emitted by the reporter is absorbed by the quencher (fluorescent resonance energy transfer). However, when the probe is cleaved by the 5′-nuclease activity of the Taq polymerase during amplification, the dyes are separated, resulting in detectable fluorescence related to the amount of product amplified. The increase in reporter fluorescence is monitored with real-time amplification, using a combined thermal cycle and fluorescence reader (e.g., ABI PRISM TM 7900). These units are costly and are outside the reach of most diagnostic laboratories. However, costs are falling and portable units are now available that can give diagnosis in the field in less than 1 hour. Real-time PCR has advantages over classical PCR: it gives quantitative data and requires neither agarose gel electrophoresis to detect the amplified product nor interpretation by eye. Real-time PCR should be suited to the large-scale detection of virus and viroids as it offers advantages over other detection methods in terms of sensitivity, the use of multiplex assays, the possibility of automation using liquid-handling robots, and quantification. Real-time methods have been developed for detection of PSTVd (Boonham et al., 2004) and for direct tuber testing for viruses (Klerks et al., 2001). However, the large-scale extraction of nucleic acid has still to be achieved, and the unit costs of real-time PCR tests for direct tuber tests are more expensive than the currently used postharvest eyeplug test. For laboratories that have the facility to conduct real-time PCR, the use of the test is mostly limited to detection of quarantine pathogens or niche markets, such as direct tuber testing of export consignments, where because of shipment deadlines a quick result may justify the extra cost.

## *MANAGEMENT*

### *Preventing Entry of New Viruses into a Country*

#### *Phytosanitary Regulation*

Potato is one of the most regulated crops moving in international trade (Ebbels, 2003) because of the ease with which pests, particularly latent pests such as viruses (and viroids), can spread from one region to another. Potatoes are moved between countries for three main reasons: planting (seed potatoes), consumption or processing (ware potatoes), and crop improvement (breeding material).

International phytosanitary activities are governed by the International Plant Protection Convention (IPPC) and the World Trade Organization's Sanitary and Phytosanitary Agreement (WTO-SPS) (Ebbels, 2003). The IPPC and WTO-SPS aim to progress global harmonization of phytosanitary measures by putting into practice the concepts of international standards, technical justification by risk assessment, transparency, consistency, equivalence, nondiscrimination, and minimal impact of measures on trade. The International Standards for Phytosanitary Measures (ISPMs) and IPPC convention can be found at http://www.ippc.int and the WTO-SPS agreement at http://www.wto.org. Regional harmonization of phytosanitary measures and adoption of sound phytosanitary policies is promoted through nine regional plant protection organizations (RPPOs), and these now cover most areas of the world (Ebbels, 2003; also see http://www.ippc.int).

For phytosanitary purposes, regulated pests are classified as quarantine or regulated nonquarantine pests (see ISPM no. 5 on the IPPC Web site for definitions). Pests (including viruses) that are officially regulated for tolerances in seed potato certification schemes may be considered as regulated nonquarantine pests if they meet the defined criteria. Contracting parties to the IPPC are required to establish, update, and publish lists of regulated pests that have been determined by pest risk analysis (PRA). PRA is the process of evaluating biological or other scientific and economic evidence to determine whether a pest should be regulated and how strong any phytosanitary measures to be applied should be. The measures must be appropriate to the risk. Measures to be applied may include imports from pest-free areas, pest-free places of production or pest-free production sites and specified treatment of the consignment prior to export. The steps that a risk analyst should take in conducting a PRA are described in ISPMs nos. 2 and 11.

If no risks are identified by PRA or phytosanitary measures are proposed the reduce the risk of pest introduction to an acceptable level, the importing NPPO (national plant protection organization) should publish the phytosanitary measures to be applied and specify any declarations to be made on the phytosanitary certificate (PC). The PC is a document issued by the exporting NPPO that certifies the plants or tubers in the consignment have been inspected or tested according to official procedures and are considered to be free from the quarantine pests specified by the importing NPPO and to conform to the current phytosanitary requirements of the importing NPPO, including those for regulated nonquarantine pests (ISPM no. 12). If the phytosanitary measures cannot reduce the risk to an acceptable level, trade of the commodity may be prohibited.

As yet there is no international standard that deals specifically with the phytosanitary aspects of trade in potatoes. NAPPO (the North American Plant Protection Organization), however, has produced a standard for Canada, Mexico and the United States (NAPPO, 2003) and EPPO (the European and Mediterranean Plant Protection Organization) has produced a standard for its member countries (EPPO, 2004c). In addition the EU allows trade between its 25 member states without quarantine barriers provided that the requirements of Plant Health Directive 2000/29/EC are fulfilled (EU, 2000).

### *Postentry Quarantine*

Postentry quarantine enables prohibited material to enter a country provided it passes a specified testing regime. Such a regime may require testing by specific methods of each individual unit of candidate material for listed pests and growing plants for a complete vegetative cycle to observe for symptoms of any pest that may have escaped detection. The most stringent quarantine programs require testing for the most important viruses by methods based on different biological principles (e.g., ELISA and bioassay) to reduce the risk of divergent serotypes or pathotypes escaping detection. Although directed primarily at quarantine pests, postentry quarantine programs may aim to ensure that the plants released are free of all pests (EPPO, 2004b).

### *Healthy Seed*

The most important virus management practice available to potato growers is the use of healthy planting material. The basic requirement for production of seed tubers free of viruses and viroids is to multiply seed potatoes for a limited number of generations under specified conditions, usually in areas of low vector pressure. In most schemes the initial pathogen or virus-free stock is derived by micropropagation or clonal selection (EPPO, 1999; Khurana and Garg, 1998). During multiplication, the spread of virus is controlled by roguing and observing official tolerance limits for viruses. Pesticides may be used to control virus vectors and in some countries the use of official dates of haulm destruction has proved an effective tool in controlling virus spread in seed potato crops. Forecasting the arrival of potential virus vectors (e.g., aphids) may be used to improve the timing of pesticide application and haulm destruction. In some schemes postharvest tuber testing of a sample of the crop is carried out to ensure that the crop meets the required tolerances.

Many countries have established certification schemes in order to supply "healthy" seed to ware growers. Seed potato certification is an administrative system for quality control of propagation and planting material that provides the purchaser with assurance that it has been produced according to the scheme requirements, particularly with regard to health and cultivar purity. Certificates or labels are awarded to material that successfully meets the scheme requirements. Certification schemes may be run by government or a designated organization that acts as the (official) certifying authority.

International and regional standards for seed potatoes are recommended by the United Nations Economic Commission for Europe (UNECE) and EPPO (EPPO, 1999; UNECE, 2003). These divide seed potatoes into categories according to quality standards. In the UNECE standard there are three categories with two classes in each category, with specific virus disease tolerances for each class (Table 11.5). Within each class, certification authorities may establish the number of years that material is kept at that class provided it meets the tolerance required for that class. In Europe, this system of categorization is widely used, but in North America seed potatoes are classified according to the generation for which specific field and tuber tolerances apply.

### *Virus and Viroid Elimination*

If no healthy tuber is available to initiate seed multiplication, two methods can be used to eliminate the virus: in vitro meristem culture of tuber-shoot tips or chemotherapy of in vitro plants. These methods can be combined with heat treatment to improve the chances of eliminating the virus.

Meristem culture of potato has been reviewed by Mellor and Stace-Smith (1977). In practice meristems comprise the meristematic dome and one or more leaf primordia (Hollings, 1965). From the meristematic dome, virus concentration increases basipetally through the leaf primordia, and thus the probability of obtaining virus-free plants is

TABLE 11. 5. UNECE virus disease tolerances.

| Seed category and class | Tolerance limit in growing crop (% virus infection) | | Tolerances crops should meet when planted |
|---|---|---|---|
| | Total virus | Severe virus | |
| A. Prebasic category | | | |
| Prebasic TC | 0 | 0 | In plants directly grown from Prebasic seed symptoms of mild or severe virus should not exceed 0.1% |
| Prebasic seed | 0.1 | 0.1 | |
| B. Basic category | | | |
| Basic seed I | 0.4 | 0.2 | In plants directly grown from Basic seed symptoms of mild or severe virus should not exceed 4%, with not more than 2% showing severe virus disease |
| Basic seed II | 0.8 | 0.4 | |
| C. Certified category | | | |
| Certified seed I | 2 | 1 | In plants directly grown from Certified seed symptoms of severe virus should not exceed 10% |
| Certified seed II | 10 | 2 | |

inversely related to the size of the explant used. Conversely, the successful development of an explant into a new plant is directly proportional to its size. Some viruses (e.g., PLRV, PVA, PVY) can be eliminated from relatively large meristems (0.8 mm long with three to four leaf primordia), whereas other viruses (e.g., PVS, PVX) require excision of smaller meristems (0.3 mm long with two leaf primordia). Successful plant regeneration from meristems and virus elimination can depend also on the cultivar, the physiological condition of the donor plant, and the virus strain (Khurana and Sane,

1998). Heat treatment of tubers may increase the number of meristems testing negative (Stace-Smith and Mellor, 1968). Detailed procedures for potato meristem culture have been described elsewhere (Wang, 1985). Isometric viruses such as PLRV are particularly sensitive to heat and can be eliminated by treatment of tubers at 37°C for 8 weeks without the need for meristem culture (Kassanis, 1950). However, heat causes tubers to deteriorate.

A more effective and simple method that does not involve excision of meristems is chemotherapy of in vitro propagated plants using the antiviral synthetic riboside, ribavirin (Cassells and Long, 1982) combined with thermotherapy. The method is effective against a wide range of potato viruses. Infected plantlets are established on MS medium (Murashige and Skoog, 1962) at 22-25°C, with a photoperiod of 16 hours' light plus 8 hours' darkness. Nodal cuttings are then subcultured into medium supplemented with 20 mg/L ribavirin and, after establishment, heat treatment is initiated. A 4 hour alternating 35°C light and 31°C dark cycle for 4 weeks may be used, but for PLRV and possibly other viruses, alternating temperatures of 40°C and 25°C may be more effective, although plant mortality may increase (Jeffries, 1998). At the latter temperatures, 100 percent elimination of PLRV can be achieved without using ribavirin. If elimination is not successful after the first round of treatment, subculturing onto ribavirain- supplemented medium and thermotherapy should be continued until elimination is achieved.

The high temperatures used for virus elimination cannot be used to eliminate PSTVd. Instead, a low-temperature method that reduces viroid replication has been developed (Lizarraga et al., 1980).

### *Sanitation*

In order to prevent or minimize infection of clean stocks by mechanically transmitted viruses, cultural practices and sanitation are important. Early-generation clean stocks should be kept separate from later stocks and particularly ware potatoes. Similarly, separate machinery (e.g., tractors, graders) and equipment (e.g., storage boxes) should be used for the different stocks or cleaned and then disinfected. To avoid transfer of mechanically transmitted viruses by personnel during the growing season, roguers and inspection staff should disinfect clothing (leggings) and footware between stocks. For disinfecting hands and small implements such as knives, undiluted skimmed milk containing at least 3.5 percent protein, 3 percent trisodium orthophosphate (Broadbent, 1963), and 10 percent trisodium phosphate plus detergent (Mulholland, 1962) can be effective. For cutting knives, dipping in alcohol and flaming may also be effective for many mechanically transmitted viruses, although not completely effective for PSTVd (Roistacher et al., 1969). Buildings, boxes, and machinery should be cleaned using high-pressure steam or water (above 50°C) washers. This may not be completely effective in killing the pathogen but will reduce virus concentration on the surface and the presence of organic material that may affect the efficacy of disinfectants. Disinfectants include those based on glutaraldeyde (e.g., Ben-glucid, Glucid, Horticide), phenol (e.g., Panacide, Sudol), amphoterics/surfactant compounds (e.g., Tego), and sodium hypochlorite (e.g., Chloros, also effective against PSTVd). Newer oxidizing disinfectants such as Virkon S and Menno Florades (also effective against PSTVd) can also be used.

### *Volunteer Inoculum*

In countries where frost penetration or temperature is insufficient to kill tubers that remain in the ground after harvest, volunteer potato plants that appear in rotation crops are a potential source of inoculum for potato viruses. Crop rotation and good crop husbandry including roguing are important in the control of volunteer potato plants. Potato dumps may also be a source of virus (Frazer, 1987).

For viruses such as alfalfa mosaic virus (AMV), PYVV and TSWV, which are not primarily dependent on the potato for survival and spread (Table 11.1), alternative hosts are important in the infection of "clean" potato stocks. Soilborne viruses are particularly difficult to control as they may be able to survive in perennial weed hosts and are transmitted through botanical seed (e.g., TRV) or may survive for long periods in the resting stage of the vector (e.g., PMTV).

### *Cropping Pattern*

Changing cropping patterns can affect vector populations and potential virus incidence. Expansion in cropping of canola *(Brassica napus oleifera)*, a favorable host for *M. persicae*, is thought to be a major factor in increasing aphid pressure in Minnesota and North Dakota and the spread of PLRV and PVY. Canola is thought to act as a bridging host for spring migrant aphids from the southern overwintering sites before potatoes emerge (Ragsdale et al., 2001). In France, $PVY^{O}$ epidemics have been attributed to large flights of cereal aphids, particularly *Rhopalosiphum padi*, as a result of increased maize cultivation and the changes in cereal cultivars (Robert et al., 2000). On the other hand, soil has been cleared of viruliferous nematodes by growing alfalfa, a nonhost of TRV, over a 3-year period (Thomas et al., 2001). A dense alfalfa crop prevents weed establishment and so prevents nonviruliferous and juvenile nematodes from acquiring the virus. In the United Kingdom, the incidence of TRV induced spraing was substantially reduced in potato when the potato crop followed 3 years of barley (French and Wilson, 1976). However, although growing alfalfa and barley may be effective in the short term, the virus may still be present in any viable residual weed seeds and so may reestablish at a later date.

### *Chemicals*

#### *Insecticides*

In general, insecticides that are systemic in action are successful in diminishing field spread of persistent viruses such as PLRV because of the latent period required before the aphids are able to transmit virus. Insecticides can be applied at planting or with foliar applications, or both. Although much PLRV is spread within fields, particularly in the early part of the season by nonwinged (apterous) forms, virus can also be carried into fields by migrating winged (alate) aphids moving long distances. In these cases, insecticides may not kill aphids quickly enough to prevent transmission. A possible exception is the insecticide pymetrozine, which is highly selective; once aphids acquire a toxic dose they cease to feed (Harrewijn and Kayser, 1997).

Nonpersistently aphid transmitted viruses such as PVY are thought to be acquired and transmitted largely within a field by probing of winged aphids flying from plant to plant. Because of the short time of probing required to become infective, insecticides that may be effective against persistent viruses have little or low value in preventing the spread of nonpersistent viruses (Khurana and Garg, 1998). However, insecticides are much used by growers in the hope that they will suppress PVY spread, and possibly their greatest effect is in killing developing populations of colonizing aphids before they move to new plants by walking or flying.

Chemical control measures for nonpersistent viruses are largely focused on those chemicals that kill the vector rapidly, repel the vector, or modify vector behavior to prevent probing (Perring et al., 1999). Pyrethroids possess many of these properties but work mainly by causing aphid hyperactivity followed by rapid incapacitation prior to death (Gibson et al., 1982).

Constant and frequent applications of insecticide can cause problems because natural enemies and predators of aphids may be killed and insecticide resistance may develop rapidly, particularly toward the end of the season. Some *M. persicae* genotypes are resistant to a number of major insecticide classes (e.g., organochlorines, organophosphates, carbamates, and pyrethroids), although they can be controlled by newer insecticides (e.g., pymetrozine, imidacloprid, and methemidophos). The arrival and spread of *M. persicae* resistant to MACE (modified acetyl cholinesterase) is a consequence of insecticide use. In some aphid clones, MACE resistance is combined with all of the other known resistance mechanisms (Foster et al., 2000). However, there is evidence that some insecticide-resistant *M. persicae* clones may be less adapted for cold temperature survival than nonresistant clones in the United Kingdom (Foster et al., 1997). Imidacloprid has helped reduce the incidence of whitefly-transmitted viruses, but resistance is now developing in whitefly (Polston and Anderson, 1997).

### *Nematicides*

Soil fumigants such as dazomet, 1,3-dicloropropene, and methyl bromide provide good short-term control of nematode-transmitted viruses

(Harrison, 1977), but control is not complete or permanent because nematodes can survive at depths below the reach of the fumigants (Weingartner et al., 1983). Nonvolatile nematicides, such as the carbamates aldicarb and oxamyl are nematostats that paralyze the nematodes for a period of time rather than killing them, thus preventing them from feeding and transmitting the virus. Although the application of such chemicals allows farmers to grow a crop with a reasonable degree of protection, when the chemical is denatured and ceases to affect the nematodes, the previous level of infestation will return.

*Mineral Oils*

Mineral oils provide good control of PVY, but not PLRV. Efficacy, however, is dependent on maintaining adequate coverage as the plants grow (Boiteau and Wood, 1982). In addition, plants treated with mineral oils are more difficult to rogue and inspect than untreated plants and overapplication can lead to phytotoxic effects (Peters, 1987). The mechanism involved in control of PVY by mineral oils is not clearly understood. No difference in probing or feeding behavior has been observed (Powell, 1992), so the effect may occur after this. Combining oils and pyrethroids gave better control of PVY than mineral oils alone (Gibson and Rice, 1986).

***Repellents, Physical Barriers, and Crop Separation***

The repellent neem oil applied every 7-10 days from emergence until the end of the peak of aphid flight (Handizi et al., 2001) may reduce the incidence of aphid-transmitted viruses. Fleece application is also effective provided it remains intact during the time of aphid flight (Harrewijn et al., 1991). Fleece may be more effective than oil plus pyrethroid applications but is a more expensive treatment (Legorburu, 1998).

Shortwave light reflecting materials, such as straw mulch applied to the ridges and furrows, have been shown to reduce aphid infestation and PVY incidence significantly (Doring and Saucke, 2002). Mulching is most effective early in the growing season when vector pressure is high and young plants are most susceptible to virus infection. The effectiveness of the mulch decreases with increasing crop ground cover and the reducing area of visible mulch. Border crops of forage sorghum, winter wheat, or soybean, several meters wide, planted immediately next to the potatoes and with a fallow area to the outside have significantly reduced spread of PVY in seed potato crops. This planting pattern is thought to be effective because on landing at the interface of fallow ground and a green crop, aphids become disturbed and are unable to locate the host. Immigrating viruliferous aphids feed on the barrier crop and are cleaned of virus before continuing flight (DiFonzo et al., 1996). Increasing the fallow areas in crops is thought to increase the "visibility" of crops to immigrating aphids.

Seed and ware potatoes should not be grown on the same farm, and there should also be minimum separation distances between seed crops. For whitefly- and thrip-transmitted viruses the potato crop should not be grown near tomato or other vector and potato virus hosts. However, control of long-distance spread may not be possible in conditions where vectors move with the prevailing winds (Polston and Anderson, 1997).

### *Forecasting*

Models have been developed using biological and environmental data to predict aphid flight and changes in virus incidence that may be used by certification authorities to improve the timing of virus control measures such as haulm destruction and pesticide application (Robert et al., 2000).

These models may take into account variables such as the number of vector species caught and their populations, relative efficiency factors, the time of crop infestation, the number of infective sources (whether within or outside the crop), and cultivar and mature plant resistance (Sigvald, 1998). The relative efficiency factor for an aphid species is the efficiency of transmission for each aphid species relative to the most efficient vector *M. persicae,* and these factors have been calculated for a number of viruses including $PVY^O$ and $PVY^N$ (Sigvald, 1986; De Bokx and Piron, 1990). Forecasting the spread of potyviruses such as PVY in potato is much more difficult than for PLRV because they are spread by many different aphid species, including vagrant species with varying levels of efficiency (Sigvald,

1986). Weed species and other crops may also act as a source of inoculum, particularly in warmer climates. In contrast, PLRV is primarily spread by *M. persicae* and has few alternative host plants in countries with temperate climates.

Aphid numbers can be monitored using traps for use in the models. Since the 1960s a network of around 70 Rothamsted Insect Survey suction traps has been developed across Europe. Standardized trapping now operates in many countries producing seed potatoes. The traps sample airborne aphids at 12.2 m above the ground, a height chosen to give a good estimate of the numbers of small wind-dispersed insects. Such traps may give a good regional indication of aphid flight but may not always predict aphid flight locally (Woodford et al., 1977). For this, sticky traps and water-filled yellow trays situated in crops provide a better indicator of alate aphids. Crop-specific monitoring can help to refine decision making to reduce sprays in periods of low risk or increase sprays in periods of high risk (Northing et al., 2004).

### *Host-Mediated Resistance*

Virus resistance in potato was reviewed by Valkonen (1994) and by Solomon-Blackburn and Barker (2001). Several different types of natural resistance have been identified, including extreme resistance (ER), hypersensitive resistance (HR), resistance to virus infection, resistance to virus accumulation, and resistance to virus movement. Tolerance is frequently confused with resistance.

A plant with ER to a virus shows no symptoms or very limited necrosis (pinpoint lesions or flecks) when inoculated. Virus cannot be detected in inoculated plants or is detected only at extremely low levels using sensitive diagnostic methods. There are several well-described genes that confer ER (e.g., gene *Rx* against PVX and $Ry_{sto}$ against PVA, PVV and PVY). Although a strain of PVX was found that would overcome the resistance conferred by gene *Rx* (Moreira et al., 1980), little other evidence has been found that such genes break down and new resistance-breaking isolates develop. Robinson and Dale (1994) identified resistant cultivars that are immune to TRV (i.e., no spraing symptoms develop and no virus can be recovered). Despite the obvious value of genes for immunity to virus infection, only gene *Rx* has been incorporated into cultivars in several breeding programs.

A plant with HR will usually show local necrotic lesions in manually inoculated leaves where infection is localized and/or systemic. Virus is usually detectable and HR is often strain specific. Resistance can also be affected by environmental conditions or the physiology of the host plant. Although ER is generally regarded as the most useful form of resistance to incorporate, HR also provides useful protection and is more widely available than ER. ER and HR are both dominant monogenic resistances. However, resistance to mechanical inoculation with PVS is controlled by the recessive gene $ss_{tbr}$. Major-gene resistance (ER and HR) to viruses in potato has proved durable, in contrast to major-gene resistance to late blight *(Phytophthora infestans).* Although there has been some evolution of resistance-breaking strains (Jones, 1985), this is a relatively slow process.

Resistance to virus infection is the term used by Barker (1987) where the likelihood of infection (by natural means) is reduced (e.g., quantitative resistance to PLRV whereby fewer plants of resistant clones or cultivars become infected through aphids in field conditions) (Davidson, 1973). It was called field resistance by Cooper and Jones (1983). It is usually determined by field exposure where the virus is spread from infector plants by aphids. Its inheritance appears to be polygenic (Davidson, 1973).

Plants with resistance to virus accumulation can be infected, but the virus reaches only a relatively low concentration in the plant. This type of resistance has been shown for PLRV in a range of *S. tuberosum* breeding lines and other potato material (Van den Heuvel et al., 1993). The most resistant tetraploid clones had PLRV concentrations only 1-5 percent of those found in susceptible clones. The advantage of this type of resistance is that virus is less likely to be acquired and spread to other plants by aphids (Barker and Woodford, 1992).

Resistance to virus movement has long been recognized for PLRV (Hutton and Brock, 1953) where fewer than expected infected daughter plants are obtained from an infected plant. It may or may not be associated with phloem necrosis (Hutton and Brock, 1953). Resistance to phloem transport has been demonstrated by Wilson and

Jones (1992), who found it to be separate from resistance to accumulation and resistance to infection.

Although tolerance is not a form of virus resistance, this trait has been both deliberately and inadvertently selected in many breeding programs. Tolerant cultivars can be defined as those that show no symptoms or few obvious symptoms when plants are infected. Although tolerance may be seen as a valuable trait, it has the major disadvantage that tolerant cultivars can act as virus reservoirs. For example, Robinson and Dale (1994) have identified cultivars and clones that are susceptible to TRV infection but show few or no spraing symptoms.

### *Pathogen-Derived Resistance*

An alternative to host-mediated resistance is the introduction of pathogen-derived resistance through genetic engineering, and this has been reviewed for potato by Solomon-Blackburn and Barker (2001) and Ghosh et al. (2002), and for PLRV by Barker and Waterhouse (1999). The most common pathogen-derived transgene used for resistance is the one encoding the virus CP from which is derived coat protein–mediated resistance (CPMR). There are numerous examples of this type of resistance incorporated in potato, particularly for viruses such as PLRV, PVX, and PVY.

In some cases, but not all, the resistant transgenic plants produce CP. The resistance varies in degree and type but often acts against virus accumulation. For example, Kawchuk et al. (1991) observed PLRV titers as low as 1 percent of the titer found in control plants at the stage of primary infection. *Nicotiana benthamiana* (a model species for studying PMTV) has been transformed to express a translatable version of the CP gene from PMTV. In glasshouse tests all transgenic lines were immune or highly resistant to PMTV (Reavy et al., 1995), and resistance to PMTV was more effective in leaves than in roots (Germundsson et al., 2002): although roots became infected following inoculation with viruliferous *S. subterranea,* no systemic movement of virus from roots to leaves occurred. The same transgene construct for PMTV has been used to produce transgenic lines of the potato cultivars Saturna and Pentland Marble (Barker et al., 1998). When tested for resistance to PMTV by growing in pots with soil known to be infested with viruliferous *S. subterranea* in a screenhouse, all lines were highly resistant to PMTV and very few tubers became infected or developed spraing symptoms. However, when the transgenic Saturna lines were planted in a field in southern Sweden, the resistance was not as strong, although one line remained uninfected (Germundsson et al., 2002). This indicates that it is important to test the resistance of transgenic plants in several locations because environment may influence the expression of resistance.

CPMR is often strain specific but sometimes not. Resistance is conferred by nontranslatable CP genes (e.g., antisense sequences) or by truncated versions of the viral CP gene. In recent years the underlying mechanisms of the most useful forms of transgenic resistance have become apparent, and this is well illustrated by work with PVY. Thus, several reports have indicated that transformation of tobacco with the $PVY^N$ CP gene sequence confers resistance to $PVY^N$ (Sudarsono et al., 1995). It has also been shown that resistance can be induced in potato by transformation with PVY CP gene sequence expressed in sense, frameshift, and antisense forms (Smith et al., 1994). The mechanism of transgenic resistance in these cases was proposed to be an RNA-mediated defense (RMD) that is based on a post-transcriptional RNA-mediated gene-silencing response (PTGS) (Smith et al., 1995). The different forms of RMD probably induce resistance through a common pathway involving dsRNA (Di Serio et al., 2001). PTGS is similar to related mechanisms such as RNA interference in animals (Hannon, 2002) and seems to act by inducing sequence-specific RNA degradation of transgene transcript RNA and homologous RNA in the plants, a process that is triggered by activation of host ribonucleases (Metzlaff et al., 1997).

A highly efficient way to confer virus resistance is to transform plants with transgenes that simultaneously express sense and antisense mRNA and hairpin loop structures cloned from virus sequences (Levin et al., 2000). The transgene transcript RNA is thought to produce a partially double-stranded structure that induces a strong form of virus resistance through PTGS (Baulcombe, 1996). These mechanisms may provide a universal means of inducing resistance to all viruses, so that by combining or "stacking" individual transgenes, multifunctional transgenes will become available.

Although potato viruses are largely well characterized and detection methods for most are well established, problems of virus control in various crops remain. In developed agricultural systems these largely concern the difficulties in controlling virus vectors and lack of epidemiological knowledge. In developing countries, problems are largely of providing the detection and control methods at a realistic cost. In both situations, improved plant resistance through the use of host genes or genetic manipulation in breeding programs may offer a solution, particularly if there is pressure to reduce pesticide use on food crops. Resistance will largely benefit ware growers and, paradoxically, will remove one of the reasons specialist seed growers exist.

## REFERENCES

Abdel-Salam, A.M. (1991). Tomato yellow leaf curl virus in Egypt. *Bulletin of the Faculty of Agriculture, University of Cairo 42:*521-532.

Adam, G., Winter, S., and Lesemann, D.-E. (1990). Characterisation of a new strain of tobacco necrosis virus isolated from nutrient feeding solution. *Annals of Applied Biology 116:*523-536.

Adams, M.J. (1991). Transmission of plant viruses by fungi. *Annals of Applied Biology 118:*479-492.

Anonymous (2003). *MAF Biosecurity Authority Standard 155.02.06. Importation of Nursery Stock.* Ministry of Agriculture and Forestry, New Zealand.

Arif, M., Torrance, L., and Reavy, B. (1995). Acquisition and transmission of potato mop-top furovirus by a culture of *Spongospora subterranea f. sp. subterranea* derived from single cystosorus. *Annals of Applied Biology 126:*493-503.

Babos, P. and Kassanis, B. (1963). Serological relationships and some properties of tobacco necrosis strains. *Journal of General Microbiology 32:*135-144.

Banttari, E.E. and Goodwin, P.H. (1985). Detection of potato viruses S, X and Y by ELISA on nitrocellulose membrane (dot-ELISA). *Plant Disease 69:*202-205.

Barker, H. (1987). Multiple components of the resistance of potatoes to potato leafroll virus. *Annals of Applied Biology 111:*641-648.

Barker, H., Reavy, B., and McGeachy, K.D. (1998). High level of resistance in potato mop-top virus induced by transformation with the coat protein gene. *European Journal of Plant Pathology 104:*737-740.

Barker, H. and Waterhouse, P.M. (1999). The development of resistance to luteoviruses mediated by host genes and pathogen-derived transgenes. In Smith, H.G. and Barker, H. (Eds.), *The Luteoviridae* (pp. 169-210). Wallingford, UK: CABI Publishing.

Barker, H. and Woodford, J.A.T. (1992). Spread of potato leafroll virus is decreased from plants of potato clones in which virus accumulation is restricted. *Annals of Applied Biology 121:*345-354.

Baulcombe, D.C. (1996). Mechanisms of pathogen-derived resistance to viruses in transgenic plants. *The Plant Cell 8:*1833-1844.

Beltz, B.S. and Burd, G.D. (1989). *Immunocytochemical Techniques: Principles and Practices.* Oxford, UK: Blackwell.

Blanco-Urgoiti, B., Sanchez, F., de San Roman, C.P., Dopazo, J., and Ponz, F. (1998). Potato virus Y group C isolates are a homogeneous pathotype but two different genetic strains. *Journal of General Virology 79:*2037-2042.

Boiteau, G. and Wood, F.A. (1982). Persistence of mineral oil spray deposits on potato leaves. *American Potato Journal 59:*55-63.

Boonham, N., Perez, L.G., Mendez, M.S., Peralta, E.L., Blockley, A., Walsh, K., Barker, I., and Mumford, R.A. (2004). Development of a real-time RT-PCR assay for the detection of *Potato spindle tuber viroid. Journal of Virological Methods 116:*139-146.

Boonham, N., Walsh, K., Madagan, K., Graham, I., and Barker, I. (2003). Detection of potato viruses using microarray technology: Towards a generic method for plant viral disease diagnosis. *Journal of Virological Methods 108:*181-187.

Broadbent, L. (1963). The epidemiology of tomato mosaic. III. Cleaning virus from hands and tools. *Annals of Applied Biology 52:*225-232.

Brown, J.K. and Bird, J. (1992). Whitefly-transmitted geminiviruses and associated disorders in the Americas and the Caribbean basin. *Plant Disease 76:*220-225.

Brunt, A.A., Crabtree, K., Dallwitz, M.J., Gibbs, A.J., and Watson, L. (Eds.) (1996). *Viruses of Plants. Descriptions and lists from the VIDE Database.* Wallingford, UK: CAB International.

Calvert, E.L. (1968). The reaction of potato varieties to potato-mop virus. *Record of Agricultural Research, Northern Ireland 17:*31-40.

Cassells, A.C. and Long, R.D. (1982). The elimination of potato viruses X, Y, S and M in meristem and explant cultures of potato in the presence of virazole. *Potato Research 25:*165-173.

Chrzanowska, M. (2002). Importance of different strains of PVY in potato production and breeding program in Poland. *Potato Research 45:*101-102 (abstract).

Clark, M.F. and Adams, A.N. (1977). Characteristics of the microplate method of enzyme linked immunosorbent assay for the detection of plant viruses. *Journal of General Virology 34:*475-483.

Cockerham, G. and McGhee, T.M.R. (1953). Potato stunt disease. *Scottish Plant Breeding Station Annual Report* (pp. 46-49).

Cooper, J.I. and Harrison, B.D. (1973). The role of weed hosts and the distribution and activity of vector nematodes in the ecology of tobacco rattle virus. *Annals of Applied Biology 73:*53-66.

Cooper, J.I. and Jones, A.T. (1983). Responses of plants to viruses: Proposals for the use of terms. *Phytopathology 73:*127-128.

Copeland, R. (1998). Assaying levels of plant virus by ELISA. In Foster, G.D. and Taylor, S. (Eds.), *Plant Virus Protocols. From Virus Isolation to Transgenic Resistance* (pp. 455-460). Totowa, NJ: Humana Press.

Cordero, M., Ramos, P.L., Hernandez, L., Fernandez, A.I., Echemendia, A.L., Peral, R., Gonzalez, G., Garcia, D., Valdes, S., Estevez, A. et al. (2003). Identification of *Tomato mottle begomovirus* stains in Cuban potato fields. *Phytoparasitica 31:*478-489.

Creamer, R., Luque-Williams, M., and Howo, M. (1996). Epidemiology and incidence of beet curly top geminivirus in naturally infected weed hosts. *Plant Disease 80:*553-535.

Dale, M.F.B., Robinson, D.J., Griffiths, D.W., Todd, D., and Bain, H. (2000). Effects of tuber-borne M-type strain of tobacco rattle virus on yield and quality attributes of potato tubers of the cultivar Wilja. *European Journal of Plant Pathology 106:*275-282.

Dale, M.F.B. and Solomon, R.M. (1988). A glasshouse test to assess the sensitivity of potato cultivars to tobacco rattle virus. *Annals of Applied Biology 112:*225-229.

Davidson, T.M.W. (1973). Assessing resistance to leafroll in potato seedlings. *Potato Research 16:*99-108.

De Bokx, J.A. and Piron, P.G.M. (1990). Relative efficiency of a number of aphid species in the transmission of potato virus $Y^N$ in the Netherlands. *Netherlands Journal of Plant Pathology 96:*237-246.

Debrot, E.A. and Centeno, F. (1985). Infeccion natural de la papa en Venezuela con el Mosaico amarillo del tomate, un geminivirus transmitado por moscas blancos. *Agronomia Tropical 35:*125-138.

DiFonzo, C.D., Ragsdale, D.W., Radcliffe, E.B., Gudmestad, N.C., and Secor, G.A. (1996). Crop borders reduce potato virus Y incidence in seed potato. *Annals of Applied Biology 129:*289-302.

Di Serio, F., Schob, H., Iglesias, A., Tarina, C., Bouldoires, E., and Meins, F. (2001). Sense- and antisense-mediated gene silencing in tobacco is inhibited by the same viral suppressors and is associated with accumulation of small RNAs. *Proceedings of the National Academy of Sciences (USA) 98:*6506-6510.

Dolby, C.A. and Jones, R.A.C. (1987). Occurrence of the Andean strain of potato virus S in imported potato material and its effects on potato cultivars. *Plant Pathology 36:*381-388.

Doring, T.F. and Saucke, H. (2002). Straw mulch for control of virus vectors in organic seed potatoes. In *Abstracts of Papers and Posters of European Association for Potato Research Conference* (p. 168). Hamburg, Germany: EAPR.

Ebbels, D.L. (2003). *Principles of Plant Health and Quarantine.* Wallingford, UK: CAB International.

EPPO. (1999). EPPO Standards. PM 4/28 (1). Certification Schemes. Seed Potatoes. *EPPO Bulletin 29:*253-267.

———. (2004a). EPPO Standards. PM 7/33 (1). *Potato spindle tuber pospiviroid. EPPO Bulletin 34:*257-269.

EPPO. (2004b). EPPO Standards. PM3/21 (2). Post-entry quarantine for potato. *EPPO Bulletin 34:*443–454.

———. (2004c) EPPO Standards. PM 8/1. Commodity-specific phytosanitary measures. Potato. *EPPO Bulletin 34:*463–477.

EU. (2000). Council Directive 2000/29/EC of 8 May 2000 on protective measures against the introduction into the Community of organisms harmful to plants or plant products and against their spread within the Community. *Official Journal of the European Communities* L *169:*1-112.

Foster, S.P., Denholm, I., and Devonshire, A.L. (2000). The ups and downs of insecticide resistance in peach-potato aphids *(Myzus persicae)* in the UK. *Crop Protection 19:*873-879.

Foster, S.P., Harrington, R., Devonshire, A.L., Denholm, I., Clark, S.J., and Mugglestone, M.A. (1997). Evidence for a possible fitness trade-off between insecticide resistance and the low temperature movement that is essential for survival of UK populations of *Myzus persicae* (Hemiptera: Aphididae). *Bulletin of Entomological Research 87:*573-579.

Franc, G.D. and Banttari, E.E. (1984). The transmission of potato virus S by the cutting knife and retention time of infectious PVS on common surfaces. *American Potato Journal 61:*253-260.

Frazer, B.D. (1987). Aphid and virus management in British Columbia. In Boiteau, G., Singh, R.P., and Parry, R.H. (Eds.), *Potato Pest Management in Canada* (pp. 23-29). Ottawa, Canada: Agriculture Canada.

French, N. and Wilson, W.R. (1976). Influence of crop rotation, weed control and nematicides on spraing in potatoes. *Plant Pathology 25:*167-172.

Fuentes, S. (1991). *Identificacion, caracteristicas y distribucion de un virus baciliforme aislado de papa (Solanum tuberosum).* M.Sc Thesis, Universidad Nacional Agraria La Molina. Lima, Peru.

Fulton, J.P., Gergerich, R.C., and Scott, H.A. (1987). Beetle transmission of plant viruses *Annual Review of Phytopathology 25:*111-123.

Garg, I.D. and Khurana, S.M. Paul (1992). Factors governing immune electron microscopy of flexuous potato viruses. *Acta Virology 36:*435-442.

Garg, I.D., Khurana, S.M. Paul, Kumar, Shiv, and Lakra, B.S. (2001). Association of a geminivirus with potato apical leaf curl in India and its immuno-electron microscopic detection. *Journal of Indian Potato Association 28:*227-232.

Germundsson, A., Sandgren, M., Barker, H., Savenkov, E.I., and Valkonen, J.P.T. (2002). Initial infection of roots and leaves reveals different resistance phenotypes associated with CP gene-mediated resistance to Potato mop-top virus. *Journal of General Virology 83:*1201-2109.

Ghosh, S.B., Nagi, L.H.S., Ganapathi, T.R., Khurana, S.M. Paul, and Bapat, V.A. (2002). Cloning and sequencing of potato virus Y coat protein gene from an Indian isolate and development of transgenic tobacco for PVY resistance. *Current Science 82:*885-859.

Gibson, R.W. and Rice, A.D. (1986). The combined use of mineral oils and pyrethroids to control plant viruses transmitted non- and semi-persistently by *Myzus persicae. Annals of Applied Biology 109:*465-72.

Gibson, R.W., Rice, A.D., and Sawicki, R.M. (1982). Effects of the pyrethroid deltamethrin on the acquisition and inoculation of viruses by *Myzus persicae. Annals of Applied Biology 100:*49-54.

Handizi, A., Serra, J.A., and Legorburu, F.J. (2001). Escaping from PVY infection in the field: Planting dates, aphid repellents and resistance inducers. *Proceedings of the 11th Virology Section Meeting of the European Association of Potato Research* (p. 25). Havlickuv Brod-Trest, Czech Republic: EAPR.

Hannon, G.J. (2002). RNA Interference. *Nature 418:*244-251.

Harper, K., Toth, R.L., Mayo, M.A., and Torrance, L. (1999). Properties of a panel of single chain variable fragments against Potato leafroll virus obtained from two phage display libraries. *Journal of Virological Methods 81:*159-168.

Harrewijn, P., Den Ouden, H., and Piron, P.G.M. (1991). Polymer webs to prevent virus transmission by aphids in seed potatoes. *Entomologia Experimentalis et Applicata 58:*101-107.

Harrewijn, P. and Kayser, H. (1997). Pymetrozine, a fast-acting and selective inhibitor of aphid feeding. In-situ studies with electronic monitoring of feeding behaviour. *Pesticide Science 49:*130-140.

Harrison, B.D. (1977). Ecology and control of viruses with soil-inhabiting vectors. *Annual Review of Phytopathology 15:*331-360.

Harrison, J.G., Searle, R.J., and Williams, N.A. (1997). Powdery scab disease of potato—a review. *Plant Pathology 46:*1-25.

Hess, K.R., Zhang, W., Baggerly, K.A., Stivers, D.N., and Coombes, K.R. (2001). Microarrays: Handling the deluge of data and extracting reliable information. *Trends in Biotechnology 19:*463-468.

Hill, S.A. (1984). *Methods in Plant Virology.* London: Blackwell Scientific.

Hollings, M. (1965). Disease control through virus-free stock. *Annual Review of Phytopathology 3:*367-396.

Horvath, J., Lesemann, D.E., Koenig, R., Weidemann, H.-L., and Hoekstra, R. (1993) Detection of a tymovirus from *Solanum hannemannii. Quartech* (A Newsletter for European Community Potato Quarantine Stations) *1:*2-3.

Horvath, J., Wolf, I., Bosze, Z., Pocsai, E., and Kazinczi, G. (1996). Is tobacco rattle tobravirus transmissible by true potato seeds. *Proceedings of the 9th EAPR Virology Section Meeting* (pp. 71-72). Bled, 1995.

Hull, R. (2002) *Matthews' Plant Virology,* Fourth edition. London: Academic Press.

Hutton, E.M. and Brock, R.D. (1953). Reactions and field resistance of some potato varieties and hybrids to the leaf-roll virus. *Australian Journal of Agricultural Research 4:*256-263.

Jain, R. K., Khurana, S. M. Paul, Bhat, A. L., and Chaudhuary, V. (2004). Nucleocapsid protein gene sequence studies suggest that potato stem necrosis disease is caused by strain of *Groundnut bud necrosis virus. Indian Phytopathology 57:69–173.*

Jeffries, C.J. (1993) Transmission of tomato black ring virus in potato seed. *Quartech* (A Newsletter for European Community Quarantine Stations) *1:*3.

Jeffries, C.J. (1998). *FAO/IPGRI Technical Guidelines for the Safe Movement of Germplasm, No. 19. Potato.* Food and Agriculture Organization of the United Nations/International Plant Genetic Resources Institute, Rome. (Virus descriptions at http://www.ecoport.org).

Jeffries C.J., Chard, J.M., and Brattey, C. (1993). Coping with plant health risks posed by gene bank collections of potato. In *Plant Health and the European Single Market* (pp. 145-156). *Monograph, No 54*, British Crop Protection Council.

Jeffries, C. and James, C. (2005). Development of an EU protocol for the detection and diagnosis of *Potato spindle tuber pospiviroid. EPPO Bulletin 35:*125-132.

Jones, R.A.C. (1982). Tests for transmission of four potato viruses through potato true seed. *Annals of Applied Biology 100:*315-320.

———. (1985). Further studies on resistance-breaking strains of potato virus X. *Plant Pathology 34:*182-189.

———. (1988). Epidemiology and control of potato mop-top virus. In, Cooper, J.I. and Asher, M.J.C. (Eds.), *Developments in Applied Biology II. Viruses with Fungal Vectors* (pp. 255-270). Wellesbourne, UK: Association of Applied Biologists.

———. (1990). Strain group specific and virus specific hypersensitive reactions to infection with potyviruses in potato cultivars. *Annals of Applied Biology 117:*93-105.

Jones, R.A.C. and Harrison, B.D. (1972). Ecological studies on potato mop-top virus in Scotland. *Annals of Applied Biology 71:*47-57.

Jordan, W. (2005). DAS ELISA. In Burns, R. (Ed.), *Immunochemical Protocols*, Third edition (pp. 207-214). Totowa, NJ: Humana Press.

Karande, A.A., Savithri, H.S., and Khurana, S.M. Paul (1998). Monoclonal antibodies—Production and application for the detection and diagnosis of potato viruses. In Khurana, S.M. Paul, Chandra, R., and Upadhya, M. (Eds.), *Comprehensive Potato Biotechnology* (pp. 162-186). New Delhi, India: Malhotra Publishing.

Kassanis, B. (1950). Heat inactivation of leaf roll virus in potato tubers. *Annals of Applied Biology 37:*339-41.

Kawchuk L.M., Martin, R.R., and McPherson, J. (1991). Sense and antisense RNA-mediated resistance to potato leaf roll virus in Russet Burbank potato plants. *Molecular Plant–Microbe Interactions 4:*247-253.

Kemeny, D.M (1991). *A Practical Guide to ELISA.* Oxford, UK: Permagon Press.

Khurana, S.M. Paul (2004). Potato viruses and their management. In Naqvi, S.A.M.H. (Ed.), *Diseases of Fruits and Vegetables* (Vol. II, pp. 389-440). Dordrecht, The Netherlands: Kluwer Academic Publishers.

Khurana, S.M. Paul and Garg, I.D. (1998). Present status of controlling mechanically and non-persistently aphid transmitted potato viruses. In Hadidi, A., Khetarpal R.K., and Koganezawa, H. (Eds.), *Plant Virus Disease Control* (pp. 593-615). St. Paul, MN: APS Press.

Khurana, S.M. Paul, Garg, I.D., Singh, M.N., and Kumar, S. (1993). Storage and preservation affect immuno electron microscopy of flexuous potato viruses. *Asian Potato Journal 3:*33-36.

Khurana, S.M. Paul and Sane, A. (1998). Apical meristem culture—A tool for virus elimination. In Khurana, S.M. Paul, Chandra, R., and Upadhya, M. (Eds.), *Comprehensive Potato Biotechnology* (pp. 207-232). New Delhi, India: Malhotra Publishing.

Klein, M. (2001). Transmission of viruses by leafhoppers and thrips. In Loebenstein, G., Berger, P.H., Brunt, A.A. and Lawson, R.H. (Eds.), *Virus and Virus-like Diseases of Potatoes and Production of Seed-potatoes* (pp. 227-236). Dordrecht, The Netherlands: Kluwer Academic.

Klerks, M.M., Leone, G.O.M., Verbeek, L.M., van den Heuvel, J.F.J.M., and Schoen, C.D. (2001). Development of a multiplex Amplidet RNA for the simultaneous detection of Potato leafroll virus and Potato virus Y in potato tubers. *Journal of Virological Methods 93:*115-125.

Kohler, G. and Milstein, C. (1975). Continuous cultures of fused cells secreting antibody of predefined specificity. *Nature 256:*495-497.

Kumar, S. and Singh, S. (1999). Sensitivity of polyvalent DOT-ELISA in comparison to DOT and DAS ELISA for detection of potato viruses. *Journal of Indian Potato Association 26:*58-62.

Lambert, D.H., Levy, L., Mavrodieva, V.A., Jhonson, S.B., Babcock, M.J., and Vayda, M. E. (2003). First report of potato mop-top virus on potato from the United States. *Plant Disease 87*:872.

Legorburu, F.J. (1998). Control of PVY transmission in the field. *Proceedings of the 10th Virology Section Meeting of the European Association of Potato Research* (p. 9). Baden, Austria: EAPR.

Levin, J.Z., de Framond, A.J., Tuttle, A., Bauer, M.W., and Heifetz, P.B. (2000). Methods of double-stranded RNA-mediated gene inactivation in *Arabidopsis* and their use to define an essential gene in methionine biosynthesis. *Plant Molecular Biology 44:*759-775.

Lizarraga, R.E., Salazar, L.F., Roca, W.M., and Schilde-Rentschler, L. (1980). Elimination of potato spindle tuber viroid by low temperature and meristem culture. *Phytopathology 70:*754-755.

Manzer, F.E. and Merriam, D. (1961). Field transmission of the potato spindle tuber virus and virus X by cultivating and hilling equipment. *American Potato Journal 38:*346-352.

Martinez-Soriano J.P., Galindo-Alonso, J., Maroon, C.J.M., Yucel, I., Smith, D.R., and Diener, T.O. (1996). Mexican papita viroid: Putative ancestor of crop viroids. *Proceedings of the National Academy of Sciences (USA) 93:*9397-9401.

Mellor, F.C. and Stace-Smith, R. (1977). Virus-free potatoes by tissue culture. In Reinert, J. and Bajaj, Y.P.S. (Eds.), *Applied and Fundamental Aspects of Plant Cell, Tissue, and Organ Culture* (pp. 616-735). New York: Springer-Verlag.

Metzlaff, M., O'Dell, M., Cluster, P.D., and Flavell, R.B. (1997). RNA-mediated RNA degradation and chalcone synthase A silencing in Petunia. *Cell 88:*845-854.

Morales, F.J., Lastra, R., de Uzcátegui, R.C., and Calvert, L. (2001). Potato yellow mosaic virus: A synonym of Tomato yellow mosaic virus. *Archives of Virology 146:*2249-2253.

Moreira, A., Jones, R.A.C., and Fribourg, C.E. (1980). Properties of a resistance-breaking strain of potato virus X. *Annals of Applied Biology 95:*93-103.

Mulholland, R.I. (1962). Control of the spread of mechanically transmitted plant viruses. *Commonwealth Phytopathological News 8:*60-61.

Murashige, T. and Skoog, F. (1962). A revised medium for rapid growth and bioassays with tobacco tissue cultures. *Physiologia Plantarum 15:*473-497.

NAPPO (2003). Requirements for importation of potatoes in a NAPPO. member country *Regional Standard for Phytosanitary Measures no 3.* Ottawa, Canada: North American Plant Protection Organisation.

Nie, X. and Singh, R.P. (2000). Detection of multiple potato viruses using an oligo(dT) as a common cDNA primer in multiplex RT-PCR. *Journal of Virological Methods 86:*179-185.

Nisbet, C., Butzonitch, I., Colavita, M., Daniels, J., Martin J., Burns R., George, E., Mulholland V., and Jeffries C. J. (2005). Characterisation of potato rough dwarf virus and potato virus P. *Abstracts of Papers and Posters of the 16th Triennial Conference of the European Association of Potato Research* (pp. 301-303), Bilbao, Basque Country: EAPR.

Northing, P., Walters, K.F.A., Barker, I., and King, L. (2004). Crop specific virus risk assessment: A cost effective management tool for the seed potato growing industry. *Proceedings of Conference of Crop Protection in Northern Britain* (pp. 297-302). Dundee, UK: Scottish Crop Research Institute.

Orlandi, R., Gussow, D.H., Jones, P.T., and Winter, G. (1989). Cloning immunoglobulin variable domains for expression by the polymerase chain reaction. *Proceedings of the National Academy of Sciences (USA) 86:*3833-3837.

Owens, R.A., Khurana, S.M. Paul, Smith, D.R., Singh M.N., and Garg, I.D. (1992). A new mild strain of potato spindle tuber viroid isolated from wild *Solanum* spp. in India. *Plant Disease 76:*527-529.

Perring, T.M., Gruenhagen, N.M., and Farrar, C.A. (1999). Management of plant viral diseases through chemical control of insect vectors. *Annual Review of Entomology 44:*457-481.

Peters, D. (1987). Control of virus spread. In De Bokx, J.A. and van der Want, J.P.H. (Eds.), *Viruses of Potatoes and Seed-potato Production* (pp. 171-74). Wageningen, The Netherlands: Pudoc.

Polston, J.E. and Anderson, P.K. (1997). The emergence of whitefly-transmitted geminiviruses in tomato in the western hemisphere. *Plant Disease 81:*1358-1369.

Powell, G. (1992). The effect of mineral oil on stylet activities and potato virus-Y transmission by aphids. *Entomologia Experimentalis et Applicata 63:*237-242.

Querci, M., Owens, R.A., Bartolini, I., Lazarte, V., and Salazar, L.F. (1997). Evidence for heterologous encapsidation of potato spindle tuber viroid in particles of potato leafroll virus. *Journal of General Virology 78:*1207-1211.

Querci, M. and Salazar, L. (1998). Molecular probes for detection of viruses/viroid in potato. In, Khurana, S.M. Paul, Chandra, R., and Upadhya, M. (Eds.), *Comprehensive Potato Biotechnology* (pp. 187-206). New Delhi, India: Malhotra Publishing.

Radcliffe, E.B., Ragsdale, D.W., and Flanders, K.L. (1993). Management of aphids and leafhoppers. In Rowe, R.C. (Ed.), *Potato Health Management* (pp. 117-126). St Paul, MN: APS Press.

Ragsdale, D.W., Radcliffe, E.B., and DiFonzo, C.D. (2001). Epidemiology and field control of PVY and PLRV. In Loebenstein, G., Berger, P.H., Brunt, A.A., and Lawson, R.H. (Eds.), *Virus and Virus-like Diseases of Potatoes and Production of Seed-potatoes* (pp. 237-270). Dordrecht, The Netherlands: Kluwer Academic.

Reavy, B., Arif, M., Kashiwazaki, S., Webster, K.D., and Barker, H. (1995). Coat protein-mediated immunity to potato mop-top virus in *Nicotiana benthamiana* is effective against fungal inoculation. *Molecular Plant–Microbe Interactions 8:*286-291.

Reisner, D. (1987). Structure formation (physical-chemical properties). In Diener, T.O. (Ed.), *The Viroids* (pp. 63-98). New York: Plenum Press.

Ribeiro, S.G., Inoue-Nagata, A.K., Daniels, J., and de Ávila, A.C. (2005). Potato deforming mosaic disease is caused by an isolate of *Tomato yellow vein streak virus. New Disease Reports.* http://www.bspp.org.uk/ndr/jan2006/2005-70.asp.

Robert, Y. and Bourdin, D. (2001). Aphid transmission of potato viruses. In Loebenstein, G., Berger, P.H., Brunt, A.A., and Lawson, R.H. (Eds.), *Virus and Virus-like Diseases of Potatoes and Production of Seed-potatoes* (pp. 195-225). Dordrecht, The Netherlands: Kluwer Academic.

Robert, Y., Woodford, J.A.T., and Ducray-Bourdin, D.G. (2000). Some epidemiological approaches to the control of aphid-borne virus diseases in seed potato crops in northern Europe. *Virus Research 71:*33-47.

Roberts, I.M. (1986). Practical aspects of handling, preparing and staining samples containing plant virus particles for electron microscopy. In Jones, R.A.C. and Torrance, L. (Eds.), *Developments and Applications in Virus Testing* (pp. 213-243). Wellesbourne, UK: Association of Applied Biologists.

Robinson, D.J. and Dale, M.F.B. (1994). Susceptibility, resistance and tolerance of potato cultivars to tobacco rattle virus and spraing disease. *Aspects of Applied Biology 39:*61-66.

Robinson, D.J. and Harrison, B.D. (2003). Tobacco rattle virus. *Descriptions of Plant Viruses no 346.* Wellesbourne, UK: Association of Applied Biologists.

Roenhorst, J.W., Butot, R.P.T., van Der Heijden, K.A., Hooftman, M., and van Zaayen, A. (2000). Detection of *Chrysanthemum* stunt viroid and Potato spindle tuber viroid by return-polyacrylamide gel electrophoresis. *EPPO Bulletin 30:*453-456.

Roistacher, C.N., Calavan, E.C., and Blue, R.L. (1969). Citrus exocortis virus-chemical inactivation on tools, tolerance to heat and separation of isolates. *Plant Disease Reporter 53:*333-336.

Rongai, D., Cerato, C., Martelli, R., and Ghendini, R. (1998). Aspects of insecticide resistance and reproductive biology of *Aphis gossypii* Glover on seed potatoes. *Potato Research 41:*29-37.

Salaman, R.N. (1926). *Potato Varieties.* Cambridge, UK: Cambridge University Press.

Salazar, L.F. (1996a). *Potato Viruses and their Control.* Lima, Peru: International Potato Center.

———. (1996b). Potato yellow vein disease: Evidence for the involvement of an unusual virus. *CIP Circular (Sept. issue)*: 2-4.

Salazar, L.F., Muller, G., Querci, M., Zapata, J.L., and Owens, R.A. (2000). Potato yellow vein virus: Its host range, distribution in South America and identification as a crinivirus transmitted by *Trialeurodes vaporariorum. Annals of Applied Biology 137:*7-19.

Salazar, L.F., Querci, M., Bartolini, I., and Lazarte, V. (1995). Aphid transmission of potato spindle tuber viroid assisted by potato leafroll virus. *Fitopatologia 30:*56-58.

Saldarriaga, A., Alvarez, A.M. and Jaramillo, J.E. (1988). Efecto del amarillamiento devenas transmitido por *Trialeurodes vaporariorum* (Westwood) en papa. *Revista Colombiana de Entomologia 14:*3-8.

Schumacher, J., Meyer, N., Riesner, D., and Weidemann, H.-L. (1986). Diagnostic procedure for detection of viroids and viruses with circular RNAs by "return" gel electrophoresis. *Journal of Phytopathology 115:*332-343.

Sigvald, R. (1986). Forecasting the incidence of potato virus Y. In McLean, G.D., Farret, R.G., and Ruesink, W.G (Eds.), *Plant Virus Epidemics—Monitoring, Modelling and Predicting Outbreaks* (pp. 419-441). Australia: Academic Press.

———. (1998). Forecasting aphid borne virus diseases. In Hadidi, A., Khetrapal R.K., and Koganezawa, H. (Eds.), *Plant Virus Disease Control* (pp. 172-187). St. Paul, MN: APS Press.

Singh, R.P. (1970). Seed transmission of potato spindle tuber virus in tomato and potato. *American Potato Journal 47:*225-227.

Singh, R.P., Boucher, A., and Somerville, T.H. (1992). Detection of potato spindle tuber viroid in the pollen and various parts of potato plant pollinated with viroid-infected pollen. *Plant Disease 76:*951-953.

Slack, S.A. (1983). Identification of an isolate of the Andean strain of potato virus S in North America. *Plant Disease 67:*786-789.

Smith, H.A., Powers, H., Swaney, S., Brown, C., and Dougherty W.G. (1995). Transgenic potato virus Y resistance in potato: Evidence for an RNA-mediated cellular response. *Phytopathology 85:*864-870.

Smith, H.A., Swaney, S.L., Parks, T.D., Wernsman, E.A., and Dougherty, W.G. (1994). Transgenic plant-virus resistance mediated by untranslatable sense RNAs—expression, regulation, and fate of nonessential RNAs. *Plant Cell 6:*1441-1453.

Solomon-Blackburn, R.M. and Barker, H. (2001). A review of host major-gene resistance to potato viruses X, Y, A and V in potato: Genes, genetics and mapped locations. *Heredity 86:*8-16.

Stace-Smith, R. and Mellor, F.C. (1968). Eradication of potato viruses X and S by thermotherapy and axillary bud culture. *Phytopathology 58:*199-203.

Stevenson, W.R., Loria, R., Franc, G.D., and Weingartner, D.P. (Eds.) (2001). *Compendium of Potato Diseases.* St Paul, MN: APS Press.

Sudarsono, Young, J.B., Woloshuk, S.L., Parry, D.C., Hellmann, G.M., Wernsman, E.A., Lommel, S.A., and Weissinger, A.K. (1995). Transgenic burley and flue-cured tobacco with resistance to 4 necrotic isolates of potato virus Y. *Phytopathology 85:*1493-1499.

Thomas, P.E and Mink, G.I. (1979). Beet curly top virus. *AAB Descriptions of Plant Viruses.* (http://www.dpvweb.net/dpv/showdpv.php?dpvno=210).

Thomas, P.E., Mojtahedi, H., Crosslin, J.M., and Santo, G.S. (2001). Reduction and elimination of tobacco rattle virus from soils in alfalfa cultures. *American Potato Journal 78:*487 (abstract).

Toth, R.L., Harper, K., Mayo, M.A., and Torrance, L. (1999). Fusion proteins of single-chain variable fragments derived from phage display libraries are effective reagents for the routine diagnosis of Potato leafroll virus infection in potato. *Phytopathology 89:*1015-1021.

UNECE. (2003). UNECE Standard S-1 concerning the certification and commercial quality control of seed potatoes. (http://www.unece.org/trade/agr/standard/).

Usharani, K.S., Surendranath, B., Garg, I.D., Khurana, S.M. Paul, and Malathi, V.G. (2004). Potato leaf curl—a new disease of potato in northern India caused by a strain of Tomato leaf curl New Delhi virus. *Plant Pathology 53:*235.

Valkonen, J.P.T. (1994). Natural genes and mechanisms for resistance to viruses in cultivated and wild potato species (*Solanum* spp.). *Plant Breeding 112:*1-16.

Valkonen, J.P.T., Pehu, E., and Watanabe, K. (1992). Symptom expression and seed transmission of alfalfa mosaic virus and potato yellowing virus (SB-22) in *Solanum brevidens* and *S. etuberosum. Potato Research 35:*403-410.

Van de Graaf, P., Lees, A.K., Wale, S.J., and Duncan, J.M. (2005). Effect of inoculum level and environmental factors on potato powdery scab caused by *Spongospora subterranean. Plant Pathology 45:*22-28.

Van den Heuvel, J.F.J.M., Dirven, J.A.A.M., Van Os, G.J., and Peters, D. (1993). Acquisition of potato leafroll virus by *Myzus persicae* from secondarily-infected potato plants of different genotypes. *Potato Research 36:*89-96.

Van Dijk, P. and Cuperus, C. (1989). Reactions of *Nicotiana* species to potato viruses A, X and Y and tobacco mosaic virus in relation to their taxonomy and geographical origin. *Netherlands Journal of Plant Pathology 95:*343-356.

Verhoeven, J.T.J. and Roenhorst, J.W. (2003). Detection of a broad range of potato viruses in a single assay by mechanical inoculation of herbaceous test plants. *EPPO Bulletin 33:*305-311.

Verhoeven, J.T.J., Jansen, C.C.C., Willemen, T.M., Kox, L.F.F., Owens, R.A., and Roenhorst, J.W. (2004). Natural infections of tomato by *Citrus exocortis viroid, Columnea latent viroid, Potato spindle tuber viroid* and *Tomato chlorotic dwarf viroid. European Journal of Plant Pathology 110:453-456.*.

Verma, Y., Khurana, S.M. Paul, Mohan, Jitendra, Sood, Shivali, and Mukherjee, Krishanu (2003). Development of a non-radioactive probe for detection of potato virus Y. *Journal of Indian Potato Association 30:*111-112.

Vunsh, R., Rosner, A., and Stein, A. (1990). The use of the polymerase chain reaction (PCR) for the detection of bean yellow mosaic virus in gladiolus. *Annals of Applied Biology 117*:561-569.

Wang, P. (1985). Potato tissue culture and its applications in agricultural. In Paul, H.L (Ed.), *Potato Physiology* (pp 503-577). London: Academic Press.

Weidemann, H.-L. (1988). Importance and control of potato virus $Y^N$ in seed potato production. *Potato Research 31*:85-94.

Weingartner, D.P., Shumaker, J.R., and Smart, G.C. (1983). Why soil fumigation fails to control corky ringspot disease in Florida. *Plant Disease 67*:130-134.

Wetzel, T., Candresse, T., Macquaire, G., Ravelonandro, M., and Dunez, J. (1992). A highly sensitive immunocapture polymerase chain reaction method for plum pox potyvirus detection. *Journal of Virological Methods 39*:27-37.

Wilson, C.R. and Jones, R.A.C. (1992). Resistance to phloem transport of potato leafroll virus in potato plants. *Journal of General Virology 73*:3219-3224.

Wolf, I. and Horvath, S. (2001). Occurrence of Potato Y (PVY) potyvirus strains in potato fields in Hungary. Novenyvedelem *36*:449-455.

Woodford, J.A.T., Shaw, M.W., McKinlay, R.G., and Foster, G.N. (1977). The potato aphid spray warning scheme in Scotland. *Proceedings of a Symposium on Problems of Pest and Disease Control in Northern Britain* (pp. 28-31). Dundee, UK.

Xenophontos, S., Robinson, D.J., Dale, M.F.B., and Brown, D.J.F. (1998). Evidence for persistent, symptomless infection of some potato cultivars with tobacco rattle virus. *Potato Research 41*:255-265.

Younkin, S.G. (1942). Weed sucepts of the potato yellow dwarf virus. *American Potato Journal 19*:6-10.

Ziegler, A., Torrance, L., Macintosh, S.M., Cowan, G.H., and Mayo, M.A. (1995). Cucumber mosaic cucumovirus antibodies from a synthetic phage display library. *Virology 214*:235-238.

# Chapter 12

# Insects, Nematodes, and Integrated Pest Management

Hanafi Abdelhaq

Despite technological advances leading to tremendous yield increase, potato production continues to face pest threats. Potato is host to many insects and nematodes, which often vary depending on the region, the growing season, and the production system. For instance, for potato seed production, the most significant insects are aphids as vectors of viruses (Hanafi, 2000). In ware potato production, in warm climates, the key insect is the potato tuber moth *Phthorimaea operculella* (Hanafi, 1998). Certain insects and nematodes can cause tuber yield reductions of up to 50-80 percent (Radcliffe et al., 1993). Losses of this magnitude or even higher are commonly observed in potato fields heavily infested with potato tuber moth in spring and summer crops in North Africa and the Middle East (Hanafi, 1999). These key pests can be managed using pesticides. However, once farmers start using pesticides on a routine basis, often certain pest species are reduced but others show greater importance, probably because of elimination of their natural enemies or development of pesticide resistance (Radcliffe et al., 1993). For each pest a thorough understanding of the problem is required for its effective management. This chapter reviews various aspects of major insect and nematode pests infesting potato.

## *MAJOR INSECT PESTS*

### *Aphids*

In addition to the green peach aphid *Myzus persicae,* other species of worldwide importance to potato crop are *Macrosiphum euphorbiae,* buckthorn aphid *Aphis nasturtii,* and the stolon-infesting *Rhopalosiphoninus latysiphon.* Among species of regional importance are the *Aphis gossypii* Complex (including *A. frangulae, A. fabae, A. craccivora*), *Aulacorthum solani, Rhopalosiphum padi,* and *Lipaphis erysimi* (Bacon et al., 1978; Radcliffe et al., 1993; Khurana et al., 2000).

Aphids rarely cause economic damage through direct plant injury, but their role as virus vectors is the most significant (see Chapter 11 "Viruses and Viroids").

#### *Biology and Behavior*

Aphids belong to the group Homoptera and the family Aphididae. In regions with mild winters where potatoes are grown year round, aphids have a parthenogenetic life cycle, with a rapid turnover of generations resulting in a dramatic increase in their population (Hanafi, 1992). In regions with severe winters, aphids reproduce sexually and overwinter as fertilized eggs. The primary overwintering host is a woody plant (*Prunus* spp. including peach and nectarine), whereas in summer secondary hosts are often a broad range of herbaceous species. The green peach aphid is highly polyphagous, with more than 875 secondary hosts (Radcliffe et al., 1993).

Overwintering eggs hatch in spring, giving rise to apterous, parthenogenetic, viviparous mothers (fundatrices) that feed and reproduce on the primary host. Fundatrices that drop from the tree may survive on secondary hosts. Beginning with the second generation, alate morphs (spring migrants) appear. Depending on the region, migrants invading potatoes are both true spring migrants from primary host(s) and alatae produced on secondary hosts or mainly migrants from the secondary hosts (Radcliffe et al., 1993).

On potato, apterae (Figure 12.1A) are almost always the predominant morph. As the host plant ages, more alate (Figure 12.1B and C)

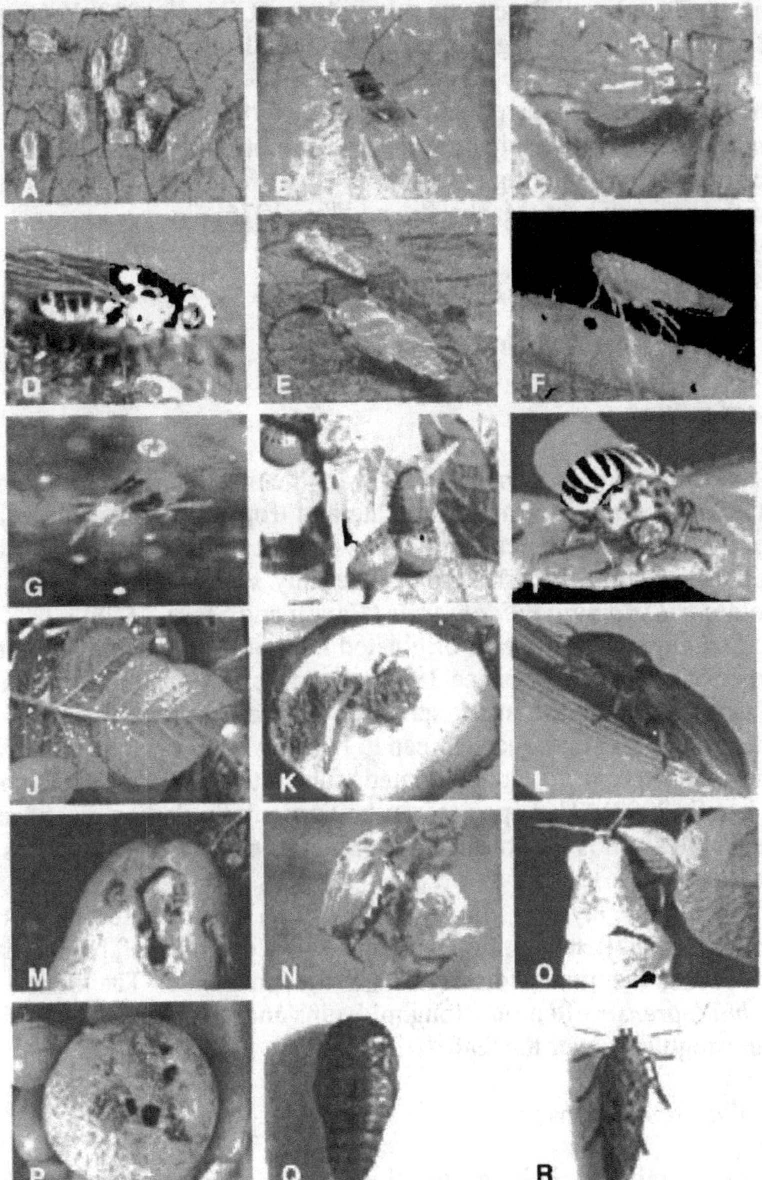


FIGURE 12.1. Important insect pests of potato. See Plate 12.1 in the color gallery.

are produced and disperse to other plants (Gibson, 1978; Ferro and Boiteau, 1993). During winter, as day length shortens, fall migrants are produced; these are both male and female. Fall migrants return to the primary host, where the female gives birth to an apterous female morph (oviparae). Oviparae mate with male fall migrants and lay fertilized overwintering eggs. These eggs are usually laid near buds of the primary host (Mackauer and Way, 1976).

*Management*

See Chapter 11 "Viruses and Viroids."

## *Leaf miners*

Leaf miner flies, *Liriomyza* spp., are the most important dipterans attacking potato in warm temperate and tropical regions. There are several species of leaf miners, which cause damage to most solanaceous and cucurbit crops. The only species that develops on potato is *Liriomyza huidobrensis* (Hanafi and Srairi, 1998).

*Liriomyza huidobrensis* originated in South and Central America, whence it has spread since 1985 to parts of the United States and Hawaii. In 1989, the insect spread to northern Europe by unknown means, and its subsequent spread to North Africa was apparently via the importation of contaminated substrate (peat) from Europe (Hanafi, 2003). This species can be particularly damaging to leafy vegetables and is the sole species of leaf miner to cause problems in potato.

The female lays eggs in the leaf tissue on the upper surface of the leaf. Mining often begins on the upper surface of the leaf, but thereafter the larvae tunnel down toward the underside. The mines of *L. huidobrensis* often run along midveins and side veins but can also run irregularly over the leaf.

*Biology and Behavior*

Leaf miners belong to the group Diptera and family Agromyzidae. *Liriomyza huidobrensis* is closely related to the tomato leaf miners *(L. trifolii and L. bryoniae)* (Hanafi, 2005). The adult of *L. huidobrensis* (Figure 12.1D) is slightly larger (1-2 mm) and darker and also has a bright yellow thoracic shield. A binocular microscope is usually necessary to distinguish between the adults of the two species. Larvae are slightly whiter than those of the tomato leaf miner (milky to yellowish white) and are at most 3.5 mm in length. Pupae are approximately 2.2 mm long. Pupae in the same crop and of the same age can nevertheless vary enormously in color, from pale yellow to black (Parella et al., 1983; van der Linden and van Achterberg, 1989).

Under favorable conditions a population of *L. huidobrensis* can quickly build up. At approximately 19°C the egg stage lasts 2 days, the larval period 6 days, and the pupal stage 8 days, with a pre-oviposition period of 1-2 days. At this and slightly warmer temperatures the life cycle lasts 2-2.5 weeks. It appears that this species is slightly more resistant to cold than the other leaf miners (Parella, 1984; van der Linden, 1993).

### *Management*

Many commonly used insecticides (deltamethrin, imidachloprid, endosulfan, etc.) are not effective against the leaf miner, although biological control can be successful. ***Diglyphus isea,*** a commonly occurring natural enemy, is quite effective against the leaf miner in many parts of the world. The mortality of the larvae of leaf miners through natural parasitism by ***Diglyphus*** can reach up to 80 percent (Hanafi and Srairi, 1998). Cultural practices aimed at eliminating favorable hosts of leaf miners from the potato field and its vicinity help reduce the population. Favorable weed species include ***Solanum nigrum*** and ***Datura stramonium.***

## ***Leafhoppers***

Leafhopper species ***Armrasca biguttula*** and ***Emposca devastans*** are important potato pests in India (Saxena et al., 1974). In North America, four species of leafhoppers are of great economic importance: *E. fabae* (Figure 12.1E, F), *E. filamenta, E. abrupta,* and *E. arida* (Delong, 1931).

In contrast to aphids, leafhoppers are important mainly because of their direct feeding damage, though some are also vectors of other pathogens. Potato plants are susceptible to injury from the potato leafhopper at all growth stages, but infestations are most damaging during early tuber bulking. Leafhoppers suck the sap of leaves, which is replaced by air, giving the cells a grayish white appearance. These grayish-white spots, about 1 mm in diameter and resembling the feeding spots of leaf miners, can be present in huge numbers, sometimes randomly distributed and sometimes in stripes. Prolonged feeding by the potato leafhopper causes a condition known as hopper burn (Prasad, 1961). The first conspicuous symptom of injury is a brown, triangular lesion at the tip of the infested leaflet. As the symptoms develop, lesions spread backward and inward from the margin, eventually destroying the entire leaflet. Damage from feeding by leafhoppers is the result of disruption of the phloem tissues responsible for transporting the photosynthates within the plant (Ladd and Rawlins, 1965).

Some species of leafhopper are important to potato health not because of direct feeding but because of their role as vectors in the transmission of certain plant pathogens. The aster leafhopper *(Macrosteles fascifrons)* is the most important in this regard. It is the principal vector of aster yellows mycoplasma-like organism (AY-MLO), which causes purple top in potato (Delong, 1971). Leafhopper generally acquires the pathogen by feeding on an infected host plant during migration from ripening small grains to potato. More than 350 plant species have been identified as hosts of AY-MLO. The pathogen must incubate for several days within the leafhopper's body before the insect can transmit it. Moreover, it takes several hours of feeding by the leafhopper to transmit the pathogen successfully (Delong, 1971). The long incubation and feeding periods required for transmission of AY-MLO afford opportunities to reduce purple top by controlling the vector. In India, another MLO, purple top roll, is transmitted by *Orosius albicinctus* (Nagaich and Giri, 1973; Khurana et al., 1988).

The clover and beet leafhoppers *(Aceratagallia sanguinolenta)* transmit potato yellow dwarf virus and beet curly top virus, respectively (see Chapter 11 "Viruses and Viroids").

### *Biology and Behavior*

The common name "leafhoppers" used here refers specifically to the family Cicadellidae. Leafhoppers, like aphids, whiteflies, and scale insects, belong to the suborder Homoptera of the order Hemiptera. Within this suborder, aphids, whiteflies, and scale insects constitute the group Sternorrhyncha, and the cicadas and the leafhoppers form the group Auchenorrhyncha. The former group has long, threadlike antennae, whereas the latter has short antennae and relatively large eyes situated at the sides of the head.

Leafhoppers pass through three life stages: egg, nymph (in five instars), and adult. On potato, they usually complete several generations per year. The population density is dependent upon the date of arrival on the crop and the temperature. In summer, potato leafhopper adults live for 30-40 days, and some even cross over 90 days, with the result that generations overlap. The females lay an average of 2 or 3 eggs per day and as many as 200 over a lifetime. At summer temperatures (above 28°C), eggs hatch in approximately 10 days. The nymphs are yellowish green when young but closely resemble the adult leafhopper when older. They require approximately 12 days to develop into adults. New adults begin laying eggs when they are 6 days old (Radcliffe et al., 1993).

### *Management*

The size of potato leafhopper population required to cause economic damage varies with the cultivar, the stage of plant growth, and the environment. Early-maturing cultivars are generally more susceptible to the potato leafhopper, but these cultivars bulk more rapidly and their yield may actually be affected less. Another common assumption is that the potato crop is more susceptible to leafhopper injury if it is under drought, and hence it is more important to control leafhoppers under water stress (Radcliffe et al., 1993).

If leafhopper populations exceed locally accepted action thresholds, insecticides (imidachloprid for instance) provide the only effective means of controlling these pests on potato. Potato leafhopper populations are relatively easy to monitor. Adults are best sampled by sweep net and nymphs by examining randomly picked midplant leaves. The number of samples required to estimate the size of a population satisfactorily depends on the population density. A sample of 35 picked leaves per location generally gives population estimates of sufficient accuracy for management decisions. In the United States, insecticide application is recommended when the population density exceeds ten nymphs per 100 leaves (Radcliffe, 1982; Hanafi et al., 1995).

Soil systemics (aldicarb) applied in furrow at planting or side-dressed at plant emergence give 6-8 weeks of control and can essentially prevent the transmission of leafhopper-borne pathogens. However, for reasons of cost, systemic insecticides are probably seldom used specifically for this purpose. The standard industry practice for leafhopper control on fresh-market potatoes is to apply foliar sprays of methamidophos. Insecticide resistance is not a problem with the potato leafhopper, and almost any registered insecticide provides good control of this pest (Radcliffe et al., 1993).

### *Psyllids*

The psyllid *(Paratrioza coekerelli)* is an occasional potato pest. Most of the damage is caused by nymphs, which inject a toxin while feeding. As a result, immature leaves turn red or purple. Older leaves become unusually thick, roll upward, and turn yellow, and the internodes become shortened. Affected plants produce excessive numbers of small, distorted tubers, which sprout without any dormancy. Small aerial tubers may also appear in leaf axils. Tubers produced in a heavily infested field are generally too small to be marketable and also not suitable for seed purposes (Radcliffe et al., 1991).

The severity of damage caused by a psyllid infestation is determined by several factors, including the size of the colonizing population, the growth stage of the potato crop at the time of colonization, prevailing temperatures, and the extent to which natural enemies are present. The greatest yield reduction occurs when plants are infested early in the season. As few as 3 or 4 psyllids per plant can produce psyllid yellows on plants in prebloom stage under favorable conditions, but 15 or more psyllids per plant may be required to cause damage to plants in bloom (Boiteau et al., 1987).

Larger plants that provide more shade are most likely to become heavily infested. The availability of suitable host plants when migrants move into an area is an important factor in maintaining populations in a region. For this reason, early-planted potatoes can be heavily damaged.

*Biology and Behavior*

Colonizing adults lay eggs along the leaf margins of host plants. The light-yellow eggs are spindle-shaped and suspended on short stalks from leaves on which they are deposited by the female, which resembles a tiny cicada. There are five nymphal stages, and the life cycle is completed in approximately 4 weeks. The nymphs are flattened, scale-like, and pale yellow-green.

The potato psyllid is very sensitive to temperature, which to a large extent determines its movement and distribution. Rising temperatures in the overwintering areas induce the production of migrants. Cool weather is ideal for holding populations within an area, and outbreaks often follow cool periods in early summer.

*Management*

It is important to monitor psyllid populations before initiating any control measures as psyllid infestations are sporadic. Early scouting should concentrate on the lower leaves of potato plants or on Solanaceous weeds growing near potato fields. Yellow pan traps can be used for early detection, but there are no good correlations between trap catches and resultant damage. When a field is scouted, samples comprising 50 leaflets randomly picked up throughout the field should be carefully examined. Treatment should commence if there is more than one nymph per ten leaflets on young plants (prebloom) or three to four nymphs per leaflet on plants in full bloom or older plants. Feeding by adult psyllids has no effect on potato yields at any time.

If psyllid populations reach or cross these action thresholds, insecticides such as abamectine provide the only consistent control measure. It is important to cover the canopy thoroughly with the spray to control nymphs feeding on the undersides of leaves. As potato psyllid infestations are sporadic and are initiated by migrating adults, cultural practices such as crop rotation and planting time have little effect.

### *Spider Mites*

Spider mites *(Tetranychus urticae, Polyphagotarsonemus latus, Aculops lycopersici)* are pests of many crops throughout the world. The two-spotted spider mite, also known as the red spider mite *(Tetranychus urticae)* (Figure 12.1G), is by far the most important species in potato (Jeppson et al., 1975). Owing to their great reproductive capacity they are capable of destroying plants very rapidly. Larvae, nymphs, and adults cause damage to the potato plant by feeding on plant tissue and sap. They mainly occur on the underside of leaves, where they pierce the cells and suck out the contents. These dead cells become yellow, and in many plants the damage can also be seen on the upper surface of leaves as small yellow dots. As the damage increases, whole leaves turn yellow, and as the chlorophyll is reduced, the leaf, and eventually the whole plant, may die. In the event of a severe attack of spider mite, transpiration is reduced, resulting in a higher temperature and a lower humidity at leaf surface. This change in the microclimate at the leaf surface suits the spider mite and rapidly promotes its population growth. The nymphs and adults also produce webs, and plants can be completely covered with such webs, in which the mites swarm.

The two-spotted spider mite can spread through a crop in various ways. If plants are heavily infested, mites fall to the ground and move over to other plants or can easily migrate to new plants along crop rows. They also secrete silk threads on which they are easily dispersed over great distances by air currents. Mites can also be dispersed mechanically by machinery, clothing, and other objects.

#### *Biology and Behavior*

Spider mites belong to the class Arachnida (along with spiders), the subclass Acarina or Acari (mites), the order Acariformes, and the family Tetranychidae. This family contains many harmful plant-pest species. Of the family Tetranychidae, *T. urticae* remains the most polyphagous species and is an important pest of many ornamental and food crops throughout the world (Jeppson et al., 1975).

The adult female *T. urticae* has an oval body that is round at the posterior. The color can vary from orange, light yellow, or light green to dark green, red, brown, or almost black. The male is smaller and more active than the female, with a body form that is narrower and more pointed and a color that varies from light yellow or orange to dark yellow or brown depending on the season. Both males and females usually have two large black spots, hence the common name "two-spotted spider mite." The spots can vary both in form and size. A similar species often encountered in the literature is *T. cinnabarinus*, the carnation spider mite. However, many researchers are convinced that this is not a separate species but rather one of a complex of different races or biotypes of the two-spotted spider mite that vary in host plant preference, color (green and red), and form (oval). As there is no correlation between various features of different races or biotypes, the different populations are now all regarded as a *T. urticae* complex and not as a separate species (Zhang, 1995).

Population growth in two-spotted spider mite depends on temperature and relative humidity, as well as the variety, age, and quality of the crop. Of these, temperature is certainly the most important. Different biotypes of spider mite may also show differing growth rates. Development does not occur at temperatures below 12°C; the population growth increases with temperature up to 30°C but declines again at higher temperatures. Atmospheric humidity also has an effect on population growth, with *T. urticae* laying more eggs and developing faster at lower humidity (<80 percent) (Malais and Ravensberg, 2003). The spider mite inhabits a microhabitat provided by the layer of air at the surface of the leaf. Temperature and humidity within this boundary layer can differ significantly from the conditions prevailing in ambient air.

Under summer temperatures, a spider mite population can develop extremely rapidly. As a result of this rapid population growth and the frequent use of chemical pesticides, resistance can also develop very rapidly (Jeppson et al., 1975).

When environmental conditions change adversely, female spider mites enter diapause. Such adverse conditions include decreasing day length, falling temperature, and decline or deterioration in food supply. Females entering diapause become orange-red within 3-5 days of becoming adult. After being fertilized, they overwinter in a dormant condition, hidden within various host plants. During this period they eat nothing, lay no eggs, and are less susceptible to chemical pesticides. They are also more difficult for predatory mites to find, not only because they hide in concealed places but also because the cessation of feeding activity means they no longer cause the release of attractive volatile compounds from damaged plant tissue. As soon as conditions become more favorable in the spring, the females become active again and resume egg laying. A single mating is sufficient to fertilize all the eggs. Fertilized females produce both male and female progeny, whereas unfertilized females produce only males. The sex ratio in a population of *T. urticae* is usually one male to three females (Popov et al., 1992).

Two-spotted spider mites pass through five developmental stages: egg, larva, first-stage nymph (protonymph), second-stage nymph (deutonymph), and adult mite. In the larval and nymphal stages, an active period and a resting period of roughly equal duration can be distinguished. Eggs are usually found on the underside of leaves. They are round, with an average diameter of 0.14 mm, and are transparent when freshly laid, later turning opaque. By hatching time they are straw-colored (Popov et al., 1992).

The larvae have three pairs of legs, and when hatched they are virtually colorless with two dark red eyes. Once they begin feeding their color changes to light green, brownish yellow, or even dark green. Two dark spots also develop in the middle of the body. Once sufficient food has been consumed, they become inactive and withdraw their legs until they develop into a protonymph.

Protonymphs have four pairs of legs and are somewhat larger than the larvae. Their color varies from light to dark green. The two body spots are larger and clearer than at the larval stage. Here, too, a period of feeding is succeeded by a resting period that terminates in a molt, out of which develops the deutonymph. This second-stage nymph is larger than the protonymph but has the same coloration. Male and female individuals can usually be distinguished at the deutonymph stage, the male being elongate and slightly smaller than the female, which is rounder in form (Malais and Ravensberg, 2003).

### *Management*

Creating a more humid environment around the plants will help prevent spider mite infestation. Spider mites often appear in a potato crop in specific places, mainly at the edges and in rows facing the wind direction. Such places are often dusty and have a drier, warmer climate more favorable for spider mites. Spraying water helps prevent infestation of spider mites.

The widespread use of chemical pesticides has led to many two-spotted spider mite populations developing resistance to a range of pesticides (Helle and Sabelis, 1985). In many instances, biological control of this pest occurs naturally. In fact the predatory mite *Phytoseiulus persimilis* occurs in high numbers in fields where pesticides are not used heavily. Other natural enemies of spider mites are the predatory bug *Macrolophus caliginosus* and the predatory mites *Amblyseius cucumeris* and *A. degenerans.* The lacewing *Chryperla carnea* also preys on spider mites (Sharaf, 1984).

### *Colorado Beetles*

The Colorado beetle, *Leptinotorsa decemlineata,* originally comes from the eastern slopes of the Rocky Mountains in the west of the United States. The beetle spread throughout the United States and Canada and was first reported in Europe (in Germany) in 1877 (Hsiao, 1978). The Colorado beetle later spread to Mexico, parts of Central America and the Caribbean region, most of Europe, and parts of Asia (Radclifffe, 1982).

Colorado beetles have a preference for potatoes but also feed on many other Solanaceous species, such as brinjal (eggplant), sweet pepper, and tomato (Hsiao, 1978). Both adult and larvae damage plants by eating the leaves. They are such prolific feeders that plants can be rapidly defoliated, leaving only bare stalks and skeletonized foliage.

### *Biology and Behavior*

*Leptinotorsa decemlineata* belongs to the order Coleoptera and the family Chrysomelidae. This is a large family with more than 25,000 species, almost all of which are leaf eaters. Most leaf beetles are small and attractive, with shiny metallic coloration. The adult Colorado beetle (Figure 12.1I) is approximately 1 cm long, with an oval-shaped body. The color is pale yellow to brown-yellow with irregular black spots on the head and thorax and five black longitudinal stripes on the wing covers. The legs are predominantly yellow (Jacques, 1988).

The beetles come to the ground in spring or early summer (depending on the climate), and if the temperature is high enough (>10°C) they can fly. A female is able to lay several hundred eggs. Larvae hatch from the eggs after approximately 7 days and begin to feed immediately. Larval development takes approximately 2-3 weeks and pupation another 2-3 weeks. There are four larval instars. The young larva is red-brown, and older larvae are orange to red. Larvae have a conspicuous head with biting mouthparts, three pairs of legs and a humpbacked body with two rows of black dots on either side (Figure 12.1H). When fully grown they fall to the ground and bury themselves in order to pupate. The life cycle is completed when the adult beetle develops from this pupa. Adult beetles overwinter, digging into the ground to a depth of 25-40 cm in order to hibernate (Jacques, 1988).

### *Management*

The bacterium *Bacillus thuringiensis tenebrionis* can be used effectively as a control agent against several species of beetle, including *L. decemlineata.* The larvae are killed by ingesting the bacteria along with leaf material. Another strain of this bacterium, *Bacillus kurstaki,* is also efficient against this pest. In seed potato, aldicarb application in furrow at planting gives adequate control. In ware potatoes, other safe insecticides such as fenvalerate, oxamyl, and endosulfan are recommended (Radcliffe, 1982).

### *Whiteflies*

The most common species of whitefly affecting potatoes are the glasshouse whitefly *Trialeurodes vaporariorum* and the tobacco whitefly *Bemisia tabaci.* These two species are widespread and broadly comparable in outline. Adult whiteflies are mostly found on the underside of young leaves (Figure 12.1J). If whitefly populations reach a high level, they can cause plant death by feeding. Sooty mould on honeydew secretion by whiteflies leads to indirect damage to the plants by decreasing photosynthesis.

### *Biology and Behavior*

Whiteflies belong to the family Aleyrodidae, order Hemiptera (true bugs), suborder Homoptera, and division Sternorrhyncha. Both whitefly species that affect potato, *T. vaporariorum* and *B. tabaci,* belong to the subfamily Aleyrodinae. Whiteflies have six distinct developmental stages: the egg; first, second, third, and fourth larval instars; and adult. The late fourth larval stage is frequently referred to as a "pupa," but strictly speaking it is not a true pupal instar because there is no separate pupal molt, even though the external appearance of the insect changes markedly during this stage (Cock, 1986).

The eggs are elliptical and often deposited in circles on the lower surface of leaves. The young larvae, which have well-developed legs and antennae on hatching, are known as "crawlers." In the second and third larval stages, the legs and antennae are reduced to one or two segments and are no longer visible. The growth of hairs on the insect depends on the hairs of the leaf—the longer the hairs on the leaf, the longer are the hairs on the insect. In other words, the larva adapts itself to the structure of the leaf on which it lives. In the fourth larval stage, the insect finally becomes flattened and the cuticle hardens. As soon as the red eyes of the adult whitefly are visible in the fourth larval instar, it is referred to as a "pupa." *Trialeurodes vapárariorum* and *B. tabaci* can be distinguished by several features during the pupal stage. In *T. vapárariorum* a wreath of short waxy threads can be seen around the body, projecting outward. This is lacking in *B. tabaci,* whose pupae are broad, flat, and transparent or yellowish. Pupae of *T. vapárariorum* are small, white, and oval (Cock, 1986).

The adult emerges from the hardened cuticle of the pupa through a characteristic T-shaped opening. Newly emerged whiteflies have two pairs of transparent wings; later on wings and the body are covered with a white, waxy powder that gives the insect its characteristic appearance.

*Management*

Natural enemies of whiteflies help in checking its population buildup in potato crop. The two parasitoids most commonly found on whiteflies are *Encarsia formosa* and *Eretmocerus erimicus* or *E. moundus.* The bugs *Macrolophus caligonisus* and *Nesidiocorris tenuis* occur naturally in many parts of the world and play an important role as predators of whiteflies not only in potato but also in most other field crops (Hanafi and Schnitzler, 2004). In addition to the parasitoids and predators, the entomopathogenic fungi *Verticillium lacanii* can also infect and kill whiteflies (Hanafi, 2003). These beneficial organisms, however, are quite sensitive to most of the insecticides and fungicides used on potatoes to control pests and diseases. The use of broad-spectrum pesticides often upsets the action of the beneficial organisms, resulting in dangerous levels of whitefly population (Bouharroud et al., 2003).

*Wireworms*

Many species of wireworms are pests of potato worldwide. *Agriotes lineatus, A. obscurus,* and *A. spectator* are the species of economic importance (van der Zaag et al., 1996). Though they inhabit most regions, none of the species can be regarded as a key pest; yet collectively they represent a serious problem, especially in areas of the world with heavy soils. Early in the growing season, wireworm larvae bore seed tubers (Figure 12.1K). These feeding sites sometimes become infected with fungi or bacteria and result in weak shoots and gappy stands. Later in the season, wireworms bore bulking tubers, resulting in holes that may be shallow or deep, usually characterized by a clean, round entry (2-3 mm diameter). This type of injury does not reduce yields, but it makes the tubers unmarketable.

### *Biology and Behavior*

Wireworms belong to the order Coleoptera and the family Elateridae. Although different wireworm species have similar biology, life cycles range from 1 to 5 years depending on species and soil temperature. The pest can overwinter in all life stages, and generally several different larval instars can be found at any one time during the summer. For this reason, the relative abundance of wireworms in different life stages may not be synchronous from one season to the next (Wolfgang and Rieckmann, 1991).

As most of the life cycle is spent in the larval stages, the larvae predominate in a field. They move up and down in the soil profile in response to changes in temperature and moisture. In temperate areas, the larvae remain in the top few centimeters of soil during the growing season. They burrow down as deep as 60 cm to overwinter. In the spring, when soil temperatures become favorable, they move back to the upper layer to resume feeding. Most damage to potato is caused by larvae in their second and third years of development.

Last-instar larvae either pupate immediately or overwinter in the pupal cells and pupate in spring. In the spring and summer, the adults (Figure 12.1L) come to the surface, and if adequate host plants are not present, they fly to lay their eggs in cracks or crevices or burrow into the soil to lay eggs near host plants. Young larvae feed on roots of cereal crops, pasture grasses, and weeds.

### *Management*

Application of soil insecticides (Mocap, metam sodium, diazinon) is the only viable management option. Insecticides can be banded or broadcast before planting, or they can be side-dressed after plants begin to emerge. For broadcasting, any labeled granular or tank-mixed product is spread uniformly over the entire field and mixed into the soil immediately to the depth specified on the label. For in-furrow applications, placing the granules in a narrow band of 5-10 cm below the seed piece at planting is recommended. To determine whether a side-dress treatment is needed after planting, seed pieces should be dug at random throughout the field as soon as shoots begin to emerge to check for wireworm injury. This is also an excellent way to determine the effectiveness of a preplant treatment. There is no well-established action threshold for this approach. If a second application is necessary, an insecticide different from that used in the pre-plant or pre-emergence treatment can be applied. Crops such as alfalfa can be used for rotation with potato in fields prone to wireworm infestation (Rowe, 1993).

### *White Grubs*

White grubs or chafer grubs are the larvae of such beetles as the cockchafer. The most important species on potato and other horticultural crops is *Melolontha melolontha* (Figure 12.1M, N).

#### *Biology and Behavior*

Whitegrubs belong to the order Coleoptera and the family Scarabeidae. The development from the egg to the adult beetle takes 1 year. At the beginning of the warm season, the female beetle lays its eggs in the soil. The larvae at full development measures 5 cm and is curved toward its white abdomen. The 3-5 cm long chafer grubs are white to yellow and have a red or brown head. They are usually curved toward the abdomen. The larvae of white grubs usually remain in the soil for 3-5 years and overwinter deep in the soil.

#### *Management*

The behavior of white grub larvae in the soil is similar to that of wireworms. Thus management practices used for wireworms are also applicable to white grubs.

### *Cutworms*

Numerous species of cutworm attack potato. Among the most cosmopolitan and destructive are *Agrotis* spp. Larvae cut the stalks of young growing plants. After tuberization, cutworm damage is largely confined to the tubers. Crevice caused by larvae can easily be confused with white grub damage.

### *Biology and Behavior*

Cutworms belong to the order Lepidoptera and family Noctuidae. They are the larvae of various species of noctuid moths. Cutworm larvae are nocturnal, living just below the ground surface. They are 3-5 cm long caterpillars dirty gray in color. The young caterpillars first feed on leaves and then burrow into the soil and feed on roots and tubers. The caterpillars when fully grown pupate near the soil surface and develop into moths (Figure 12.1O) (van der Zaag et al., 1996).

### *Management*

In plots where there is a risk of damage as a result of cutworms, the application of approved insecticides in the soil before planting is required.

## *Potato Tuber Moth*

The potato tuber moth (PTM), *Phthorimaea operculella* (Zeller), is one of the most damaging pests of potatoes in the field and in storage and is generally of greater importance in warmer climates (Haines, 1977; Hanafi, 2005). Although quarantine is commonly practiced against infested tubers, tuber moths continue to spread to new areas. The PTM is assumed to have originated in the humid subtropics of South America as its principal host plants, potato and tobacco *(Nicotiana tabacum)*, are indigenous to that area, and the parasite complex of this pest is rich.

In the field, the PTM can damage up to 25 percent of tubers, but in storage damage can be 100 percent (Raman, 1987; Hanafi, 2005). In fields, larvae bore into leaves, stems, and tubers. Although the damage caused to plant foliage usually does not reduce yields, infestation in tubers can lower their market value. In the field, most damage occurs just prior to harvest, especially when vines are dying either naturally or upon cutting. PTM larvae leave the dying vines, and if the soil is dry and cracked they move down the cracks and infest tubers.

In storage, larvae tunnel through tubers (Figure 12.1P), filling the tunnels with the frass and webbing and allowing easy entry and development of fungi such as *Fusarium* and bacteria *Erwinia.* Besides potato, tuber moths also feed on some other Solanaceous plants, including tomato, eggplant, and pepper (Von Arx et al., 1987).

### *Biology and Behavior*

The PTM belongs to the order Lepidoptera and the family Gelechidae. The life cycle of the PTM lasts 20-30 days at 28°C. Between 2 and 12 generations occur per year depending on the climate (Raman, 1988). The eggs are white-cream (0.5 mm) and are deposited individually on the lower leaf surface, stems, tubers, storage bags or boxes, and even dust in potato stores (Hanafi, 1999, 2005). The larva, which is the only stage that causes damage, emerges in 5 days. At the end of its development, the larva measures 10 mm in length. The color of the larva varies from white-green to gray-yellow with a brown head. The larval stage lasts approximately 14 days (Von Arx et al., 1987).

The pupa (Figure 12.1Q) is brown, measures 6 mm, and is enclosed in a white-gray cocoon found near the soil surface, on tubers, on bags or boxes, and on soil dust in stores. The pupal stage lasts approximately 8 days (Hanafi, 1999). The adult PTM measures 15 mm; its anterior wings are gray-brown with small dark spots, whereas posterior wings are white-cream (Figure 12.1R). The longevity of the adult is 10-15 days. Adults are active at night and are difficult to find during the day but are easily recognizable by their short flight when disturbed (Hanafi, 1999).

Tuber moths lay eggs directly on leaves and on tubers left uncovered in the field or in storage. On leaves, eggs are deposited on dorsal surfaces and emerging larvae mine leaflets. As leaflets are consumed, larvae may leave their tunnels and move on to the adjacent leaflets. Sometimes they even mine stems, causing the death of plant terminals. From foliage, larvae move to feed on tubers by crawling down through cracks in the soil. They do not bore tubers via the stems. When ready to pupate, larvae spin cocoons on the soil surface or on debris under plants. They rarely pupate in tubers. Larvae that have fed on potatoes in storage can crawl considerable distances to find protected sites for pupation, such as crevices in store walls, floors, and the crates (Von Arx et al., 1987).

Temperatures between 20 and 25°C are the most ideal for PTM development. During the hot season, the PTM can complete a generation in approximately 3 weeks. Larval growth and development slow down at cooler temperatures but continue as long as temperatures are higher than 11°C. During winter, the PTM may take as long as 7 months to complete a generation.

*Management*

Some of the general cultural practices useful in checking the population buildup of the PTM, and hence reducing crop damage, are as follows (Hanafi, 2005):

- PTM-infested seed tubers are not used for planting.
- Potato fields are rotated and the current potato crop is located as far away from the previous potato fields as possible. Any volunteer potato plants growing in uncultivated areas and in other fields are destroyed.
- Tubers are well covered. The moths cannot reach tubers covered with more than a few centimeters of soil, unless it is dry and deeply cracked.
- Furrow irrigation is avoided in areas with heavy soil as this leads to cracks. Sprinkler and drip irrigation help to keep the soil surface sealed, especially in fine-textured soils, which tend to develop deep cracks upon drying.
- Extremely important is to harvest tubers as soon as they have matured. PTM infestations increase sharply near crop maturity.
- Newly dug potatoes are not covered with green vines, because as soon as vines wilt, tuber moth larvae leave them and infest the tubers.
- After harvest, discarded tubers are not left in the field, as they provide food for developing larvae throughout the winter months.
- Harvested produce should be stored below 10°C to prevent eggs hatching and larvae feeding on tubers.

All these practices are designed to reduce PTM infestation in the field prior to tuber storage (Hanafi, 1999, 2005). Sex pheromones provide a useful tool for monitoring populations of PTMs under both field and storage conditions. Various trap designs have been used, but a combination of pheromone and water traps works best. Mass trapping using sex pheromones has demonstrated the feasibility of direct control. In field studies, one pheromone trap per 25 $m^2$ reduced tuber damage by 45 percent compared with plots without pheromone traps. In storage, both water pan and tunnel traps were effective in reducing tuber moth damage (Hanafi, 2005).

Monitoring traps can be made from pans approximately 20 cm in diameter filled with water to a depth of at least 7 cm. The pheromone lure is placed beneath a piece of metal sheet bent to form a tent-shaped cover over the pan. The lure should be suspended 2.5 cm above the water surface so that male moths attracted to it fall into the water. A few drops of soap are added to the water to break the surface tension and prevent the moths from escaping. The traps are placed between plants, on the tops of beds approximately 15 m from the edge of the field. The moths are counted twice a week in each trap and water is replenished.

The threshold level of PTM population for taking up chemical control measures may vary depending on the region, cropping season, type of soil, and type of irrigation system. As a general rule for fresh-market potatoes, insecticide spray should be started when the number of trapped moths reaches 15-20 at any one time the traps are checked. It is extremely important to control this pest prior to crop harvest. Hence, insecticides (deltamethrin, endosulfan, methamidophos) should be sprayed toward the newly hatched larvae (Hanafi, 2005).

For controlling the PTM damage in rustic stores, chemicals recommended are pyrethroids or carbamates, if tubers are to be used later for seed. If tubers are to be used for consumption, biological insecticides such as *Bacillus thuringiensis* and Granulosis virus, which are entomopathogens of the PTM, are recommended. These have the important advantage of posing no health risk to applicators, consumers, or for that matter other species of animals.

## *MAJOR NEMATODE PESTS*

Plant-parasitic nematodes are a severe constraint for potato production in sandy soil, especially in regions where more than one crop is grown per year. In temperate zones, at intermediate to high altitude, with high rainfall, snow, and cold winters and dry cool summers, the cyst nematode *Globodera rostochiensis* is quite prevalent as an infestation of potato. In moderately warm zones at intermediate altitude, with cool rainy winters and hot dry summers, especially in coastal belts, the root-knot nematodes such as *Meloidogyne hapla* occur commonly in potato fields. In warmer and subtropical zones at low altitude, with cool winters, little rainfall, and very hot summers, rot nematodes *M. javanica, M. incognita,* and *Ditylenchus destructor* are quite common.

Although many nematodes are capable of inflicting severe damage directly to growing plants, the degree of injury to potato depends on the level of soil infestation by the dominant species, new races or pathotypes, and the prevailing stress conditions such as drought, waterlogging, salinity, and inadequate plant nutrition.

Many chemicals available in the market for the control of nematodes are too expensive for potato growers. This reality demands that nematode management should become an integrated program of practices that includes alternative measures to the use of chemicals and provides sustainable nematode control systems. No standard formula exists, and such an integrated program must be designed to fit local requirements and options available affordably within a specific region. Modified cultural practices, nonhost rotation crops, host resistance or tolerance, biorational nematicides, nematode antagonists (*Tagetes* spp.), cover crops, and better nematode identification and enumeration are the areas of improvement for a nematode management program (van der Zaag et al., 1996).

### *Northern Root-Knot Nematode*

The northern root-knot nematode *(Meloidoygyne hapla)* affects almost all dicotyledons, including potato (Figure 12.2A) and almost all broad-leaf weeds. The nematode occurs mainly in light sandy soils. Infested crop shows localized areas of stunted growth and greater sensitivity to drought, the degree of which depends on the nematode density in the soil, cultivar, and water shortage. On affected roots small galls are formed, which typically have several lateral rootlets attached. The roots do not function normally, and if nematode

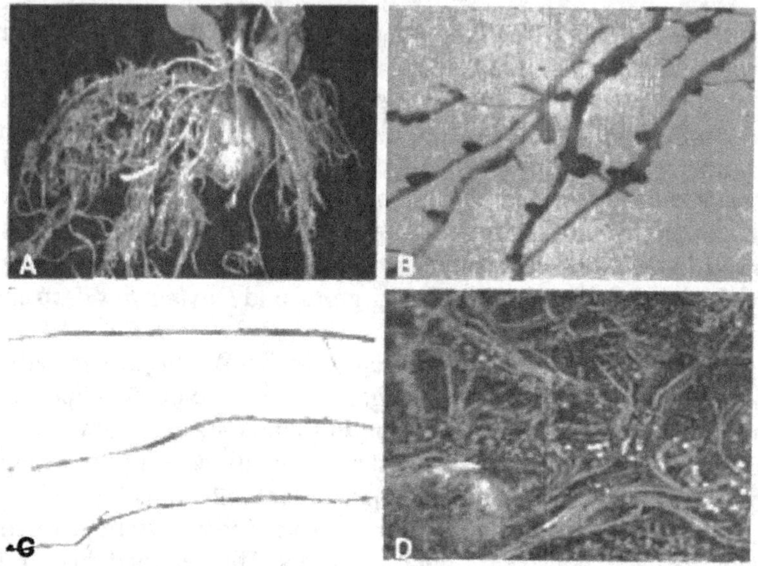


FIGURE 12.2. Some nematode pests of potato. See Plate 12.2 in the color gallery.

densities are high, the root system remains underdeveloped. Tubers are generally not affected, unless produced in areas with soil temperature above 20°C.

## *Biology and Behavior*

*Meloidoygyne hapla* belongs to the order Tylenchida and family Tylenchidae. Nematodes survive crop-free periods as eggs in a gelatinous matrix or as free larvae. They become active at soil temperatures above 8°C. Only the slender, mobile, second-stage larva that hatches from the egg is infectious. They usually invade just behind the root cap and move between the cortical cells to the zone of cell differentiation, where they arrive at the initial vascular bundle. Here a few cortical cells are stimulated to form giant cells on which the nematodes feed. As a result of additional cell division in the cortex, the root tissue swells, leading to the formation of the typical root knots.

As nematodes develop into adults the females swell and become immobile, whereas males are slender. Each female lays 200-500 eggs in a gelatinous matrix. After some time the egg packets turn brown and can be seen as brown "pinheads" on the root knots. When the female dies, the egg packet often becomes detached from the root. At soil temperature of 8°C, the life cycle is completed in approximately 8 weeks (van der Zaag et al., 1996).

### *Management*

Monocotyledons such as cereals and grasses are nonhosts. By cultivating such crops or leaving the land fallow (with proper weed control), nematode densities can be reduced by 80-90 percent per year. The greatest reduction is achieved in moist and bare land. Late planting after tillage helps to eliminate the initial population by starvation. *Meloidoygyne hapla,* however, is most effectively controlled by treating the soil with nematicides (metam sodium, Mocap).

### *Rot Nematodes*

Rot nematode *(Ditylenchus destructor)* damages potatoes mainly in sandy soils. Primarily it attacks stolons and tubers in the soil, penetrating via lenticels and eyes and leaving roots unaffected. Initial symptoms are gray to chalky white spots that can be seen when the potatoes are peeled. The skin takes on a paperlike appearance, shrivels, and consequently cracks. Later, gray cavities and channels develop that extend up to the vascular ring. This is followed by the formation of dark brown areas associated with fungal and bacterial rot organisms. The initial symptoms closely resemble those of the injury caused by the stem and bulb nematode *(Ditylenchus dipsaci).* However, contrary to stem and bulb nematodes, rot nematodes rarely affect stems. Damage is severe if contaminated tubers are stored above 5°C as the nematodes continue to reproduce and, if the tubers are wet, may spread from one tuber to another. As the nematodes produce the enzyme amylase, which converts starch into sugar, the initial dry rot caused by the nematode may develop into soft rot.

### Biology and Behavior

*Ditylenchus destructor* belongs to the order Tylenchida and the family Tylenchidae. Adults are slender and 0.8-1.4 mm long. A single female is able to lay 200-250 eggs. At 6-10°C the life cycle is completed in 68 days and at 15-20°C in 20-26 days. This nematode cannot survive under adverse conditions. However, it can survive on decaying plant remains, weeds, and fungi, and damp soils are more favorable for survival than dry ones (van der Zaag et al., 1996).

### Management

Stocks of potato infested with potato rot nematodes must be processed as soon as possible. If the potatoes are to be stored, they must be dried and stored below 5°C. At low levels of infestation no symptoms are seen. Hence, it is not possible to remove all infected potatoes from a contaminated stock. Potatoes from such stocks should not be used as seed.

*Ditylenchus destructor* has a very wide host range, including green manure crops such as red clover, many weeds, and many common soil-inhabiting fungi. Because of its wide host range, this polyphagous nematode is difficult to control by crop rotation. One of the few effective measures is to cultivate cereals and combine them with complete weed control.

## Columbia Root-Knot Nematode

The Columbia root-knot nematode *(Meloidogyne chitwoodi)* was identified as a potato pest organism recently and described as a new nematode species in 1980. It has been found in several regions of Europe, North and South America, and South Africa but is probably more widely spread than is recognized at present.

Stunted growth and late flowering have been seen in potato crops grown in fields highly infested with this nematode. Severely affected roots show small, slightly elongated, club-shaped knots or galls (Figure 12.2B). Lateral root growth from knots, a typical feature for the northern root-knot nematode, does not occur. Under warm growing conditions, when subsequent generations develop tubers are also affected severely. If three generations develop, more than 80 percent of the tubers may be infested. Affected tubers are marked by galls that contain both the nematodes and their eggs. These galls are easily confused with those of tropical root-knot nematode species *(Meloidogyne arenaria, M. incognita,* and *M. javanica)* and of powdery scab (van der Zaag et al., 1996).

Tuber infestation is favored by moisture (irrigation) at the time the eggs of the preceding generation hatch on the roots. The nematodes tend to penetrate quite deeply into the tuber, and affected tubers are not acceptable for processing. In general, the damage is mainly qualitative. Only if initial densities are very high may serious yield losses occur, and the specific gravity of the tubers of affected crops is reduced. In the United States, stocks containing more than 10 percent infested tubers cannot be sold as ware potatoes.

*Biology and Behavior*

During crop-free periods, the nematodes survive in soil as eggs and free larvae. They also survive in the roots of other host crops in all stages of development. The nematode becomes active at 5°C. The slender nematode emerges as a second-stage larva from the eggshell; this is the only stage in which the nematode is able to migrate. The larvae often invade roots where they branch. The larva stimulates the plant to form four large nutritive cells (giant cells), on which it feeds. The distinctive "root knots" are actually clusters of larvae and their associated giant cells. The nematodes develop into adults through a series of three molts, during which they gradually swell. The spherical female ruptures the root tissue and deposits her eggs in a gelatinous substance. After the final molt, the slender males arise from the bottle-shaped fourth larval stage. At the time females rupture the root tissue with their posterior, males leave the roots to search for female mates. However, *M. chitwoodi* also multiplies asexually. At soil temperature of 20°C, a generation is completed in 56-57 days and at 10°C in 187-189 days. Tuber invasion takes place through lenticels and the nonsuberized parts of the epidermis. Although this species of nematode is able to penetrate deep into tuber tissue, most nematodes are located between the epidermis and vascular ring. Surrounding cells in the tuber's cortical and parenchymatous tissues are disintegrated, making room for the developing females and their eggs. These surrounding cells then suberize, turning brown and constituting a protective layer around the eggs. This process results in characteristic brown pustules or galls on the tuber surface and the necrotic spots in cortex and flesh. The larvae of the new generation can leave the tuber only via wounds or when affected tissues disintegrate. As the nematodes inside the tuber can survive as long as the tuber itself, they are readily spread by infested seed tuber stocks.

#### *Management*

In addition to cereals and grasses, a wide range of dicotyledons can serve as hosts for the Columbia root-knot nematode. Crop husbandry practices such as rotation are still being evaluated. *Meloidogyne chitwoodi* is very sensitive to such soil fumigants as methyl bromide and metam sodium.

### ***Root-Lesion Nematode***

The root-lesion nematode *Pratylenchus penetrans* has a very wide host range (more than 160 hosts have been described). It occurs mainly in sandy soils but is also found in loamy sand, sandy loam, and peat soil. In addition to *P. penetrans*, several other root-lesion nematode species exist but have so far not been shown clearly to cause economic damage. The most common species are *P. crenatus, P. neglectus, P. scribneri*, and *P. brachyurus*. There is evidence that *P. crenatus* may cause damage if the soil pH is particularly low. Other lesion nematodes, such as *P. scribneri* and *P. brachyurus*, may affect tubers and cause severe tuber damage. Though potato is only moderately sensitive to *P. penetrans*, yield losses of more than 50 percent have been reported. Some varieties have proved to be more susceptible to damage than others. *Pratylenchus penetrans* enhances infection of *Verticilium dahilae*, and *P. neglectus* is believed to enhance infection of both *V. dahliae* and *Rhizoctonia solani*.

After normal emergence, affected plants develop slowly, remain small, have a dull and dark foliage, flower late, and often senesce prematurely. Because of their poorly developed root system, affected plants often show deficiency symptoms and are highly sensitive to drought. Affected roots show elongated, yellowish brown to dark brown and black necrotic lesions (Figure 12.2C). The plant endeavors to replace the damaged roots, which results in the root system taking on a bushy appearance. Roots rot at the site of the lesions and then become breakable, especially late in the season. *Pratylenchus penetrans* is primarily a root pathogen, usually affecting only cortical tissues; damage is caused by direct feeding.

### *Biology and Behavior*

Root-lesion nematodes belong to order Tylenchida and family Pratylenchidae. They are 0.3-0.8 mm long and have a very simple life cycle. They live mainly inside the root cortex, where fertilized females lay one to two eggs per day. Occasionally eggs are laid in the soil. One female may lay up to 40 eggs during its entire life cycle. The first molt takes place inside the eggshell, and the second-stage larva emerges from the egg. After three more molts the larva develops into an adult nematode. At 18°C, the life cycle is completed in 7 weeks, which means that in each growing season the nematode may have three to six generations. Larvae of all stages are slender and mobile and are able to penetrate roots, usually in the root hair zone. Stolons may also be invaded and develop lesions. The nematodes survive crop-free periods in soil or in affected plant tissues. In soil the fourth-stage larva is the most persistent one. Provided soil moisture content is sufficiently high, populations decrease fastest in soil at high temperatures. In dry soils the nematode may survive in a quiescent state.

### *Management*

Owing to the wide host range of *P. penetrans*, crop rotation is not an effective control. However, beet is an important nonhost crop and populations usually decrease faster when beet are grown than under fallow. The root-lesion nematode is most effectively controlled by soil treatment with volatile nematicides (methyl bromide and metam sodium) that spread through the soil in the gas phase. Soil-sampling analysis provides an estimate of initial population densities, enabling a risk analysis with respect to possible crop damage and yield loss.

### Stem and Bulb Nematode

The stem and bulb nematode *(Ditylenchus dipsaci)* occurs mainly in sandy clay and clay soils. Several races are known. Races that infest potatoes can also multiply on peas, broad beans, rye, and onions.

In potato, these nematodes can affect both foliage and tubers. When foliage is affected, the stem shows swellings and bends; galls may also develop in leaf axils. The leaves become frizzled and the main vein of the leaflets is often distorted. Symptoms of tuber infection are grayish brown depressions overlain by a papery skin. Such affected areas deepen; tissues turn rusty brown and granular and may develop dry rot.

#### *Biology and Behavior*

Stem and bulb nematodes belong to order Tylenchida and family Aguininidae. Both the larvae and the adults of these nematodes are slender and mobile. The first molt takes place inside the egg, after which the second-stage larva hatches. Fourth-stage larvae penetrate the plant from the soil. After one more molt they develop into adults. The fourth-stage larvae are resistant to adverse conditions and can survive long periods in the soil and in plant debris. They are resistant to drought if desiccation takes place slowly. In mature seed (e.g., onion seed) or desiccating plant tissue, nematodes enter into a quiescent state (anabiosis) and can survive for many years. All stages of the nematode can survive in infested tubers. The nematodes become active at 1-2°C and start to reproduce at 5°C. At 15°C, the life cycle is completed in 19-23 days. Because a single female lays 200-500 eggs during its entire life cycle and many generations are completed during a growing season, these nematodes can reproduce explosively under favorable conditions.

#### *Management*

Heavy soils favored by stem and bulb nematodes are difficult to disinfect. Moreover, a small residual population (e.g., only one nematode per kilogram of soil) can still cause damage. Affected stocks of tubers must be processed as soon as possible. If such stocks have to be stored, they must be dried and stored below 5°C as quickly as possible to slow down or prevent population increase. Seed potatoes must be free of these nematodes. The risk of infestation and damage to potatoes can be avoided by analyzing soil samples for the presence of the nematodes.

### *Potato Cyst Nematodes*

Two species of potato cyst nematodes are known: *Globodera rostochiensis* (the golden nematode) and *Globodera pallida* (the white nematode). The first sign of injury in the field consists of small areas of stunted plants. Infested plants are dark in color and generally flower later than healthy plants. The root system is bushy because infested roots tend to produce more branches than healthy ones. The damage caused depends on the potato variety and the population density of nematodes. High nematode populations can cause crop loss of more than 50 percent (van der Zaag et al., 1996).

#### *Biology and Behavior*

The cyst nematode belongs to order Tylenchida and family Heteroderidae. It is a highly specialized root parasite that develops only on potato. The cyst nematode penetrates the root at the growing point and moves through the root's cortical cells to the vascular bundles. The young spherical females erupt through the root's epidermis and can be seen as white cysts on the roots (Figure 12.2D). The young swollen female cysts usually become visible on the roots approximately 6-8 weeks after planting and are golden yellow (van der Zaag et al., 1996).

One female produces between 100 and 500 eggs and may be fertilized by more than one male. The fertilized eggs are actually not laid but remain inside the hardening cuticle of the dying female. Dead females containing the eggs are called cysts. The mature cysts detach from the root and remain in the soil after potatoes have been harvested. The eggs and larvae are well protected inside the cyst. Cysts may still contain viable eggs 8-10 years after the last potato crop. Cyst nematodes become active at temperatures above 10°C and reach optimum development at 25°C (van der Zaag et al., 1996).

*Management*

Legislative quarantine measures must be enforced to prevent the introduction of cyst nematodes through potato seed tubers to new areas. If the nematode becomes established in an area, sanitary measures should be taken to prevent its spread. Disinfection of machinery is of paramount importance when moving from infested fields into clean ones. Only long rotation (approximately 7 years) can provide adequate control of the cyst nematode. It is advisable wherever possible to use resistant or tolerant varieties. If soil becomes heavily infested it is necessary to disinfect it using a chemical fumigant such as methyl bromide.

## *INTEGRATED PEST MANAGEMENT*

Integrated pest management (IPM) is a sustainable and environmentally sensitive approach to pest management that relies on a combination of biological, cultural, mechanical, physical, and chemical practices. IPM programs use available comprehensive information on the life cycles of pests and their interaction with the environment. This information, in combination with available pest control methods, is used to manage pest damage by the most economical means and with the least possible hazard to users, consumers, and the environment. IPM has attracted most interest in the developed world.

Even though the idea for using multiple tactics and strategies for pest control was introduced in the late nineteenth century, the concept of IPM did not become popular until the 1970s and 1980s. As public knowledge on pest resistance and the dangers caused by conventional chemical pesticides to human health and the environment grew, certain pesticides (e.g., DDT [dichloro-diphenyl-trichloroethane] and dieldrin) were removed from the market. The deployment of sustainable and effective pest management strategies that did not rely on chemical control became an absolute necessity. These strategies included the prevention or avoidance of pest and pathogens, decreased pest populations, decreased host susceptibility, and pest eradication. These strategies were appealing not only for their environmental and ecological protection but also for their economic benefits. In practicing IPM, a four-tier approach is followed.

### *Set Action Thresholds*

Before any pest control action is taken, IPM first sets an action threshold, a point at which pest populations or environmental conditions indicate that pest control action must be taken. Sighting a single pest does not always mean control is needed. The level at which pests will become an economic threat is critical to guiding pest control decisions.

### *Monitor and Identify Pests*

Not all insects, weeds, and other living organisms require control. Many organisms are innocuous, and some are even beneficial. IPM programs work to monitor for pests and identify them accurately, so that appropriate control decisions can be made in conjunction with action thresholds. This monitoring and identification removes the possibility that pesticides will be used when they are not really needed or that the wrong kind of pesticide will be used.

### *Prevention*

As a first line of pest control, IPM programs work to manage the crop to prevent pests from becoming a threat. In a potato crop, this may mean using cultural methods, such as rotation between different crops, hilling potatoes, selecting pest-resistant varieties, and planting pest- and disease-free seed. These control methods can be very effective and cost efficient and present little risk to people or the environment. This applies particularly to the PTM and nematodes, as discussed above.

### *Control*

Once monitoring, identification, and action thresholds indicate that pest control is required and preventive methods are no longer effective or available, IPM programs then evaluate the proper control method for both effectiveness and risk. Effective, less risky pest controls are chosen first, including highly targeted chemicals (e.g., pheromones to disrupt pest mating) and mechanical control (e.g., trapping and weeding). If further monitoring, identification, and action thresholds indicate that less risky controls are not working, then additional pest control methods need to be employed.

### *Use of Biopesticides*

Biopesticides are pesticides derived from natural materials such as animals, plants, bacteria, and certain minerals. For example, canola oil and baking soda have pesticidal applications and are considered biopesticides. At the end of 2004, there were approximately 324 registered biopesticide active ingredients and close to 1,000 products. Biopesticides fall into three major classes.

#### *Microbial Pesticides*

Microbial pesticides have a microorganism (e.g., a bacterium, fungus, virus, or protozoan) as the active ingredient. Microbial pesticides can control many kinds of pests, although each separate active ingredient is relatively specific for its target pest(s). The most widely used microbial pesticides are subspecies and strains of *Bacillus thuringiensis,* or Bt. Each strain of this bacterium produces a different mix of proteins and specifically kills one or a few related species of insect larvae. Some Bts control moth larvae found on plants; others are specific for the larvae of flies (mosquitoes) or beetles (Colorado beetle). The target insect species are determined by whether the particular Bt produces a protein that can bind to a larval gut receptor, thereby causing the insect larvae to starve.

#### *Biochemical Pesticides*

Biochemical pesticides are generally naturally occurring substances with a nontoxic mode of action against the target pest. Other characteristics may include low application rates or volumes and a narrow range of target species. The nontoxic mode of action against the target pest does not assume a lack of toxicity to humans or nontarget organisms. Furthermore, the "natural origin" of a pesticidal active ingredient does not imply a nontoxic mode of action. For example, pyrethrins, active ingredients produced by plants in the genus *Chrysanthemum,* have a toxic mode of action as neurotoxins (fast-acting paralyzing agents). Similarly, antibiotics from microorganisms, if used as pesticides, would not be considered biologicals because, by definition, they act via a toxic mode of action. Nontoxic modes of action would be (1) growth and development changes; (2) lures, attractants, and repellents; (3) irritants (e.g., capsaicin from red pepper, although irritation can be considered a toxic effect in regard to human health effects); and (4) suffocation (e.g., vegetable oil).

### *Plant-Incorporated Protectants*

Plant-incorporated protectants are produced from genetic material that has been added to a plant. For example, scientists can take the gene for the Bt pesticidal protein and introduce it into the plant's own genetic material. The plant then manufactures the substance that kills the pest. Several potato cultivars have been transformed genetically for resistance to Colorado beetle and PTM. However, transgenic potatoes with resistance to PTM are not yet available commercially (Gressel et al., 2004).

Biopesticides in IPM could be conceptualized as an extension of the natural pest controls. These products provide some unique benefits over conventional chemicals because of their selectivity to a target pest, lower toxicity to beneficial insects and workers, and shorter reentry intervals. Biopesticides are used for prophylactic tactics, but recently increasing numbers of new products have been developed to provide curative properties rather than preventive ones.

We should not see IPM as a control strategy that will completely replace chemical control. It is a powerful option and can be applied on a much larger scale than at present. For the chemical industry, IPM may result in extended use of products because of slower development of resistance and a more positive public perception of the role of the pesticide industry. In order to serve potato, as well as the environment and human health, we should use the best prevention and reduction strategies to develop effective IPM methods. Designing such environmentally safer IPM programs is a challenge for our profession, though it has been met to a certain extent for the PTM (Hanafi, 2005).

## REFERENCES

Bacon, O.G., Burton, V.E., and Wyman, J.A. (1978). Management of insect pests of potatoes. *California Agriculture 32:*26-27.

Boiteau, G., Singh, R.P., and Parry, R.H. (1987). Potato pest management in Canada. *Proceedings of Symposium on Improving Potato Production*, Fredericton, New Brunswick.

Bouharroud, R., Hanafi, A., Murphy, B., and Serghini, M.A. (2003). New developments in Bemisia tabaci resistance to insecticides in greenhouse tomato in Morocco. In Castané, C. and Hanafi, A. (Eds.). *Integrated Control in Protected Crops: Mediterranean Climate. Bulletin International Organisation for Biological and Integrated Control of Noxious Animal and Plants* (Vol. 26, pp. 21-25). Rue Sully, France: INRA.

Cock, M.J.W. (1986). *Bemisia Tabaci: A literature Survey on the Cotton Whitefly with an Annotated Bibliography.* Berks, Kenya: CAB International Institute of Biological Control and Food & Agricultural Organization of the United Nations.

Delong, D.M. (1931). Distribution of the potato leafhopper (*Emposca fabae* Harris) and its close relatives. *Journal of Economic Entomology 24:*475-479.

———. (1971). The bionomics of leafhoppers. *Annual Review of Entomology 16:*179-210.

Ferro, D.N. and Boiteau, G. (1993). Management of insect pests. In Randall C. Rowe (Ed.), *Potato Health Management* (pp. 103-115). St. Paul Mn., USA: APS Press.

Gibson, R.W. (1978). Pests aspects of potato production. In Harris, P.M. (Ed.), *The Potato Crop* (pp. 470-502). New York, USA: Chapman and Hall.

Gressel, J., Hanafi, A., Head, G., Marasas, W., Obilana, B., Ochanda, J., Souissi, T., and Tzotzos, G. (2004). Major heretofore intractable biotic constraints to African food security that may be amenable to novel biotechnological solutions. *Crop Protection 23:*661-689.

Haines, C.P. (1977). The potato tuber moth, *Phthorimaea operculella* (Zeller): A bibiliography of recent literature and review of its biology and control on potatoes in the field and in stores. *Report Tropical Production Institute,* G112.

Hanafi, A. (1992). *Spread and Control of Potato Leafroll Virus in Potato: Implications for Management of Seed Potatoes in the Souss Valley, Morocco.* Ph.D. Dissertation, University of Minnesota, USA.

———. (1998). Integrated pest management of potato tuber moth (PTM) *Phthorimaea operculella*, in potato field and in storage in the Souss Valley of Morocco. In Hanafi, A., Achouri, M., and Baudoin, W. (Eds.), *Proceedings of the international symposium on integrated production and protection in horticultural crops* (pp. 452-459). Agadir, Morocco: IAV Academic Publisher .

———. (1999). Integrated pest management of potato tuber moth in field and storage. *Potato Research 42:*373-380.

———. (2000). An overview of the management of insect transmitted viruses in potato. In Adipala, E., Nampala, P., and Osiru, M. (Eds.), *Proceedings* of *African Potato Association Conference* (Vol. 5, pp. 281-283). Kampala, Uganda.

———. (2003). Integrated production and protection today and in the future. In Lamalfa, G., Lipari, V., Noto, G., and Leonardi, C. (Eds.), *Greenhouse Crops in the Mediterranean Region. ACTA Horticulturae 614:*755-765.

———. The sanitary quality of substrates: A key component in IPP and its potential role in the spread of invasive species of pests and diseases of greenhouse crops. In Abouhadid, A.F. (Ed.), The horizon of using organic matter and substrates in horticulture. *ACTA Horticulturae 608:*21-24.

———. (2005). Invasive Pests and Diseases: A Challenge to IPM in Greenhouse Crops. Guest Editorial. *Phytoparasitica 33:*423-426.

———. (2005). Integrated management of potato tuber moth in field and storage. In Haverkort, A.J. and Struick P.C. (Eds.), *Potato in Progress: Science Meets Practice* (pp. 203-210). Wageningen Academic Publisher .

Hanafi, A., Radcliffe, E.B., and Ragsdale, D.W. (1989). Spread and control of Potato Leafroll Virus in Minnesota. *Journal of Economic Entomology 82:*1201-1206.

———. (1995). Spread and control of Potato Leafroll Virus in the Souss Valley of Morocco. *Crop Protection 14:*145-153.

Hanafi, A. and Schnitzler, W.H. 2004. Integrated production and protection in greenhouse tomato in Morocco. *ACTA Horticulturae 659:*323-330.

Hanafi, A. and Srairi, I. (1998). Lutte intégrée contre la mineuse de la tomate *Liriomyza bryoniae* en culture de tomate sous serre dans le Souss. In Hanafi, A., Achouri, M., and Baudoin, W. (Eds.), *Proceedings of the International Symposium on Integrated Production and Protection in Horticultural Crops* (pp. 50-59). Agadir, Morocco: IAV Academic Publisher .

Helle, W. and Sabelis, M.W. (Eds.) (1985). *World Crop Pests*, Amsterdam, The Netherlands: Elsevier.

Hsiao, T.H. (1978). Host plant adaptations among geographic populations of the Colorado potato beetle. *Applied Experimental Entomology 24:*237-247.

Jacques, R.L., Jr. (1988). The potato beetles. The genus *Leptinotarsa* in North America (Coleoptera: Chrysomelidae). *Flora & Fauna Handbook no.3:*144.

Jeppson. L.R., Keifer, H.H., and Baker, E.W. (1975). *Mites Injurious to Economic Plants.* Berkeley, USA: University of California Press.

Khurana, S.M. Paul, Singh, R.A., and Kaley, D.M. (1988). Mycoplasma associated potato diseases and their control in India. In Maramorosch, K. and Raychaudhari, S.P. (Eds.), *Mycoplasma Diseases of Crops: Basic and Applied Aspects* (pp. 285-316). New York, USA: Springer-Verlag.

Khurana, S.M. Paul, Verma, K.D., and Chandla, V.K. (2000). Potato virus vectors and their management. In Khurana, S.M. Paul, Shekhawat, G.S., Singh, B.P., and Pandey, S.K. (Eds.), *Potato: Global Research and Development* (Vol.1, pp. 352-362). Shimla, India: Indian Potato Association.

Ladd, T.L., Jr. and Rawlins, W.A. (1965). The effects of feeding of the potato leafhopper on photosynthesis and respiration in the potato plant. *Journal of Economic Entomology 58:*623-628.

Mackauer, M. and Way, M.J. (1976). *Myzus persicae* Sulz an aphid of world importance. In Delucchi, V.L. (Ed.), *International Biological Programme 9. Studies in Biological Control* (pp. 51-119). London, UK: Cambridge University Press.

Malais, M.H. and Ravensberg, W.J. (Eds.) (2003). *Knowing and Recognizing: The Biology of Glasshouse Pests and their Natural Enemies.* Doetinchem, The Netherlands: Koppert B.V. Reed Business Information.

Nagaich, B.B. and Giri, B.K. (1973). Purple top roll diseases of potato. *American Potato Journal 50:*79-85.

Parella, M.P. (1984). Effect of temperature on oviposition, feeding and longetivity of *Liriomyza trifolii* (Diptera: Agromyzidae). *The Canadian Entomologist 116:*85-92.

Parella, M.P., Robb, K.L., and Bethke, J. (1983). Influence of selected host plants on the biology of *Liriomyza trifolii* (Diptera: Agromyzidae). *Annals of the Entomological Society of America 76:*112-115.

Popov, N.A., Zabudskaya, I.A., Shiiko, E.A.S., Mencher, E.M., and Shevchenco, V.B. (1992). Development of glasshouse pests of cucumber under different conditions, in relation to possibilities for their biological control. *EPPO Bulletin 22:*529-536.

Prasad, S.K. (1961). Quantitative estimation of damage to potatoes caused by potato leafhopper, *Emposca fabae* (Harris). *Indian Potato Journal 3:*105-107.

Radcliffe, E.B. (1982). Insects Pests of Potato. *Annual Review of Entomology 27:*173-204.

Radcliffe, E.B., Flanders, K.L., Ragsdale, D.W., and Noetzel, D.M. (1991). Potato insects- pest management systems for potato insects. In Pimentel, D. (Ed.), *Handbook of Pest Management,* Second edition (pp. 587-621). Boca Raton, FL: CRC Press.

Radcliffe, E.B., Ragsdale, D.W., and Flanders, K.L. (1993). Management of aphids and leafhoppers. In Randall C. Rowe (Ed.), *Potato Health Management* (pp. 117-126). St Paul, Mn, USA: APS Press.

Raman, K.V. (1987). Survey of diseases and pests in Africa: Pests. *Acta Horticulturae 213:*143-150.

———. (1988). Integrated pest management for potatoes in developing countries. *CIP Circular 16:*1-8.

Rowe, R.C. (1993). *Potato Health Management.* Plant health management series. St. Paul, Mn., USA: APS Press.

Saxena, K.N., Gaudhi, J.R., and Saxena, R.C. (1974). Patterns of relationship between certain leafhoppers and plants. I Responses to plants. *Applied Experimental Entomology 17:*303-318.

Sharaf, N.S. (1984). Studies on natural enemies of tetranychid mites infesting eggplant in the Jordan Valley. *Journal of Applied Entomology 98:*527-533.

van der Linden, A. (1993). Overwintering of *Liriomiza bryoniae* and *Liriomyza huidobrensis* (Diptera: Agromyzidae) in the Netherlands. *Proceedings of Experimental and Applied Entomology 4:*145-150.

van der Linden, A. and van Achterberg, C. (1989). Recognition of eggs and larvae of the parasitoids of *Liriomyza* spp. (Diptera: Agromyzidae; Hymenoptera: Braconidae and Eupholidae). *Entomologische Berichten 49:*138-140.

van der Zaag, D.E., Asscheman, E., Bokx, J.A., Brinkman, H., Bus, C.B., Hotsma, P.H., Meijers, C.P., Mulde, A., Scholte, K., Turkensteen, L.J., and Wustman, R. (Eds.) (1996). *Potato Diseases, Pests and Defects.* den Haag, The Netherlands: Netherlands Potato Consultative Institute (NIVAA).

Von Arx, R., Guider, J., Cheikh, M., and Ben temine, A. (1987). Integrated control of potato tuber moth in Tunisia. *Applied Insect Science 8:*989-994.

Wolfgang R. and Rieckmann, W. (Eds.) (1991). *Maladies et Ravageurs de la Poome de Terre.* Nyon, Suisse : Th. Mann. Gelsenkirchen-Buer.

Zhang, Z.Q. (1995). *Mites of Glasshouses and Nurseries: Identification, Biology and Control.* London, UK: CAB International Institute of Entomology.

# Chapter 13

# Storage

Klaus Gottschalk
R. Ezekiel

In most countries, only one potato crop is raised each year, and as a result fresh potatoes are available for only 2-3 months. Therefore, potatoes need to be stored to meet the requirement during the remainder of the year. Potatoes are used for three purposes: seed, ware, and processing. Storage requirements vary depending upon the purpose for which potatoes are to be used. Seed potatoes are required only at the time of planting; therefore, they need to be stored for longer—approximately 6 months in countries where the potato crop is grown during summer and approximately 8 months in countries where the potato crop is grown during fall/winter. Generally, seed potatoes are stored at low temperatures (2-4°C). Ware and processing potatoes are in demand throughout the year; hence both short- and long-term storage are needed. Ware and processing potatoes are stored at higher temperatures (8-12°C) to reduce sugar accumulation.

The purpose of storage is to preserve the quality of potatoes for as long as possible with minimum storage losses. An important factor reducing quality during storage is physiological weight loss as a result of respiration and transpiration. Respiratory loss is minimized through temperature control and transpiratory loss is reduced by maintaining high relative humidity. During storage many physiological and biochemical changes take place, influenced by a number of factors. Knowledge of these changes and factors is necessary for successful storage of potatoes.

## *PHYSIOLOGICAL AGE AND DORMANCY*

### *Physiological Age*

Physiological age is defined as the physiological state of the tuber, which influences its productive capacity. It is determined by chronological age and environmental conditions during plant growth and storage. Chronological age is measured from the time of tuber initiation, expressed in days, weeks, or months, without reference to environmental conditions. Of the environmental conditions, especially during storage, temperature has a strong influence on physiological age.

The physiological age of a potato tuber can be divided into two main periods: dormancy and incubation (Reust, 1982, 1983). Dormancy begins with the initiation of a tuber and ends when the tuber begins to sprout. The incubation period starts with sprouting of the tuber and ends when a new plant is produced in the field or premature tuber formation occurs during storage. Changes in the physiological activity of a tuber during aging, which are associated with biochemical changes, are shown in Figure 13.1 (Müller, 1975, 1978, 1983).

During storage, the tuber passes through different phases of physiological activity. At harvest, when the tuber is fully mature, it is absolutely dormant, characterized by low level of sugars (Müller, 1978). In this phase, the tuber is physiologically unstable and is susceptible to stresses of different kinds (e.g., changes in temperature and relative humidity). The next phase is dormancy. During this phase, physiological activities are minimal and the tuber is relatively less susceptible. At the end of the dormancy period, physiological activities increase again as the sprout growth begins, and the tuber becomes more susceptible to stresses. This phase is also characterized by substantial

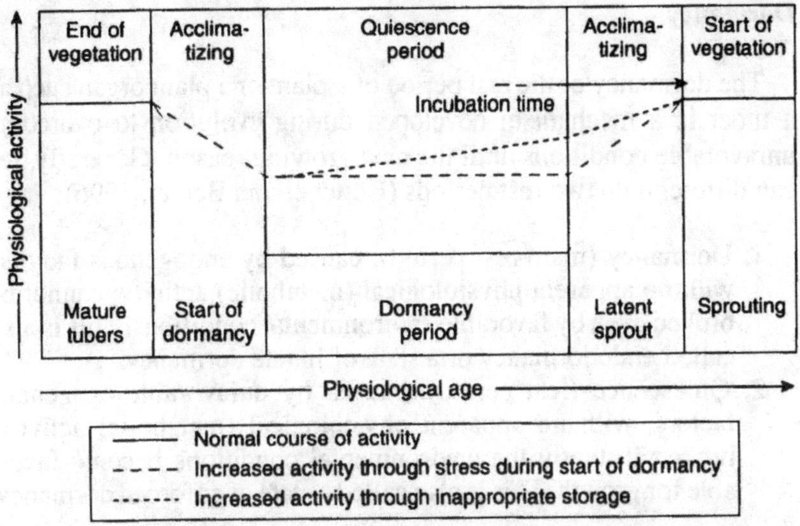


FIGURE 13.1. Schematic representation of changes in physiological activity of tubers during vegetative growth and storage. *Source:* After Müller (1975, 1978, 1983).

biochemical changes such as an increase in sugars, which are required for sprout growth. The lengths of these consecutive periods (shown in Figure 13.1) vary depending on the environmental conditions and may be a few weeks shorter under warmer storage conditions.

The physiological age of tubers can be determined in several ways. Sprouting capacity (expressed as grams of sprouts per 100 g tuber weight) (Krijthe, 1962) and accumulated day-degrees (the sum of daily average temperatures above 4°C) from the end of the dormancy period (Allen and O'Brien, 1986) have been found to be useful measures of physiological age. With increases in the number of day-degrees, tuber yields have been reported to increase at early harvest and decrease at final harvest (Ezekiel and Singh, 2003). Another measure of physiological age is incubation period, expressed as accumulated day-degrees above 0°C (Reust, 1983). (For more information on physiological aging, index of aging, and yield prediction modeling, see Sands, 1989; Coleman, 2000; Ezekiel and Singh, 2003.)

### Dormancy

The dormancy or the rest period of a plant or a plant organ such as a tuber is a mechanism developed during evolution to overcome unfavorable conditions until the next growing season. Generally, we can differentiate two rest periods (Böttcher and Belker, 1996):

1. Dormancy (main rest period): caused by endogenous factors, with no apparent physiological (metabolic) activity; cannot be broken even by favorable environmental conditions. This is also called endodormancy or a state of innate dormancy.
2. Quiescence (rest period): caused by unfavorable exogenous factors, with no apparent physiological (metabolic) activity; broken instantly the environmental conditions become favorable for growth. This is also called a state of enforced dormancy.

During the dormancy period, even if there is no apparent physiological activity, the bud is alive and has the potential to grow. During this period, physiological and biochemical changes are actually at their lowest, but it would be wrong to say that dormancy is a state in which no changes take place. Burton (1989, page 470) states "a tuber with no metabolic change would be dead, not just dormant." A clear understanding of the terms "dormancy" and "quiescence" is helpful in storage management. For example, in long-term storage of ware potatoes, the dormancy period is prolonged through appropriate storage conditions, and once the natural dormancy period is over, the tubers are kept in a quiescent state either by maintaining environmental conditions unfavorable for growth or by using sprout suppressants. In the short-term storage of seed potatoes, after the natural dormancy period, sprout growth is promoted by setting up a storage environment favorable for growth.

The point at which the dormancy of a potato tuber begins and ends has been much debated. Different definitions have been put forward, and so the duration of dormancy has been found to vary. Wright and Peacock (1934, page 20) state "the rest period is that period immediately following harvest during which the tubers will not sprout even under favorable growing conditions." Burton (1989) considers tuber initiation to be the beginning of dormancy. For practical purposes, two definitions of dormancy can be used (van der Schild, 1990):

1. the time (in days or weeks) between haulms cutting or harvest (digging) and the initiation under favorable conditions of sprout growth
2. the time (in days or weeks) between tuber initiation and the initiation under favorable conditions of sprout growth

Although the second definition is more accurate, the first definition is practically more useful.

The duration of dormancy depends on variety, tuber maturity, weather conditions during crop growth, irrigation, and soil conditions. Of these factors, variety is the most important, and significant differences have been observed among potato genotypes in the length of dormancy (Joseph and Gopal, 1994). Generally, the dormancy period of cultivated varieties is 5-9 weeks. Among the environmental factors, temperature and relative humidity are most important. Cold and wet weather during crop growth prolong the dormancy period, whereas warm and dry conditions shorten it (Burton, 1966). Storage at low temperatures prolongs dormancy. Higher relative humidity favors sprout growth, and this effect is more pronounced at higher temperatures.

Initiation of sprout growth marks the end of dormancy. Gibberellic acid (GA) and abscisic acid (ABA)—the two endogenous growth regulators—have been reported to regulate dormancy and sprout growth. At the time of tuber initiation (beginning of dormancy), the balance between growth-promoting and growth-inhibiting substances is stable (Burton, 1989). The balance of GA and ABA changes during tuber development and dormancy. The ABA concentration is high and the GA concentration very low during dormancy (Sonnewald, 2001). There is a sharp decline in ABA content (10-100 times) at the end of dormancy and initiation of sprout growth, with a concomitant increase in GA content (Figure 13.2). Though ABA and GA are considered the main regulators of dormancy, the underlying process is quite complex and probably involves other growth regulators (Gopal et al., 2004).

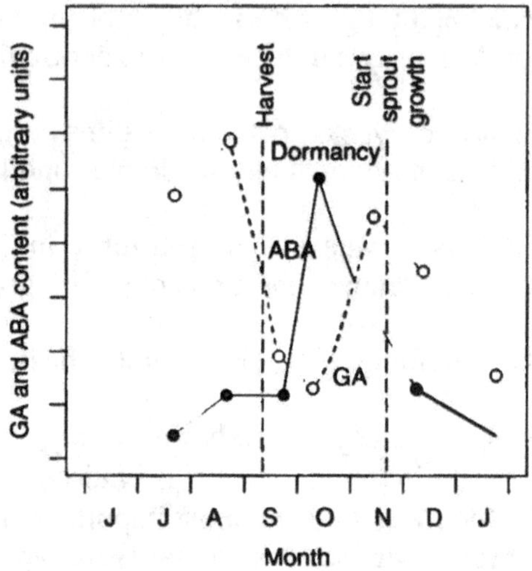


FIGURE 13.2. Schematic representation of changes in abscisic acid (ABA) and gibberellic acid (GA) concentrations in tubers during storage. *Source:* After Bialek and Bielinska-Czarnecka (1978); Burton et al. (1992); van der Plas (1987).

After the end of dormancy as a result of sprouting, water loss increases significantly and storage management becomes difficult. Sprouting in ware potatoes can cause severe quality loss if the weight loss is unacceptably high. In seed potatoes limited sprouting may not cause any loss in quality, but the economic loss could be significant. Excessive sprouting increases the physiological age of the seed tubers, resulting in yield reduction (Reust, 1983).

## *CONDITION OF POTATOES BEFORE STORAGE*

Not all potatoes are suitable for long-term storage. The primary requirement is that tubers should be mature, with firm skin. The periderm of immature tubers is not completely suberized to form a protective shield against mechanical impacts and diseases. During lifting, transporting, and loading operations, the tubers may be damaged by mechanical impacts, which can lead to development of black discoloration (black spots) and wounds. Further, potatoes that go into store should be clean, as adherent soil and plant residues can interfere with the ventilation of the tubers. Therefore, dry cleaning (brushing) is required immediately after the lifting of potatoes, but this has to be done carefully to avoid mechanical damage and consequent considerable lowering of quality. Washing potatoes before storage is not recommended for two reasons: (1) it increases the risk of spreading microbial diseases, and (2) additional care has to be taken to ensure complete drying of tubers before they are stored. If the surface moisture is not eliminated, microorganisms can multiply and cause rotting. Since the drying process influences the evaporation of water and heat transfer, it should be done in such a way that water loss from the tuber is minimized. Even a small quantity of diseased tubers in a lot raises a potential risk that the disease could spread and the whole lot be spoiled during storage. Tubers with latent infection appear healthy. The infection could have occurred because of crop cultivation in a soil infested by fungal or bacterial pathogens. Therefore, a test of tubers for freedom from pathogens is recommended before filling the store.

The quality requirements mentioned above need to be met if potatoes are to be sold to quality-conscious consumers and processors after a long storage period. The European Association for Potato Research (EAPR) has established working groups to define quality standards, and international quality standards are regulated by the United Nations/Economic Commission for Europe (UN/ECE).

## *STORAGE CONDITIONS*

In the early nineteenth century, potatoes were stored in unventilated clamps. Ventilated clamps came later. Although unventilated clamps have become obsolete in developed countries, they are still in use in developing countries for the short-term storage of ware potatoes. For long-term storage of large quantities, potato stores were constructed in the developed countries in the 1930s. In these stores, outside-air cooling systems were used to maintain the desired temperature.

Potato stores with electronically controlled environments have been developed since the 1960s.

### *Nonrefrigerated Storage*

Refrigerated storage remains the best option for the long-term storage of potatoes. For short-term storage (3-4 months), nonrefrigerated, on-farm storage methods are cheap alternatives, particularly in tropical and subtropical countries. Diffused-light storage for seed potatoes and storage in simple piles or clamps for ware potatoes have been found to be effective methods in South America (Booth and Shaw, 1981). In India, a nonrefrigerated store run on passive evaporative cooling was suggested as an alternative storage method (Kaul and Sukumaran, 1984), but it was not adopted by the farmers owing to its high initial cost compared with the benefits obtained. On-farm storage of potatoes in heaps (unventilated clamps) and pits is common in some parts of India (Ezekiel et al., 2002). Potatoes stored in heaps and pits contain low levels of reducing sugars and are therefore suitable for processing.

### *Refrigerated Storage*

Optimal storage conditions are based on the purpose for which the potatoes are stored and the cultivar. These optimal conditions should be reached during the cooling process and maintained throughout the storage period. For ware potatoes, storage time can be up to 12 months; hence sprout growth has to be checked to prevent excessive weight loss leading to shriveling. Cold storage (2-4°C) helps not only in controlling sprout growth but also in maintaining a low respiration rate. Cold storage is recommended for seed potatoes, for which cold-induced sweetening is not a problem. However, low-temperature storage is not suitable for potatoes meant for chips (crisps) or French fries because low temperatures cause sweetening as a result of excessive sugar accumulation and discoloration during frying. However, "4°-varieties" resistant to cold-induced sweetening have been developed (Putz, 1997). Storing potatoes at temperatures above 7°C is called warm storage; it is practiced for potatoes meant for processing into chips and French fries.

TABLE 13.1. Temperatures for long-term storage.

| Purpose | Temperature range (°C) |
|---|---|
| Seed potatoes | 2-4 |
| Ware potatoes | 4-6 |
| Industrial potatoes | 6-8 |
| Potatoes for crisping (chips) | 7-12 |
| Potatoes for French fries (crisps) | 6-8 |

The storage temperatures shown in Table 13.1 can generally be used for long-term storage of potatoes for different purposes. When potatoes are stored at temperatures above 4°C, sprout growth occurs. Therefore, processing potatoes need to be treated with sprout inhibitors. The relative humidity should be maintained at approximately 95 percent. The oxygen concentration should be in the normal range of approximately 21 percent and carbon dioxide concentration should be less than 1 percent (Rastovski, 1987).

## *STORAGE PROCESS*

The potato storage operation can be said to begin with the harvesting of the tubers and end when the store is unloaded. The entire storage process can be divided into several stages: (1) loading the store with potatoes, (2) drying the tubers to remove surface moisture, (3) wound healing, (4) cooling the stack (lowering the temperature gradually to the desired level), (5) maintaining the temperature and relative humidity at the desired level, (6) warming the stack (in preparation for unloading the store), and (7) unloading the store.

### *Loading*

In a modern potato store, potatoes are stored either in bulk or in bins. Bulk storage is suitable for storing potatoes of one variety for a specific purpose; they can be loaded or unloaded in one go. Generally, potatoes can be stacked in bulk up to a height of 4-6 m, but this can cause bruises or pressure spots as the result of pressure exerted by the upper layer of potatoes on the lower layer (Figure 13.3). Bruises can affect the quality

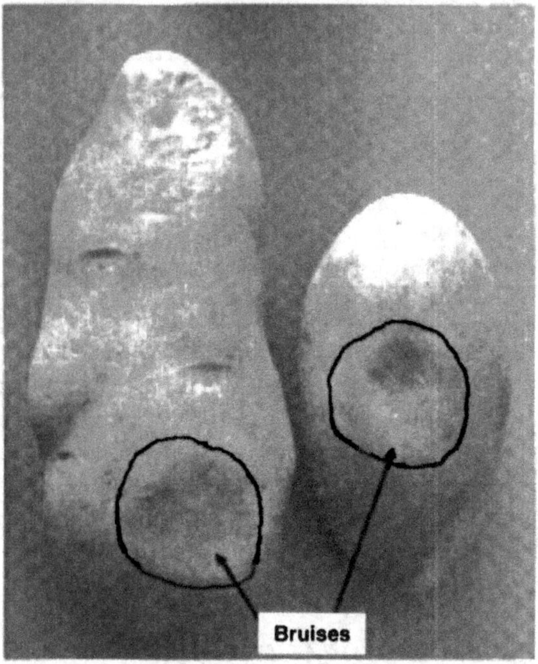


FIGURE 13.3. Development of bruises or pressure spots as a result of pressure exerted on tubers in the lower layer by tubers in the upper layer. See Plate 13.3 in the color gallery.

of stored potatoes considerably. Bruises or pressure spots can be avoided by using bin boxes in which potatoes are filled to a height of 1-1.5 m. Further, bin storage facilitates storing potatoes of different varieties and of different origins for different purposes. Storing potatoes in bin boxes offers several other advantages: the boxes are freely maneuverable to allow flexible configuration of the stack, drying and cooling are easy, sprout suppressants can be applied more effectively, and unloading of different lots is easy. However, in using bin boxes, maintaining uniform temperature and relative humidity in the stacks becomes difficult since airflow around the stack is influenced by both storehouse geometry and air duct system geometry (Gottschalk, 1999b). The investment costs are also relatively high for the boxes. The air duct system is, however, simpler and less expensive than the underfloor duct system of a bulk store. It has been shown that even a huge bin box store of 15,000 tonne capacity can run with a free-convection ventilation system without fans for long-term storage of ware potatoes (Kern et al., 2002).

### *Drying*

During the drying process, the thin water film on the tuber surface evaporates into the ambient air and the air in the boundary layer close to the tuber surface is saturated with water vapor while the water film exists. Vapor transport through the boundary layer is equivalent to mass transport during the transpiration process, as described below. The drying "velocity" (i.e., the vapor loss per unit of time) is proportional to the partial pressure difference (the pressure deficiency) between the saturated humid air close to the tuber surface and the ambient humid air. The drying velocity can be increased by increasing the flow rate of the air passing through the heap, boxes, or bulk and by increasing ambient air temperature. When the water film is dried from the surface, evaporation continues as a result of transpiration, removing bounded water transported from inside the tuber to the surface. Mass loss will start, and this must be stopped by properly controlling the drying process. The point at which a dry surface is achieved can be seen from a sharp increase in air temperature (Figure 13.4) or by measuring the potatoes' conductivity.

### *Wound Healing*

The use of machines during harvest, transport, grading, and filling causes various types of mechanical damage, such as bruising, abrasions, and cracks on the tuber surface. The wounds become easy entry points for microbes that cause diseases; therefore, faster wound healing is necessary to prevent infections. As living organs, tubers are capable of reforming damaged tissues, even repeatedly. This process of wound healing starts immediately after storage, that is, during the drying phase. The climate within the store should be controlled in such a way that the tissue on the wounded surface forms the periderm quickly and completes the secondary protective tissue of the periderm by forming cork, cork cambium, and phelloderm. Formation of suberin (cork) begins within a few hours and a permanent

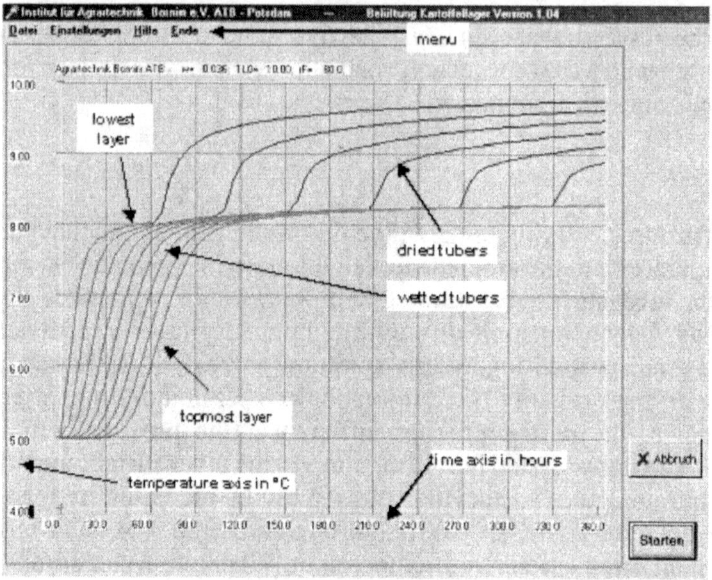


FIGURE 13.4. Temperature courses during surface drying in a heap of potatoes stored at 5°C. Drying with air temperature of 10°C, 80 percent relative humidity, and air velocity of 0.036 m·s$^{-1}$ through the heap. Results from a simulation program. *Source:* Gottschalk (1999c).

barrier is formed, preventing the entry of fungi and bacteria and loss of moisture. This process of suberization is slow at lower temperatures and stops at 2°C. Under suitable environmental conditions, a few layers of wound periderm similar to normal periderm in structure are formed in 5-7 days. Phellogen (cork cambium) is the basis for the formation of periderm. The ideal conditions for wound healing are higher temperatures (18-20°C) and high relative humidity (95 percent and above) with free access to fresh air (oxygen). As temperatures above 15°C and high relative humidity are also conducive to the spread of soft rot caused by *Erwinia* spp., there is a high risk of total loss within a few days if the wound healing is not complete. At temperatures below 10°C, relative humidity less than 80 percent, and oxygen concentration less than 10 percent, periderm development is poor, and when the relative humidity is less than 30 percent, wound healing stops (Burton, 1974). It takes approximately 2 weeks for complete wound healing, and this time is critical because of the risk of infection of the tuber by microorganisms and higher water loss from the tuber. Therefore, ventilation should be limited to a few hours daily to remove surface moisture and to keep oxygen concentration at approximately 20 percent. When fresh air flows through potatoes stored in bulk, it is quickly saturated to 100 percent relative humidity. In freshly harvested tubers, especially immature tubers, water loss can be quite high. Thus, within the first few weeks of storage, a considerable amount of water evaporates from the tubers, resulting in maximum weight loss at the beginning of storage period. Heavily wounded and mechanically shaken tubers respire intensely. The rate of respiration in tubers with unhealed wounds is much higher than that in mature and well-cured tubers.

### *Cooling*

After drying and wound healing comes the main storage phase. Where the climate is moderate, for example in central European countries, outside air is used for cooling or warming, and thus temperature maintenance is economical. Ventilation is from bottom to top through an air duct system constructed at the floor of the store or through underfloor air ducts, which are more expensive.

The main problem during this period is excessive weight (water) loss through evaporation. The evaporation process (i.e., mass transfer from the tuber surface to flowing ambient air) is indirectly proportional to the temperature difference between the potato bulk and the inlet air temperature. Actually, the mass flux density for the evaporated water (vapor) from the surface into the air is proportional to the partial pressure difference between the humid air close to the surface of the tuber (the boundary layer) and the humid air that passes by the ambience of the tuber. The partial pressure difference is the "driving" force for the vapor transfer in humid air. The partial pressures themselves are strongly dependent on the temperature and the air humidity.

Potatoes should be cooled down to the desired temperature gradually by decreasing the temperature by 0.5-1 K per day, either with fresh air or using a mechanical refrigeration system, depending upon local conditions. In temperate countries, the former practice is common; in tropical countries, the latter. The cooling velocity (degrees Kelvin per unit of time) can be accelerated by applying a higher airflow rate (air velocity) through the bulk. Approximate cooling time can be calculated using Fourier's heat equation for a solid sphere, involving the heat transfer coefficient. The tuber can be regarded as an ideal sphere. The boundary condition for the heat exchange across the boundary layer from the tuber surface to the ambient flowing air is of "mixed" type; that is, it couples heat flux through the boundary layer with the "driving" temperature difference.

$$\alpha\,(T_{tuber} - T_{ambient}) = -\lambda(\partial T_{tuber}/\partial n)_{boundary}$$

where $\lambda$ is the heat conductivity of potato tissue, $\alpha$ is the heat transfer coefficient, $T$ is temperature, and $n$ is the normal vector perpendicular to the tuber surface (boundary).

The dependency of the heat transfer coefficient on the airflow velocity is described (Rastovski, 1981a) as

$$\alpha = 7.8w^{0.6}/D^{0.4}\ (\mathrm{W{\cdot}m^{-2}{\cdot}K^{-1}}),$$

where $w$ is the air velocity in meters per second and $D$ is the potato tuber diameter in meters.

Using $D = 0.05$m, for example, analytic calculation of cooling a potato with variant airflow velocities and constant ambient air temperature is shown in Figure 13.5, using the normalized temperature $T_n = (T - T_{ambient})/(T_{tuber} - T_{ambient})$.

In this simple analytical model, evaporation or condensation effects have not been taken into consideration. More complex models taking these factors into consideration have been developed (e.g., Maltry and Winter, 1992). A numerical model of this kind has been developed and validated by Gottschalk (1996) for calculating tuber temperature, tuber moisture, air temperature, and air relative humidity over time and along the height of a ventilated heap. A comparison of simulated heap cooling and measurements is shown in Figure 13.6 (Gottschalk, 1999a).

After storage at low temperatures, potatoes should be warmed up to a temperature of 12-15°C before they are unloaded and

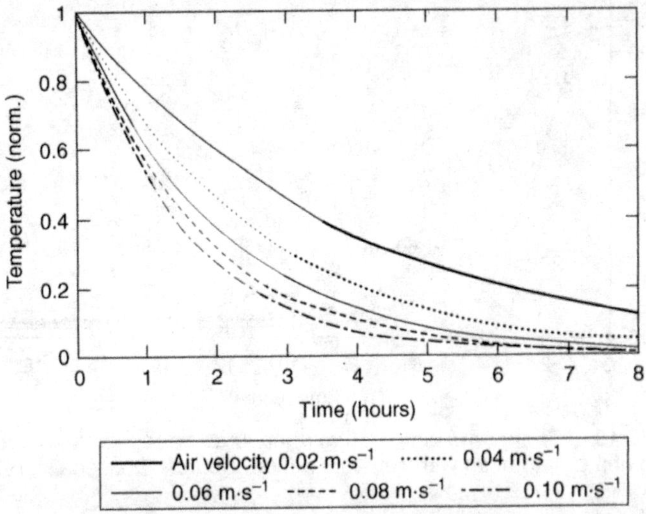


FIGURE 13.5. Analytical time course of temperature during cooling of a spherical tuber (5 cm diameter) using normalized temperature.

packed. Potatoes are most susceptible to mechanical impacts leading to discoloration at low temperatures. However, warming should be done carefully as there is a risk of condensation of water on the tuber surface when cold tubers are heated with warm and humid air. If condensation occurs, the potatoes should be dried well before packing to prevent rotting. Theoretical interpretations of condensation effects have been given by Hylmö et al. (1976). Condensation effects have also been taken into consideration in the simulation model for bulk ventilation developed by Gottschalk (1996).

A number of measures to prevent condensation problems have been recommended (Pringle, 1996). Some of the most important are to (1) stop ventilation when the dew point is reached; (2) seal the store to prevent the entry of warm air; (3) mix inlet air with store air to avoid large temperature differences (above 3 K); (4) use forced air ventilation or refrigeration if sprouts start to elongate.

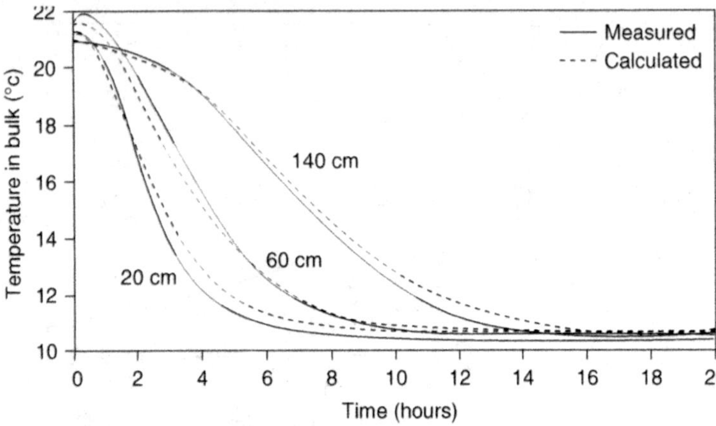


FIGURE 13.6. Changes in temperature (measured and simulated values) during cooling of a potato heap with constant airflow. *Source:* Gottschalk (1999a).

## *Respiration*

As living organisms, tubers respire continuously. During respiration oxygen ($O_2$) is taken up and carbon dioxide ($CO_2$), water, and heat are released. Some of the water released during respiration is retained in the tuber, and the remainder is lost through evaporation. The $CO_2$ released during respiration, if allowed to accumulate in stores, can affect the quality of stored potatoes. Therefore, periodical ventilation is necessary to remove the accumulated $CO_2$ and to supply fresh $O_2$.

Potatoes are rich in starch, which is hydrolyzed to glucose, which forms the substrate for respiration according to the following reaction:

$$C_6H_{12}O_6 + 6O_2 = 6CO_2 + 6H_2O + \text{energy (heat)}.$$

During respiration, for 1 g of glucose, 1.07 g of oxygen is taken up and 1.47 g of $CO_2$, 0.6 g of water, and 16 kJ of energy are produced. Only 32 percent of the energy produced (i.e., 5.1 kJ) is fixed in adenosine triphosphate (ATP) as metabolic energy and the remainder (68 percent, i.e., 10.9 kJ) is released as heat (van Es and Hartmans,

1981). A complete sequence of enzymatic reactions is involved in the aerobic respiration process in which adenosine diphosphate (ADP) and inorganic phosphate ($P_i$) are used to form ATP, the energy carrier in the cell. The overall reaction can be written as below (Burton et al., 1992).

$$C_6H_{12}O_6 + 6O_2 + 38ADP + 38P_i = 6CO_2 + 6H_2O + 38ATP.$$

The release of heat and water during the respiration process causes the loss of dry matter in tubers during storage. The loss is about 1.5 percent of fresh weight during 6 months of storage at 5-10°C. The total dry matter loss is 6-7 percent but converted water remains in the tuber to offset this (Rastovski, 1981b).

The uptake of oxygen is an index of oxidative activity in the tuber, and the production of $CO_2$ during respiration is an index of the breakdown of carbohydrates. The molecular ratio of $O_2$ and $CO_2$ is known as the respiratory quotient (RQ):

$$RQ = \frac{\text{Volume of } O_2 \text{ absorbed per hour}}{\text{Volume of } CO_2 \text{ released per hour}}$$

The RQ should ideally be 1, if the $O_2$ supply is not limiting. However, the RQ for stored potatoes has been found to vary depending upon the age of the tubers. The $CO_2$ to $O_2$ ratio was found to be approximately 0.8 during the early period of storage and approximately 1.3 when sprout growth started (Burton, 1989). In senescent tubers, it was as high as 4 because of an excess of $CO_2$ released over $O_2$ absorbed (Burton, 1963; Isherwood and Burton, 1975). There is another type of respiration, called residual respiration, which is an enzymatic process involved in wound healing. During this, $O_2$ is taken up and $CO_2$ and heat are produced, but there is no ATP production (DeEll et al., 2003).

Storage temperature has a strong influence on the rate of respiration. There is a rise in the respiration rate at temperatures below 5°C and above 15°C. The rate of respiration is minimized between 5 and 10°C (Figure 13.7). At lower temperatures, the high concentration of reducing sugars formed by the breakdown of starch causes an increase in the rate of respiration. The activity of the enzyme invertase,

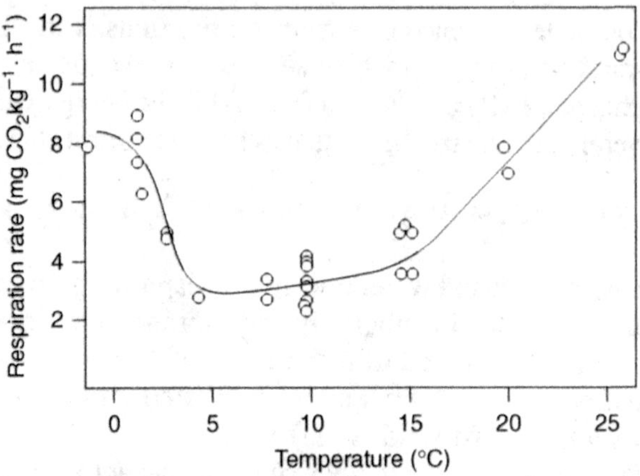


FIGURE 13.7. Rate of respiration of potato tubers after approximately 1 month of storage at various temperatures. *Source:* Redrawn from Burton (1989).

which hydrolyses sucrose into glucose and fructose, is also high at lower storage temperatures, leading to the accumulation of reducing sugars. At higher storage temperatures, the sugars are partly reconditioned (up to 80 percent) and converted back into starch.

The $O_2$ uptake is generally high in immature tubers. It was found to be 40 ml·kg$^{-1}$·h$^{-1}$ in immature tubers immediately after harvest and 3-5 ml·kg$^{-1}$·h$^{-1}$ in mature tubers, at 10°C in normal air (Burton, 1974). During storage it further decreased to 1.5-2.5 ml·kg$^{-1}$·h$^{-1}$. The rate of respiration was influenced by variations in the $O_2$ concentration at the beginning and end, but not during the middle, of the storage period. Respiration activity was reported to increase after 2 weeks of storage and then to decrease again, but the saccharine and reducing sugar content continued to increase. This indicates that respiration is not necessarily related to sugar content. The rate of respiration was determined by Schippers (1977b) in a number of varieties stored at different temperatures. The respiration rate was 1.5-4 ml·kg$^{-1}$·h$^{-1}$ at 2.5°C and 2.5-12.5 ml·kg$^{-1}$·h$^{-1}$ at 15°C. The rates were higher at the start of the storage period, especially at storage temperatures of 2.5 and 5°C. The rate of respiration was high immediately after harvest, and it decreased to 2.5-12.5 ml·kg$^{-1}$·h$^{-1}$ after 3-6 weeks of storage.

(For more information on the factors influencing the rate of respiration, see Schippers, 1977a.)

Since gas exchange takes place through the lenticels, the presence of a water film on the tuber surface obstructs the exchange of gases and consequently the respiration process. If the tuber surface is not properly dried, anaerobic conditions are created, leading to suffocation of the tuber.

### *$O_2$ and $CO_2$ in the Storage Atmosphere*

The concentration of gases in the storage atmosphere affects tuber quality. Excessive accumulation of $CO_2$ may cause black heart, and affected tubers eventually rot. Higher $CO_2$ concentrations are also known to affect the processing quality of stored potatoes. Accumulation of $CO_2$ up to 3 percent in well-sealed stores is reported to result in unacceptable fry color. Therefore, regular ventilation of stores for 2-4 hours per day is recommended (Veerman and van Loon, 1997).

Low $O_2$ concentration in the storage atmosphere stimulates the end of dormancy. The optimal $O_2$ concentration for promoting sprout growth has been found to be 3-5 percent (Hartmans, 1993). Increased $CO_2$ concentration (2-4 percent) is also known to stimulate sprout growth. Breaking of dormancy through the controlling of gas concentrations in the storage atmosphere can be an alternative to treatments with toxic chemicals such as rindite or carbon disulphide. However, it is expensive.

Controlled-atmosphere storage is much used in the storage of fruits and vegetables but not for potatoes so far. Coleman and McInerney (1997) have developed a computerized controlled-atmosphere storage method for seed potatoes by using different concentration of nitrogen, carbon dioxide, oxygen, and traces of ethylene.

### *Water Loss*

During storage, water is lost from the tubers at a rate proportional to the vapor pressure deficit ($P_i - P_a$) between the intercellular space and the ambient humid air. The mass transport takes place within the boundary layer close to the tuber surface (Gottschalk and Linke, 2002). The partial vapor pressure of tuber tissue is dependent on the temperature and the molar concentration of the soluble matter in the intercellular water. The moisture that diffuses through the boundary layer around the tuber surface is transported away from the tuber by air flowing past the tuber surface. Above a certain air velocity, the overall mass transfer coefficient increases with increasing air velocity and tends asymptotically to an almost constant value (Rastovski, 1978).

According to Fick's law of diffusion, the water evaporation rate (mass flux density) $j$ is proportional to the pressure deficiency with a proportional factor $k$, the transpiration coefficient; hence:

$$j = k(P_i - P_a)\ (\text{kg}\cdot\text{m}^{-2}\cdot\text{s}^{-1}).$$

The transpiration coefficient $k$ is also related to the reciprocal of the mass diffusion resistance ($r_d$):

$$r_d = 1/k\ (\text{kg}^{-1}\cdot\text{m}^2\cdot\text{s}\cdot\text{Pa}).$$

The total mass (moisture) loss $\Delta m$ as a result of evaporation from the tuber surface $A$ is therefore

$$\Delta m = jA\Delta t$$

for constant $j$ over time interval (e.g., storage time) $\Delta t$.

However, $j$ is in practice not constant for all time $\Delta t$ because $k$ (or $r_d$) changes with the change in the permeability of the periderm during its suberization.

Average evaporative loss was measured in two potato cultivars stored at 10°C with a water vapor pressure deficit for the store air of 2.1 mbar. It varied from 6.2 to 6.9 $\mu g\cdot cm^{-2}\cdot h^{-1}\cdot mbar^{-1}$ in variety Majestic and from 9.9 to 10.9 $\mu g\cdot cm^{-2}\cdot h^{-1}\cdot mbar^{-1}$ in variety King Edward. This corresponds to average mass losses of 0.1 and 0.15 percent per week per millibar, respectively. The evaporation loss increases rapidly in sprouted tubers since the sprout surface is 100-150 times more permeable to water vapor than the periderm of the tuber per unit area and time (van Es and Hartmans, 1987b). This leads to considerable weight loss and shriveling of tubers. The periderm of the tuber forms an effective barrier to water loss. The importance of the periderm can be judged from the fact that removal of the tuber skin resulted in an immediate 300-500 fold increase in evaporation (Burton, 1989).

### *Store Ventilation and Environment Control*

During long-term storage, metabolically or physically produced heat, accumulated gases such as $CO_2$ and ethylene, and moisture should be removed regularly from the tuber stack. Ventilation of the store with fresh air at a comparatively low airflow rate is normally sufficient. For uniform temperature distribution, heat and moisture must be removed regularly from the tuber surface. When the stack is being cooled and during the wound healing period, a stronger airflow of sufficiently high rate is necessary. Insufficient ventilation gives rise to the accumulation of moisture and self-heating, leading to an increased risk of spread of microbial diseases, which can result in very high storage losses. Especially when potatoes are stored in bin boxes, sufficient airflow inside the boxes must be ensured. In fact, the common problem in such cases is that the ventilation is not sufficient to remove the heat and moisture from the "heart" of the stack, although tubers near the periphery of the stack are sufficiently ventilated. During cooling and wound healing, a relatively high ventilation rate is necessary to accelerate these two processes. However, during subsequent storage, a lower ventilation rate is recommended. A lower airflow rate at regular intervals is sufficient to remove metabolically produced heat and the $CO_2$ produced during respiration. Airflow improvements can be made using computational fluid dynamics simulations (Gottschalk, 1999b).

Developing control strategies for biological products such as potatoes is difficult because of strong nonlinearity and changing behavior during storage. In model-based controllers, mathematical equations are used to describe the process under control. Simple environment feedback systems are not able to respond accurately to a real biological system (Farkas, 2003). Nevertheless, such control systems are being used widely in controlled-atmosphere stores. Knowledge-based adaptive control strategies are found in the form of expert systems, fuzzy logic control systems (Gottschalk et al., 2003), and parameter-adaptive optimal control systems using artificial intelligence strategies as artificial neuronal network systems, neuro-fuzzy systems, genetic algorithms, and evolutionary algorithms (Gottschalk, 1995).

## SPROUT INHIBITION

A number of chemicals have been tried for inhibiting sprout growth in potatoes. Among these, only two, maleic hydrazide (MH) and isopropyl N-(3-chlorophenyl) carbamate (CIPC), are in use on a large scale. MH has been used extensively for sprout inhibition and has been found to be very effective in suppressing sprout growth when applied as a preharvest foliar spray. MH does not affect tuber quality for ware and processing purposes (Ezekiel et al., 1984; Sawyer and Malagamba, 1987). If applied too early, MH can cause tuber abnormalities; if applied too late, it may not be effective. Although MH is widely used in the United States, it is prohibited in the European Union.

The most commonly used sprout suppressants are isopropyl N-phenylcarbamate (IPC) and CIPC. A mixture of IPC and CIPC is also used. CIPC should be applied only to healthy and dry, nonsprouted tubers. The tubers should be mature, with a well-developed periderm and fully healed wounds. The chemical is deposited on the tuber surface and prevents the growth of sprouts. The sprout inhibition effect may not be complete after the first application. Therefore, a second application is recommended after approximately 6 weeks.

Seed potatoes should not be placed in a store in which sprout inhibitors have been applied within the past few weeks. Boyd et al. (1982) observed that residues of less than 0.3 $mg \cdot kg^{-1}$ delayed mean emergence time and reduced the yield. With higher residues (0.3-1 $mg \cdot kg^{-1}$), the effect was more drastic, but varieties differed in their response.

### Natural Sprout Inhibitors

Environmental and health concerns have led to an increasing tendency among consumers to avoid chemically treated foods. As a result, attempts are being made to find and use naturally occurring substances with sprout inhibition properties as alternatives to chemical sprout inhibitors. The sprout suppression effect of several volatile oils extracted from aromatic plants is being investigated. Among the important aromatic plants under investigation are *Carum carvi* (caraway seed), *Lavendula angustifolia* (lavender), *Mentha pulegium* (mint), *M. spicata* (spearmint), *Origanum onites* (Turkish oregano), *Origanum vulgare* (Greek oregano), *Rosmarinus officinalis* (rosemary), and *Salvia fruticosa* (sage). Oils extracted from these plants (except oregano) showed sprout suppression effects. Greek oregano showed a strong inhibiting effect against *Erwinia carotovora,* which causes soft rot (Vokou et al., 1993). Application of crude extracts instead of oils had similar effects, and this might be an attractive and economical proposition in countries to which these aromatic plants are native. Neomenthol and menthone can also be used as sprout suppressants (Coleman et al., 2001).

Of the several essential volatile oils trialed for their sprout suppression effect, the monoterpene S-(+)-carvone, derived from the essential oil of caraway seed *(Carum carvi)* (consisting of 50-60 percent carvone), has been found to be the most effective and useful. As it is volatile, carvone evaporates quickly in the storage environment; hence repeat applications are needed.

Carvone-treated tubers showed better results than tubers treated with a conventional IPC/CIPC mixture (Hartmans et al., 1995). Most of the carvone residues could be recovered from the peel, and the level of residue was significantly lower than the level of IPC/CIPC residue in the treated potatoes. Further, carvone did not affect the color of fries prepared from stored tubers. Kalt et al. (1999) compared the sprout suppression effects of carvone, dimethylnaphthalene (DMN), ethylene, and CIPC and determined fry color and sugar content during 25 weeks of storage at 9°C. All treatments were effective, but the degree of sprout suppression varied. Maximum sprout suppression was observed with CIPC, followed by carvone, ethylene, and DMN. Ethylene treatment affected the fry color adversely. Sorce et al. (1997) found carvone to be an effective sprout suppressant even for seed potatoes. Carvone at a concentration of 7.15 $mmol \cdot mol^{-1}$ was effective in suppressing sprout growth temporarily, with subsequent sprout growth being normal. A lower concentration was not effective in suppressing sprout growth, and a higher concentration not only suppressed sprout growth but also adversely affected subsequent sprout development, thus lowering the seed quality.

In addition to being an effective sprout suppressant, carvone is also known to inhibit the growth of certain fungi and microorganisms and to act as an insect repellent (Vokou et al., 1993).

### *Light*

Light is not relevant for the storage of ware and processing potatoes as they are stored in the dark to prevent greening from the development of chlorophyll. However, exposure to light might be useful in suppressing the sprouting of seed potatoes, especially under nonrefrigerated storage. Light at the lower and upper ends of the visible spectrum has a positive effect on sprout inhibition. Diffused-light storage of seed potatoes was developed based on this principle, and it has been found to be quite useful in many developing countries with moderate temperatures during the storage period.

Exposure to light also causes accumulation of glycoalkaloids, which are toxic at higher concentrations. Griffiths et al. (1994) investigated the effect of light on glycoalkaloid development and found significant varietal difference in glycoalkaloid and chlorophyll contents when tubers were exposed continuously for a period of 48 hours to a radiant flux density of 140 $\mu mol \cdot s^{-1} \cdot m^{-2}$.

### *Irradiation*

Irradiation with beta, gamma, or X-rays effectively inhibits sprouting (Burton, 1989). However, irradiation prevents cell division; thus wound healing is affected and the risk of rotting increases, especially at higher storage temperatures of 17-30°C (Ezekiel, Singh et al., 2003). It is also reported to increase after-cooking discoloration in some varieties. Irradiation up to 100 Gy was reported not to affect suberization (Thomas, 1982). Irradiation of potatoes is very costly and is prohibited in some countries.

## *BIOCHEMICAL CHANGES DURING STORAGE*

Among the important biochemical constituents of tubers that are influenced by long-term storage are starch, sugars, protein, amino acids, vitamins, alkaloids, and organic acids. Storage is also known to affect the activity of enzymes.

The average dry matter in tubers ranges from 15 to 35 percent. Generally, the dry matter content of potatoes increases during storage as the loss of water through evaporation is much more than the loss of dry matter owing to respiration (Smith, 1987). Changes in dry matter during storage are influenced by storage temperature, relative humidity in the storage atmosphere, and storage duration.

There is a strong relationship ($r = 0.956$) between the dry matter content and starch content of potatoes (Burton, 1989):

$$\text{starch content (\%)} = 17.55 + 0.891(\text{dry matter content [\%]} - 24.18).$$

Starch content can also be derived from underwater weight using conversion tables (van Es and Hartmans, 1987a; Ezekiel, Singh, et al., 2003). However, care should be taken when using these methods; probable discrepancies are discussed by Samotus et al. (1986). Near-infrared spectroscopy has also been used for the rapid determination of the starch and dry matter content of potatoes (Putz and Weber, 1996).

The total carbohydrate content of tubers is not known to change significantly during storage, but changes in starch and sugar contents do occur and have been the subject of investigation since 1882 (Müller-Thurgau, 1882). Generally, at low storage temperatures, starch decreases and sugars increase as a result of the conversion of starch into sugars. At high temperatures sugars decrease and starch increases because of the conversion of sugars into starch and the utilization of sugars in respiration (Smith, 1987). Sucrose, fructose, and glucose are the three major sugars in potato tubers, and the content of each varies with storage temperature. Fructose content was found to be high in tubers stored at low temperatures, and glucose content was high at high temperatures (Arreguin-Lozano and Bonner, 1949). Reducing sugar accumulation was high in potatoes stored at 4.4°C, but there was little change in reducing sugars at 21°C (Hyde and

Morrison, 1964). The accumulation of sugars at low temperatures results in a sweet taste. When the total sugar content of a tuber exceeds 5 percent of fresh weight, it imparts a noticeable sweet taste and such potatoes are considered unacceptable for ware and processing purposes. The acceptable limit of reducing sugars content in a tuber for processing is 0.25-0.5 percent of fresh weight. When it exceeds this limit chips (crisps) from such tubers develop a brown color. The accumulation of sugars at low temperatures is called cold-induced sweetening, and the quantity of sugars that accumulates varies with cultivar. This process starts slowly at approximately 9°C, reaches a maximum at approximately 3°C, and stops at approximately −1.7°C, the freezing point of the tuber. Cold-induced sweetening is reversible to a large extent by storing the potatoes at a higher temperature of 10-20°C. During such reconditioning, some 80 percent of the reducing sugars is converted back into starch and a part of the remaining sugars is used up in respiration. However, after prolonged storage at low temperatures, reconditioning does not help in restoring the sugar concentration to the prestorage level. In other words, the sugar level in reconditioned tubers may not be low enough for processing. Sugar accumulation is minimum when potatoes are stored at 10-13°C, and this temperature range is considered optimum for the storage of potatoes to be processed. However, sugar accumulation can occur after prolonged storage even at this temperature.

Most researchers have found only a slight change in the total nitrogen content during storage, but significant changes in free amino acids and amides have been observed (Lisinska, 1989). Proteins are broken down into free amino acids during prolonged storage (Habib and Brown, 1957; Ezekiel, Mehta et al., 2003). Protein synthesis may also occur during storage, although low temperature reduces the rate of this process and enhances protease activity. Changes in free amino acids were found to be greater at 21.1°C than at 12.8°C (Sweeny et al., 1969). Storage studies at 5 and 10°C up to 40 weeks have shown that the free amino acid content of stored tubers is controlled by a combination of factors, including proteinase activity, protein synthesis, and the overall balance of nitrogen flow (Ezekiel, Mehta et al., 2003). Nitrate content is also affected by storage conditions, and a decrease in nitrate content after six months of storage at 4°C has been reported (Augustine et al., 1977).

The important glycoalkaloids present in potato tuber are alpha solanine (approximately 40 percent) and alpha chaconine (approximately 60 percent). Significant losses in glycoalkaloid content was observed after 6 months of storage (Haase, 2002). During long-term (9 months) storage at 4°C, Olsson (1995, 1996) observed an inconsistent trend in glycoalkaloid content. There was an increase at the beginning of the storage period followed by a decrease. Glycoalkaloid content increased again during sprout growth, decreased subsequently, and increased again at the end of the storage period. Varieties with higher glycoalkaloid contents at the beginning of storage showed greater variations during the storage period.

Vitamin C content changes considerably during the storage period, and dehydro ascorbic acid is irreversibly converted to 2,3-dioxo-L-gulonic acid. Ascorbic acid content generally decreases during storage, and the decrease is rapid (approximately 10 percent) within 3-4 weeks during storage at 10°C (Burton, 1989). Augustine et al. (1978) also observed decreases in ascorbic acid, niacin, and folic acid contents but an increase in vitamin $B_6$ during low-temperature storage. There was little difference in thiamin and riboflavin content.

Changes in organic acids have also been reported. Sweeney et al. (1969) found an increase in citric acid and a decrease in malic acid contents at 12.8 and 21.1°C. Shekhar and Iritani (1979) observed an increase in both acids during low-temperature storage. An increase in fatty acid content and a decrease in triglyceride and phospholipid content in tubers stored at 5.6°C have been reported (Cheng and Muneta, 1978).

The enzymes present in the potato tuber are numerous and form a major part of the potato protein. Most of the studies carried out on changes in enzyme activities during storage are in relation to changes in starch and sugar contents.

Starch degradation can begin within a few days of transferring tubers to low temperature (Hill et al., 1996), and this breakdown is believed to be regulated by the enzymes amylases, maltases, and phosphorylases (Davies, 1998). Amylase activity is low at low temperatures, but phosphorylase activity is high, indicating that the degradation of starch during cold storage (4°C) is principally phosphorylitic. At high temperatures phosphorylitic activity is low because of the formation of an inhibitor of phosphorylase (Smith, 1987). Phosphorylase activity has been found to be higher at 4.4°C than at 21.1°C (Hyde and Morrison, 1964). The increase in amylase during cold sweetening has been found to be less in cold sweetening–resistant genotypes than in cold sweetening–susceptible genotypes (Cochrane et al., 1991). The accumulation of reducing sugars could be attributable to the hydrolytic action of invertase on sucrose. No invertase activity was observed in freshly harvested tubers, but significant activity was observed after cold storage (Schwimmer et al., 1961). Reducing sugar accumulation was not proportional to invertase activity (Pressey, 1969). Changes in the activity of sucrose synthetase and sucrose phosphate synthase during storage have been reported (Pressey, 1970). Sucrose synthase activity decreases sharply after harvest and remains low thereafter during warm storage but increases slowly during cold storage up to 30 weeks. At low temperatures, sucrose accumulation can precede accumulation of glucose and fructose. Attempts have been made to restrict the production of sucrose by down-regulating the expression of uridine diphosphoglucose (UDPG) phosphorylase, the enzyme which forms the UDPG required for sucrose phosphate synthase (Davies, 1998). The level of sucrose synthase, the primary sucrose cleaving enzyme in developing tubers, is usually low in stored tubers. In contrast, invertase, which is practically nonexistent in developing tubers, is the principle sucrose degrading enzyme in stored tubers.

Proteolytic activity increases markedly around the buds during the break of dormancy (Burton, 1989). Polyphenoloxidase activity is reported to increase in eyes and buds but to decrease in other parts of the tuber (Smith, 1987). Peroxidase activity is positively correlated with storage temperature, and the highest polyphenoloxidase activity was observed in tubers stored at 10°C. The activity of a low-oxygen-affinity terminal oxidase system decreases during storage and increases with the increase in respiration after the start of sprouting (Burton, 1974).

In conclusion, various physical and biochemical changes in potato tubers after harvest can lead to substantial postharvest losses. Weight loss as a result of evaporation is the major storage loss, and it can be minimized through proper control of the storage environment. Ventilation control is of utmost importance to ensure minimum weight loss, optimum air circulation, and prevention of $CO_2$ accumulation in the stores. Effective storage management keeps the sugar level within the acceptable limit. Although simple storage methods involving minimal investment and running costs can be very useful for short-term storage of potatoes in many parts of the world, new storage strategies make use of information generated by simulation models. All these developments have resulted in precise storage environment control, which has facilitated the maintenance of the desired quality in stored tubers.

## REFERENCES

Allen, E.J. and O'Brien, P.J. (1986).The practical significance of accumulated day-degrees as a measure of physiological age of seed potato tubers. *Field Crops Research 14:*141-151.

Arreguin-Lozano, B. and Banner, J. (1949). Experiments on sucrose formation by potato tubers as influenced by temperature. *Plant Physiology 24:*720-738.

Augustine, J., Johnson, S.R., Teitzel, C., Toma, R.B., Shaw, R.L., True, R.H., Hogan, J.M., and Deutsch, R.M. (1978). Vitamin composition of freshly harvested and stored potatoes. *Journal of Food Science 43:*1566-1574.

Augustine, J., McDole, R.E., and Painter, C.G. (1977). Influence of fertilizers, irrigation and storage treatments on nitrate-N content of potato tubers. *American Potato Journal 54:*125-136.

Bialek, K. and Bielinska-Czarnecka, M. (1978). Gibberellin pattern in potato tubers in relation to cultivar specificity and weather conditions. *Bulletin de l'Académie Polonaise des Sciences / Série des Sciences biologiques,* Polska Akademia Nauk, Warshaw *26:*505-512.

Booth, R.H. and Shaw, R.L. (1981). *Principles of Potato Storage.* Lima Peru: International Potato Center.

Böttcher, H. and Belker, N. (1996). *Frischhaltung und Lagerung von Gemüse* (pp. 11-36). Stuttgart, Germany, Ulmer.

Boyd, I.M.G., Dalziel, J., and Duncan, H.J. (1982). Studies on sprout suppressants. 5. The effect of chlorpropham contamination on the performance of seed potatoes. *Potato Research 25:*51-57.

Burton, W.G. (1963). The effect of stage of maturity and storage on the respiratory quotient of potato tubers. *European Potato Journal 6:*268-270.

———. (1966). *The Potato.* Wageningen, The Netherlands: Veenman & Zonen.

Burton, W.G. (1974). Oxygen uptake and carbon dioxide output of stored potato tubers. *Potato Research 17:*113-137.

———. (1989). *The Potato,* Third edition. New York, USA: Wiley.

Burton, W.G., van Es, A., and Hartmans, K.J. (1992). The physics and physiology of storage. In Harris, P.M. (Ed.), *The Potato Crop* (pp. 608-727). London, UK: Chapman & Hall.

Cheng, F.C. and Muneta, P. (1978). Lipid composition of potatoes as affected by storage and potassium fertilization. *American Potato Journal 55:*441-448.

Cochrane, M.P., Duffus, C.M., Allison, M.J., and Mackay, G.R. (1991). Amylolytic activity in stored potato tubers. 2. The effect of low temperature storage on the activities of α and β-amylase and α-glucosidase in potato tubers. *Potato Research 34:*333-341.

Coleman, W.K. (2000). Physiological ageing of potato tubers: A review. *Annals of Applied Biology 137:*189-199.

Coleman, W.K., Lonergan, G., and Silk, P. (2001). Potato sprout growth suppression by menthone and neomenthol, volatile oil components of *Minthostachys, Satureja, Bystropogon,* and *Mentha* species. *American Journal of Potato Research 78:*345-354.

Coleman, W.K. and McInerney, J. (1997). Enhanced dormancy release and emergence from potato tubers after exposure to a controlled atmosphere. *American Potato Journal 74:*173-182.

Davies, H.V. (1998). Prospects for manipulating carbohydrate metabolism in potato tuber. *Aspects of Applied Biology 52:*245-254.

DeEll, J.R., Prange, R.K., and Peppelenbos, H.W. (2003). Post harvest physiology of fresh fruits and vegetables. In Chakravarty, A., Majumdar, A.S., Vijaya Raghavan, G.S., and Ramaswamy, H.S. (Eds.), *Handbook of Post Harvest Technology. Cereals, Fruits, Vegetables, Tea, and Spices* (pp. 455-483). New York, USA: Dekker.

Ezekiel, R., Dahiya, P.S., and Shekhawat, G.S. (2002). *Traditional Methods of Potato Storage in the Malwa Region of Madhya Pradesh.* Technical Bulletin No. 57, Shimla, India: Central Potato Research Institute.

Ezekiel, R., Daniel, E.V., and Sukumaran, N.P. (1984). Effect of maleic hydrazide on storage behaviour and chemical composition of potatoes in Nilgiris. *Journal of Indian Potato Association 11:*1-11.

Ezekiel, R., Mehta, A., Dinesh Kumar, and Manish Das. (2003). Potato storage, In Khurana, S.M. Paul, Minhas, J.S., and Pandey, S.K. (Eds.), *The Potato: Production and Utilization in Sub-tropics* (pp. 323-335). New Delhi, India: Mehta Publishers.

Ezekiel, R. and Singh, B. (2003). Seed physiology. In Khurana, S.M. Paul, Minhas, J.S., and Pandey, S.K. (Eds.), *The Potato: Production and Utilization in Subtropics* (pp. 323-335). New Delhi, India: Mehta Publishers.

Ezekiel, R., Singh, B., and Gopal, J. (2003). Relationship between under water weight and specific gravity, dry matter and starch content of potatoes grown in India. *Journal of Indian Potato Association 30:*7-14.

Farkas, I. (2003). Control aspects of post harvest technologies. In Chakravarty, A., Majumdar, A.S., Vijaya Raghavan, G.S., and Ramaswamy, H.S. (Eds.), *Handbook of Post Harvest Technology: Cereals, Fruits, Vegetables, Tea, and Spices* (pp. 845-866). New York, USA: Dekker.

Gopal, J., Chamail, A., and Sarkar, D. (2004). In vitro production of microtubers for conservation of potato germplasm: Effect of genotype, abscisic acid and sucrose. *In Vitro Cellular and Developmental Biology-Plant 40:*485-490.

Gottschalk, K. (1995). Adaptive Parameteroptimierung unscharfer Regler mittels Genetischer Algorithmen zur Klimatisierung von Kartoffellagern. *Agrartechnische Forschung 1 H.* 2, 95-101.

———. (1996). Mathematical modelling of the thermal behaviour of stored potatoes and developing of fuzzy control algorithms to optimise the climate in store houses. *Proceedings 2nd International IFAC/ISHS Workshop on Mathematical and Control Applications in Agriculture and Horticulture,* Silsoe, UK: *Acta Horticulturae 406:*331-339.

———. (1999a). Belüftung von Kartoffelschüttungen. *Landtechnik 54:*338-339.

———. (1999b). Die Durchströmung von Kistenlagern für Kartoffeln. *Der Kartoffelbau 50:*375-377.

———. (1999c). Ein Computer Programm zur Berechnung der Temperaturverläufe im Kartoffellager. *Abstracts 14th Triennial Conference of the European Association for Potato Research* (pp. 674-675). Sorrento, Italy: EAPR.

Gottschalk, K. and Linke, M. (2002). Modelling post harvest transpiration and mass loss of horticultural produce. *Proceedings 6th International Symposium on Fruit, Nut and Vegetable Production Engineering* (pp. 369-374). Institut für Agrartechnik Bornim e.V., Potsdam, Germany.

Gottschalk, K., Nagy, L., and Farkas, I. (2003). Improved climate control for potato stores by using fuzzy controllers. *COMPAG Computers and Electronics in Agriculture 40:*127-140.

Griffiths, D.W., Dale, M.F.B., and Bain, H. (1994). The effect of cultivar, maturity and storage on photo-induced changes in the total glycoalkaloid and chlorophyll contents of potatoes *(Solanum tuberosum). Plant Science 98:*103-109.

Haase, N. (2002). Veränderung der Inhaltsstoffe durch Lagerung and Verarbeitung. *Der Kartoffelbau 53:*284-289.

Habib, A.T. and Brown, H.D. (1957). Role of reducing sugars and amino acids in the browning of potato chips. *Food Technology 11:*85-89.

Hartmans, K.J. (1993). Influence of low oxygen conditions on dormancy and sprout growth of six seed potato cultivars. *Proceedings 12th EAPR Conference* (pp. 89-90). Paris, France: EAPR.

Hartmans, K.J., Diepenhorst, P., Bakker, W., and Gorries, L.G.M. (1995). The use of carvone in agriculture: Sprout suppression of potatoes and antifungal activity against potato tuber and other plant diseases. *Industrial Crops and Products 4:*3-13.

Hill, L.M., Reimholz, R., Shroder, R., Nielsen, T.H., and Stitt, M. (1996). The onset of sucrose accumulation in cold-stored potato tubers is caused by an increased rate of sucrose synthesis and coincides with low levels of hexose-phosphates, an activation of sucrose phosphate synthase and the appearance of a new form of amylase. *Plant Cell and Environment 19:*1233-1237.

Hyde, R.B. and Morrison, J.W. (1964). The effect of storage temperature on reducing sugars, pH and phosphorylase enzyme activity in potato tubers. *American Potato Journal 41:*163-168.

Hylmö, B., Persson, T., and Wikberg, C. (1976). Bulk storing of potatoes. Interpretation of a condensation problem. *Acta Agriculturæ Scandinavia 26:*99-102.

Isherwood, F.A. and Burton, W.G. (1975). The effect of senescence, handling, sprouting and chemical sprout suppression upon the respiratory quotient of stored potato tubers. *Potato Research 18:*98-104.

Joseph, T.A. and Gopal, J. (1994). Variation in tuber dormancy of *andigena* potato *(Solanum tuberosum* ssp. *andigena). Indian Journal of Agricultural Sciences 64:*247-248.

Kalt, W., Prange, R.K., Daniels-Lake, B.J., Walsh, J., Dean, P., and Coffin, R. (1999). Alternative compounds for the maintenance of processing quality of stored potatoes *(Solanum tuberosum). Journal of Food Processing and Preservation 23:*71-81.

Kaul, H.N. and Sukumaran, N.P. (1984). *A Potato Store Run on Passive Evaporative Cooling.* Technical Bulletin No. 11, Shimla, India: Central Potato Research Institute.

Kern, A., Krumbiegel, D., Pötke, E., and Schuhmann, P. (2002). Qualitätsschonende Lager- und Lüftungssysteme. *Der Kartoffelbau 53:*226-233.

Krijthe, N. (1962). Observations on the sprouting of seed potatoes. *European Potato Journal 5:*316-333.

Lisinska, G. (1989). Potato storage. In Lisinska, G. and Leszczynski, W. (Eds.), *Potato Science and Technology* (pp. 129-165). London, UK: Elsevier.

Maltry, W. and Winter, U. (1992). Testing the response of stored potatoes to climatic storehouse conditions by mathematical simulation. *International Conference on Agricultural Engineering, AgEng* (pp. 1-14). Uppsala, Sweden.

Müller, K. (1975). Kennzeichnung des Vegetations- und Lagerungsverlaufs der Kartoffel. *Der Kartoffelbau 26:*166-167.

———. (1978). Changes in composition of the tuber as a criterion of its physiological age. *Potato Research 21:*55-57.

———. (1983). Zur Frage der qualitätserhaltenden Lagerung von Kartoffeln. *Der Kartoffelbau 34:*376-379.

Müller-Thurgau, H. (1882). Ueber Zuckeranhäufung in Pflanzentheilen in Folge niederer Temperatur. (Sugar accumulation in portions of plants at low temperatures). *Landwirtschaftliches Jahrbuch 11:*751-828.

Olsson, K. (1995). Changes in glycoalkaloid content of potato during long-term storage. *Abstracts EAPR Joint Meeting of the Agronomy and Utilization Sections* (p. 49). Lissabon, Portugal: EAPR.

———. (1996). Vorkommen von Glykoalkaloiden in Kartoffelknollen. Bericht über die 18. *Kartoffel-Tagung Arbeitsgemeinschaft Kartoffelforschung e.V. Band 18:*45-53.

Pressey, R. (1969). Role of invertase in the accumulation of sugars in cold-stored potatoes. *American Potato Journal 46:*291-297.

———. (1970). Changes in sucrose synthase and sucrose phosphate synthetase activities during storage of potatoes. *American Potato Journal 47:*245-251.

Pringle, R.T. (1996). Storage of seed potatoes in pallet boxes. 2. Causes of tuber surface wetting. *Potato Research 39:*223-240.

Putz, B. (1997). Erste 4°C-Typen bei Kartoffeln aus deutscher Züchtung. Bericht über die 19. *Kartoffeltagung. Arbeitsgemeinschaft Kartoffelforschung e.V., Detmold. 19:*25-30.

Putz, B. and Weber, L. (1996). NIT für die Rohstoffbewertung. Bericht über die 18. *Kartoffeltagung Arbeitsgemeinschaft Kartoffelforschung e.V. Band 18:*28-34.

Rastovski, A. (1978). Potato storage and storage environment. *Survey Papers, 7th Triennial Conference EAPR* (pp. 97-108). Warsaw, Poland, EAPR.

———. (1981a). Cooling of potatoes and water loss. In Rastovski, A. and van Es, A. (Eds.), *Storage of Potatoes* (pp. 175-208). Wageningen, The Netherlands: PUDOC.

———. (1981b). Storage losses, In Rastovski, A. and van Es, A. (Eds.), *Storage of Potatoes* (pp. 169-172). Wageningen, The Netherlands: PUDOC.

———. (1987). Storage conditions. In Rastovski, A., van Es, A., Buitelar, N., de Haan, P.H., Hartmans, K.J., Meijers, C.P., van der Schild, J.H.W., Sijlring, P.H., Sparenberg, H., van Zwol, B.H., and van der Zaag, D.E. (Eds.), *Storage of Potatoes* (pp. 181-182). Wageningen, The Netherlands: PUDOC.

Reust, W. (1982). Contribution à l'appréciation de l'àge physiologique des tubercules de pommes de terre (*Solanum tuberosum* L.) et ètude de son importance sur le rendement. *PhD Dissertation, Zürich.*

———. (1983). Das physiologische Alter der Kartoffelknollen und seine Wirkung auf den Ertrag. *Der Kartoffelbau 34:*13-15.

Samotus, B., Leja, M., Scigalski, A., Dulinski, J., and Siwanowicz, R. (1986). Analytical remarks to explain some discrepancies in the determination from potato tubers density. *Starch/Stärke 38:*14-19.

Sands, P.J. (1989). A model of the development and bulking of potatoes *(Solanum tuberosum)* VI. Predicting the time of emergence. *Field Crops Research 20:*165-174.

Sawyer, R.L. and Malagamba, J.P. (1987). Sprout inhibition. In Talburt, W.F. and Smith, O. (Eds.), *Potato Processing* (pp. 183-202). New York, USA: Van Nostrand Reinhold.

Schippers, P.A. (1977a). The rate of respiration of potato tubers during storage. 1 Review of literature. *Potato Research 20:*173-188.

———. (1977b). The rate of respiration of potato tubers during storage. 2. Results of experiments in 1972 and 1973. *Potato Research 20:*189-206.

Schwimmer, S., Makower, R.U., and Rorem, E.S. (1961). Invertase and invertase inhibitor in potato. *Plant Physiology 36:*313-316.

Shekhar, V.C. and Iritani, W.M. (1979). Changes in malic and citric acid contents during growth and storage of *Solanum tuberosum L. American Potato Journal 56:*87-94.

Smith, O. (1987). Transport and storage of potatoes. In Talburt, W.F. and Smith, O. (Eds.), *Potato Processing* (pp. 203-286). New York, USA: Van Nostrand Reinhold.

Sonnewald, U. (2001). Control of potato tuber sprouting. *Trends in Plant Science 6:*333-335.

Sorce, C., Lorenzi, R., and Ranalli, P. (1997). The effects of (S)-(+)-carvone treatments on seed potato tuber dormancy and sprouting. *Potato Research 40:*155-161.

Sweeney, J.P., Hepner, P.A., and Libeck, S.Y. (1969). Organic acid, amino acid and ascorbic acid content of potatoes as affected by storage conditions. *American Potato Journal 46:*463-469.

Thomas, P. (1982). Wound-induced suberization and periderm development in potato tubers as affected by temperature and gamma irradiation. *Potato Research 25:*155-164.

van der Plas, L.H.W. (1987). Potato tuber storage: Biochemical and physiological changes. In Bajaj, Y.P.S. (Ed.), *Biotechnology in Agriculture and Forestry: Potato* (pp. 109-135). Berlin, Germany: Springer.

Van der Schild, J.H.W. (1990): *Kartoffellagerung*. Hamburg, Germany: Behr's.

Van Es, A. and Hartmans, K.J. (1981). Respiration. In Rastovski, A. and van Es, A. (Eds.), *Storage of Potatoes* (pp. 120-127). Wageningen, the Netherlands: PUDOC.

———. (1987a). Structure and chemical composition of the potato, In Rastovski, A. and van Es, A. (Eds.), *Storage of Potatoes* (pp. 15-78). Wageningen, The Netherlands: PUDOC.

———. (1987b). Water balance of the potato tuber. In Rastovski, A. and van Es, A. (Eds.), *Storage of Potatoes* (pp. 141-147). Wageningen, The Netherlands: PUDOC.

Veerman, A. and van Loon, C.D. (1997). Einfluss von $CO_2$ auf die Fritierfarbe während der Lagerung. *Bericht über die 19. Kartoffeltagung. Arbeitsgemeinschaft Kartoffelforschung e.V. 19:*8-9.

Vokou, D., Vareltzidou, S., and Katinakis, P. (1993). Effects of aromatic plants on potato storage: Sprout supression and antimicrobial activity. *Agriculture, Ecosystems and Environment 47:*223-235.

Wright, R.C. and Peacock, W.M. (1934). *Influence of the Storage Temperatures on the Rest Period and Dormancy of Potatoes* (pp. 1-21). Technical Bulletin. No. 424. Washington D.C., USA: US Department of Agriculture.

# Chapter 14

# Processing and Value Addition

Xiu-Qing Li
Martin G. Scanlon
Qiang Liu
Warren K. Coleman

Potato tubers are traditionally used as food after baking or boiling. The commercial value of potatoes is, however, considerably increased when they are processed into French fries, chips, other frozen or dehydrated products, and starch. Processing of potatoes appears to have originated in pre-Columbian South America. The need to preserve the potato tubers for use during unfavorable seasons led to rudimentary but effective freeze-drying of tubers as well as storage of living material (Salaman, 1985; Talburt and Smith, 1987).

The top three countries for potato production are China, Russia, and India, which produced 66.8, 36.7, and 23.2 million metric tons of potatoes in 2003, respectively (FAO, 2004). However, the top three potato-processing countries are the United States, Canada, and the Netherlands. The world's leading exporter of frozen potato products is Canada, followed by the United States and the European Union (USDA, 2004). In 2002, the share of frozen potato product exports was approximately 43 percent for Canada, 29 percent for the United States, and 15 percent for the EU (USDA, 2004). More than 50 percent of potatoes produced in Canada are processed, mostly into

French fries (AAFC, 2001). In northern Ontario (Canada), fries and chips are the number one contributor of food energy for females and number three (after white bread and condensed milk) for males (Wolever et al., 1997). Similarly, 57 percent of the potatoes grown in the United States are processed (USDA, 2001), including approximately 28 percent into fries and 11 percent into chips or shoestrings. In addition, potato has potential to produce medical vaccines and other highly valuable proteins (molecular farming) (Dai et al., 2000).

This chapter reviews the industrial technology in use for the production of potato chips, French fries, and potato starch. Current knowledge on the use of peels and waste and the production of pharmaceutical proteins using potato cells is also summarized. Additional information on specific processing operations can be found in Talburt and Smith (1987), Lisinska and Leszczynski (1989), Andersson et al. (1994), Krokida et al. (2001), Mellema (2003), Atkins (2003), and Somsen et al. (2004).

## *FOOD VALUE OF POTATOES*

Potato has food properties between vegetables and cereals in terms of energy, vitamins, minerals, amino acids, and dietary fibers. Table 14.1 summarizes the major food components of potatoes. These components vary considerably, however, depending on the variety, the growth stage, and storage conditions. For example, the activity of sucrose synthase, an important enzyme affecting sucrose and starch contents in storage organs, varies depending on the degree of tuber maturity and growth conditions (Li, 2001; Li and Zhang, 2003). Although potatoes also have some antinutritive components such as glycoalkaloids, these components can be minimized through appropriate handling or processing.

### *Dry Matter, Amino Acids, Antioxidants, and Vitamins*

The dry matter content of tubers was reported to vary from 19.9 to 25.6 percent in a field test in the United States (Love et al., 1998). In India, the dry matter content in tubers of potato germplasm lines ranged from 15 to 28 percent (Jai Gopal, personal communication).

TABLE 14.1. Chemical composition of potatoes.

| Component | Approximate content | Reference |
|---|---|---|
| Dry matter | 19.9-25.6%, f.w. | Love et al. (1998) |
| Sucrose at 4°C | 0.13-0.68%, f.w. | Love et al. (1998) |
| Glucose at 4°C | 0.05-0.58%, f.w. | Love et al. (1998) |
| Glucose at 10°C | 0.01-0.2%, f.w. | Love et al. (1998) |
| Protein | 0.6-1.2%, f.w. | Burton (1989) |
| Asparagines (free) | 110-529 mg/100 g, f.w.[a] | Talley et al. (1970) |
| Glutamine (free) | 23-409 mg/100 g, f.w.[a] | Talley et al. (1970) |
| Proline (free) | 2-209 mg/100 g, f.w.[a] | Talley et al. (1970) |
| Other amino acid (free) | 0.2-117 mg/100 g, f.w. | Talley et al. (1970) |
| Lipid | 0.075-0.126%, f.w. | Burton (1989) |
| Vitamin C | ~20 mg/100 g, f.w. | Burton (1989) |
| Vitamin E | ~0.1 mg/100 g, f.w. | Paul and Southgate (1978) |
| Folic acid | 0.01-0.02 mg/100 g, f.w. | Augustin et al. (1978) |
| Riboflavin | 0.04-0.05 mg/100 g, f.w. | Augustin et al. (1978) |
| Nitrogen (total) | 0.2-0.4%, f.w. | Talburt and Smith (1987) |
| Potassium | 564 mg/100 g, f.w.[a] | Burton (1989) |
| Phosphorus | 30-60 mg/100 g, f.w.[a] | Burton (1989) |
| Calcium | 6-18 mg/100 g, f.w.[a] | Burton (1989) |
| Dietary fiber | 1-2%, f.w. | Paul and Southgate (1978) |
| Glycoalkaloids | <20 mg/100 g, f.w. | Dale et al. (1998) |

[a] Converted from milligrams per 100 g dry weight, assuming the dry matter content is 20 percent.

Glucose content varies between 0.05 and 0.58 percent. The most abundant amino acids are asparagine and glutamine. Blue-flesh potatoes are rich in antioxidants. Potato is rich in vitamin C, but lipid content is low. The vitamin A level in fresh potatoes is very low (not included in Table 14.1), and the calcium content is also low (approximately 6-18 mg per 100 g on fresh weight basis).

### *Dietary Fiber*

Potato cell wall contains dietary fiber at a level of 1-2 percent (w/w) in boiled potatoes (Paul and Southgate, 1978). In contrast to lignified fibers, which are not effective in lowering cholesterol levels, the potato dietary fibers (mainly unlignified) can effectively reduce cholesterol levels (Lazarov and Werman, 1996). Potato dishes provide 24 percent of the total daily dietary fiber in an average Swedish diet (Wirfält et al., 2002).

### *Antinutritive or Toxic Components*

Glycoalkaloids, found in all parts of potato plants, are the best known and most studied toxins in potatoes. Glycoalkaloids in potato plants are mainly glycosides of the aglycone solanidine. The total glycoalkaloid levels in the tubers of four varieties tested ranged from 0.7 to 17.8 mg per 100 g on a fresh weight basis (Cieslik and Praznik, 1998). Glycoalkloid content is highest in the skin area and lowest in the flesh. The glycoalkaloid content increases significantly when tubers are damaged by bruising (Dale et al., 1998), and it usually increases when tubers are exposed to light. However, there is no metabolic link between the light-induced accumulation of chlorophyll and glycoalkaloids (Edwards and Cobb, 1997; Edwards et al., 1998). Glycoalkaloid content also increases during storage, but the increase is lower at 10°C than at 4°C (Cieslik and Praznik, 1998). In many countries, such as Canada and the United Kingdom, potato clones can be registered as commercial varieties only if they have glycoalkaloid levels of less than 20 mg per 100 g in fresh tubers. Recently, however, glycoalkaloids and their metabolites have also been reported to have beneficial effects for human health such as inhibiting the growth of human-colon- and liver- cancer cells (Lee et al., 2004).

Acrylamide can be formed in small but variable amounts during cooking (frying and baking) at high temperatures (above 100°C) in potato-based and most other starch-based foods, such as wheat bread. At high temperatures (e.g., 180°C), acrylamide is formed as a result of the Maillard reaction mainly between asparagine (a nutrient amino acid) and reducing sugars (mainly glucose). Other sugars, such as sucrose, fructose, and lactose, also lead to the formation of acrylamide (Stadler et al., 2002). As the Maillard reaction causes the formation of acrylamide and browning of fried or baked products, the color appearance in fried or baked potato products is now an important safety and quality issue. However, no excess risk for various cancers has been observed in consumers of 14 different food items, including potato fries and chips, where acrylamide contents ranged between 30 and 1,200 $\mu g \cdot kg^{-1}$ (Mucci et al., 2003). (For a review of acrylamide as a potential food contaminant, see Friedman, 2003).

Protease inhibitors in raw potatoes can be toxic at high concentrations. In cooked potatoes (e.g., boiled, baked, or fried), protease inhibitors are denatured by high temperature and become normal protein nutrients.

## *FACTORS IMPORTANT TO THE PRODUCTION OF POTATOES FOR PROCESSING*

### *Variety*

Varieties with increased tuber solids and decreased glucose content are suitable for processing. Tuber morphology is another important criterion in selecting a variety for processing into chips and French fries. Varieties recommended for chipping are "round," so that chips have a more equidimensional morphology. In contrast, varieties recommended for french fry production fall into the "very long" category. Tubers with shallow eyes can reduce the peeling cost (Li et al., 2005).

Since the early 1960s, directed breeding for such traits as low reducing sugars and the ability of tubers to chip directly out of a 4°C storage environment (i.e., "cold chippers") has led incrementally to significant improvements in processing quality for both French fries (e.g., Shepody, released in 1980) and potato chips (e.g., Atlantic, released in 1976; Snowden, released in 1990). (Profiles of potato varieties currently used for processing can be found on the Web; e.g., www.ume.maine.edu/PAA/Varieties; www.oregonstate.edu/ potatoes; www.europotato.org; www.dpw.wau.nl/pv/query.asp.) Progress in genetic improvement for quality traits is described in Chapter 5 "Quality Improvement."

### *Field Management*

Selection of an appropriate growing site (Agblor and Scanlon, 2002; Mazurczyk et al., 2003) as well as specific site management (Carter et al., 2003) is important for the production of tubers suitable for processing. Site-specific optimization of seed age, seed size, and planting density in conjunction with levels of fertilizers containing nitrogen, phosphorus, and potassium (Coleman and Coleman, 1986; Tai and Coleman, 1999; Coleman, 2000) can help in achieving high tuber specific gravity along with desired tuber shape and size in the produce. In addition, resistance or avoidance of abiotic (McGarry et al., 1996; Kouwenhoven et al., 2003) and biotic stresses can have a pronounced effect on final tuber quality.

### *Tuber Harvest*

Acceptable chemical maturity (based on sucrose content) of tubers at harvest (Gould, 1999) is essential for successful processing. Determination of tuber maturity using a sucrose rating test (Sowokinos and Preston, 1988) has become an established tool in the management of tubers destined for processing. In order to minimize the increase in glucose and fructose levels during storage, freshly harvested mature tubers should have low sucrose levels (below 1 mg sucrose per gram on a fresh weight basis). As bruising affects the processing quality of potatoes, tools such as the "electronic potato," an electronic device containing accelerometers, data manipulators, and wireless communicators (Van Canneyt et al., 2003) have been developed to assess damage from bruising during rough handling.

### *Storage*

The choice of storage temperature depends on the ability to suppress sprout growth and inhibit accumulation of reducing sugar in tubers. If the storage temperature is too low (less than 7-8°C), sugars will accumulate in tubers ("cold-induced sweetening"). On the other hand, higher temperature promotes sprouting. Once sprouting begins after completion of dormancy, the processing quality of tubers decreases rapidly as reducing sugars begin to accumulate ("senescence sweetening"). In addition, tubers become flaccid because of water loss and unfit for processing by rapid mechanical systems. Although sprout suppressants (both natural and synthetic) can prevent sprout development (Coleman et al., 2001), they often have no effect on reducing sugar accumulation and can lead to higher sugar levels within the tuber (see Chapter 13 "Storage").

## *PROCESSING POTATOES INTO FRENCH FRIES AND CHIPS*

Generic potato processing lines for the production of French fries and chips are shown in Figure 14.1. Four major operations are described here to illustrate how these can be operated to attain consistent products of high quality and with good recoveries.

### *Slicing or Cutting (Size Reduction)*

Slicing or cutting of tubers follows a predetermined line using a cutting tool. In chipping, the tuber is sequentially cut in one plane to produce a series of slices. In French fry manufacture, cuts are propagated within the tuber concurrently in two planes to produce a set of strips. Strips from the outer edges of the tuber that possess five or fewer surfaces are designated "slivers" and are removed. Both the operating conditions and the structural and mechanical properties of the tuber influence the efficiency and the quality of size reduction.

Ensuring optimal alignment to effect fractures along predetermined lines parallel to the long axis of the tuber is a sophisticated practice in the cutting operation for which elaborate mechanisms have been developed (Anonymous, 1998). The usual practice is to convey pregraded tubers in a stream of water at a specific velocity to meet a cutting block (a grid of intersecting blades whose separation is consistent with fry thickness). In chip manufacture, centrifugal force is typically employed to keep one side of the tuber in a zone where a countershear moving at a specific velocity pushes the tuber onto successive blades to produce the slices of potato. The edge thickness of the blade(s) is critical to successful initiation of a cut in the tubers. Thin blades are more susceptible to blunting even though the high

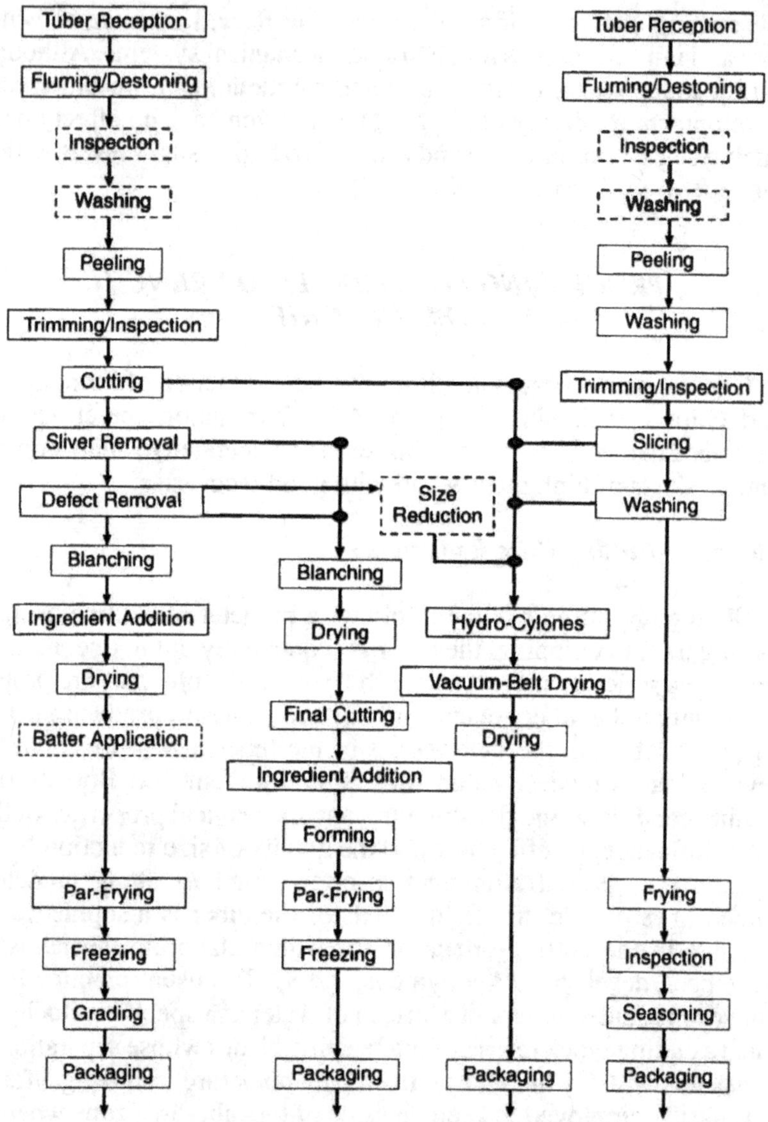


FIGURE 14.1. Generic potato processing lines for producing (from left to right) frozen par-fried French fries, frozen par-fried coproduct (hash browns or patties), potato starch, and chips. Dashed-line boxes represent optional processes.

stress at the edge of a thin blade makes cutting easier. The need to avoid blunting makes monitoring and changing of blades an ongoing challenge to processors, and a blade may last for as little as 15 min when tubers are taken straight from the field. Once fracture has been initiated, blade characteristics govern the success of cutting. Two critical parameters in this regard are the thickness of the blade away from the edge (when the blade is angled to create a thin edge) and the surface characteristics. The former determines the strain energy in the tuber in the vicinity of the blade edge, and the latter affects the coefficient of friction between the sides of the growing strip and the sides of the blades. Both parameters modify the stress distribution at the tip of the fracture. An undesirable outcome of not attaining an optimal stress distribution is the development of brittle fractures running obliquely into the body of the strip. When the strips are fried, these regions desiccate faster and separate from the main body of the strip, giving rise to an undesirable "feathered" appearance (Anonymous, 1998).

The physical properties of the tubers also affect the size reduction operation. In addition to the coefficient of friction, the yield stress, fracture toughness, and elastic modulus of the potato tissue will all affect process efficiency and product quality. These physical properties are in turn affected by the physiological condition of the tuber. Freshly harvested tubers typically possess a strong turgor pressure, and this affects the strain energy within the tuber as fractures are propagated. Low temperature also stiffens the tuber, thereby exacerbating the difficulty of avoiding brittle fractures. In some regions, when tubers are harvested under cool conditions and fed directly into the plant, high turgor and low temperature alter properties so much that tubers "shatter" when they meet the cutting grid. This is due to catastrophic brittle fractures running along the whole thickness of the strips. To overcome this problem, some processors preheat the tubers, typically at 50-60°C for periods commensurate with tuber size. This not only "softens" the tubers owing to the temperature rise but also eliminates the effect of turgor pressure (Laza et al., 2001).

### *Defect Removal*

Numerous defects can occur in potato strips or slices. All arise from the tuber—from an interaction between physiology and either agronomic practices (hollow heart, greening), mechanical handling (bruises), or pests (e.g., wireworm, common scab and rot). Defective portions of tubers should be removed as soon as possible to prevent unnecessarily transporting material that does not contribute to product yield and to prevent cross-contamination that may give rise to bad flavor or the appearance of defects in the product. Although a manual inspection is performed, usually after peeling, internal defects, such as internal blackspot and hollow heart, will not be apparent until after size reduction. The vast number of strips to inspect at this point precludes manual inspection as an option and thus requires sophisticated automatic inspection and defect removal operations to improve recovery from strips containing these defects.

In an automated defect removal system for fry manufacture, the flow of strips is adjusted so that individual strips are conveyed in a number of lanes to the inspection station. The strips are inspected by cameras, which ideally inspect 360° around the longitudinal axis of the strips as they are propelled in free flight toward a receiving conveyor. Those strips in which the camera/computer system detects a defect are diverted from their original trajectory by a jet of compressed air onto a separate conveyor. The defective strips are scanned by a second set of cameras controlling a trimming device that precisely excises the defects from the strip; the defective ones are routed to the starch recovery line. Portions of the strips that are defect-free are recovered, either joining the fry line again (if of sufficient length) or being diverted to the coproduct line.

### *Blanching*

For food products, quality as assessed by the consumer is usually identified with good appearance, texture, and flavor. As remarked by Burton (1989), attaining good appearance and texture simultaneously can pose a considerable challenge to the processor. Although they are absent from conventional chip production lines, blanching and drying operations considerably enhance the scope for independent adjustment of product appearance and texture beyond that achievable by frying alone. The blanching operation is important because it eliminates undesirable enzymatic activities in the frozen product and largely controls fry appearance and fry texture (Agblor and Scanlon, 2000). A primary role of the blancher is to remove sugars, the limiting component in the Maillard reaction (Garrote et al., 1984), and consequently to improve the appearance and reduce the acrylamide content in the fried products.

To reduce the potential for development of reducing sugars in the tuber, good tuber storage practice is essential. However, it may not be possible always, so tubers from storage may have sugar concentrations that are higher than acceptable. The onus is thus on the processor to reduce the sugar content of the strips. By placing the strips in the blancher and maintaining a concentration difference in sugars between the extracting medium and the strip, sugar concentration is reduced. Although the cell membranes normally prevent the free diffusion of sugars from potato parenchyma cells, the membranes are quickly denatured in the blancher and so free diffusion is thought to exist (Garrote et al., 1984; Kaymak and Kincal, 1994). Figure 14.2 illustrates how time-temperature conditions in the blancher affect the sugar concentration of strips (shoestring fries) and their potential product appearance after final frying.

Free diffusion of sugars from the strips is an ideal situation that is perhaps rarely attained. Because the cellular structure of hot potato parenchyma is complex at a number of different length scales, impediments to free diffusion are expected (Andersson et al., 1994). Examples are the degree of breakdown and hydration of the pectins, the starch content of the cells, the size of cells, and the presence of complexing substances within the strips that may bind with the sugars and hinder their free passage. Each of these factors results from an interaction between the structure and composition of the tubers, but the magnitude of its effects depends on the processing conditions employed. For example, pectin degradation and a concomitant opening up of the strip structure are markedly affected by the time-temperature treatment received in the blancher, the blancher's pH, and the presence or absence of calcium ions (Andersson et al., 1994). Therefore, maintaining adequate control of strip texture while achieving the optimal extraction of sugars for good product appearance (as indicated in Figure 14.2) is still a major challenge for potato processors.

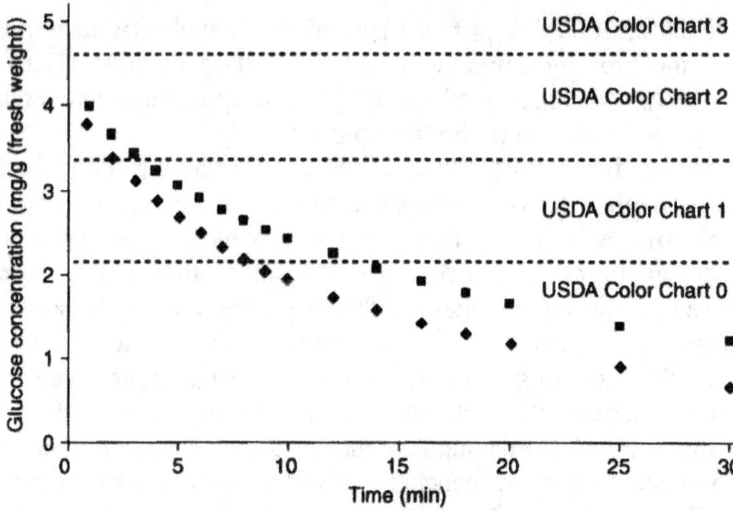


FIGURE 14.2. Change in glucose concentration of shoestring fry strips (thickness of 0.284 in.) from blanching in fresh water at 68°C (squares) and 98°C (diamonds), assuming a glucose in potato diffusion coefficient of $9 \times 10^{-9}$ $m^2 \cdot s^{-1}$ at 98°C and $5.6 \times 10^{-9}$ $m^2 \cdot s^{-1}$ at 68°C (the latter based on changes in the viscosity of water and absolute temperature relative to those at 90°C). The fry colors of the strips at different blanching times are based on the results of Pritchard (1993).

## *Frying*

The frying process is a complex heat and mass transfer operation (Singh, 1995). In chip manufacture it is carried to completion. In the manufacture of frozen fries and other coproducts, the frying operation at the plant is only partial. This par-frying operation has two important functions: it partially cooks the products so that food preparation time is reduced and it imparts an oil-rich layer on the exterior of the products so that they do not agglomerate during the freezing operation. Failure to maintain individual units of the product would have disastrous quality implications at fast-food restaurants.

Because frying removes moisture, the solids content of the incoming tubers is important in the manufacture of all types of potato products. Moisture content needs to be less than 2 percent to ensure stability in chips and less than 70 percent in par-fried fries to meet customer specifications. If these targets are not met, chips can lose their crispness, and the fully finished fries can exhibit limpness or separation of the interior and the crust. However, even if the product leaving the manufacturing plant has the correct mean moisture content, there is no guarantee that the product is meeting all quality specifications. For instance, some customers even stipulate the maximum number of limp fries that can be present in a sample of the fries. The processor must then ensure that not only are frying conditions configured to attain the correct mean moisture but tubers entering the process do not display an inordinate spread in solids content that would confound reduction in the number of defective units. The inherent radial distribution in solids content across the tuber also contributes to heterogeneity in moisture content between strips and hence to texture variability (Agblor and Scanlon, 2000).

Increasing solids content in the tubers can reduce fat content in the product because there is essentially an inverse relation between moisture loss during frying and fat uptake (Mellema, 2003). Although fat is an essential component in the flavor of chips, fries, and other potato products, it is under increasing scrutiny by health-conscious consumers. As a result, processing strategies are focused on limiting fat uptake during frying (Gould, 2001; Mellema, 2003). One of the easiest routes (especially for fry processors) is to lose more moisture during drying and less during frying. Chip processors have had some success in reducing fat content by using air impingement strategies to remove a portion of the water prior to frying. A coating to restrict fat uptake is also a feasible strategy (Mellema, 2003). A third processing strategy, mentioned by Gould (2001), is to fry at high temperatures, so that target moisture contents are attained in a shorter time, and therefore there is less time for oil uptake. However, such a strategy cannot be divorced from consideration of other quality factors. This is apparent when considering that at times processors may need to maximize production in order to meet customer orders. This may stretch the capability of the fryer to deliver sufficient thermal power to evaporate water and perform adequate precooking. In such situations the fryer may experience large temperature deviations from its set point. One parameter that can be affected in such a scenario is the texture of the product, with textural quality being lost in proportion to the temperature deviation associated with the fry load. Another consideration working against the use of high oil temperatures is that the types of oils that are stable at relatively high temperatures are saturated or trans-fat (saturated vegetable oil) oils, which are not favored by consumers.

## *POTATO STARCH*

Starch content in potatoes varies with variety and plant growth stage and ranges between 66 and 80 percent on a dry weight basis. Year-to-year variation in the content and functional properties of the starch of the same variety has also been reported (Weisenborn et al., 1994; Svegmark et al., 2002). Potato starch is a polysaccharide composed of a mixture of two molecular entities: amylose (a linear fraction) and amylopectin (a highly branched fraction). The amylose content in starch from various varieties ranges between 17 and 25 percent. Potato starch granules are generally oval (Figure 14.3A) and the largest (20-100 μm) of the most commonly used starches. Under a microscope with polarized light, potato starch shows a typical birefringence cross, indicating a high degree of molecular orientation within the granule (Figure 14.3B). Potato starch also contains higher phosphorus content than other plant starches (approximately 0.08% in potato starch, 0.02% in corn starch, 0.01% in waxy maize starch, and 0.01% in tapioca starch). Phosphorus content in starch is also affected by fertilization and variety (Liu, 1997). Internal lipids and proteins are virtually absent from potato starch granules. Native potato starch gives a B-type X-ray diffraction pattern with approximately 28 percent crystallinity (Zobel, 1988). Potato starch granules consist of amorphous and ordered (i.e., crystalline) areas. The ordered areas are formed from the short branch chains of amylopectin molecules arranged in a cluster, whereas the amorphous areas are composed of branching points with α-1,6 linkage of amylopectin, associated with amylose molecules. The granular structure of the starch is essentially determined by genetic factors that govern starch biosynthesis (Guilbot and Mercier, 1985).

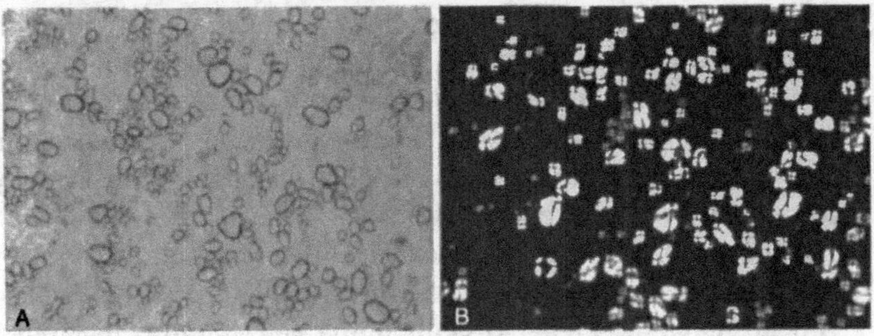


FIGURE 14.3. Potato starch granules observed under light microscope *(A)* without polarized light and *(B)* with polarized light.

Potato starch has numerous useful functional properties such as thickening, coating, gelling, adhesion, and encapsulation. Some of these functionalities are unique to the polymer as a result of the structure of amylose and amylopectin and their organization. Swelling, gelatinization, pasting, and retrogradation, which underlie starch functionality, are important phenomena in potato starch applications.

### *Isolation of Starch*

The main extraction process is wet-milling, including tuber washing, peeling, soaking, disintegrating, centrifuging, and drying. Soaking in an aqueous solution of sodium bisulfite and citric acid prevents biochemical reactions such as discoloration of the tuber. Disintegration and centrifugation are used to separate starch from other components. Potato starch can be isolated and purified by washing, sedimentation, and centrifugation. The wet starch is then either dried, chemically modified, or enzymatically hydrolyzed to manufacture new starch derivatives and to improve its functional properties.

### *Starch Gelatinization*

Potato starch granules are insoluble in cold water but swell in warm water. When starch granules are heated to a higher temperature (above 60°C) in the presence of excess water, an order-disorder phase transition occurs. Starch gelatinization is the collapse (disruption) of molecular orderliness within the starch granule along with irreversible changes in properties such as granular swelling, loss of birefringence, crystallite melting, viscosity development, and solubilization. The gelatinization characteristics of potato starch are influenced by potato variety, moisture content (Liu et al., 2002), and environmental conditions such as pressure, mechanical damage, and presence of small molecular solutes as well as hydrophilic hydrocolloids.

Potato starch has higher swelling power and solubility than other starches. These effects are due mainly to a higher content of phosphate groups on the potato starch molecule. The repulsion between phosphate groups on adjacent chains increases hydration by weakening the extent of bonding within the crystalline domains (Galliard and Bowler, 1987).

A variety of analytical techniques have been employed to understand the mechanism of gelatinization. These include viscometry, optical microscopy, electron microscopy, differential scanning calorimetry, X-ray diffraction, nuclear magnetic resonance spectroscopy, Fourier transform infrared spectroscopy, and, most recently, simultaneous X-ray scattering.

### *Starch Retrogradation*

Retrogradation describes changes in physical behavior following starch gelatinization. Starch retrogradation is the process that occurs when starch molecules reassociate and form an ordered structure during storage. In an initial step, two chains may associate. Ultimately, under favorable conditions, a crystalline order appears and physical phase separation occurs (Atwell et al., 1988). The retrograded potato starch, which shows a B-type X-ray diffraction pattern, contains both crystalline and amorphous regions. Starch retrogradation is influenced by the potato variety, the fine structure of amylopectin, amylose content, moisture content, and other foreign components (Liu, 1997). Retrogradation is important in industrial uses of potato starch as it can be a desired end point in certain applications, but it also causes instability of starch pastes.

### *Starch Rheology*

Starch rheology is the study of stress-deformation relationships of starch. The rheological properties of potato starch are important to both food and industrial processing. During processing, starch dispersions are subjected to combined high heating and shear rates that affect the rheological properties as well as the final characteristics of the product. Starch gelatinization, especially granular swelling, changes the rheological properties of starch. The subsequent retrogradation will further modify the rheological properties of starch. Depending on the starch concentration, the final structure of starchy products will give a thickened solution or a gelled structure. When starch is cooked, the flow behavior of the granule slurry changes markedly as the suspension becomes a dispersion of swollen granules, partially disintegrated granules, dissolved amylose, and a number of intermediate molecules. The cooked product is called a starch paste. In general, a starch paste can be described as a two-phase system composed of a dispersed phase of swollen granules and a continuous phase of leached amylose. If the amylose phase is continuous, aggregation with linear segments of amylopectin on cooling will result in the formation of a strong gel.

Rheological properties of potato starch have also been widely investigated using a Brabender Visco Amylograph (BVA) and Rapid Visco Analyzer (RVA). The RVA provides information on starch characteristics in a similar way to the BVA but with additional versatility of testing parameters. It has the advantages of using a small sample size, short testing time, and the ability to modify testing conditions. Potato starch shows the highest peak viscosity and the lowest pasting temperature with moderate final viscosity and lower setback compared with other commercial starches. This indicates that potato starch gelatinizes rather more easily than do other source starches, producing more viscous pastes that easily break. However, variation in the rheological properties of potato starch pastes and gels has been found among different potato varieties (Hopkins and Gormley, 2000).

### *Starch Modification*

The properties of native potato starch may not be desirable for all applications. Many approaches are available to modify the functional properties of starch to meet the requirements of a specific application. These include physical, chemical, and biotechnological modifications. Modified potato starches are used in a wide variety of food and non-food applications. Many of the modifications result in starch depolymerization, substitution, and cross-linking.

Physical methods of modifying the starch involve treating native starch granules under different temperature-moisture combinations, pressure, shear, and irradiation. The most common chemical modification processes are acid treatment, cross-linking, oxidation, substitution including esterification, etherification, and enzymatic hydrolysis.

#### *Pregelatinization*

Pregelatinized potato starch can be obtained through starch gelatinization followed by drum or spray drying. By this process, the granular integrity of starch is lost and paste viscosity is reduced.

#### *Annealing and Heat-Moisture Treatment*

Annealing is a process whereby starch granules are incubated in either excess or intermediate content (40% water or above, w/w) for a certain period of time, at a temperature above the glass transition temperature but below the gelatinization temperature. Annealing produces an ordering of polymer chains within the granule as well as in the crystallites rather than the development of new crystallites. Heat-moisture treatment is the incubation of starch granules at low moisture levels (35% water or below, w/w) for a certain period of time, at a temperature above the glass transition temperature but below the gelatinization temperature. Heat treatment of moist potato granular starches alters their crystalline structure, gelatinization temperature, and pasting characteristics.

### *Irradiation*

Irradiation is used to modify the size distribution of starch granules. Gamma irradiation, for example, is used to hydrolyze chemical bonds, thereby cleaving large molecules of starch into small fragments of dextrin that may be either electrically charged or uncharged free radicals (Ciesla et al., 1991).

### *Extrusion*

Extrusion combines several operations, including mixing, kneading, shearing, heating, cooling, shaping, and forming. The thermodynamic effects during extrusion break hydrogen bonds in starch, gelatinizing or even dextrinizing them in the process. The required energy input is often a function of starch granule size, the extent and type of crystallinity, and the purity of the starch. Starch can be hydrolyzed to varying degrees during the extrusion process. By controlling extrusion conditions to achieve the desired balance of gelatinization and molecular degradation, starch and modified starches with a wide range of cold-water solubility values can be produced.

### *Acid Treatment*

Acid modification of starch is achieved by suspending it in an aqueous solution of hydrochloric acid or sulfuric acid at a temperature between the ambient temperature and just below the pasting temperature (under 50°C) to prevent gelatinization. Acid hydrolysis of potato starch proceeds randomly, cleaving both α-1,4 and α-1,6 linkages, and releases an increasing amount of low molecular weight material. It occurs primarily in the amorphous or non-crystalline regions of starch. When the desired reduction in viscosity is obtained, the acid is neutralized and the starch recovered by filtration. The gel strength of cooled paste increases and gel clarity improves in acid-modified potato starch.

### *Oxidation*

Starch oxidation takes place in the presence of oxidants such as hydrogen peroxide, chlorine as sodium hypochlorite, and potassium permanganate. The process results in the oxidation of hydroxyls. Upon oxidation, carboxyl (COOH) or carbonyl (C=O) groups are introduced into the starch molecule, depending on the type of oxidant used. The bulkiness of carboxyls and carbonyls reduces the tendency of starch to retrograde. In addition, the carbonyl group interacts with the surface of food, promoting adhesion. Both amylose and amylopectin are involved, but the introduction of carboxyl or carbonyl groups into amylose chains is the main factor in reducing the tendency of starch to retrogradation and gelling.

### *Cross-Linking*

When chemicals have more than two functional groups to attack starch molecules (e.g., sodium trimetaphosphate and epichlorohydrin), reaction at two different hydroxyls is possible, resulting in cross-linking between hydroxyls on the same molecule or on different molecules. Cross-linking treatment is intended to add chemical bonds at random locations within granules and to provide more structural integrity by bridging one starch molecule to another within the starch granule. These chemical bonds stabilize and strengthen the granules. Cross-linked potato starch pastes are more viscous and heavy and are less likely to break down upon extended cooking, increased acid treatment, or severe agitation.

### *Substitution*

Substitution is the introduction of chemicals into potato starch by reacting with some of the starch hydroxyl groups. Chemical substitution enables the modification of starch functional properties. For example, to increase the water resistance of starch, the hydrophobicity can be increased by acetylation of the hydroxyl groups. Substituted starch can be obtained by esterification using acetic anhydride, vinyl acetate, orthophoric acid, sodium or potassium orthophosphate or sodium tripolyphosphate, sodium trimetaphosphate (cross-linking), phosphorus oxychloride, or urea. Substituted starch can also be obtained by etherification using ethylene oxide, propylene oxide, monochloroacetic acid, or quarternary amines (cationization). When chemical groups such as acetate or hydroxypropyl are introduced into the molecule, the tendency for intermolecular association is reduced or restricted. Increasing the bulkiness of amylose molecules or introducing repelling electrical charges results in an increased gelatinization rate, high paste viscosity, inhibited gelation, and freeze-thaw stabilization in substituted starch.

*Biochemical Modification*

Enzymatic hydrolysis is also widely used to modify potato starch. It is much more specific than acid hydrolysis. Enzymes used in potato starch modification include α-amylase, β-amylase, glucoamylase (amyloglucosidase), pullulanase, and isomerase. A much wider range of products has become possible following the introduction of starch-degrading enzymes at different alcohols (Kim and Robyt, 1999). Potato starch hydrolysates include maltodextrins, malto-oligosaccharides, syrup, maltose, and glucose.

***Potato Starch for Food and Nonfood Applications***

Through effective chemical and physical modifications, potato starch has been applied in a wide variety of industrial products, including food ingredients, sizing agents for paper and textiles, and starch-based plastics. The major consumers of potato starch are the food industries, where the texture, viscosity, and color that starch granules contribute to products are of primary interest. Modified potato starch is used to replace the fat in flour tortillas while maintaining a soft texture and foldability. It is also used in the fat-free processed meat industry. Its clean flavor and high water-binding capabilities make potato starch ideal for baking low-fat and nonfat hotdogs. Extruded potato starch can also be used to make food products for children, nonallergenic formulas, and various functional foods. Convenience foods such as instant puddings can be manufactured using pregelatinized potato starch. Acid-modified potato starch is used in confectionery products such as starch jelly candies. Thus, modified potato starch can be labeled as "modified food starch" and provides an economical alternative to maize and wheat starches.

Modified potato starch, as a raw material, is the second largest component in papermaking by volume (Maher and Cremer, 1986). In the paper industry, native and modified starches are used to increase dry paper strength and improve surface properties. The process of starch modification by enzyme conversion is well established in the paper industry. Enzyme-converted starch pastes can be used in most paper surface sizing and coating applications. Oxidized starch has also been widely used for paper sizing and coating because of its excellent film-forming and binding properties (Bramel, 1986). Cationic starch is an important industrial starch with superior adhesive properties that has been widely used as a wet end additive in the manufacture of paper to improve sheet strength and enhance the retention of pigments, dye, and filler (Solarek, 1986). In the textile industry potato starch is used in adhesive and building product applications to increase weaving efficiencies and printing ability. For uses in the pulp and paper industry modified starches with low degrees of substitution are generally preferred. However, for uses such as hydrophobic coatings and adhesives and as blend compatibilizers, highly substituted hydrophobic products may be of value. Modified potato starch can be used in deep-well drilling for oil and gas to control fluid loss. It can also be used in the purification process for drinking water as a flocculation agent.

In addition to these applications, potato starch has been applied in the development of biodegradable plastic. Biodegradable plastic is a new generation of material that has significantly reduced environmental impact. This material performs as a traditional plastic and is completely degradable in a composting cycle through the action of living organisms. To obtain starch-based plastic materials, granular potato starches are processed with the aid of additives and plasticizers using conventional techniques. Glycerol and water are good plasticizers for starch (Fishman et al., 2000). Biodegradable packaging has the potential to replace synthetic, nonbiodegradable packaging for use by the food industry because of its high oxygen permeability. However, the effect of moisture on the stability of such packaging limits its usefulness in other industries. To overcome this problem, starch and polyethylene, polystyrene or polyvinyl alcohol (PVOH), and polyacetic acid were blended to yield thermoplastic materials with properties superior to those of starch alone. Some starch-PVOH blends appear to have potential to replace LDPE (low-density polyethylene) films in applications where mechanical properties are critical for the intended use and good moisture barrier properties are not necessary. Other starch-PVOH blends are being explored as replacements for polyethylene in disposable food-service items (Krochta and De Mulder-Johnston, 1997). This approach avoids the safety concerns related to longer-term food packaging application.

## *POTATO PEELS, CULLS, AND WASTEWATER*

The loss (peels, culls) during potato tuber processing varies depending on peeling methods (steam, mechanical, etc.) and tuber quality. In the case of mechanical peeling, the peeling loss may be as high as 22 percent (Weaver et al., 1979). Still higher loss up to 27 percent was reported for potato varieties suffering sunburn greening (Morris and Lee, 1984). In order to maximize the use of both starchy and nonstarchy components and to reduce environmental impact of potato processing, research efforts are under way to convert potato peels, culls, and peeling-generated wastewater to other coproducts.

### *Fermentation to Produce Enzymes, Proteins, Wine, and Chemicals*

Potato peels and wastewater from steam peeling have been found to be excellent media for fermentation that can produce useful enzymes, proteins, and chemicals. Fermentation by *Bacillus subtilis* using potato peel substrates, prepared by blending the peels and removal of large particles by filtration, produced useful enzymes such as α-amylase, neutral and alkaline proteases, and polygalacturonate-lyase (Mahmood et al., 1998). The production of these enzymes in potato-based media was found to be similar to that in glucose-based culture. Potato peels were successfully used in producing glucoamylase enzyme in fermentation with *Aspergillus niger* (Fadel, 2001). The enzyme, whose function is to hydrolyse starch to glucose, was active in a wide range of pH values (3.0-7.5) and temperatures (50-75°C).

Yeast culture using the effluent from the steam peeling of potatoes produced yeast solids, which can be used as a protein supply in feed (Karim and Sistrunk, 1985). Fermentation with potato waste can also effectively produce several useful chemicals such as ethanol, acetone, and butanol (Grobben et al., 1993; Kobayashi et al., 1998) or lactic acid (Khalaf, 2001). The highest production of lactic acid was about 75 g per liter of culture.

Potatoes can also be used, at least theoretically, for producing wine (Rink, 1997). Potato wine can be used in the same way as vodka but has considerably less alcohol content.

### *Potato Peels as Dietary Fiber*

Potato peels from steam or abrasion peeling can potentially be used as dietary fiber as the glycoalkaloid content in most varieties is within the limits prescribed for food safety (Sotelo and Serrano, 2000). The effluent from the abrasion peeling method has more starch and less dietary fiber than does the effluent from the steam peeling procedure (Camire et al., 1997). When used in biscuits as a source of dietary fiber, potato peels were better than wheat bran in terms of diversity of particle size (Abdel, 1991). After 4 weeks, rats fed with potato peels showed a 40 percent decrease in plasma cholesterol and 30 percent reduction in hepatic fat cholesterol compared with rats fed with cellulose. The data indicate that potato peels might be considered as a source of dietary fibers for human consumption to lower cholesterol levels (Lazarov and Werman, 1996).

## *MOLECULAR FARMING FOR VALUE-ADDED PROTEINS IN POTATO*

Molecular farming is the use of plant cells to express recombinant genes and produce value-added products, usually proteins. When the products are for medical use in human and animals, molecular farming is often called molecular pharming. Potato was the first plant used in molecular farming and is one of the most commonly used (Ma et al., 1997; Arakawa et al., 1997).

. The main advantages of using the potato for molecular farming are its easy genetic transformation, high productivity, suitability for storage, vegetative propagation, and lower degree of outcrossing. Further, it is relatively easy to process and purify the transgenic protein from potato tubers because they have thin cell walls and lack chlorophyll.

### *Vaccines and Other Medical Proteins Produced in Potatoes*

Potato has been used successfully for developing vaccines or medicines against cholera, other bacteria, cancers, hepatitis B, rabbit hemorrhagic virus, enteric virus, Norwalk virus, foot and mouth disease virus, and diabetes (Table 14.2). These human or animal proteins produced in potatoes have been found to be functional, though some potential modifications such as glycosylations need to be checked case by case. The use of potato can dramatically reduce the production cost of medical proteins compared with producing them in animal cells. To express the proteins effectively in potato cells, each transgene should be under the control of a strong promoter such as the 35S promoter of cauliflower mosaic virus or the Rubisco (ribulose-1,5-bisphosphate carboxylase) gene promoter. Rubisco promoters express proteins mainly in the green parts of plants. Recently, another strong promoter, tCUP promoter (tobacco-constitutive promoter), has been confirmed to express in all types of living cells in the dicotyledonary plants (Malik et al., 2002; Wu et al., 2003). This promoter may be a suitable candidate to express molecular farming proteins in both the leaves and tubers of potato plants.

The content of value-added proteins expressed in potato cell cytoplasm is usually less than 0.1 percent. The content of a secreted protein of the human serum albumin can be significantly improved to 0.2 percent if the protein transgene is in fusion with a protease inhibitor signal sequence (Farran et al., 2002). When an endoglucanase E1 protein was expressed in the cells and located in chloroplasts, the E1 protein accumulated up to 2.6 percent of total leaf soluble proteins (Dai et al., 2000). The transgenic E1 content in leaves was higher in potato than in tobacco bearing the same transgene constructs (Dai et al., 2000).

TABLE 14.2. Molecular "pharming" using potato cells as factories.

| Function of transgenic protein | Against disease | Transgene in potato | Reference |
|---|---|---|---|
| Antibacterial vaccine | Cholera | Cholera toxin subunits | Arakawa et al. (1997); Yu and Langridge (2001) |
| Antibacterial vaccine | *Escherichia coli* | Enterotoxin B subunit | Mason et al. (1998) |
| Antibacterial medicine | Multiple bacteria | Human antimicrobial lactoferrin | Chong and Langridge (2000) |
| Anticancer drug | Cancer | Human interleukin-2 | Park and Cheong (2002) |
| Anticancer drug | Cancer | Human tumor necrosis factor-alpha | Ohya et al. (2002) |
| Anticancer drug | Cancer | Human interleukin-2 | Park and Cheong (2001) |
| Anticancer drug; antiviral drug | Cancer | Human alpha-interferon | Ohya et al. (2001) |
| Antiviral vaccine | Hepatitis B | Hepatitis B surface antigen | Richter et al. (2000); Kong et al. (2001) |
| Antiviral vaccine | Rabbit calicivirus | Rabbit haemorrhagic virus VP60 protein | Castanon et al. (1999) |
| Antiviral vaccine | Enteric virus | Murine rotavirus capsid protein (VP60) | Yu and Langridge (2003) |
| Antiviral vaccine | Norwalk virus | Norwalk virus capsid protein | Tacket et al. (2000) |
| Antiviral vaccine | Foot and mouth disease virus | VP1 protein of foot and mouth disease virus | Carrillo et al. (2001) |
| Anticancer drug | Cancers | Single-chain Fv (scFv) and derivatives | Savelyeva et al. (2001) |
| Serum albumin | Blood-related | Human serum albumin | Farran et al. (2002) |
| Antidiabetes vaccine | Diabetes | Glutamic acid decarboxylase (GAD) | Ma et al. (1997) |

In conclusion, processing has significantly enriched potato products and increased their value. Frozen products and chips are currently the major trading products from potato processing. The need for continuous improvement of the quality of potato products and the need to address consumer perceptions about their healthiness will inevitably affect the methods of potato processing. The environmental impact of potato processing can be minimized by converting potato peels, culls, and wastewater to useful coproducts. Additional challenges in processing will center on reducing plant inputs and acquiring and using real-time information at critical points in the process flow. Potato starch is another value-added component that can be processed from potato tubers. The diverse properties and options for inexpensive modifications of potato starch allow it to be used in many industrial applications. The use of potato cells for producing vaccines presents many challenging opportunities. No doubt potato is a wholesome food; further research can make it still healthier. There is a definite need to integrate genetic improvement and processing technologies to achieve this goal.

## REFERENCES


AAFC (Agriculture & Agri-Food Canada) (2001). 2000/2001 *Canadian Potato Situation and Trends.* Ottawa, Ontario K1A 0C5, Market Industry & Services Branch, http://www.agr.ca.

Abdel, M.M.M. (1991). Effect of dietary fibre of potato peel on the rheological and organoleptic characteristics of biscuits. *Egyptian Journal of Food Science 19:*293-300.

Agblor, A. and Scanlon, M.G. (2000). Processing conditions influencing the physical properties of French fried potatoes. *Potato Research 43:*163-178.

———. (2002). Effect of storage period, cultivar and two growing locations on the processing quality of French fried potatoes. *American Journal of Potato Research 79:*167-172.

Andersson, A., Gekas, V., Lind, I., Oliveira, F., and Öste, R. (1994). Effect of preheating on potato texture. *Critical Reviews in Food Science and Nutrition 34:*229-251.

Anonymous (1998). Profitable size reduction. *Potato Business World 6:*21-30.

Arakawa, T., Chong, D.K.X., Merritt, J.L., and Langridge, W.H.R. (1997). Expression of cholera toxin B subunit oligomers in transgenic potato plants. *Transgenic Research 6:*403-413.

Atkins, A.G. (2003). Modelling metal cutting using modern ductile fracture mechanics: Quantitative explanations for some longstanding problems. *International Journal of Mechanical Sciences 45:*373–396.

Atwell, W.A., Hood, L.F., Lineback, D.R., Varriano-marston, E., and Zobel, H.F. (1988). The terminology and methodology associated with basic starch phenomena. *Cereal Food World 33:*306-311.

Augustin, J., True, R.H., Hogan, J.M., and Deutsch, R.M. (1978). Vitamin content of freshly harvested and stored potatoes. *Journal of Food Science 43:*1566-1570.

Bramel, G.F. (1986). Modified starches for surface coating of paper. *Tappi 69:*54-56.

Burton, W.G. (1989). *The Potato,* Third edition. Harlow, U.K.: Longman Scientific & Technical.

Camire, M.E., Violette, D., Dougherty, M.P., and McLaughlin, M.A. (1997). Potato peel dietary fibre composition: Effects of peeling and extrusion cooking processes. *Journal of Agricultural and Food Chemistry 45:*1404-1408.

Carrillo, C., Wigdorovitz, A., Trono, K., Dus Santos, M.J., Castanon, S., Sadir, A.M., Ordas, R., Escribano, J.M., and Borca, M.V. (2001). Induction of a virus-specific antibody response to foot and mouth disease virus using the structural protein VP1 expressed in transgenic potato plants. *Viral Immunology 14:*49-57.

Carter, M.R., Kunelius, H.T., Sanderson, J.B., KImpinski, J., Platt, H.W., and Bolinder, M.A. (2003). Productivity parameters and soil health dynamics under long-term 2-year potato rotations in Atlantic Canada. *Soil and Tillage Research 72:*153-168.

Castanon, S., Marin, M.S., Martin, A.J.M., Boga, J.A., Casais, R., Humara, J.M., Ordas, R.J., and Parra, F. (1999). Immunization with potato plants expressing VP60 protein protects against rabbit hemorrhagic disease virus. *Journal of Virology 73:*4452-4455.

Chong, D.K.X. and Langridge, W.H.R. (2000). Expression of full-length bioactive antimicrobial human lactoferrin in potato plants. *Transgenic Research 9:*71-78.

Ciesla, K., Zoltowski, T., and Mogilevsky, L.Y. (1991). Detection of starch transformation under gamma irradiation by small-angle X-ray scattering. *Starch 43:*11.

Cieslik, E. and Praznik, W. (1998). Changes of glycoalkaloid content in potato tubers of selected varieties during vegetation and storage. *Polish Journal of Food and Nutrition Sciences 7:*417-422.

Coleman, W.K. (2000). Physiological ageing of potato tubers: A review. *Annals of Applied Biology 137:*189-199.

Coleman, W.K. and Coleman, S.E. (1986). The effects of bromoethane and ethanol on potato *(Solanum tuberosum)* tuber sprouting and subsequent yield responses. *American Potato Journal 63:*373-377.

Coleman, W.K., Lonergan, G., and Silk, P. (2001). Potato sprout growth suppression by menthone and neomenthol, volatile oil components of Minthostachys, Satureja, Bystropogon, and Mentha species. *American Journal of Potato Research 78:*345-354.

Dai, Z., Hooker, B.S., Anderson, D.B., and Thomas, S.R. (2000). Improved plant-based production of E1 endoglucanase using potato: Expression optimization and tissue targeting. *Molecular Breeding 6:*277-285.

Dale, M.F.B., Griffiths, D.W., and Bain, H. (1998). Effect of bruising on the total glycoalkaloid and chlorogenic acid content of potato *(Solanum tuberosum)* tubers of five cultivars. *Journal of the Science of Food and Agriculture 77:*499-505.

Edwards, E.J. and Cobb, A.H. (1997). Effect of temperature on glycoalkaloid and chlorophyll accumulation in potatoes (*Solanum tuberosum* L. cv. King Edward) stored at low photon flux density, including preliminary modeling using an artificial neural network. *Journal of Agricultural and Food Chemistry 45:*1032-1038.

Edwards, E.J., Saint, R.E., and Cobb, A.H. (1998). Is there a link between greening and light-enhanced glycoalkaloid accumulation in potato (*Solanum tuberosum* L.) tubers? *Journal of the Science of Food and Agriculture 76:*327-333.

Fadel, M. (2001). Utilization of agro-industrial wastes in the production of fungal glucoamylase under solid state fermentation system. *Egyptian Journal of Microbiology 35:*155-172.

FAO (2004). *FAOSTAT Agriculture Data.* http://apps.fao.org.

Farran, I., Sanchez, S.J.J., Medina, J.F., Prieto, J., and Mingo, C.A.M. (2002). Targeted expression of human serum albumin to potato tubers. *Transgenic Research 11:*337-346.

Fishman, M.L., Coffin, D.R., Konstance, R.P., and Onwulata, C.I. (2000). Extrusion of pectin/starch blends plasticized with glycerol. *Carbohydrate Polymers 41:*317-325.

Friedman, M. (2003). Chemistry, biochemistry, and safety of acrylamide. A review. *Journal of Agricultural and Food Chemistry 51:*4504-4526.

Galliard, T. and Bowler, P. (1987). Morphology and composition of starch. In Galliard, T. (Ed.), *Starch: Properties and Potential* (p. 55). New York: John Wiley & Sons.

Garrote, R.L., Bertone, R.A., and Silva, E.R. (1984). Effect of soaking-blanching conditions on glucose losses in potato slices. *Canadian Institute of Food Science and Technology Journal 17:*111-113.

Gould, W.A. (1999). *Potato Production, Processing and Technology.* Timonium, Maryland: CTI Publications, Inc.

———. (2001). Potatoes and potato chips. In Lusas, E.W. and Rooney, L.W. (Eds.), *Snack Foods Processing* (pp. 227-246). Lancaster, U.K.: Technomic Publishing Co., Inc.

Grobben, N.G., Eggink, G., Cuperus, F.P., and Huizing, H.J. (1993). Production of acetone, butanol and ethanol (ABE) from potato wastes: Fermentation with integrated membrane extraction. *Applied Microbiology and Biotechnology 39:*494-498.

Guilbot, A. and Mercier, C. (1985). Starch. In Aspinall, G.O. (Ed.), *Polysaccharide* (p. 209). New York: Academic Press, Inc.

Hopkins, S. and Gormley, R. (2000). Rheological properties of pastes and gels made from starch separated from different potato cultivars. *Lebensmittel-Wissenschaft und-Technologie 33:*388-396.

Karim, M.I.A. and Sistrunk, W.A. (1985). The use of selected strains of yeasts in the treatment of processing waste water from lye-peeled and steam-peeled potatoes. *Journal of Food Processing and Preservation 8:*175-190.

Kaymak, F. and Kincal, N.S. (1994). Apparent diffusivities of reducing sugars in potato strips blanched in water. *International Journal of Food Science and Technology 29:*63-70.

Khalaf, S.A. (2001). Lactic acid production by interspecific hybrids of *Rhizopus* strains from potato processing peel waste. *Egyptian Journal of Microbiology 36:*89-102.

Kim, Y.K. and Robyt, J.F. (1999). Enzyme modification of starch granules: In situ reaction of glucoamylase to give complete retention of D-glucose inside the granule. *Carbohydrate Research 318:*129-134.

Kobayashi, F., Sawada, T., Nakamura, Y., Ohnaga, M., Godliving, M., and Ushiyama, T. (1998). Saccharification and alcohol fermentation in starch solution of steam-exploded potato. *Applied Biochemistry and Biotechnology 69:* 177-189.

Kong, Q., Richter, L., Yang, Y.F., Arntzen, C.J., Mason, H.S., and Thanavala, Y. (2001). Oral immunization with hepatitis B surface antigen expressed in transgenic plants. *Proceedings of the National Academy of Sciences, USA 98:*11539-11544.

Kouwenhoven, J.K., Perdok, U.D., Jonkheer, E.C., Sikkema, P.K., and Wieringa, A. (2003). Soil ridge geometry for green control in French fry potato production on loamy clay soils in The Netherlands. *Soil and Tillage Research, 74:* 125-141.

Krochta, J.M. and De Mulder-Johnston, C.L.C. (1997). Agricultural materials as renewable resources. *Food Technology 51:*61-74.

Krokida, M.K., Oreopoulou, V., Maroulis, Z.B., and Marinos-Kouris, D. (2001). Deep fat frying of potato strips—quality issues. *Drying Technology 19:*79-935.

Laza, M., Scanlon, M.G., and Mazza, G. (2001). The effect of tuber pre-heating temperature and storage time on the mechanical properties of potatoes. *Food Research International 34:*659-667.

Lazarov, K. and Werman, M.J. (1996). Hypocholesterolaemic effect of potato peel as a dietary fibre source. *Medical Science Research 24:*581-582.

Lee, K.R., Kozukue, N., Han, J.S., Park, J.H., Chang, E.Y., Baek, E.J., Chang, J.S., and Friedman, M. (2004). Glycoalkaloids and metabolites inhibit the growth of human colon (HT29) and liver (HepG2) cancer cells. *Journal of Agricultural and Food Chemistry 52:*2832-2839.

Li, X.Q. (2001). Potato EST sequencing and analysis: Which genes involved in carbohydrate metabolism are most active in immature potatoes? *American Journal of Potato Research 78:*467 (abstract).

Li, X.Q., De Jong, H., De Jong, D.M., and De Jong, W.S. (2005). Inheritance and genetic mapping of tuber eye depth in cultivated diploid potatoes. *Theoretical and Applied Genetics 110:* 1068-1073.

Li, X.Q. and Zhang, D.P. (2003). Gene expression activity and pathway selection for sucrose metabolism in developing storage root of sweet potato. *Plant and Cell Physiology 44:*630-636.

Lisinska, G. and Leszczynski, W. (1989). *Potato Science and Technology.* Barking, England: Elsevier.

Liu, Q. (1997). *Characterization of Physico-chemical Properties of Starch from Various Potatoes and Other Sources.* Ph.D. thesis. University Laval.

Liu, Q., Charlet, G., Yelle, S., and Arul, J. (2002). A study of phase transition in the starch-water system I. Starch gelatinization at high moisture level. *Food International 35:*397-407.

Love, S.L., Pavek, J.J., Thompson-Johns, A., and Bohl, W. (1998). Breeding progress for potato chip quality in North American cultivars. *American Journal of Potato Research 75:*27-36.

Ma, S.W., Zhao, D.L., Yin, Z.Q., Mukherjee, R., Singh, B., Qin, H.Y., Stiller, C.R., and Jevnikar, A.M. (1997). Transgenic plants expressing autoantigens fed to mice to induce oral immune tolerance. *Nature Medicine 3:*793-796.

Maher, S.L. and Cremer, C.W. (1986). Use of Modified Starches: Paper Industry. In Wurzburg, O.B. (Ed.), *Modified Starch: Properties and Uses* (pp. 213-227). Boca Raton, FL: CRC Press.

Mahmood, A.U., Greenman, J., and Scragg, A.H. (1998). Orange and potato peel extracts: Analysis and use as *Bacillus* substrates for the production of extracellular enzymes in continuous culture. *Enzyme and Microbial Technology 22:*130-137.

Malik, K., Wu, K., Li, X.Q., Martin-Heller, T., Hu, M., Foster, E., Tian, L., Wang, C., Ward, K. et al. (2002). A constitutive gene expression system derived from the tCUP cryptic promoter elements. *Theoretical and Applied Genetics 105:*505-514.

Mason, H.S., Haq, T.A., Clements, J.D., and Arntzen, C.J. (1998). Edible vaccine protects mice against *Escherichia coli* heat-labile enterotoxin (LT): Potatoes expressing a synthetic LT-B gene. *Vaccine 16:*1336-1343.

Mazurczyk, W., Lutomirska, B., and Wierzbicka, A. (2003). Relation between air temperature and length of vegetation period of potato crops. *Agricultural and Forest Meteorology 118:*169-172.

McGarry, A., Hole, C.C., Drew, R.L.K., and Parsons, N. (1996). Internal damage in potato tubers: A critical review. *Postharvest Biology and Technology 8:*239-258.

Mellema, M. (2003). Mechanism and reduction of fat uptake in deep-fat fried foods. *Trends in Food Science and Technology 14:*364-373.

Morris, S.C. and Lee, T.H. (1984). The toxicity and tetragenicity of Solanaceae glycoalkaloids, particularly those of the potato *(Solanum tuberosum). Food Technology Australia 36:*118-124.

Mucci, L.A., Dickman, P.W., Steineck, G., Adami, H.O., and Augustsson, K. (2003). Dietary acrylamide and cancer of the large bowel, kidney, and bladder: Absence of an association in a population-based study in Sweden. *British Journal of Cancer. 88:*84-89.

Ohya, K., Itchoda, N., Ohashi, K., Onuma, M., Sugimoto, C., and Matsumura, T. (2002). Expression of biologically active human tumor necrosis factor-alpha in transgenic potato plant. *Journal of Interferon and Cytokine Research 22:*371-378.

Ohya, K., Matsumura, T., Ohashi, K., Onuma, M., and Sugimoto, C. (2001). Expression of two subtypes of human IFN-alpha in transgenic potato plants. *Journal of Interferon and Cytokine Research 21:*595-602.

Park, Y. and Cheong, H. (2001). The effect of the recombinant human interleukin-2 gene in potato (*Solanum tuberosum* cv. Superior). *Journal of Plant Biology 44:*193-198.

———. (2002). Expression and production of recombinant human interleukin-2 in potato plants. *Protein Expression and Purification 25:*160-165.

Paul, A. and Southgate, D.A.T. (1978). *The Composition of Foods.* Fourth edition London, UK: HMSO.

Pritchard, M.K. (1993). Relationships of sugars to colour of processed potatoes. *Horticulture Vegetable News,* Manitoba Agriculture, Winnipeg, MB.

Richter, L.J., Thanavala, Y., Arntzen, C.J., and Mason, H.S. (2000). Production of hepatitis B surface antigen in transgenic plants for oral immunization. *Nature Biotechnology 18:*1167-1171.

Rink, J. (1997). Potato wine. *American Wine Society Journal 29:*23-25.

Salaman, R.N. (1985). *The History and Social Influence of the Potato.* Cambridge and New York: Cambridge University Press.

Savelyeva, N., Munday, R., Spellerberg, M.B., Lomonossoff, G.P., and Stevenson, F.K. (2001). Plant viral genes in DNA idiotypic vaccines activate linked CD4+ T-cell mediated immunity against B-cell malignancies. *Nature Biotechnology 19:*760-764.

Singh, R.P. (1995). Heat and mass transfer in foods during deep-fat frying. *Food Technology 49:*134-137.

Solarek, D.B. (1986). Cationic starches. In Wurzburg, O.B. (Ed.), *Modified Starches: Properties and Uses* (p. 113). Boca Raton, FL: CRC Press, Inc.

Somsen, D., Capelle, A., and Tramper, J. (2004). Manufacturing of par-fried French fries. Part 3: A blueprint to predict the maximum production yield. *Journal of Food Engineering 61:*209-219.

Sotelo, A. and Serrano, B. (2000). High-performance liquid chromatographic determination of the glycoalkaloids alpha-solanine and alpha-chaconine in 12 commercial varieties of Mexican potato. *Journal of Agricultural and Food Chemistry 48:*2472-2475.

Sowokinos, J.R. and Preston, D.A. (1988). Maintenance of potato processing quality by chemical maturity monitoring (CMM). *Minnesota Agricultural Experiment Station Bulletin 586:*1-11.

Stadler, R.H., Blank, I., Varga, N., Robert, F., Hau, J., Guy, P.A., Robert, M.C., and Riediker, S. (2002). Acrylamide from Maillard reaction products. *Nature 419:*449-450.

Svegmark, K., Helmersson, K., Nilsson, G., Nilsson, P.O., Andersson, R., and Svensson, E. (2002). Comparison of potato amylopectin starches and potato starch—influence of year and variety. *Carbohydrate Polymers 47:*331-340.

Tacket, C.O., Mason, H.S., Losonsky, G., Estes, M.K., Levine, M.M., and Arntzen, C.J. (2000). Human immune responses to a novel Norwalk virus vaccine delivered in transgenic potatoes. *Journal of Infectious Diseases 182:*302-305.

Tai, G.C.C. and Coleman, W.K. (1999). Genotype x environment interaction of potato chip colour. *Canadian Journal of Plant Science 79:*433-438.

Talburt, W.F. and Smith, O. (1987). *Potato Processing,* Fourth edition. New York: Avi, Van Nostrand Reinhold.

Talley, E.A., Fitzpatrick, T.J., and Porter, W.L. (1970). Chemical composition of potatoes. VIII. Effect of variety, location and year of growth on the content of nitrogen compounds. *American Potato Journal 47:*231-244.

USDA. (2001). *Frozen Potato Fries Situation and Outlook.* January 2001, http://www.fas.usda.gov/htp/horticulture/potato.html

———. (2004). *Frozen Potato Products Situation and Outlook for Selected Countries.* www.fas.usda.gov/htp/horticulture/Proc.Veg/04-01-04%20Spud%20 Situation.pdf.

Van Canneyt, T., Tijskens, E., Ramon, H., Verschoore, R., and Sonck, B. (2003). Characterisation of a potato-shaped instrumented device. *Biosystems Engineering 86:*275-285.

Weaver, M.L., Ng, K.C., and Huxsoll, C.C. (1979). Sampling potato tubers to determine peel loss. *American Potato Journal 56:*217-224.

Weisenborn, D.P., Orr, P.H., Casper, H.H., and Tacke, B.K. (1994). Potato starch paste behaviour as related to some physical/chemical properties. *Journal of Food Science 59:*644-648.

Wirfält, E., Mattisson, I., Johansson, U., Gullberg, B., Wallström, P., and Berglund, G. (2002). A methodological report from the Malmö Diet and Cancer study: Development and evaluation of altered routines in dietary data processing. *Nutrition Journal 1:*3.

Wolever, T.M.S., Hamad, S., Gittelsohn, J., Hanley, A., Logan, A., Harris, S., and Zinman, B. (1997). Nutrient intake and food use in an Ojibwa-Cree community in Northern Ontario assessed by 24H dietary recall. *Nutrition Research 17:*603-618.

Wu, K., Hu, M., Martin, T., Wang, C.M., Li, X.Q., Tian, L., Brown, D., and Miki, B. (2003). The cryptic enhancer elements of the *tCUP* promoter. *Plant Molecular Biology 51:*351-362.

Yu, J. and Langridge, W.H.R. (2001). A plant-based multicomponent vaccine protects mice from enteric diseases. *Nature Biotechnology 19:*548-552.

———. (2003). Expression of rotavirus capsid protein VP6 in transgenic potato and its oral immunogenicity in mice. *Transgenic Research 12:*163-169.

Zobel, H.F. (1988). Molecules to granules: A comprehensive starch review. *Starch 40:*44-50.

# Chapter 15

# True Potato Seed

Jai Gopal
R. Ortiz

Traditionally, potato crop is grown from tubers, often called seed tubers. This vegetative method of potato production has many advantages, including ease of planting, vigorous plant growth, and high yields of uniform produce. However, low multiplication rate, high storage and transportation costs, carryover of pathogens, and physiological degeneration are among the constraints associated with the use of seed tubers. Inevitably, crop performance declines as viruses accumulate, so that an elaborate and complex system of disease-free seed production is necessary. The costs of healthy planting material are therefore higher for a potato crop than for any other staple crop: an average of 20-30 percent of total production costs (Almekinders et al., 1996). In warm and developing regions of the world, the constraints of seed tuber–based potato production systems are more serious as quality seed tubers are either not easily available or priced too highly, and stocks degenerate quickly because of systemic as well as tuber-borne diseases (Malagamba and Monares, 1988). Healthy seed tubers under these conditions account for 50-70 percent of total production costs (Horton and Sawyer, 1985).

Sexual or botanical seed of potato (Figure 15.1A), commonly called true potato seed (TPS), is a radical alternative to seed tubers for

FIGURE 15.1. Raising potato crop from true seeds. See Plate 15.1 in the color gallery.

raising a commercial crop, and it appears promising in the warm tropics of the developing world, especially where lack of planting materials precludes the expansion of potato production (Ortiz, 1997). Compared with seed tubers, TPS offers several advantages, such as low cost of planting material, reduced transmission of pathogens and pests, and convenience as well as inexpensiveness of storage and transport. With rare but important exceptions (e.g., potato spindle tuber viroid), TPS is free of virtually all diseases, including systemically transmitted viruses (see Chapter 11 "Viruses and Viroids"). So, TPS provides cheap, healthy planting materials, thereby saving on the production costs (10-60 percent) of imported tubers (CIP, 1985; Ortiz, 1997). Moreover, only 200-500 g of TPS is required for transplanting 1 ha of 60,000 seedlings (Bofu et al., 1987), which reduces transportation by the 2 tons of tubers required to plant the same land area. However, potato production from TPS needs a longer growing season for tuber maturation (approximately 30-40 extra days) than a crop raised from tuber propagules and is more susceptible to weeds, water stress, and biotic stresses during the early growth stages. Nevertheless, by using TPS, a considerable amount of valuable potato stock that is used as seed can be saved for consumption as food.

The use of TPS for potato seed tuber production, sometimes with the purpose of disease elimination, stock rejuvenation, and creation of new cultivars, probably originated in the Andes before the time of the Incas (Umaerus, 1987; Malagamba and Monares, 1988; Quiros et al., 1992). The possibility of raising commercial crops from TPS was first explored in India in the late 1940s. The initial attempts were not, however, successful because efforts were mainly made to use self-seeds of the cultivar Phulwa, which flowers naturally under the short days when potato crop is grown during winter on Indian plains. The produce of this population was highly heterogeneous for most of the economic characteristics and had low yields as a result of inbreeding depression (Ramanujam, 1950, 1954). Renewed interest and formal research efforts to exploit TPS as a low-cost alternative for potato production in developing countries came from the International Potato Center (CIP) in 1977 (Mendoza, 1984). In the 1990s more than 40 national potato programs in the developing world assessed potato production from TPS and at least 8 countries produced potato commercially using first-generation tubers (or tuberlets) from TPS (Pallais, 1991). There are few potato-producing countries in the developing world where TPS has not been tried (Almekinders et al., 1996). This chapter reviews the developments in TPS technology in terms of its problems and prospects.

## *USE OF TPS*

An important characteristic of the use of TPS in a potato production system is the flexibility of the technology. TPS can be sown directly in the field or in nursery beds, and seedlings from TPS can either be left in place (Figure 15.1B) or transplanted (Figure 15.1C) (Wiersema, 1986a; Malagamba, 1988). Seedling tubers (Figure 15.1D; i.e., tubers harvested from TPS seedlings) can either be sold as ware potatoes or used as seed tubers. Since plants grown from TPS generally produce more and smaller tubers per stem than plants from seed tubers, seedling tuber production is especially suitable to form the basis of a cheap, localized seed tuber system.

### *Direct Seeding*

TPS seedlings are small and rather delicate and thus make severe demands on horticultural skills. Raising the crop from direct seeding is prone to risks of failure owing to poor germination and high mortality of seedlings. The problems of seedling emergence with TPS are more pronounced under warm conditions (Malagamba, 1988). Direct seeding is thus rarely recommended for raising a commercial potato crop. However, under experimental conditions direct sowing of 200 g TPS and an emergence of 75 percent in well-controlled field conditions (250,000-300,000 seeds per hectare) was reported to produce a yield of 28-30 tons of seedling tubers with an average weight of 25-30 g (Renia, 1995).

### *Seedling Transplants*

Raising seedlings in nursery beds and transplanting them to the field is an alternative way of using TPS for potato production. Nursery beds allow proper water and soil management and ensure rapid germination and seedling growth. Sowing in nursery beds also permits off-season production as nursery beds can be protected against extreme temperatures and watered optimally. This also shortens the growing period of the crop in the field. Transplanting requires good field conditions and assured irrigation, and it carries considerable labor costs. Direct seeding and transplanting for ware tuber production seem to have potential only in areas where consumers accept small tubers.

### *Seedling Tubers*

Seedlings can be allowed to grow to maturity in the nursery beds to produce seedling tubers. Seedling tuber production in nursery beds or otherwise well-controlled environments is the most feasible way of using TPS at present. Maximum yields of seedling tubers in nursery beds are obtained at plant densities of 80-100 seedlings per square meter, resulting in seedling tubers of 1 to more than 40 g with an average size of 10-15 g (Wiersema, 1984, 1986b). Lower plant densities reduce yields but allow for higher average tuber weight

(Wiersema, 1986b). The possibility of manipulating seedling tuber size is important as the weight of seed tubers influences subsequent crop performance (Wiersema and Cabello, 1986). Smaller seedling tubers show slower ground cover by the foliage and later crop maturity. Only 750 kg of seed tubers of an average tuber size of 15 g suffice to plant 1 ha of potatoes, compared with 2-3 tons of 40-60 g seed tubers (Wiersema, 1985). This means that 6-7.5 g of TPS (0.60-0.75 mg per seed, 80 percent emergence) and 80 m$^2$ of nursery bed can produce seedling tubers for a 1 ha potato crop.

The use of seedling tubers or later-generation tubers from TPS populations is culturally similar (in seed rate, initial crop development, and number of tubers per stem, among other traits) to the use of tubers from conventional cultivars. In addition, the yield potential of seedling tubers and later generations of selected TPS populations compares well with that of the clonal cultivars (Wiersema, 1984; Love et al., 1994; Benz et al., 1995). The variation in performance of the tubers derived from selected TPS populations is not significantly larger than that of tubers from a clonal cultivar (Love et al., 1994).

## *PRODUCTION OF TPS*

Production of a large quantity of quality TPS at a low cost is of prime importance for the adoption of TPS technology on a commercial scale. Thus it is important that the parental genotypes flower over a sufficient length of time and that the flowers do not drop but develop into berries with enough seeds. Genotype, day length, and temperature are the main factors that determine flowering and fruiting in potato.

A number of other factors, such as inflorescence position (Almekinders and Wiersema, 1991), plant/stem density (Almekinders, 1991), competition between flower and tuber (Thijn, 1954), precipitation (Jaühnl, 1954), nutrient level (Bamberg and Hanneman, 1988; Otazu and Amoros, 1991), and date of planting (Jaühnl, 1954), also are known to influence production of flowers and fruits in potato. Although flower primordia of potato can arise in total darkness (Jones and Borthwick, 1938), a photoperiod of 14-18 hours and night temperatures of 15-20°C favor flower production and berry setting in potato (Clarke and Lombard, 1939; Bodlaender, 1963; Almekinders, 1992). In the tropics and subtropics, conditions conducive to flowering and fruiting are available only at high altitudes (more than 1500 m above sea level) where the crop is grown during summer. Under these conditions, too, not all genotypes flower (Gopal, 1994). A number of methods (Thijn, 1954; Sadik, 1983; Gopal and Rana, 1988) have been suggested to improve flowering and berry setting in potato. Pallais (1986) could produce as many as approximately 117 berries per plant in a potato crop where plants were not hilled, stolons were pruned, and vines were trained on a trellis system.

Potato flower has a multiovular gynoecium, and an optimum pollen load is therefore necessary for maximum seed set. On average a berry has 100-200 seeds, though in certain genotypes fully grown berries may have up to 400 seeds (Upadhya et al., 1984; Sharma and Choudhary, 1985). With an increase in berry weight, the number of seeds increases and average seed weight decreases. Significant variations have been reported in berry and seed characteristics for both open-pollinated (OP) and hybrid seed production (Upadhya et al., 1984; Sharma and Choudhary, 1985; Pallais, 1986; Malagamba, 1988; Almekinders et al., 1995). Open-pollinated seed production can be improved simply by enhancing flowering and berry setting, whereas hybrid seed production also depends on the pollinating efficiency and the physiological and genetic makeup of the female and male parents. Combining-ability differences have been found for the success of hybridization among both females and males (Gopal et al., 2004). Thus selection of a good general combiner for the production of a high number of seeds per pollinated flower is important for improving the production of hybrid seed. For large-scale production of hybrid seeds, emasculation of the flower is avoided; as well as being cumbersome and costly, this results in fewer seeds per berry than from hand-pollinated nonemasculated flowers (Gopal et al., 2004). Selection of male-sterile females is particularly useful for the production of hybrid seeds without emasculation. Male-sterile lines are quite common in cultivated potatoes (Gopal, 1994). It might be possible to incorporate male sterility to facilitate natural crossing in the field under insect visitation without the necessity of hand-pollination

(Ortiz and Peloquin, 1991). Use of apomixis is more appealing, but this could not be easily incorporated (Hermsen, 1980).

## *TPS QUALITY*

The quality of TPS is determined by genetic constitution of the parents, stage of seed development, nutritional status of mother plant, and biochemical components of the seed. In warm tropical locations, a high level of seed germination, fast emergence, and early vigorous growth of seedlings are required for commercial production (Malagamba, 1988). Seed vigor, generally defined as the superior performance of a genotype after planting compared with the same genotype or other genotypes under specific experimental conditions, is a good indicator of the seed quality (McDaniel, 1973). Bold seeds are known to result in vigorous and high-yielding seedlings (Dayal et al., 1984).

Longer on-plant maturation of berries is known to produce more vigorous TPS (Pallais et al., 1989). The formation and growth of the seed is also strongly affected by environmental and climatic conditions such as temperature, water, light, and the kind and quantity of available nutrients (Delouche 1980; Siddique and Goodwin, 1980; Gray and Thomas, 1982; Pet and Garretsen, 1983; Evenari, 1984). Nitrogen application during and after pollinations is known to result in dormancy enhancement, delayed plant senescence, and increased TPS vigor, size, and weight (Pallais, 1987). In potato, moderately cool weather, especially at night, favors the setting of seed, and a gradually falling temperature with a moderate amount of moisture is potentially favorable. Dry locations free of major diseases are generally suited for production of high-quality seed (Delouche, 1980). Light intensity and photoperiod during seed development have significant effects on the subsequent seed performance of many crop species (Gutterman, 1982).

Upadhya et al. (1981) described an additional quality component of TPS: the embryo shape of the seed (Figure 15.2). Seeds with a circinnate and inverted "U"–shaped embryos are heavier than seeds with other types of embryos (rod-shaped, aborted, etc.) and give a

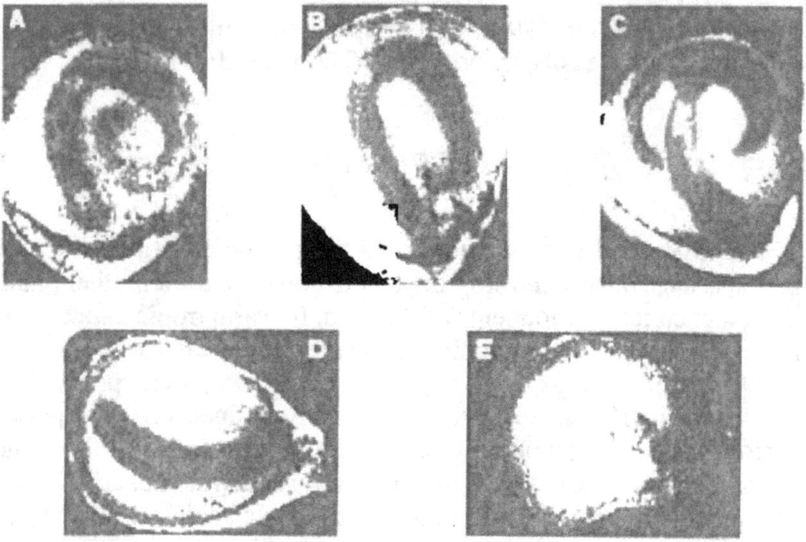

FIGURE 15.2. Embryo types in true potato seed. *(A)* Circinnate. *(B)* Inverted "U"–shaped. *(C)* Curved cotyledons in opposite direction. *(D)* Rod-shaped. *(E)* Aborted. *Source:* after Upadhya et al. (1981).

higher percentage of germination and more vigorous seedlings with higher yield. It was found that more than 90 percent of seeds of both open and hybrid families have circinnate and inverted "U"–shaped embryos (Table 15.1), and these two classes of embryos were equivalent for various seedling characteristics (Gopal, 2003). If the poor-vigor seedlings are eliminated before transplanting, the seeds with undesired embryos are eliminated, improving the productivity of the seedlings.

Seed deterioration begins in the field, probably soon after seed development has ended on the mother plant (Roberts, 1979). Stored TPS is assessed for viability on the basis of germination under controlled conditions. The decrease of seed germinability during storage is, however, preceded by a decrease in seed vigor, expressed by a slower rate of germination and seedling growth (Harrington, 1971; Van Staden et al., 1982; White, 1983a). Stored TPS is, however, remarkable in its ability to germinate even after 25 years at mild

TABLE 15.1. Percentage of different types of embryos in cross- and open-pollinated families.

| Family type | Family | Embryo type (%) | | | | |
|---|---|---|---|---|---|---|
| | | A | B | C | D | E |
| Cross-pollinated | CP1207 × EX/A680-16 | 84 | 14 | 0 | 0 | 2 |
| | CP1351 × EX/A680-16 | 68 | 29 | 0 | 0 | 3 |
| | CP1525 × EX/A680-16 | 85 | 13 | 0 | 1 | 1 |
| | CP1548 × EX/A680-16 | 88 | 7 | 0 | 3 | 2 |
| | CP1985 × EX/A680-16 | 77 | 17 | 0 | 2 | 4 |
| | CP2378 × EX/A680-16 | 51 | 44 | 0 | 0 | 5 |
| | CP2384 × EX/A680-16 | 75 | 19 | 0 | 3 | 3 |
| | CP2386 × EX/A680-16 | 52 | 28 | 7 | 5 | 8 |
| | CP3204 × EX/A680-16 | 64 | 29 | 0 | 2 | 5 |
| | CP3275 × EX/A680-16 | 88 | 4 | 1 | 1 | 6 |
| | CP3276 × EX/A680-16 | 51 | 44 | 0 | 3 | 2 |
| | CP3281 × EX/A680-16 | 92 | 4 | 0 | 1 | 3 |
| | CP3358 × EX/A680-16 | 78 | 17 | 0 | 3 | 2 |
| | CP3379 × EX/A680-16 | 79 | 8 | 0 | 11 | 2 |
| | CP3281 × Kufri Jyoti | 73 | 24 | 0 | 0 | 3 |
| | CP3362 × Kufri Jyoti | 82 | 15 | 0 | 1 | 2 |
| | CP3164 × CP3290 | 84 | 14 | 0 | 1 | 1 |
| | CP3281 × CP3290 | 93 | 3 | 0 | 1 | 3 |
| | CP1525 × CP2334 | 84 | 10 | 0 | 2 | 4 |
| | CP2378 × CP2334 | 64 | 29 | 1 | 3 | 3 |
| | CP3281 × CP2334 | 86 | 5 | 0 | 3 | 6 |
| | CP3358 × CP2334 | 69 | 23 | 0 | 4 | 4 |
| | CP1162 × Kathadin | 76 | 18 | 0 | 0 | 6 |
| | CP3281 × Kathadin | 89 | 8 | 0 | 0 | 3 |
| | CP2378 × Kufri Jawahar | 80 | 16 | 0 | 3 | 1 |
| | CP3164 × Kufri Jawahar | 70 | 11 | 0 | 16 | 3 |
| | CP3281 × Kufri Jawahar | 66 | 30 | 0 | 0 | 4 |

TABLE 15.1 *(continued)*

| Family type | Family | Embryo type (%) | | | | |
|---|---|---|---|---|---|---|
| | | A | B | C | D | E |
| | CP1525 × Kufri Pukhraj | 40 | 55 | 0 | 2 | 3 |
| | CP2378 × Kufri Pukhraj | 80 | 19 | 0 | 0 | 1 |
| | CP3275 × Kufri Pukhraj | 72 | 25 | 0 | 2 | 1 |
| | CP3276 × Kufri Pukhraj | 38 | 55 | 0 | 4 | 3 |
| | Mean ± SE | 73.5 ±2.6 | 20.5 ±2.5 | 0.29 ±0.22 | 2.48 ±0.59 | 3.19 ±0.31 |
| Open-pollinated | CP1406 | 61 | 34 | 0 | 1 | 4 |
| | CP1458 | 48 | 36 | 0 | 9 | 7 |
| | CP1704 | 55 | 40 | 0 | 2 | 3 |
| | CP1824 | 73 | 21 | 0 | 4 | 2 |
| | CP2132 | 75 | 22 | 0 | 0 | 3 |
| | CP2184 | 72 | 24 | 0 | 1 | 3 |
| | CP2298 | 80 | 14 | 0 | 4 | 2 |
| | CP3192 | 74 | 18 | 0 | 2 | 6 |
| | D-150 | 78 | 14 | 0 | 1 | 7 |
| | EB/C543 | 45 | 37 | 0 | 4 | 14 |
| | Kufri Pukhraj | 80 | 16 | 0 | 2 | 2 |
| | TPS-67 | 72 | 24 | 0 | 0 | 4 |
| | Mean ± SE | 67.3 ±3.6 | 25.0 ±2.7 | 0±0 | 2.50 ±0.72 | 4.75 ±0.99 |

*Source:* Gopal (2003).

*Note:* A, circinnate; B, inverted "U"; C, curved cotyledons in opposite direction; D, rod-shaped; E, aborted.

temperatures; 40 years or more of viability can be expected if the seeds are thoroughly dried (Howard, 1980). High temperature and humidity are the key factors responsible for seed deterioration in the tropics (D. J. Thompson, 1971; Gregg, 1984). It is generally accepted that in seed containing 5-14 percent moisture, successive 1 percent decreases of seed moisture double the storage life of seed (Harrington, 1971; Gregg, 1984). Rapid drying to safer moisture levels after extraction decreases seed quality reductions before and during storage. Large amounts of forced air may be required to dry TPS in high-humidity environments as temperatures in excess of 25°C have to be avoided (White, 1983a, 1983b). TPS dried immediately after extraction to moisture levels of 2-5 percent should be sealed and stored at 2-5°C (Hegarty, 1975) to maintain its quality over longer periods. In a laboratory, TPS can also be dried over silica gel. In the tropics, parched rice serves well to dry seed sufficiently for short-term storage (Sadik and White, 1982). TPS treated for 24 hours with 1,000-1,500 ppm of gibberellic acid can be stored for up to 2 years without any deleterious consequences (D'Antonio and McHale, 1988).

## *TPS DORMANCY*

A seed is considered dormant when it is unable to achieve germination under favorable conditions. TPS is dormant at harvest, a condition that generally disappears after 6 months, with traces remaining for 18 months (Simmonds, 1963a). Seed dormancy sometimes results in erratic seed germination even after prolonged storage (Simmonds 1963a, 1968). The problems of seedling emergence in TPS are greatly increased under warm conditions (Clark and Stevenson, 1943; Ghate et al., 1983; Malagamba, 1988). The optimum temperature for seed germination ranges from 15 to 20°C (Gallagher and Navbi, 1985; Lam, 1968), and emergence from soil can be severely inhibited above 25°C. Pallais et al. (1991) reported that old (22 month) seed consistently emerged sooner at high temperature and produced seedlings with higher dry weight than new (10 month) seed. Seed presowing techniques such as osmotic priming have demonstrated the potential to increase seedling emergence under unfavorable temperatures in many species (Bradford, 1986). Priming of TPS in a solution of $KNO_3$ plus $K_3PO_4$ at –1.0 MPa after 6 months of seed storage has been shown to be effective in enhancing seedling emergence (Pallais, 1989; Pallais et al., 1990). Traditionally, before sowing, TPS is soaked for 24-48 hours in a 1,000-2,000 ppm solution of gibberellic acid (Spicer, 1961); nondormant seed is often treated as well. Pallais et al. (1991) showed that seed priming is more effective than the standard gibberellic acid treatment. Pallais (1989) also found gibberellic acid to have a detrimental effect on subsequent seedling growth, whereas seed priming improved seedling vigor (Pallais et al., 1991).

The physiology of TPS dormancy is not well understood, but the inhibitory factors are believed to be under phytochrome control (Taylorson, 1982). They are considered to be in both the testa and the embryo (Simmonds, 1963a, 1963b). A general observation is that seed produced in long days is heavier and more dormant than seed produced in short days (Taylorson, 1982). Gutterman (1982) attributed the phenomenon of longer dormancy in seeds produced under long days to increased dry matter accumulation causing a thickening of the testa and imposing a physical restraint on germination. He also suggested that long days induce changes in endogenous hormone levels such as an increase in concentration of abscisic acid (germination inhibitor). Leaves of potato plants grown under short days were shown to contain significantly lower levels of endogenous gibberellins (germination promoters) than those grown under long days (Railton and Wareing, 1973).

Dry storage intensifies the dormancy of TPS but does not reimpose it on nondormant seed: cool storage (0-5°C) keeps dormancy constant, whereas air storage causes a steady decline (Simmonds, 1963a, 1963b). TPS progenitors often vary at the species level in terms of dormancy (Mendoza, 1984), and Simmonds (1964) found wide differences in the intensity of dormancy between different *Solanum* spp. Therefore, it is possible that similar physiological control of dormancy is not present in all TPS progenies.

Potato tubers that are dormant at harvest are an objective of most breeding programs. Tuber dormancy has been associated with TPS dormancy (Simmonds, 1964). Therefore, to select TPS progenies with little or no seed dormancy that produce tubers with moderate levels of dormancy appears difficult. However, some variability may occur in correlation between the tuber and TPS dormancy (Simmonds, 1964).

## BREEDING TPS POPULATIONS

The dominant difference between potato crops grown from TPS and those from normal sized tubers is the slow initial development of plants grown from TPS. As a result, a TPS crop develops ground cover slowly. The establishment of the crop in the field is also longer and more vulnerable. Although progenies vary, typically, a stem from TPS produces a larger number of tubers with a lower average tuber weight than a stem from a normal sized tuber (Wiersema, 1984; Benz et al., 1995). This characteristic is further accentuated when a larger number of stems per unit area are used in a crop grown from TPS to compensate for the slower ground cover. The identification of TPS progenies with quick seed germination and fast seedling growth speeds up seedling establishment and the development of ground cover, thereby mitigating the most important drawback in the use of TPS.

### *Genetics and Breeding of TPS*

TPS is not unknown to potato breeders, who use true seed for the genetic enhancement of the potato genome through sexual hybridization. Nonetheless, the TPS breeding strategy must consider a sexually propagated crop that is grown for the harvest of its vegetative part: the tuber. The cultivated potato is a complex tetraploid that exhibits tetrasomic inheritance (Howard, 1970). Most potato cultivars, especially in high-input agricultural systems, are homogeneous highly heterozygous genotypes showing great uniformity owing to vegetative propagation. As a result, the segregating population ensuing from crosses between potato cultivars is a heterogeneous collection of genotypes for various characteristics, which poses a problem for potato production from TPS. The remedy to this drawback is to breed for parents from which relatively homogeneous TPS offspring may ensue. Thus selection of TPS offspring with an acceptable level of plant-to-plant variability, particularly for tuber characteristics, is the major objective in developing productive TPS cultivars. Likewise, because top-yielding TPS offspring for each location should be locally adapted, cultivars or advanced selections from the target area must be the parental sources of TPS offspring (Ortiz and Golmirzaie, 2004a).

Although inbreeding gives the highest level of gametic uniformity and homogeneity, resulting in completely uniform progenies, it also causes significant yield reductions and poor performance stability (Howard, 1970; Atlin, 1985b; Mendoza and Marca, 1986). The theory of inbreeding and crossbreeding is more complex in tetraploids than in diploids. The primary effect of inbreeding is to reduce the frequency of heterozygous genotypes. According to Jackson et al. (1984), the most common explanation of inbreeding depression is the fixation of unfavorable or deleterious recessives with an increase in homozygosity. The deleterious effect of inbreeding is also due to the fact that most of characteristics in potato, including tuber yield, are determined by genes with nonadditive effects (Plaisted et al., 1962; P. G. Thompson et al., 1983; Gopal, 1998), and increased homozygosity results in the loss of such interactions, including epistasis.

Golmirzaie, Ortiz et al. (1998) used ten Andigena clones as parental material to derive hybrid ($S_0$), inbred ($S_1$ and $S_2$), and OP ($OP_1$ and $OP_2$) generations. Significant differences among generations were found for pollen production, pollen viability (as determined by its stainability with acetocarmine glycerol), number of flowers and berries per plant, number of seeds per berry, weight of 1,000 seeds, and tuber yield per plant. The parental populations were significantly different for most of the traits, but not for flower production and berry weight. The authors indicated that all traits evaluated except seed weight showed a strong inbreeding depression, and the OP progenies had intermediate values between the $S_0$ and the $S_1$, which demonstrates that open pollination in potatoes is not exclusively the product of selfing; it also results from outcrossing. Similarly, correlation coefficients were calculated between the level of inbreeding and performance of different traits in selfed and sib-mated generations of five heterogeneous TPS hybrid parental sources (Golmirzaie, Bretschneider, et al., 1998). Inbreeding depression was expressed mainly by pollen viability ($r = -0.912$, $P < 0.01$) and tuber yield ($r = -0.837$, $0.01 < P < 0.05$). Furthermore, the TPS parental sources responded quite differently to inbreeding depression. They did not show the same amount of depression for yield as they did for the characteristics connected with fertility. The high tuber-yielding parental sources displayed a greater inbreeding depression for tuber yield than the lower-yielding ones.

### *Hybrid versus Open-Pollinated TPS*

Heterosis, or hybrid vigor, is the converse of inbreeding depression. As a result of this phenomenon, potato hybrid families are reported to be more productive than OP ones (Mendoza, 1984; Kidane-Mariam et al., 1985a).

For the production of hybrids, Tuberosum × Andigena crosses are particularly useful as heterosis for yield is spectacular in such families (Glendinning, 1969; Cubillos and Plaisted, 1976; Ortiz, 1997; Simmonds, 1997; Gopal et al., 2000). For temperate conditions, however, *andigena* clones need to be preselected for adaptations to long days. Long-day-adapted *andigena* has been named Neotuberosum (Simmonds, 1976).

Hybrid families are produced, by controlled pollination, from parents that have good general combining ability for agronomic and reproductive characteristics. Open-pollinated families on the other hand are the simplest types of genetic material resulting from natural pollination (predominately self-pollination) (Atlin, 1985b). Nevertheless, the proportion of hybrid offspring may be increased in OP families as a variable rate of natural outcrossing has been reported in cultivated tetraploid potato (Brown, 1993). The cost of OP seed is significantly lower than that of hybrid seed as its production does not require specialist skills. OP families from some genotypes can be as productive as the best hybrid families (Golmirzaie et al., 1994, 2004). This suggests that some parental sources might be able to "resist" the inbreeding effect (Golmirzaie, Bretschneider et al., 1998). Further, OP seed in potatoes is not exclusively the product of selfing; it also results from outcrossing (Golmirzaie, Ortiz et al., 1998), thus offering the possibility of using it in a TPS breeding program with favorable results. By selecting early at the seedling stage for the most vigorous individuals, it is possible to increase the performance of OP progenies for agronomic and reproductive characteristics (Atlin, 1985a; Atlin and Wiersema, 1988; Kidane-Mariam et al., 1985b; Golmirzaie and Mendoza, 1986; Golmirzaie and Ortiz, 2002a), thus partly offsetting the deleterious effects of inbreeding.

Cytoplasmic male sterility (CMS) could be one of the main options for reducing TPS production costs, but the male sterile flowers should still attract bumblebees for natural pollination to replace hand-pollination. Tetrad pollen sterility, which provides a source of male sterility, has some potential for TPS breeding (Golmirzaie et al., 2003). Potato cybrids arising from the fusion between cytoplasm and nuclear gene donors can be a source of such a male sterility system. However, reciprocal crosses may perform significantly differently for tuber characteristics, suggesting the importance of cytoplasmic effects in potato crop. Nevertheless, experimental results suggest that cytoplasmic effects in TPS breeding populations are more important for reproductive characteristics such as pollen production than for tuber yield (Golmirzaie and Ortiz, 2003). Male sterility in potato occurs as a result of the interaction between sensitive cytoplasm and dominant nuclear genes (Iwanaga et al., 1991; Ortiz et al., 1993).

### *Synthetics*

According to Golmirzaie et al. (1994) a promising method for producing low-cost TPS is to create a synthetic population. A synthetic is a mixture of agronomically uniform TPS families, produced by bulking equal amounts of seed from various cross-combinations or OP families (Golmirzaie and Ortiz, 2002b). In both cross- and OP-families, parents with good general combining ability for agronomic and reproductive characteristics need to be identified (Golmirzaie and Ortiz, 2004; Ortiz and Golmirzaie, 2003b, 2004b). Golmirzaie (1987) showed that by selecting the right parental lines with high general combining ability, it is possible to obtain high-performance TPS in a naturally produced synthetic population with as few as only two parents. Theoretically, however, one would expect a single best OP or cross-pollinated population to do better than a mixture of two or more such populations (Golmirzaie and Ortiz, 2002b).

### *Use of Diploids*

Among potatoes of different ploidy levels, tetraploids have the highest yield potential (Maris, 1990). However, production of 2*n* gametes (or gametophytes with the sporophytic chromosome number) by diploids can be exploited in $4x \times 2x$ crosses to produce productive and uniform tetraploid TPS families (Peloquin, 1983). The occurrence of $2n$ gametes in diploids results from modified meiosis, which is genetically equivalent to first-division restitution (FDR) or second division restitution (SDR) (Iwanaga et al., 1989). FDR gametes transmit approximately 83 percent of their heterozygosity and are genetically more uniform than SDR gametes (Peloquin, 1979). The importance of $2n$ gametes is not only to achieve maximum heterozygosity in $4x \times 2x$ progenies but also to produce adequate homogeneity in the progenies by an appropriate choice of tetraploid parent (Golmirzaie and Mendoza, 1988; Carroll and De Maine, 1989). In one of the schemes proposed for production of $4x$ TPS from $4x \times 2x$ matings, Peloquin (1983) suggested the use of FDR gametes from selected $2x$ hybrids (with good agronomic and reproductive characteristics). Such $2x$ hybrids can be obtained by mating dihaploids (i.e., sporophytes with the gametic chromosome number or $2n = 2x = 24$) of tetraploid cultivars or advanced selections (selected for fertility and $2n$ gametes) with the diploid cultivated or wild species with 24 chromosomes (selected for $2n$ gametes formation and desired traits such as disease resistance). These selected $2x$ hybrids could be crossed to cultivars that possess good adaptation and are unrelated to the haploids involved in the $2x$ hybrid. The scheme envisages that $4x$ hybrids from such $4x \times 2x$ crosses would obtain their adaptation and desired horticultural traits from the cultivar and allelic variation from the haploid species hybrid. The scheme, though theoretically sound and with some promising experimental results, does not seem to be popular nowadays among some potato breeders, who occasionally report low fertility and low frequency of $2n$ gametes, late maturity, and poor agronomic traits of $2x$ hybrids or species. The scheme is also cumbersome and time consuming. Consequently, one very experienced potato scientist felt that the production of Tuberosum dihaploids, followed by breeding with them and finally by sexual or asexual retetraploidization was nothing but a waste of time and money. (Maris, 1990, page 32).

### Short-Duration TPS Populations

As mentioned above, TPS crop needs a longer growing duration and does not fit into the prevailing crop sequences, particularly in tropical and subtropical countries (Gaur et al., 2000), where this technology is considered to have greatest potential (Malagamba, 1988; Simmonds, 1997). In a recent review, Gaur et al. (2000) rightly highlighted the need to develop short-duration TPS cultivars. Two alternatives (i.e., early foliage maturity and early tuber bulking) exist for developing short-duration families. However, the choice depends upon the comparative TPS production potential of the parent(s) selected for this purpose.

Gopal et al. (2004) found that foliage maturity has a positive association with duration of flowering as well as intensity of flowering, indicating that late-maturing genotypes flowered profusely and for a longer duration. Very early clones generally did not flower at all. On the other hand, early bulking as judged from average tuber weight had no association with any of the flower production attributes. Thus, unlike early maturing genotypes, early bulkers may flower profusely, early, and for longer. Hence, out of foliage maturity and tuber bulking, the latter is a better criterion for the selection of parental clones in developing short-duration (early-bulking) TPS populations.

### Hardy TPS Populations

Seedling mortality after transplanting is the major risk to which the TPS transplant crop is exposed after seed germination. Gopal (2003) found that transplant survival was highly correlated with seedling vigor. Hardy seedlings thus can be indirectly selected before transplanting based on their vigor. Other characteristics—namely, leaf area, root area, root length, and seedling dry weight—were also associated with transplant survival to varying extents, but seedling vigor was a better indicator of seedling transplant survival than the root area or root length. Vigor as an indicator of hardiness also has the advantage that it can be recorded visually and effectively, without any damage to the seedlings, whereas the recording of leaf area, root area, root length, and dry weight is cumbersome and time consuming, besides being destructive.

TABLE 15.2. Mean performance of some wild and cultivated spp. accessions for seedling characteristics.

| Species | Accession number | Vigor[a] | Leaf area (cm² per seedling) | Root area (cm² per seedling) | Root length (cm per seedling) | Dry weight (mg per seedling) |
|---|---|---|---|---|---|---|
| *S. cardiophyllum* | PI283062 | 2.33 | 1.44 | 1.75 | 27.60 | 4.6 |
| | GLKS99 | 1.33 | 0.74 | 1.11 | 15.57 | 3.6 |
| *S. bulbocastanum* | PI243505 | 2.33 | 1.18 | 0.77 | 12.49 | 7.1 |
| | GLKS1741 | 1.67 | 2.19 | 2.67 | 32.22 | 6.7 |
| *S. chacoense* | PI133073 | 3.00 | 3.19 | 2.89 | 36.89 | 8.7 |
| | PI217451 | 2.67 | 2.19 | 2.53 | 34.20 | 11.0 |
| *S. demissum* | PI160208 | 3.33 | 3.02 | 3.03 | 41.88 | 3.3 |
| | PI160220 | 3.00 | 2.46 | 4.41 | 53.69 | 11.4 |
| | PI160222 | 1.33 | 0.86 | 1.43 | 26.88 | 5.1 |
| | PI160230 | 1.33 | 1.44 | 2.69 | 31.71 | 5.3 |
| | PI161149 | 1.33 | 1.14 | 4.24 | 18.88 | 5.7 |
| | PI161366 | 3.33 | 3.19 | 2.83 | 34.20 | 15.0 |
| | PI161169 | 1.67 | 2.35 | 2.53 | 30.40 | 4.0 |
| | PI175423 | 2.67 | 2.19 | 2.42 | 34.68 | 7.7 |
| | GLKS239 | 2.00 | 1.25 | 1.84 | 24.84 | 4.7 |
| | GLKS269 | 2.67 | 3.09 | 2.71 | 25.55 | 6.3 |
| | GLKS313 | 2.00 | 0.42 | 0.89 | 11.02 | 4.7 |
| *S. guerreorense* | PI161727 | 2.67 | 2.51 | 4.28 | 55.00 | 11.2 |
| | PI161730 | 2.33 | 2.35 | 3.03 | 41.80 | 7.0 |
| *S. ployadenium* | PI310963 | 2.67 | 2.96 | 4.44 | 36.96 | 4.6 |
| | PI320342 | 2.33 | 1.20 | 0.87 | 15.69 | 7.2 |
| *S. stoloniferum* | PI365398 | 2.00 | 0.52 | 0.89 | 15.23 | 5.8 |
| | GLKS530 | 1.00 | 0.62 | 1.50 | 20.17 | 13.2 |
| *S. tuberosum* ssp. *andigena*[b] | PI225629 | 4.00 | 3.38 | 3.77 | 45.85 | 12.6 |
| | PI243411 | 2.67 | 2.58 | 4.11 | 48.16 | 7.3 |
| | PI243437 | 4.33 | 1.70 | 5.21 | 40.85 | 11.6 |
| | PI243443 | 4.67 | 5.51 | 4.73 | 64.47 | 16.7 |
| *S. tuberosum* ssp. *tuberosum*[b] | CP1690 | 4.67 | 4.98 | 5.21 | 55.68 | 13.5 |
| | CP2030 | 4.33 | 3.33 | 4.58 | 48.84 | 11.5 |
| | CP2118 | 4.00 | 5.20 | 4.11 | 40.85 | 8.5 |
| | CP2284 | 3.33 | 2.85 | 3.77 | 45.28 | 9.5 |
| | LSD (0.05) | 0.99 | 1.79 | 1.37 | 14.36 | 4.2 |

*Source:* Gopal (2003). [a]Score: 1 = very poor, 5 = very good. [b]Cultivated species.

Seedling vigor and other characteristics governing hardiness could be improved genetically as enough variability was present in a population of wild and cultivated species (Gopal, 2003). Heritability and genetic advance values were also reasonably high. However, it was found that seedlings of wild species were less hardy than those of cultivated ones. Wild species, in general, also had poorer performance for all characteristics associated with transplant survival (i.e., seedling vigor, leaf area, root area, root length, and dry weight; Table 15.2). A majority (70 percent) of the tuber-bearing wild species are diploids (Hawkes, 1958). These are self-incompatible and set seeds only when crossed with some other strain possessing compatible S gene combinations. Triploid and pentaploid species are mostly sterile. Self-fertile potato species too prefer to reproduce vegetatively as a theoretically infinite number of exact replicas of the initially successful biotypes can be produced (Hawkes, 1958). The predominance of vegetative reproduction thus might have led to poor seedling hardiness in tuber-bearing wild species. On the other hand, high vigor in cultivated potatoes might arise from their being subject to selection since domestication, initially by farmers and later by potato breeders. Parental lines of cultivated species with high heritability for vigor have been identified at the CIP (P. G. Thompson et aL., 1983; Golmirzaie, 1985; Ortiz and Golmirzaie, 2003a, 2003b). Pallais et al. (1989, 1990) suggested ways to improve seedling vigor by manipulating physiological maturity, dormancy, storage conditions, and presowing treatments of true seeds. Nevertheless, the need to search for and exploit genetic variability for seedling vigor should not be underestimated if TPS is to be considered seriously as an alternative to seed tuber crop (Pallais et al., 1991).

Despite its suggested potential, sustained research, and promotion efforts, TPS has never reached a significant number of farmers (Almekinders et al., 1996). Even China, which once gained recognition for its pioneer work on TPS (Li and Shen, 1979; Bofu, 1984), now has little area under potato crop grown from TPS (Bofu et al., 1987; Gaur et al., 2000). This can be attributed to the improved availability of virus-free seed tubers following the establishment of the tuber seed farms throughout most of the potato areas of northern China after 1976 (Bofu et al., 1987). In addition to genetic, physiological, and agronomic problems associated with TPS crop, as discussed above, economic considerations are also responsible for nonadoption of this technology. Although several calculations (van der Zaag and Horton, 1983; Dayal et al., 1984; Upadhya et al., 1984; Singh and Janardanjee, 1990) suggest that economic returns are higher from TPS than from seed tuber crops, the fact remains that these calculations are too sketchy and far from being realistic (Simmonds, 1997). This technology must be made economically competitive compared with the conventional method of raising crop from seed tubers.

If we summarize the experience with this technology, we can conclude that direct seeding (drilling) as the method to raise a TPS crop has failed; transplants are better, but seedling tubers are better still. Technically, a crop raised from seedling tubers cannot be regarded as a TPS crop. In our opinion, the only practical way to raise a truly TPS crop is through seedling transplants. Further, for the production of seedling tubers too, raising a successful seedling crop either in nursery beds or as transplants in the field is mandatory. A successful crop of TPS seedlings can currently be raised at most in limited areas where small farmers grow potatoes with irrigation and intensive management in small plots. Those who have experience with raising vegetable seedlings and sufficient family labor could better manage potato seedling crops.

Breeding for hardiness and early bulking using conventional as well as nonconventional approaches is thus absolutely essential if this technology is to be popularized. On the other hand, a little variability within the produce of a TPS crop might not matter, particularly for consumers in developing countries. This might even be biologically advantageous for achieving good disease resistance. Identification of parents producing OP seeds under varying thermo-photoperiodic conditions and having reasonably uniform tuber progenies with high tuber yield is another aspect that needs to be further researched. This will help farmers produce their own seed at low cost and achieve good returns for their efforts and time spent on the production of seedling tubers and raising normal potato crop from the seedling tubers. All this needs sustained efforts to overcome various intrinsic problems associated with TPS technology. Long experience with this technology, however, makes us pessimistic about its future.

## REFERENCES

Almekinders, C.J.M. (1991). Flowering and true seed production in potato (*Solanum tuberosum* L.) 2. Effects of stem density and pruning of lateral stems. *Potato Research 34:*379-388.

———. (1992). The effect of photoperiod on flowering and TPS production in the warm tropics. *Potato Research 35:*433-442.

Almekinders, C.J.M., Chilver, A.S., and Renia, H.M. (1996). Current status of the TPS technology in the world. *Potato Research 39:*289-303.

Almekinders, C.J.M., Neuteboom, J.H., and Struik, P.C. (1995). Relation between berry weight, number of seeds per berry and 100-seed weight in potato inflorescences. *Scientia Horticulture 61:*177-184.

Almekinders, C.J.M. and Wiersema, S.G. (1991). Flowering and true seed production in potato (*Solanum tuberosum* L.) I. Effects of inflorescence position, nitrogen treatment and harvest date of berries. *Potato Research 34:*365-377.

Atlin, G. (1985a). Farmer maintenance of TPS varieties. In *Innovative Methods for Propagating Potatoes. Report of Planning Conference 1984* (pp. 39-62). Lima, Peru: International Potato Center.

———. (1985b). Inbreeding in TPS progenies: Implications for breeding and seed production strategy. In *Present and Future Strategies for Potato Breeding and Improvement, Report of Planning Conference,* 1983 (pp. 71-85). Lima, Peru: International Potato Center.

Atlin, G. and Wiersema, S.G. (1988). Selection against inbred seedlings in mixtures of inbred and hybrid true potato seed. *Potato Research 35:*105-112.

Bamberg, J.B. and Hanneman, R.E., Jr. (1988). Enhanced production of botanical seed of tuber bearing *Solanum* species via supplemental fertilizer application. *American Potato Journal 65:*470 (abstract).

Benz, J.S., Keller, E.R., and Midmore, D.J. (1995). Planting materials for warm tropic potato production growth and yield of transplanted seedlings or rooted cuttings and tuber materials in the field. *Field Crops Research 40:*179-192.

Bodlaender, K.B.A. (1963). Influence of temperature, radiation and photoperiod on development and yield. In Ivens, J.D. and Milthorpe, F.L. (Eds.), *The Growth of Potato* (pp. 199-210). London, U.K.: Butterworth.

Bofu, S. (1984). Use of true potato seed in China. *CIP Circular 12:*6-7.

Bofu, S., Qu Dong, Y., and Vander-Zaag, P. (1987). True potato seed in China: Past, present and future. *American Potato Journal 64:*321-327.

Bradford, K.J. (1986). Manipulation of seed-water relations via osmotic priming. *HortScience 21:*1105-1112.

Brown, C.R. (1993). Outcrossing rate in cultivated autotetraploid potato. *American Potato Journal 70:*725-734.

Carroll, C.P. and De Maine, M.J. (1989). The agronomic value of tetraploid F1 hybrids between potatoes of group Tuberosum and group Phureja/Stenotomum. *Potato Research 32:*447-456.

CIP (1985). *Annual Report* 1984, Lima, Peru: International Potato Center.

Clark, C.F. and Stevenson, A.J. (1943). Factors influencing the germination of seeds of the potato. *American Potato Journal 20:*247-258.

Clarke, A.E. and Lombard, P.M. (1939). Relation of length of day to flower and seed production in potato varieties. *American Potato Journal 16:*236-244.

Cubillos, A.G. and Plaisted, R.L. (1976). Heterosis for yield in hybrids between *S. tuberosum* ssp. *tuberosum* and *tuberosum* ssp. *andigena. American Potato Journal* 53:143-150.

D'Antonio, V.L. and McHale, N.A. (1988). Effect of storage temperature and extraction methods on dormancy and germination of true potato seed. *American Potato Journal 65:*573-581.

Dayal, T.R., Upadhya, M.D., and Chaturvedi, B.N. (1984). Correlation studies on 1000-true seed weight, tuber yield and other morphological traits in potato (*Solanum tuberosum* L.). *Potato Research 27:*185-188.

Delouche, J.C. (1980). Environmental effects on seed development and seed quality. *HortScience 15:*775-779.

Evenari, M. (1984). Seed physiology: From ovule to maturing seed. *Botanical Review 50:*119-142.

Gallagher, D.T.P. and Navbi, M.N. (1985). Extraction and germination of true potato seed. *New Zealand Journal of Experimental Agriculture 12:*151-154.

Gaur, P.C., Ilangantileke, S.G., and Thakur, K.C. (2000). True potato seed: Asian scenario. In Khurana, S.M. Paul, Shekhawat, G.S., Singh, B.P., and Pandey, S.K. (Eds.), *Potato, Global Research & Development* (Vol. I, pp. 656-667). Shimla, India: Indian Potato Association, Central Potato Research Institute.

Ghate, S.R., Jaworski, C.A., and Phatak, S.C. (1983). Emergence and plant stand of pregerminated true potato seed in warm climate. *American Potato Journal 60:*557-562.

Glendinning, D.R. (1969). The performance of progenies obtained by crossing Groups Andigena and Tuberosum of *Solanum tuberosum. European Potato Journal 12:*13-19.

Golmirzaie, A.M. (1985). Identification of parental lines for development of true potato seed (TPS) populations. *American Potato Journal 62:*427-428 (abstract).

———. (1987). Performance of true potato seed synthetic populations. *American Potato Journal 64:*400 (abstract).

Golmirzaie, A.M., Bretschneider, K., and Ortiz, R. (1998). Inbreeding and true seed in tetrasomic potato II. Selfing and sib-mating in heterogeneous hybrid populations of *Solanum tuberosum. Theoretical and Applied Genetics 97:*1129-1132.

Golmirzaie, A.M., Buendia, S., Espinoza, J., and Ortiz, R. (2004). Open pollinated offspring for producing potatoes from true seed. *Tropicultura. 22:* 191-198.

Golmirzaie, A.M., Malagamba, P., and Pallais, N. (1994). Breeding potatoes based on true seed propagation. In Bradshaw, J.E. and Mackay, G.R. (Eds.), *Potato Genetics* (pp. 467-498). Wallingford, U.K.: CAB International.

Golmirzaie, A.M. and Mendoza, H.A. (1986). Effect of early selection for seedling vigor on open-pollinated true potato seed. *American Potato Journal 63*:426 (abstract).

———. (1988). Breeding strategies for true potato seed production. *CIP Circular 16*:1-8.

Golmirzaie, A. and Ortiz, R. (2002a). Inbreeding and true seed in tetrasomic potato. III. Early selection on open-pollinated populations. *Theoretical and Applied Genetics 104*:157-160.

———. (2002b). Inbreeding and true seed in tetrasomic potato. IV. Synthetic cultivars. *Theoretical and Applied Genetics 104*:161-164.

———. (2003). Reciprocal effects in true potato seed breeding. *Plant Breeding 122*:372-374.

———. (2004). Genetic variation for reproductive characteristics in potato landraces and cultivars for producing true seed. *Genetic Resources and Crop Evolution 51*:759-763.

Golmirzaie, A.M., Ortiz, R., Atlin, G.N., and Iwanaga, M. (1998). Inbreeding and true seed in tetrasomic potato I. Selfing and open pollination in Andean landraces (*Solanum tuberosum* Gp. Andigena). *Theoretical and Applied Genetics 97*:1125-1128.

Golmirzaie, A.M., Tenorio, J., Serquen, F., and Ortiz, R. (2003). Cybrids and tetrad sterility for developing true potato seed hybrids. *Annals of Applied Biology 143*:231-234.

Gopal, J. (1994). Flowering behaviour, male sterility and berry setting in tetraploid *Solanum tuberosum* germpalsm. *Euphytica 72*:133-142.

———. (1998). General combining ability and its repeatability in early generations of potato breeding programmes. *Potato Research 41*:21-28.

———. (2003). True potato seed: Breeding for hardiness. *Journal of New Seeds 5*:57-73.

Gopal, J., Chahal, G.S., and Minocha, J.L. (2000). Progeny mean, heterosis and heterobeltiosis in *Solanum tuberosum* x *tuberosum* and *S. tuberosum* x *andigena* families under a short day sub-tropic environment. *Potato Research 43*:61-70.

Gopal, J., Kumar, V., and Thakur, S.S. (2004). Hybrid and open-pollinated true potato seed production: Some considerations particularly for developing short duration families. *Potato Journal 31*:13-20.

Gopal, J. and Rana, M.S. (1988). Induction of flowering in potato in North-western plains of India. *Journal of Indian Potato Association 15*:91-93.

Gray, D. and Thomas, T.H. (1982). Seed germination and seedling emergence as influenced by the position of development of the seed on, and chemical application to, the parent plant. In Khan, A.A. (Ed.), *The Physiology and Biochemistry of Seed Development, Dormancy and Germination* (pp. 81-110). New York: Elsevier.

Gregg, B. (1984). Conditioning and storage of seed in the tropics. *Extension Bulletin ASPAC Food Fertilizer Technology Center 208*:13.

Gutterman, Y. (1982). Phenotypic maternal effect of photoperiod on seed germination. In Khan, A.A. (Ed.), *The Physiology and Biochemistry of Seed Development, Dormancy and Germination* (pp. 67-79). New York: Elsevier.

Harrington, J.F. (1971). The necessity for high quality vegetable seed. *HortScience 6:*550-551.

Hawkes, J.G. (1958). Significance of wild species and primitive forms for potato breeding. *Euphytica 7:*257-270.

Hegarty, T.W. (1975). Effects of fluctuating temperature on germination and emergence of seeds in different moisture environments. *Journal Experimental Botany 26:*203-211.

Hermsen, J.G.T. (1980). Breeding for apomixis in potato: Pursuing a Utopian scheme? *Euphytica 29:*595-607.

Horton, D. and Sawyer, R.L. (1985). The potato as a world food crop with special reference to developing countries. In Li, P.H. (Ed.), *Potato Physiology* (pp. 1-34). Orlando, FL: Academic Press.

Howard, H.W. (1970). *Genetics of the Potato.* London: Logos Press.

———. (1980). Storage of true seeds of potatoes for 25 years. *Potato Research 23:*241-242.

Iwanaga, M., Jatala, P., Ortiz, R., and Guevara, E. (1989). Use of FDR 2n pollen to transfer resistance to root-knot nematodes into cultivated 4x potatoes. *Journal of the American Society of Horticultural Sciences 114:*1008-1013.

Iwanaga, M., Ortiz, R., Cipar, M.S., and Peloquin, S.J. (1991). A restorer gene for genetic-cytoplasmic male sterility in cultivated potatoes. *American Potato Journal 68:*19-28.

Jackson, M.T., Taylor, L., and Thompson, A.J. (1984). Inbreeding and true potato seed production. In *Innovative Methods for Propagating Potatoes, Report of Planning Conference* (pp. 169-179). Lima, Peru: International Potato Center.

Jaühnl, G. (1954). Zur kenntnis des Bluühverhaltens der Kartoffel. *Veroüff. Bundesanst. Alp. Landw. Admont 9:*40-74.

Jones, H.A. and Borthwick, H.A. (1938). Influence of photoperiod and other factors on the formation of flower primordia in the potato. *American Potato Journal 15:*331-336.

Kidane-Mariam, H.M., Arndt, G.C., Macaso-Khwaja, A.C., and Peloquin, S.J. (1985a). Comparisons between $4x \times 2x$ hybrid and open-pollinated true potato seed families. *Potato Research 28:*35-42.

———. (1985b). Hybrid vs. open-pollinated TPS families. In *Innovative Methods for Propagating Potatoes, Report of Planning Conference*, 1984, (pp. 25-37) Lima, Peru: International Potato Center.

Lam, S.L. (1968). Interaction of temperature and gibberellin on potato seed germination. *American Journal of Botany 55:*193-198.

Li, C.H. and Shen, C.P. (1979). Production of marketable and seed potatoes from botancial seed in the People's Republic of China. In: *Production of Potatoes from True Potato Seed. Report Planning Conference*, Manila, Philippines (pp. 21-28). Lima, Peru: International Potato Center.

Love, S.L., Thompson-Johns, A., and Werner, B. (1994). Evaluation of TPS lines for 1992 and 1993. *Report College of Agriculture* (pp. 6). Aberdeen, USA: University of Idaho.

Malagamba, P. (1988). Potato production from true seed in tropical climates. *Hortscience 23:*495-500.

Malagamba, P. and Monares, A. (1988). *True Potato Seed: Past and Present Uses.* Lima, Peru: International Potato Center.

Maris, B. (1990). Comparisons of diploid and tetraploid families derived from *Solanum phureja* x dihaploid *S. tuberosum* and their vegetatively doubled counterparts. *Euphytica 46:*15-33.

McDaniel, R.G. (1973). Genetic factors influencing seed vigor, biochemistry of heterosis. *Seed Science and Technology 1:*25-50.

Mendoza, H.A. (1984). Selection of uniform progenies to use in TPS commercial potato production: In *Innovative Methods for Propagating Potatoes. Report of Planning Conference* (pp. 5-16). Lima, Peru: International Potato Center.

Mendoza, H.A. and Marca, J.L. (1986). Performance of selfed, open-pollinated, and hybrid progenies in a *Solanum tuberosum* ssp. *andigena* population. *American Potato Journal 63:*444 (abstract).

Ortiz, R. (1997). Breeding for potato production from true seed. *Plant Breeding Abstracts 67:*1355-1360.

Ortiz, R. and Golmirzaie, A.M. (2003a). Genetic parameters for agronomic characteristics. I. Early and intermediate breeding populations of true potato seed. *Hereditas 139:*212-216.

———. (2003b). Genetic parameters for agronomic characteristics. II. Intermediate and advanced stages in a true potato seed breeding population. *Hereditas 139:*217-222.

———. (2004a). Genotype by environment interaction and selection in true potato seed breeding. *Experimental Agriculture 40:*99-107.

———. (2004b). Combining ability analysis and correlation between breeding values in true potato seed. *Plant Breeding 123:* 564-567.

Ortiz, R., Iwanaga, M., and Peloquin, S.J. (1993). Male sterility and 2n pollen in 4*x* progenies derived from 4*x* × 2*x* and 4*x* × 4*x* crosses in potatoes. *Potato Research 36:*227-236.

Ortiz, R. and Peloquin, S.J. (1991). A new method of producing 4x hybrid true potato seed. *Euphytica* 57:103-107.

Otazu, V. and Amoros, W. (1991). Potato berry blackening and premature drop due to boron deficiency. *American Potato Journal 68:*849-856.

Pallais, N. (1986). Physical relationships between berries and seeds of potato. *HortScience 21:*1359-1360.

———. (1987). True potato seed quality. *Theoretical and Applied Genetics 73:*784-792.

Pallais, N. (1989). Osmotic priming true potato seed: Effects of seed age. *Potato Research 32:*235-244.

———. (1991). TPS: Changing potato from vegetative to sexual. *HortScience 76:*239-241.

Pallais, N., Asmat, H., Fong, N., and Santos-Rojas, J. (1989). Factors affecting seedling vigor in potatoes: I. Stage of seed development. *American Potato Journal 66:*793-801.

Pallais, N.E., Espinola, N.Y., Falcon, R.M., and Garcia, R.S. (1991). Improving seedling vigor in sexual seeds of potato under high temperature. *HortScience 26:*296-299.

Pallais, N., Fong, N., Falcon, R., and Garcia, R. (1990). Factors affecting seedling vigor in potatoes: II. Genotype, after-ripening and pre-sowing treatments. *American Potato Journal 67:*109-119.

Peloquin, S.J. (1979). Breeding methods for achieving phenotypic uniformity. In *The Production of Potatoes from True Seed, Report of the Planning Conference* (pp. 151-155). Lima, Peru: International Potato Center.

———. (1983). New approaches to breeding for the potato for the year 2000. In Hooker, W.J. (Ed.), *Research for the Potato in the Year 2000* (pp. 32-34). Lima, Peru: International Potato Center.

Pet, G. and Garretsen, F. (1983). Genetical and environmental factors influencing seed size of tomato (*Lycopersicon esculentum* Mill.) and the effect of seed size on growth and development of tomato plants. *Euphytica 32:*711-718.

Plaisted, R.L., Sanford, L., Federer, W.T., Kehr, A.E., and Peterson, L.C. (1962). Specific and general combining ability for yield in potatoes. *American Potato Journal 39:*185-197.

Quiros, C.F., Ortega, R., van Raamsdock, L., Herrera-Montoya, M., Cisneros, P., Schmidt, E., and Brush, S.B. (1992). Increase of potato genetic resources in their center of diversity: The role of natural outcrossing and selection by the Andean farmer. *Genetic Resources and Crop Evolution 39:*107-112.

Railton, I.D. and Wareing, P.F. (1973). Effects of day length on endogenous gibberellins in leaves of *Solanum andigena. Physiologia Plantarum 28:*88-94.

Ramanujam, S. (1950). Problems and prospects of potato improvement in India. *Indian Journal of Genetics and Plant Breeding 10:*1-6.

———. (1954). The potato in Indian agriculture. *Journal of Soil and Water Conservation 3:*4-14.

Renia, H. (1995). True seed is a commercial reality in USA. *Potato Review 5:*48-51.

Roberts, E.H. (1979). Seed deterioration and loss of viability. In Thompson, J.R. (Ed.), *Advances in Research and Technology of Seeds,* part 4 (pp. 25-42). Wageningen, Netherlands: International Seed Association.

Sadik, S. (1983). Potato production from true seed-present and future. In Hooker, W.J. (Ed.), *Research for the Potato in the Year 2000* (pp. 18-25). Lima, Peru: International Potato Center.

Sadik, S. and White, J.W. (1982). True potato seed drying over rice. *Potato Research 25:*269-272.

Sharma, S.K. and Choudhary, S.K. (1985). Factor analysis of berry and its seed characteristics in potato. *Genetica Iberaria 37:*77-82.

Siddique, A.M.D. and Goodwin, P.B. (1980). Seed vigor in bean (*Phaseolus vulgaris* L. cv. "Apollo") as influenced by temperature and water regime during development and maturation. *Journal of Experimental Botany 31:*313-323.

Simmonds, N.W. (1963a). Experiments on the germination of potato seeds I. *European Potato Journal 6:*45-59.

———. (1963b). Experiments on the germination of potato seeds, II. *European Potato Journal 6:*69-76.

———. (1964). The genetics of seed and tuber dormancy in cultivated potatoes. *Heredity 19:*489-504.

———. (1968). Prolonged storage of potato seeds. *European Potato Journal 11:*150-156.

———. (1976). Neotuberosum and the genetic base in potato breeding. *Agricultural Research Council ARC Research Review 2:*9-11.

———. (1997). A review of potato propagation by means of seed, as distinct from clonal propagation by tubers. *Potato Research 40:*191-214.

Singh, A.N. and Janardanjee (1990). Economics of potato production from TPS and response of farmers in Bihar. In Gaur, P.C. (Ed.), *Commercial Adoption of True Potato seed Technology: Prospectus and Problems* (pp. 94-97). Shimla, India: Central Potato Research Institute.

Spicer, P.B. (1961). Use of gibberellin to hasten germination of *Solanum* seed, *Nature 89:*327-328.

Taylorson, R.B. (1982). Interaction of phytochrome and other factors in seed germination. In Khan, A.A. (Ed.), *The Physiology and Biochemistry of Seed Development, Dormancy and Germination* (pp. 323-346). New York: Elsevier.

Thijn, G.A. (1954). Observations on flower induction with potatoes. *Euphytica 3:*28-34.

Thompson, D.J. (1971). Handling seed to insure seed quality. *HortScience 6:*555-556.

Thompson, P.G., Mendoza, H.A., and Plaisted, R.L. (1983). Estimation of genetic parameters for characters related to potato propagation of true seed (TPS) in an *andigena* population. *American Potato Journal 60:*393-401.

Umaerus, M. (1987). True potato seed. In *Proceedings of the 10th Triennial Conference of EAPR* (pp. 72-102). Aalborg, Denmark: European Association for Potato Research.

Upadhya, M.D., Chandra, R., and Sharma, Y.K. (1981). Potato embryo characteristics and variation. *True Potato Seed Letter 2:*1.

Upadhya, M.D., Thakur, K.C., Juneja, A., and Kadian, M.S. (1984). True potato seed production: Flowering, quality and economics. In *Innovative Methods for Propagating Potatoes, Report Planning Conference* (pp. 117-147). Lima, Peru: International Potato Center.

van der Zaag, D.E. and Horton, D. (1983). Potato production and utilization in world perspective with special reference to the tropics and sub-tropics. *Potato Research 26:*323-362.

Van Staden J., Davey, J.E., and Brown, N.A.C. (1982). Cytokinins in seed development and germination. In Khan, A.A. (Ed.), *The Physiology and Biochemistry of Seed Development, Dormancy and Germination* (pp. 137-156). New York: Elsevier.

White, J. (1983a). Germination physiology of true potato seed. *Report CIP*, Lima, Peru: International Potato Center.

———. (1983b). Production of true potato seed. *Report CIP*, Lima, Peru: International Potato Center.

Wiersema, S.G. (1984). *The Production and Utilization of Seed Tubers Derived from True Potato Seed*. Ph.D. Thesis, UK: University of Reading.

———. (1985). Production of seed potatoes derived from true seed. *CIP Circular 13*:4.

———. (1986a). A method of producing seed tubers from true potato seed. *Potato Research 29*:225-237.

———. (1986b). The effect of density on tuber yield in plants grown from true potato seed in beds during two contrasting seasons. *American Potato Journal 63*:241-249.

Wiersema, S.G. and Cabello, R. (1986). Comparative performance of different-sized seed tubers derived from true potato seed. *American Potato Journal 63*:241-249.

# Index

9781560222729